Flow Chemistry

Volume 1: Fundamentals

Edited by
Ferenc Darvas, György Dormán, Volker Hessel,
Steven V. Ley

2nd Edition

DE GRUYTER

Editors
Dr. Ferenc Darvas
InnoStudio Inc.
Graphisoft Park
Záhony u.7
1031 Budapest
Hungary
ferenc.darvas@innostudio.org

Prof. György Dormán
ThalesNano Nanotechnology Inc.
Graphisoft Park
Zahony u. 7
Budapest 1031
Hungary
gyorgy.dorman@thalesnano.com

Prof. Volker Hessel
School of Chemical Engineering and Advanced Materials
University of Adelaide
Adelaide, Australia
and
School of Engineering
University of Warwick
Coventry CV4 7AL
UK
volker.hessel@adelaide.edu.au

Prof. Steven V. Ley
Yusuf Hamied Department of Chemistry
University of Cambridge
Lensfield Road
Cambridge CB2 1EW
United Kingdom
svl1000@cam.ac.uk

ISBN 978-3-11-069359-1
e-ISBN (PDF) 978-3-11-069367-6
e-ISBN (EPUB) 978-3-11-069377-5

Library of Congress Control Number: 2021940755

Bibliographic information published by the Deutsche Nationalbibliothek
The Deutsche Nationalbibliothek lists this publication in the Deutsche Nationalbibliografie; detailed bibliographic data are available on the Internet at http://dnb.dnb.de.

Cover image: Royalty Free Stock Illustration, ID: 1701258115; designer: jijomathaidesigners, Idukki, India
Typesetting: Integra Software Services Pvt. Ltd.
Printing and binding: CPI books GmbH, Leck

MIX
Papier aus verantwortungsvollen Quellen
FSC www.fsc.org FSC® C083411

www.degruyter.com

Ferenc Darvas, György Dormán, Volker Hessel, Steven V. Ley (Eds.)

Flow Chemistry

Also of Interest

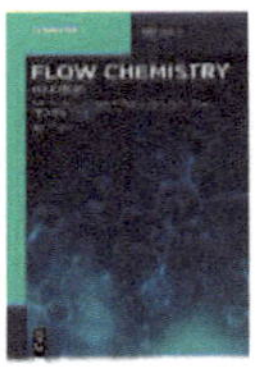

Flow Chemistry
Applications
Ferenc Darvas, György Dormán, Volker Hessel,
Steven V. Ley (Eds.), 2021
ISBN 978-3-11-069361-4, e-ISBN (PDF) 978-3-11-069369-0,
e-ISBN (EPUB) 978-3-11-069376-8
Fundamentals and *Applications*: also available as a set
Set-ISBN 978-3-11-073679-3

Basic Process Engineering Control
Paul Serban Agachi, Mircea Vasile Christea,
Emmanuel Pax Makhura, 2020
ISBN 978-3-11-064789-1, e-ISBN (PDF) 978-3-11-064793-8,
e-ISBN (EPUB) 978-3-11-064853-9

Mass Balances for Chemical Engineers
Gumersindo Feijoo, Juan Manuel Lema, Maria Teresa Moreira, 2020
ISBN 978-3-11-062428-1, e-ISBN (PDF) 978-3-11-062430-4,
e-ISBN (EPUB) 978-3-11-062431-1

Sustainable Process Engineering
Gyorgy Szekely, 2021
ISBN 978-3-11-071712-9, e-ISBN (PDF) 978-3-11-071713-6,
e-ISBN (EPUB) 978-3-11-071730-3

Industrial Separation Processes
Fundamentals
André B. de Haan, H. Burak Eral,
Boelo Schuur, 2020
ISBN 978-3-11-065473-8, e-ISBN (PDF) 978-3-11-065480-6,
e-ISBN (EPUB) 978-3-11-065491-2

Preface

In the last decade, the field of flow chemistry has advanced tremendously and a plethora of applications have been reported in different fields at an unparalleled speed. The characteristics of flow reactors are their exceptionally fast heat and mass transfer. Using so-called microreactors, virtually instantaneous mixing can be achieved for all but the fastest reactions. Similarly, the accumulation of heat, formation of hot spots, and dangers of thermal runaways can be prevented. As a result of the small reactor volumes, the overall safety of the process is significantly improved, even when harsh reaction conditions are used. Thus, this technology offers a unique way to perform ultrafast, exothermic reactions, and allows for the execution of reactions which proceed via highly unstable or even explosive intermediates. In addition, efficient telescoping of reaction sequences can be beneficial in terms of minimizing the number of unit operations and avoiding intermediate isolations, something of particular interest to the pharmaceutical industry, where complex multistep sequences often need to be performed. In contrast to only a few years ago, the flow chemistry literature is now full of publications from not only academic groups but also from scientists working in industry, reporting the results of their many different research activities in this field.

Despite the fact that there appears to be ample literature in the flow chemistry space – including several extensive monographs, books and highly cited review articles – there is a lack of suitable textbooks that can be used for teaching purposes and that can explain the fundamentals to newcomers to the field. A complaint often heard from companies is that there are not enough scientists with the unique training and skillsets of a flow chemist, that is, a person having been educated at the interface of synthetic chemistry and chemical engineering, with additional expertise – for example – in analytical chemistry and data-rich experimentation/machine learning. The first edition of the present *Graduate Textbook on Flow Chemistry*, published in 2014 was, therefore, a highly welcome and urgently needed addition to the steadily growing flow chemistry literature! Now, several years on, the second edition of this textbook is released. The original format has been kept the same, namely, a separation into two independent volumes, one dealing with fundamentals, and a second volume, more relating to the many diverse applications that can be realized with this enabling technique. Both volumes not only discuss basic theory, but also leave ample room for discussing practical considerations. The individual 22 chapters have been authored by experts in their respective fields, wisely chosen by the Editors of this textbook, now, the editorial team – in addition to the original team (Ferenc Darvas, Volker Hessel, György Dormán) – also includes Steven Ley.

It is my hope and genuine expectation that the second edition of this *Graduate Textbook on Flow Chemistry* will become the standard reference work in the field,

https://doi.org/10.1515/9783110693676-202

both at the university level and at other research institutions, where scientists have to familiarize themselves with this rapidly developing field.

October, 2021

C. Oliver Kappe
Professor
University of Graz, Austria

Acknowledgments

Since the publication of the first edition of the *Graduate Textbook on Flow Chemistry* in 2014, the field has advanced tremendously. Thus, the original editors, Ferenc Darvas, György Dormán, and Volker Hessel, joined by Steven V. Ley, have decided to write a new edition, which, besides providing a broad introduction to the subject, also covers the current state of continuousflow chemistry and also discusses practical considerations and emerging fields.

The editors would like to express their sincere gratitude to the many people who have helped to bring this book to fruition. First and foremost, the editors express their heartfelt appreciation to all authors and co-authors for their outstanding contribution, cooperation, enthusiasm, spirit, and constructive comments throughout the planning and writing of the new chapters.

A very special thanks are due to all the instrument suppliers for their contribution to the "Technology overview/Overview of the devices" chapter (AM Technology, Corning, Little Things Factory, Microinnova, Syrris, ThalesNano, Uniqsis, Vapourtec, and Zaiput).

The editors' thanks are extended to Szilvia Gilmore (Flow Chemistry Society) for her tireless efforts for coordinating and monitoring the whole project and to Réka Darvas for the great cover design, for the second time.

The editors are immensely grateful to the editorial team at De Gruyter Publishing House, especially to Nadja Schedensack, our ever-patient Project Manager, Kristin Berber-Nerlinger, for all the preliminary organization and preparation work, and Karin Sora, Vice President STEM, for the wonderful support and guidance.

Finally, the editors want to thank C. Oliver Kappe for the visionary introduction to the textbook.

https://doi.org/10.1515/9783110693676-203

Contents

About the editors

Dr. Ferenc Darvas acquired his degrees in Budapest, Hungary (medical chemistry MS, computer sciences BS, PhD in experimental biology). He has been teaching in Hungary, Spain, Austria, and the USA. Dr. Darvas has been involved in introducing microfluidics/flow chemistry methodologies for synthetizing drug candidates since the late 1990s, which led him to found ThalesNano, the inventor of H-Cube®, and the recipient of the R&D100 Award (Technical Oscar), twice.

Dr. Darvas was awarded Senator Honoris Cause by the University of Szeged, Hungary (2019) and as Fellow of the American Chemical Society (2016). Dr. Darvas is also the founder and active president of the Flow Chemistry Society, Switzerland, founder and editorial board member of the *Journal of Flow Chemistry*, founder of the Space Chemistry Consortium, organizer of the Space Chemistry Symposium series at ACS, and initiator of the world's first anti-Covid drug discovery experiments on ISS.

Prof. György Dormán obtained his PhD in organic chemistry from the Technical University of Budapest, Hungary, in 1986. Between 1982–1988 and 1996–1999, he worked at Sanofi – Chinoin in Budapest, in various research positions. In 1988–1989, he spent a post-doctoral year in the UK (University of Salford). Between 1992 and 1996, he was a visiting scientist at the State University of New York, Stony Brook. Between 1999 and 2008, he served ComGenex/AMRI as chief scientific officer. In 2008, he joined ThalesNano and worked as a director of Scientific Innovation until 2015. Since 2016, he is a consultant of InnoStudio Inc. In 2011, he became honorary professor at the University of Szeged. He is an author of 116 scientific papers and book chapters. He is a member of the editorial board of Molecular Diversity and *Mini-Reviews in Medicinal Chemistry* and member of the advisory board of *Journal of Flow Chemistry*.

Prof. Volker Hessel studied chemistry at Mainz University and received his PhD in 1993. Further career steps were as follows: 1994, Institut für Mikrotechnik Mainz/D as vice director R&D and director R&D; 2005, Eindhoven University of Technology/NL as professor; 2019, at the University of Warwick/UK as part-time professor. In 2018, he was appointed as deputy dean (research) and professor at the University of Adelaide, Australia. He is research director of Adelaide's Andy Thomas Centre of Space Resources.

Prof. Hessel's research is on microfluidic and plasma processes and their application to health, chemistry, agrifood, and space. He has published 502 peer-reviewed papers (h-index: 61, Scopus) and was authority in the Parliament Enquete Commission "Future of Chemical Industry." He received the AIChE Award "Excellence in Process Development Research" and the IUPAC-ThalesNano Prize in Flow Chemistry, as well as the ERC Advanced/Proof of Concept/Synergy and FET OPEN Grants.

https://doi.org/10.1515/9783110693676-205

Prof. Steven V. Ley obtained his PhD from Loughborough University, UK, and completed postdoctoral studies at the Ohio State University, USA, and Imperial College London, UK. He was appointed to the staff of Imperial College London, becoming professor in 1983 and head of department in 1989. He was elected to the Royal Society, London in 1990, moved to Cambridge University to the 1702 Chair of Chemistry in 1992, and was president of the Royal Society of Chemistry 2000–02.

Steve's research interests involve many aspects of organic chemistry, including synthesis, natural products, methodology, biotransformations, enabling technologies, and, in particular, his extensive work on flow chemistry. He is a recipient of numerous international awards, including the IUPAC-ThalesNano Prize in Flow Chemistry and, recently, the prestigious ACS Arthur C. Cope Award.

Contributing authors

Andrea Adamo
Zaiput Flow Technologies
Waltham, MA 02451, USA
aadamo@zaiput.com
Chapter 7

Nagaki Aiichiro
Department of Synthetic Chemistry and Biological Chemistry
Graduate School of Engineering,
Kyoto University
Kyoto, Japan
anagaki@sbchem.kyoto-u.ac.jp
Chapter 2

Lucia Anghileri
Biomolecular Systems Department
Max-Planck Institute of Colloids and Interfaces
Potsdam, Germany

and

Department of Chemistry and Biochemistry,
Freie Universität Berlin
Berlin, Germany
lucia.anghileri@mpikg.mpg.de
Chapter 5

Marcus Baumann
School of Chemistry
University College Dublin
Science Centre South
Belfield, Dublin, Ireland
marcus.baumann@ucd.ie
Chapter 4

Daniel Blanco-Ania
Educational Institute for Molecular Sciences
Radboud University
Nijmegen, The Netherlands

and

Institute for Molecules and Materials,
Radboud University
Nijmegen, The Netherlands
d.blanco@science.ru.nl
Chapter 9

Tom G. Bloemberg
Educational Institute for Molecular Sciences
Radboud University
Nijmegen, The Netherlands

and

Institute for Molecules and Materials,
Radboud University
Nijmegen, The Netherlands
t.bloemberg@fnwi.ru.nl
Chapter 9

Cormac Bracken
School of Chemistry
University College Dublin
Science Centre South
Belfield, Dublin, Ireland
cormac.bracken@ucdconnect.ie
Chapter 4

Martin Cattoen
Center for Integrated Technology and Organic Synthesis
MolSys Research Unit
University of Liège
Liège, Belgium
M.Cattoen@uliege.be
Chapter 8

Yuesu Chen
Center for Integrated Technology and Organic Synthesis
MolSys Research Unit
University of Liège
Liège, Belgium
Yuesu.Chen@uliege.be
Chapter 8

Ferenc Darvas
InnoStudio Inc.
Graphisoft Park
Záhony u.7
1031 Budapest, Hungary
ferenc.darvas@innostudio.org
Chapter 1

https://doi.org/10.1515/9783110693676-206

Benjamin J. Deadman
Centre for Rapid Online Analysis of Reactions (ROAR)
Imperial College London
Molecular Science Research Hub
London, UK
b.deadman@imperial.ac.uk
Chapter 3

Kian Donnelly
School of Chemistry
University College Dublin
Science Centre South
Belfield, Dublin, Ireland
kian.donnelly@ucdconnect.ie
Chapter 4

György Dormán
ThalesNano Inc.
Graphisoft Park
Záhony u.7
1031 Budapest, Hungary

and

Institute of Pharmaceutical Chemistry
University of Szeged
Eotvos u. 6
6720 Szeged, Hungary
gyorgy.dorman@thalesnano.com
Chapter 1

Melinda Fekete
Enzymicals AG Greifswald,
Mecklenburg-West Pomerania,
Germany
melinda.fekete@enzymicals.com
Chapter 3

Mara Di Filippo
School of Chemistry
University College Dublin
Science Centre South
Belfield, Dublin, Ireland
mara.difilippo@ucdconnect.ie
Chapter 4

Kerry Gilmore
Biomolecular Systems Department
Max-Planck Institute of Colloids and Interfaces
Potsdam, Germany

and

Department of Chemistry
University of Connecticut
Storrs, CT, USA
kerry.m.gilmore@uconn.edu

Toma Glasnov
Institute of Chemistry
University of Graz
Graz, Austria
toma.glasnov@uni-graz.at
Chapter 3

Mara Guidi
Biomolecular Systems Department
Max-Planck Institute of Colloids and Interfaces
Potsdam, Germany

and

Department of Chemistry and Biochemistry
Freie Universität Berlin
Berlin, Germany
mara.guidi@mpikg.mpg.de
Chapter 5

King Kuok (Mimi) Hii
Department of Chemistry
Imperial College London
Molecular Science Research Hub
London, UK
mimi.hii@imperial.ac.uk
Chapter 3

Haruro Ishitani
Green & Sustainable Chemistry Social Cooperation Laboratory
Graduate School of Science
The University of Tokyo
Tokyo, Japan
hishitani@chem.s.u-tokyo.ac.jp
Chapter 10

Shu Kobayashi
Department of Chemistry
School of Science
The University of Tokyo
Japan

and

Green & Sustainable Chemistry Social
Cooperation Laboratory
Graduate School of Science
The University of Tokyo
Tokyo, Japan
shu_kobayashi@chem.s.u-tokyo.ac.jp
Chapter 10

Klára Lövei
ThalesNano Inc.
Graphisoft Park
Záhony u.7
1031 Budapest, Hungary
klara.lovei@thalesnano.com
Chapter 11

Debasis Mallik
Centre for catalysis Research and Innovation
University of Ottawa
Ottawa, Canada
dmallik@uottawa.ca
Chapter 6

Lorenzo Milani
Zaiput Flow Technologies
Waltham, MA 02451, USA
lmilani@zaiput.com
Chapter 7

Jean-Christophe Monbaliu
Center for Integrated Technology and Organic
Synthesis
MolSys Research Unit
University of Liège
Liège, Belgium
jc.monbaliu@uliege.be
Chapter 8

Mathieu Morin
Department of Chemistry
Carlton University
Ottawa, Canada
mathieuandremorin@gmail.com
Chapter 6

Michael G. Organ
Centre for catalysis Research and Innovation
University of Ottawa
Ottawa, Canada

and

Department of Chemistry
York University
Toronto, Canada
Organ@uottawa.ca
Chapter 6

Floris P. J. T. Rutjes
Educational Institute for Molecular Sciences
Radboud University
Nijmegen, The Netherlands

and

Institute for Molecules and Materials,
Radboud University
Nijmegen, The Netherlands
floris.rutjes@ru.nl
Chapter 9

Peter H. Seeberger
Biomolecular Systems Department
Max-Planck Institute of Colloids and
Interfaces
Potsdam, Germany

and

Department of Chemistry and Biochemistry
Freie Universität Berlin
Berlin, Germany
peter.seeberger@mpikg.mpg.de
Chapter 5

Gellért Sipos
ComInnex Inc.
Graphisoft Park
Záhony u.7
1031 Budapest, Hungary
gellert.sipos@cominnex.com
Chapter 11

Luuk T.C.G. van Summeren
Educational Institute for Molecular Sciences
Radboud University
Nijmegen, The Netherlands
l.vansummeren@science.ru.nl
Chapter 9

Nopphon Weeranoppanant
Department of Chemical Engineering
Faculty of Engineering
Burapha University
Chon Buri, Thailand

and

School of Biomolecular Science and Engineering
Vidyasirimedhi Institute of Science and Technology (VISTEC)
Thailand
nopphon.we@eng.buu.ac.th
Chapter 7

Wenyao Zhang
Department of Chemistry
York University
Toronto, Canada
wyzhang@yorku.ca

Ferenc Darvas and György Dormán

1 Fundamentals of flow chemistry

Objective of this chapter

This chapter intends to summarize all the important basic theoretical features of flow chemistry and flow reactors. A detailed discussion of particular topics can be found in the following chapters.

1.1 General theory of flow chemistry

1.1.1 Basic features of continuous-flow operation

The typical operation sequence of traditional flask (batch) reaction in laboratory practice is shown in Fig. 1.1 [1].

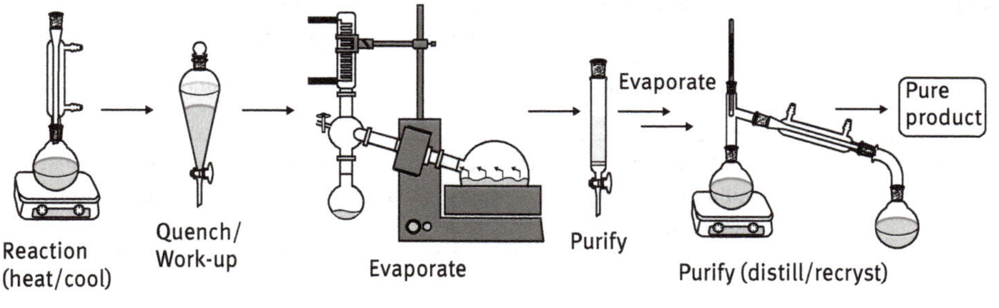

Fig. 1.1: Typical batch laboratory setup (adapted from Baxendale, I. R. (2013) The integration of flow reactors into synthetic organic chemistry. J Chem Technol Biotechnol, 88, 519–552).

The typical arrangement of continuous-flow apparatus containing reagent pumps, mixing units (T-mixer), reaction zone, quenching inlet, backpressure regulator, and product collection vessel is shown in Fig. 1.2.

Chemical synthesis in the traditional laboratory has been carried out in standardized glassware, and this has not been changed over a century [2]. There are significant differences between batch and flow processes in terms of the important measures of the reactions. In batch reaction, time is determined by how long a reaction vessel is held at a given temperature. In contrast, in continuous processes, the **residence time**, which refers to how long the reactants stay in the reactor zone, is determined by the reactor volume and the bulk flow rate (Fig. 1.3, Fig. 1.4).

https://doi.org/10.1515/9783110693676-001

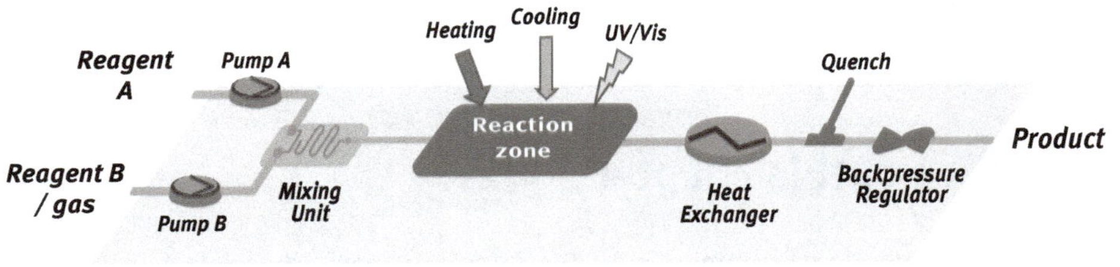

Fig. 1.2: Typical setup for flow reactions (based on Ley, S. V., & Baxendale, I. R. (2009). New tools for molecule makers: emerging technologies. Eds. M. Hicks and C. Kettner. Proc. Beilstein Symp. on Systems Chemistry, Bozen, pp. 65–85. and Gutmann, B., Cantillo, D., & Kappe, C. O. (2015). Continuous-flow technology – a tool for the safe manufacturing of active pharmaceutical ingredients. Angew Chem Int Ed, 54(23), 6688–6728).

Tab. 1.1: Comparison of the major reaction characteristics in batch and flow.

	Batch	Flow
Stoichiometry	Concentration/ratio of the molar quantities	Concentration/ratio of the flow rates
Reaction time	Time spent under the defined condition	Residence time spent in the reaction zone, depending on the flow rate and reaction volume
Reaction progress	Time spent in the flask	Distance traveled in the channel
Steady-state characteristics	It has a uniform concentration at each position within the flask at a particular moment	It has a steady but different concentration at each position throughout the length of the reactor

Stoichiometry in flow reactors is defined by the concentration of reagents and the ratio of their flow rate. Flow rate is the volume of fluid that passes through a given channel per unit time (Tab. 1.1).

In batch processes, this is defined by the concentration of chemical reagents, and the ratio of their molar quantities.

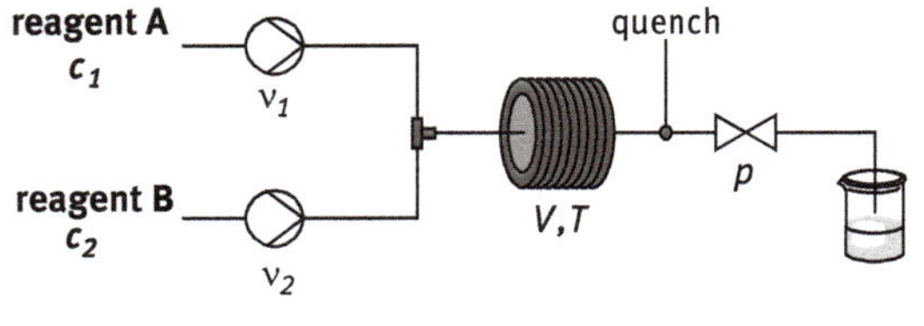

Fig. 1.3: Mixing A and B reagents under flow conditions (Plutschack, M. B., Pieber, B., Gilmore, K., & Seeberger, P. H. (2017). The hitchhiker's guide to flow chemistry, Chem Rev, 117(18), 11,796–11,893).

If a single flow rate (v1 or v2) is altered but all other parameters stay constant, it leads to changes in stoichiometry, final concentration, and residence time.

Stoichiometry can be set by the flow rates and the concentration of the reagent streams.

(For further details on this issue, please see Volume 1, Chapter 4, Title: Practical aspects of performing continuous flow Chemistry; and Chapter 5, Title: When and how to start flow chemistry?)

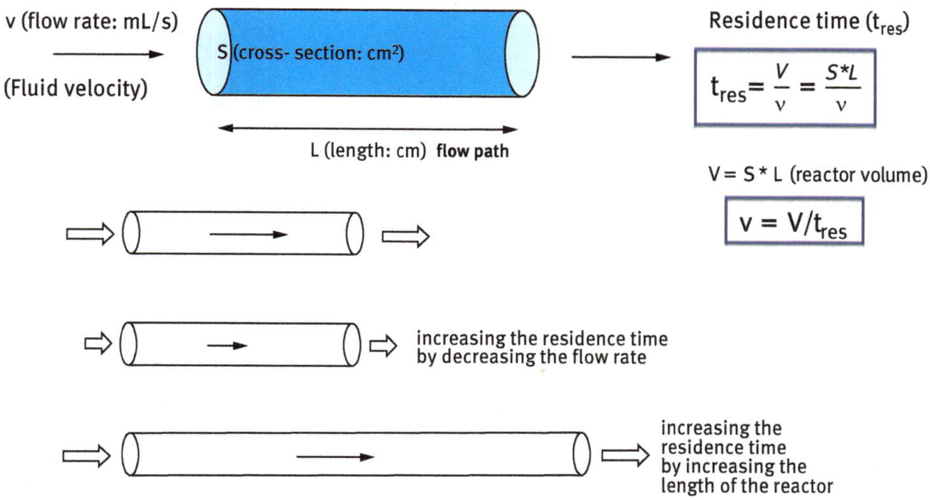

Fig. 1.4: Calculation of the residence time and its relationship with flow rate and the length of the reactor (*S* is the cross section, *L* is the length, and *V* is the flow rate).

The concentration of the reactant decays exponentially with time in the flask reactor [3], according to the rate law:

$$\text{Rate} = \frac{-d[\mathbf{A}]}{dt} = k[\mathbf{A}]$$

$$\text{Rate} = \frac{-d[\mathbf{A}]}{dt} = k[\mathbf{A}][\mathbf{B}] \text{ or Rate} = \frac{-d[\mathbf{A}]}{dt} = k[\mathbf{A}]^2$$

The reaction rate is proportional to the decrease in concentration of the reactant (**A**) in time *t*, and *k* is the rate constant for the particular reaction (first-order reaction). The rate of a second-order reaction is proportional to the concentration of two reactants, or to the square of the concentration of one.

In a microfluidic device with a constant flow rate, the concentration of the reactant decays exponentially with distance, along the reactor (Fig. 1.5). Thus, time in a flask reactor equates with distance in a flow reactor, where the reaction time is defined as the residence time between a reagent inlet and the position of the quencher

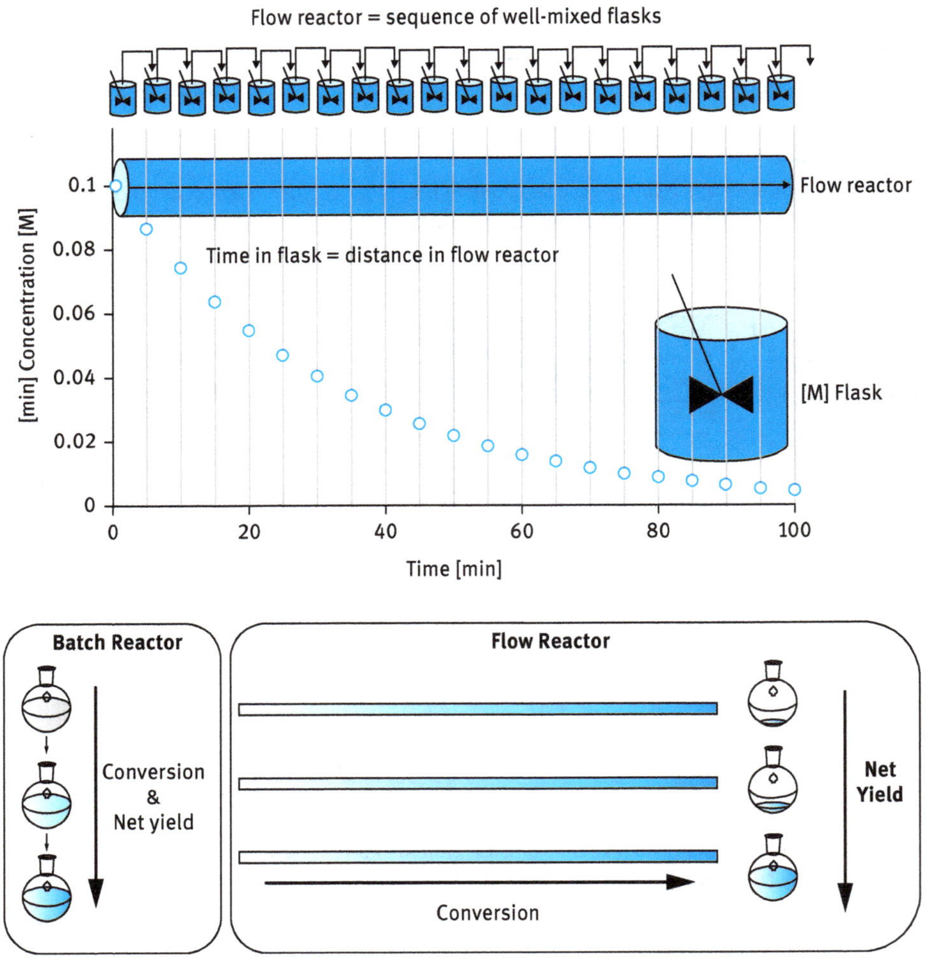

Fig. 1.5: Reactant concentration versus time (flask reactor) or distance (flow reactor) for a simple first-order reaction, under well-mixed homogeneous conditions (adapted from Valera, F. E., Quaranta, M., Moran, A., Blacker, J., Armstrong, A., Cabral, J. T., & Blackmond, D. G. (2010). The flow is the thing . . . or is it? Assessing the merits of homogeneous reactions in flask and flow, Angew Chem Int Ed, 49, 2–10; and Lummiss, J. A., Morse, P. D., Beingessner, R. L., & Jamison, T. F. (2017). Towards more efficient, greener syntheses through flow chemistry, Chem Rec, 17(7), 667–680).

inlet (Fig. 1.6) and can be precisely controlled by adjusting the length between these positions. It is also designated as reaction windows, which is the time course of the reaction between the activation (starting) and the optimal time for quenching (stopping). At a certain flow rate and volume reaction, time is equal to the residence time.

Flash chemistry is a field of chemical synthesis where extremely fast reactions involving short-lived highly reactive intermediates are conducted without deceleration,

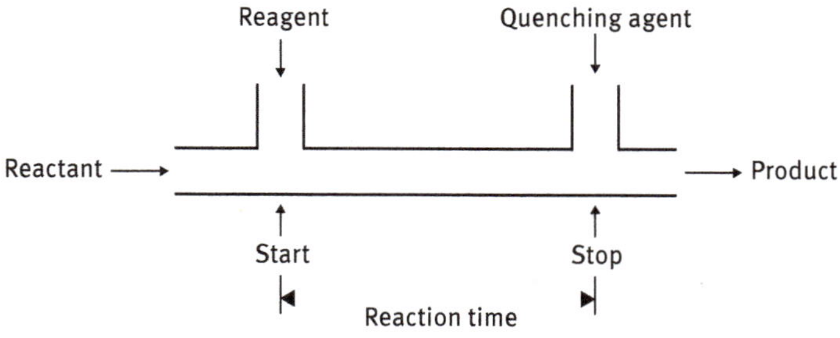

Fig. 1.6: The reaction time is defined as the residence time between a reagent inlet and the position of the quencher inlet (adapted from Yoshida, J. I. (2010). Flash chemistry: flow microreactor synthesis based on high-resolution reaction time control, Chem Rec, 10, 332–341).

in a highly controlled manner, to produce desired compounds with high selectivity by virtue of high-resolution reaction control.

(For further details on this issue, please see Volume 1, Chapter 2, Title: Principles of controlling reactions in flow chemistry) i

1.1.1.1 Basic features of flow chemistry in packed-bed microreactors

Microreactor tubes are often filled with solid-supported catalysts or reagents, and the reaction mixture passes through the stationary phase particles. There are two typical setups: packed-bed reactors and, for scale-up purposes, the trickle-bed reactors (Fig. 1.7).

The major advantageous features of the heterogeneous phase, packed-bed flow reactors are as follows: Particle movement is mainly restricted; high interfacial area facilitates better mass transfer and improved heat distribution; packed beds have improved lifetime due to decreased exposure to the environment; high reagent/catalyst excess could drive the reactions to completion; immobilized reagents are removed from reactions (placed in cartridges); and catalyst poisoning and side reactions are minimized due to the immediate removal of the product from the reactor zone (no backmixing).

Residence time depends on several factors in packed-bed reactors for liquid/solid reactions and reactions involving a gas/liquid/solid. During calculation the dead or void volume of the reactor, the difference between the total reactor volume and the volume of the stationary phase should be considered. In gas/liquid/solid reactions, the solubility of the gas in the liquid phase, the ratio of the gas/liquid phase (Fig. 1.8), particle size, and the system pressure could also influence flow rate and the residence time, and complicates the calculation; therefore, often, the simplest way is to measure the residence time by injecting a dye solution manually and monitoring its appearance at the end.

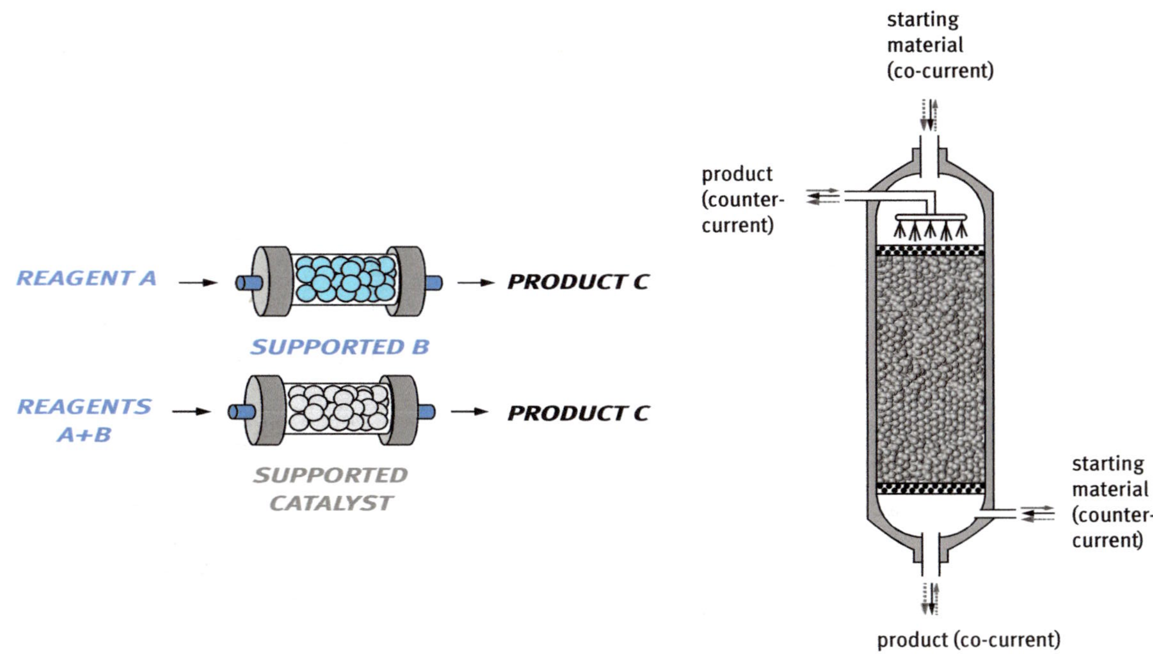

Fig. 1.7: Coley, C.W., Imbrogno, J., Mo, Y., Thomas, D.A., and Jensen, K.F. Flow chemistry system design and automation, In: Jamison, T., & Koch, G. (Eds.). (2018). Science of Synthesis: Flow Chemistry in Organic Synthesis. Georg Thieme Verlag.

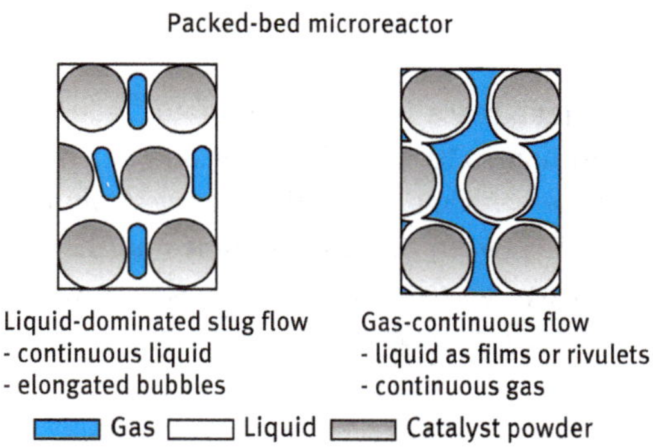

Fig. 1.8: Various flow regimes in packed reactors (adapted from Yue, J. (2018) Multiphase flow processing in microreactors combined with heterogeneous catalysis for efficient and sustainable chemical synthesis, Catal Today, 308, 3–19).

Particle size significantly influences performance. Big particles suffer from a relatively low surface-to-volume ratio, and since the reaction occurs on the surface, conversion might be inefficient. Small particles may cause high backpressure or can clog the filter unit.

There are semiempirical rules to calculate the acceptable value of the particle size.

Particle size (diameter of the particles, d_p) is dependent on the length of the catalyst bed (l) and diameter (d_r) of the tube reactor.

Rule 1: $d_p < l/25$, to avoid residence time distribution (axial dispersion).

Rule 2: $d_p < d_r/10$, to avoid "wall effect": the reactant molecules flow through without contacting the major amount of the active sites, if the particles are too large.

In packed-bed microreactors, a high resistance for fluid flow is encountered, which results in a high pressure drop over the bed. The pressure drop over the packed bed can be calculated according to Darcy's law:

$$\frac{\Delta P}{L} = \frac{\mu Q}{\kappa A}$$

where ΔP is the pressure drop, Q is the volumetric flow rate, L is the length of the bed, μ is the viscosity of the fluid, κ is permeability of the bed, and A is the cross section.

The pressure drop will increase linearly with increasing flow rates and lengths of the packed bed.

(For further details on this issue, please see Volume 1, Chapter 10, Title: Catalysis in flow and Chapter 11, Title: Gaseous reagents in flow chemistry)

1.1.2 The dimensions of flow (micro) reactors

At the molecular level, chemical reactions take place in the range of 10^{-13}–10^{-12} s, while reaction times range from minutes to hours (10^2–10^5 s) in a flask. The size of molecules is in the range of 10^{-10}–10^{-8} m, whereas the size of a flask ranges from 10^{-2} to 100 m. Hence, there is a rough correlation between the reaction time and the size of the reaction environment. In flash chemistry, we use a reactor, the size of which ranges from 10^{-6} to 10^{-3} m. The timescale for flash chemistry is milliseconds to seconds (Fig. 1.9).

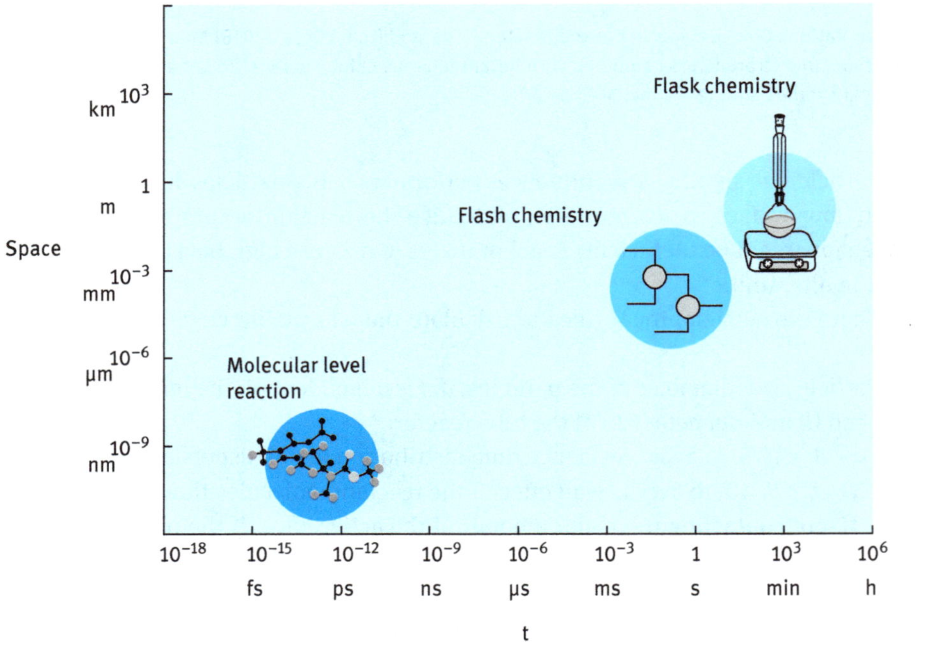

Fig. 1.9: Time/space relationship for chemical reactions (adapted from Yoshida, J. I., Nagaki, A., & Yamada, T. (2008) Flash chemistry: fast chemical synthesis by using microreactors, Chem Eur J, 14, 7450–7459).

1.1.2.1 Space–time yield

Space–time yield (STY; kg m^{-3} s^{-1}) represents the mass of a product *P* formed per volume of the reactor and time. For batch reactors, STY is less than 1, while for plug-flow reactors (PFR), it is less than 500.

Generally, volume is equal to the length cubed, while surface area is equal to length squared. When the length is shortened, surface-to-volume ratio increases. Thus, microreactors have high surface-to-volume ratio than macroreactors. As microreactors have a greater surface area per unit volume than macroreactors, heat transfer occurs rapidly in a flow microreactor, enabling fast cooling/heating and, hence, precise temperature control [4]. If the size is reduced by 100-fold, then the surface to volume increased to 100-fold (Fig. 1.10).

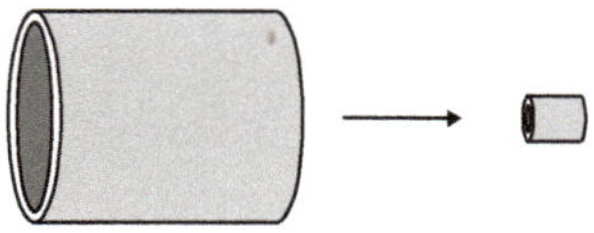

Size	1/100
Surface area	1/10000
Volume	1/1000000
Surface / Volume	100

Fig. 1.10: Comparison of the reduction ratios of the surface area and volume during 1 to 1/100 size reduction (adapted from Yoshida, J. I., Kim, H., & Nagaki, A. (2011). Green and sustainable chemical synthesis using flow microreactors, ChemSusChem 4, 331–340).

The miniaturization and large surface area allows rapid heat transfer (for slow reactions) or heat removal (for highly exothermic reactions).

1.1.3 Mixing in microreactors

Typically, to run an organic reaction with high yield and selectivity, both mass and heat transport must be carefully controlled. Continuous-flow microreactors allow rapid and homogeneous mixing because of their small dimensions (channel or capillary diameter is 0.05–0.5 cm) and, in that range, laminar flow is the predominant. This regime also recognized as PFRs if the radial diffusion is much faster than convective mass transport along the channel length. In laminar flow regime, microreactors can achieve complete mixing in microseconds, whereas classical reactors mix on a timescale of seconds or longer.

On the other hand, most of the flow chemistries operate by mixing the two mass transfer components: diffusion and convection, which could be clearly differentiated. While diffusion is a passive transport due to a concentration gradient, convection is a transport due to pressure gradient or in other words the collective movement of molecules within fluids [5]. Illustration of the differences is shown in Fig. 1.11.

In flow fast reactions can be carried out at room temperature, applying short residence time without reducing the reaction rate.

In laminar flow at steady state (as shown in Fig. 1.11), only diffusion can allow mass transfer due to the vertical concentration gradient between the two adjacent fluid layers, since convection transports mass only tangent to the velocity. However, this is only the case if the Reynolds number (see Section 1.2.4) is particularly low and pure laminar flow exists.

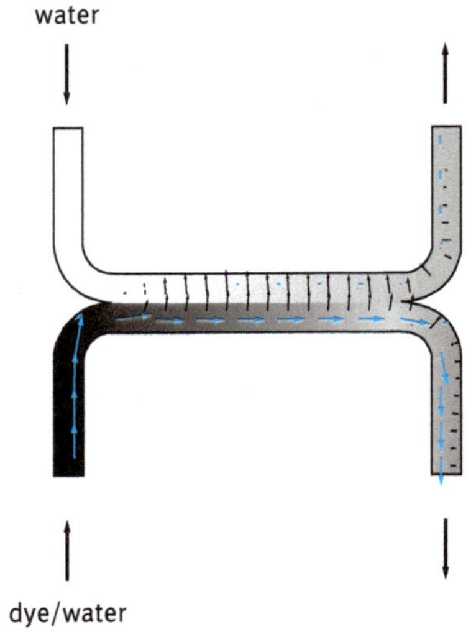

Fig. 1.11: Concentration profile (grayscale), convective flux (cyan arrows), and diffusive flux (red arrows) of a two inlet/outlet model system (adapted from Multiphysics Cyclopedia – https://www.comsol.ru/multiphysics/convection-diffusion-equation).

Fast mixing relies on the short diffusion path in microreactors. A molecule in the center of a typical microfluidic channel can reach the wall of that channel in a few seconds. The same molecule in the middle of a reaction flask (batch) would require hours to diffuse to the side wall (without mixing). A marked shortening of the diffusion path in flow (in a microreactor) results in a mixing speed that is unobtainable in batch. The time (t_d) needed for molecular diffusion is proportional to the square of the length of the diffusion path (Tab. 1.2).

Tab. 1.2: Correlation of diffusion time and size (radial diffusion distance) (adapted from Yoshida, J. I. (2015). Flash chemistry, In: Basics of Flow Microreactor Synthesis. Springer, Tokyo).

$t_d = \frac{L^2}{D}$ $D = \frac{kT}{6\pi\eta r}$	t_d is the diffusion time, L [m] is the distance over which diffusion must take place, and D [$m^2\ s^{-1}$] is the diffusion coefficient, in which k is the Boltzmann constant (1.38×10^{-23} J K^{-1}), T [K] the absolute temperature, η [kg m^{-1} s^{-1}] the absolute (solute) viscosity, and r the hydrodynamic radius [m].
Diffusion time (s)	**Size**
0.0005	1 µm
0.005	10 µm
5	100 µm
500	1 mm
Diffusion time is calculated for D (diffusion coefficient) = 1×10^{-5} cm^2 s^{-1}	

The small dimension and the short diffusion path allow rapid mixing and mass transfer, which prevents reagent accumulation and increases safety.

1.1.3.1 Mixing versus reaction rate in laminar flow

In laminar flow regime, radial mixing is strictly due to diffusion.

Damkoehler number (Da) shows if a reaction is under diffusion or kinetic control in laminar flow. It expresses the reaction rate relative to the mass transport rate (diffusion). (A chemical reaction is mixing- or diffusion-controlled, if its half-life is on the order of, or smaller than that of the relevant mixing or diffusion process):

$$\mathrm{Da} = \frac{\text{rate of reaction}}{\text{rate of diffusion}}$$

The mass transport rate in a microreactor is the diffusion rate, thus, the Damkoehler number can be written for an nth order reaction as

$$\mathrm{Da_F} = \frac{k_r C_0^{n-1}}{\left[\frac{2D_{AB}}{x^2}\right]}$$

where k_r is the specific reaction rate and C_O is the initial reagent concentration.

In general, chemical transformations are reaction rate limited when Da < 1, and mixed mass transport-reaction rate limited when Da ~ 1. When Da > 1, experimentally measured reaction rates are controlled by diffusion or mass transport; thus, flow reactors have superiority.

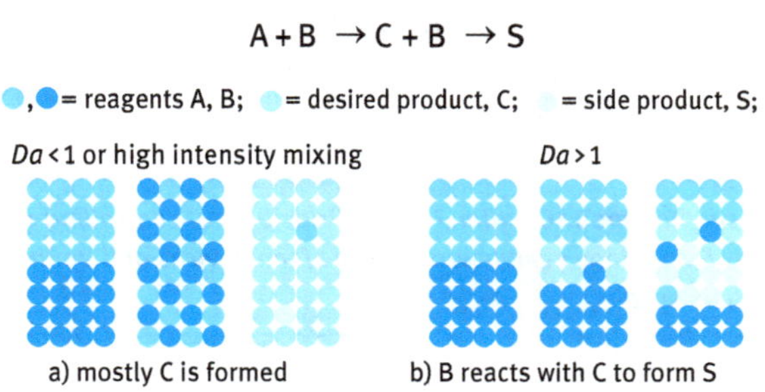

Fig. 1.12: The outcome of a consecutive reaction depending on the intensity of the mixing (Plutschack, M. B., Pieber, B., Gilmore, K., & Seeberger, P. H. (2017). The hitchhiker's guide to flow chemistry, Chem Rev, 117(18), 11,796–11,893).

The reaction outcome depends on the value of Da. If Da < 1, homogeneity (uniformity) can be achieved by diffusion/mass transfer. If Da > 1, the reaction is too fast to achieve homogeneity, and creates local concentrations of B and C, which react to give higher quantities of the side-product, S (Fig. 1.12).

There are only few options that help increase mixing or enhance diffusion in laminar flow: 1. applying narrower channel, so that the concentration gradients are larger in the vertical direction; 2. applying a longer channel or slower flow, so that the fluid takes longer to pass through the channel, and there is more time for diffusion; 3. Applying fluid having a higher diffusion coefficient, the diffusive flux will be larger.

1.1.3.2 Mixing: turning laminar flow to turbulent

Micromixers can be characterized as passive (without external energy) or active (external energy added such as oscillation or vibration). The most widely used micromixers are passive, and use mixer geometry to generate effective contact between reagents [6]. Passive mixers could act as either diffusion mixers or convection mixers.

In laminar (small channel diameter) flow devices, the rate of diffusion could be enhanced by increasing the interfacial area between fluidic layers. Typical solutions are extending the tube or dividing the flow into multiple laminates (parallel lamination, serial lamination, etc.; Fig. 1.13a, b, diffusion mixers). In laminar flow, mixing relies solely on molecular diffusion, which is often insufficient, particularly in the case of very fast, diffusion-controlled reactions ($Da > 1$). In that case, convective diffusion enhancement is needed through deformation of the lamellae arrangements by inducing a turbulent flow regime, with appropriate mixing [7]. The driving force is the pumping energy of the flowing fluid. Practically, helical or curved structures, split-recombine structures, and packed-bed reactors are used (Fig. 1.13c–e, convection mixers).

For macromixing in large-scale flow equipments, static mixers are used. Static mixers generate turbulence and intense radial mixing by applying tube inserts such as blades/helices or stretching of the fluid. Incoming fluid is split into layers and then recombines in a new sequence (Fig. 1.13f, g).

i *(For further details on this issue, please see Volume 1, Chapter 2, Title: Principles of controlling reactions in flow Chemistry, Chapter 3, Title: Technology overview/overview of the devices and Volume 2, Chapter 8, Title: Scale-up of flow chemistry system)*

1.1.4 Reaction acceleration in continuous flow

Reactions in laboratory organic synthesis usually take minutes to hours to yield the desired product [8, 9]. This duration has been considered acceptable for most practical purposes. However, more recently, there is a desire to improve efficiencies through accelerated reaction processes to enhance output and drive change in the practice of organic synthesis. However, very recently, strong demand to improve the efficiency and greenness of synthetic organic chemistry has emerged, which certainly drives a change in the laboratory routine.

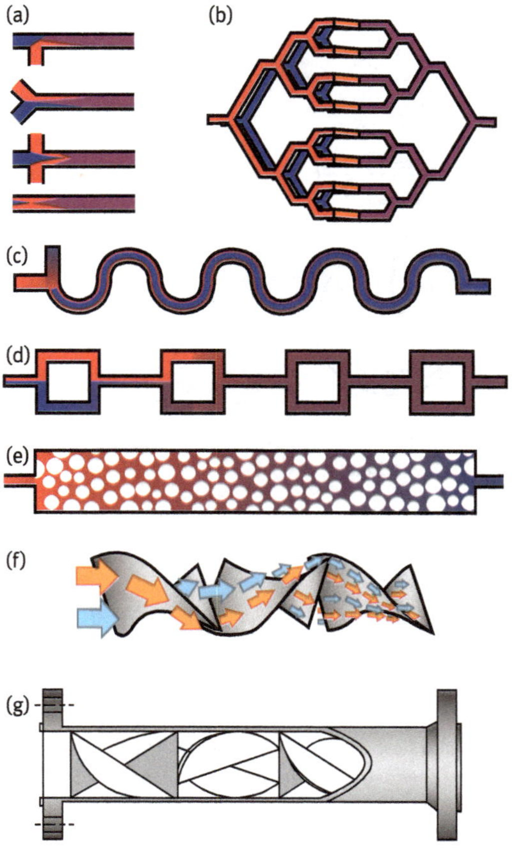

Fig. 1.13: Common types of passive and static mixers on the basis of three mixing structures: simple contacting structures (T-junction, Y-junction, cross-junction, co-flow junction, split-recombine structures), multilamination structures, helical or curved structures, and periodic static structures (Zhang, J., Wang, K., Teixeira, A. R., Jensen, K. F., & Luo, G. (2017). Design and scaling up of microchemical systems: a review, Annu Rev Chem Biomol Eng, 8, 285–305).

The reaction rate acceleration (Fig. 1.14) depends on two facts: rapid mixing (increased mass transfer) and the increased parameter space, "novel process windows" (NPW) [10]. This can be achieved by an increase in temperature, pressure, or concentration (solvent-free operation). Small dimension allows rapid heat exchange; thus, significant rate acceleration can be achieved by high temperature (>200 °C) according to the following Arrhenius equation, which expresses the dependence of the rate constant, k, of a chemical reaction on the absolute temperature T (in Kelvin), where A is the pre-exponential factor, E_a is the activation energy, and R is the universal gas constant:

$$k = Ae^{-E_a/(RT)}$$

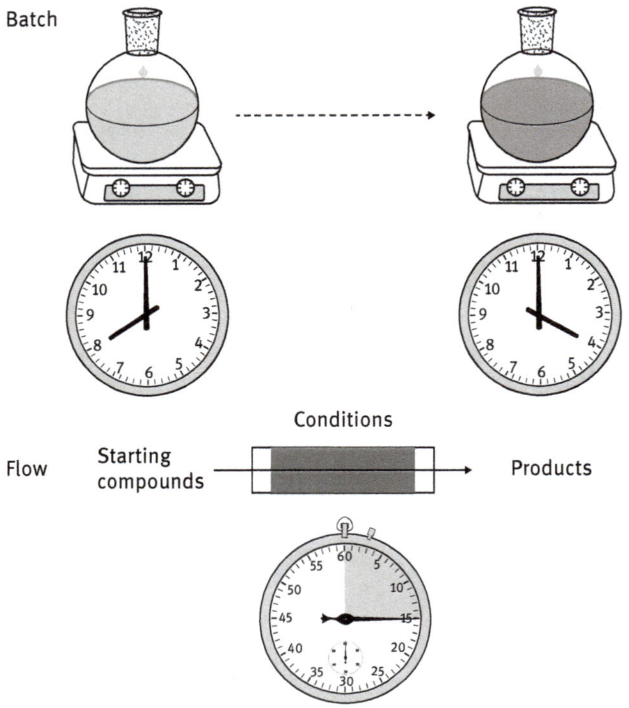

Fig. 1.14: Comparison of the reaction time in batch and flow (adapted from Wegner, J., Ceylan, S., & Kirschning, A. (2011). Ten key issues in modern flow chemistry, Chem Commun, 47, 4583–4592).

Similarly, high pressure could also increase reaction rate (conversion and yield). A combined high-temperature and high-pressure flow regime allows reaching supercritical conditions. The unique character of the supercritical fluid further accelerates many chemical reactions in flow.

1.1.5 Heterogeneous phase reactions in continuous flow [11]

Gas–liquid–solid triphasic reaction conditions that are present in heterogeneous catalytic processes involve a gas, a substrate dissolved in a solvent, and an immobilized, solid, precious metal catalyst. Such reactions require large interfacial areas (high surface-to-volume ratio) and are not attainable in normal batch systems. However, flow conditions lead to improvements in conversions, yields, and selectivities. In heterogeneous phase-flow synthesis, the immobilized reagents and catalyst are placed in cartridges (Fig. 1.15), also called fixed-bed or packed-bad reactors, separated from the reaction media. Thus, catalyst poisoning as well as side reactions are minimized due to the immediate removal of the product from the reactor zone. The

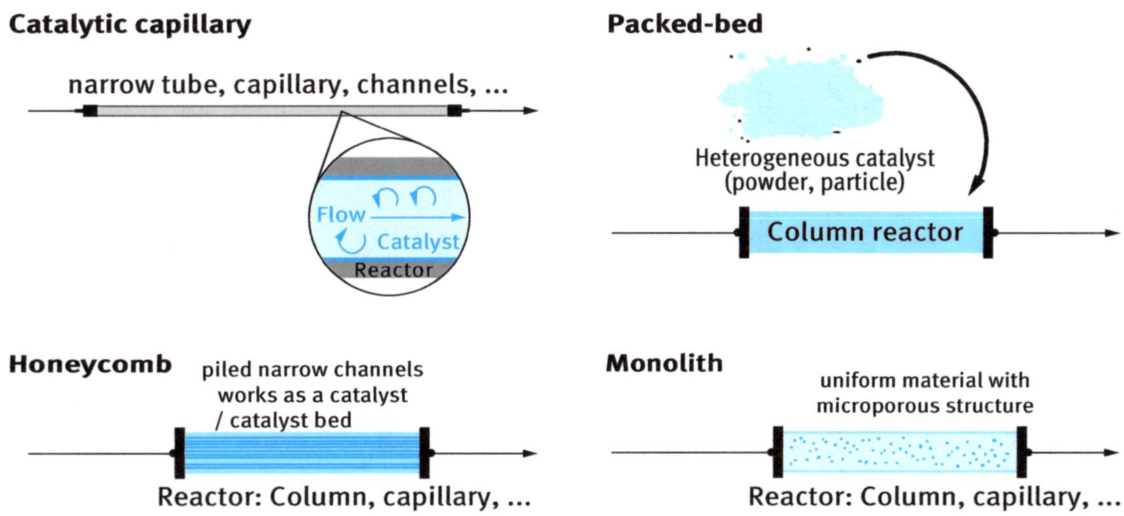

Fig. 1.15: Various setups for heterogeneous catalytic flow reactions (adapted from Masuda, K., Ichitsuka, T., Koumura, N., Sato, K., & Kobayashi, S. (2018). Flow fine synthesis with heterogeneous catalysts Tetrahedron, 74(15), 1705–1730).

reusability (recycling) of the catalysts or reagents is economical and environmentally sound. Excess reagents can be used to drive reactions to completion, without introducing difficult purification steps.

1.1.6 Multistep reactions in continuous flow

In batch, multistep synthesis proceeds through step-by-step transformation of starting materials into desired products (Fig. 1.16). Typically, after each synthetic step, products are isolated from the reaction mixture and purified to remove any undesired components that might interfere with the subsequent synthetic transformations. Single line continuous reactor network could circumvent the need for the isolation of intermediate products (telescoping) [12]. Solid-supported reagents, catalysts, and scavengers together with in-line analytics could support the single line flow synthesis [13].

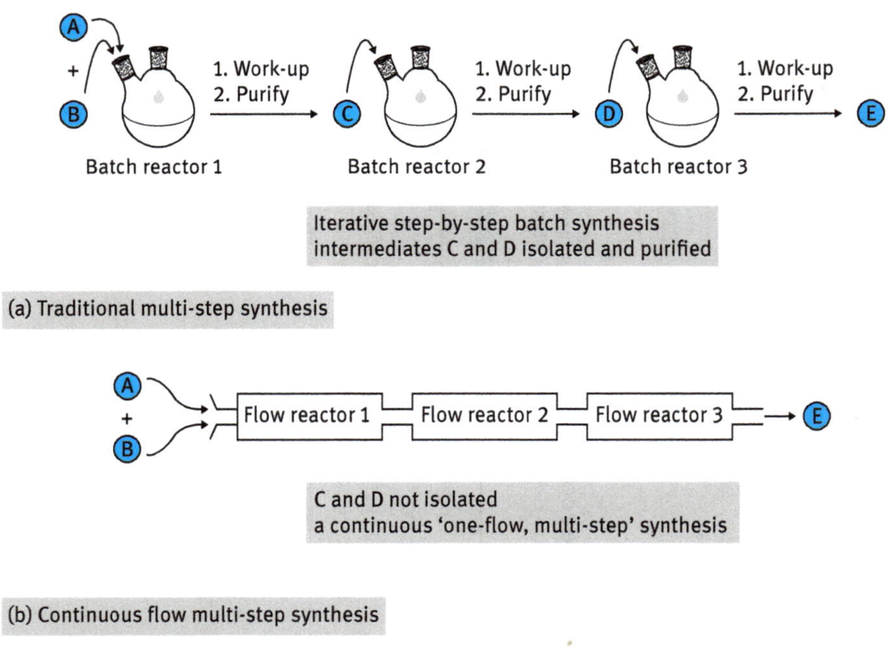

Fig. 1.16: Multistep synthesis strategies in batch and flow (adapted from Webb, D., Jamison, T. F. (2010) Continuous flow multistep organic synthesis, Chem Sci, 1, 675–680).

Definitions and relevant knowledge
- Residence time refers to how long the reactants stay in the reactor zone and is determined by the reactor volume and the bulk flow rate.
- Flow rate is the volume of fluid that passes through a given channel per unit time.

- Flash chemistry is a field of chemical synthesis, where extremely fast reactions involving short-lived highly reactive intermediates are conducted without deceleration, in a highly controlled manner, to produce desired compounds with high selectivity, by virtue of high-resolution reaction control.
- Stoichiometry in flow reactors is defined by the concentration of reagents and the ratio of their flow rate.
- Surface-to-volume ratio is the area per unit volume of the reactor.
- Diffusion is a passive transport due to a concentration gradient, while convection is a transport due to pressure gradient, or, in other words, the collective movement of molecules within fluids.
- In laminar flow regime, radial mixing is strictly due to diffusion. Damkoehler number (Da) shows if a reaction is under diffusion or kinetic control in laminar flow.
- Dispersion along the flow line emerges due to axial and radial diffusion and convection (pressure-driven fluid delivery).

(For further details on this issue, please see Volume 1, Chapter 10, Title: Catalysis in flow and Chapter 11, Title: Gaseous reagents in flow chemistry)

Further readings

- Plutschack, M. B., Pieber, B., Gilmore, K., & Seeberger, P. H. (2017). The hitchhiker's guide to flow chemistry. Chem Rev, 117(18), 11,796–11,893.
- Gutmann, B., Cantillo, D., & Kappe, C. O. (2015) Continuous-flow technology – a tool for the safe manufacturing of active pharmaceutical ingredients, Angew Chem Int Ed, 54(23), 6688–6728.
- Lummiss, J. A., Morse, P. D., Beingessner, R. L., & Jamison, T. F. (2017). Towards more efficient, greener syntheses through flow chemistry. Chem Rec, 17(7), 667–680.
- Noël, T., Su, Y., & Hessel, V. (2015). Beyond organometallic flow chemistry: The principles behind the use of continuous-flow reactors for synthesis, In: Organometallic Flow Chemistry (pp. 1–41). Springer, Cham.
- Fitzpatrick, D. E., Battilocchio, C., & Ley, S. V. (2016). Enabling Technologies for the Future of Chemical Synthesis, ACS Cent Sci, 2(3), 131–8.
- Coley, C.W., Imbrogno, J., Mo, Y., Thomas, D.A., and Jensen, K.F. Flow chemistry system design and automation, In: Jamison, T., & Koch, G. (Eds.). (2018). Science of Synthesis: Flow Chemistry in Organic Synthesis. Georg Thieme Verlag.
- Yoshida, J. I. (2015). Flash chemistry, In: Basics of Flow Microreactor Synthesis. Springer, Tokyo.
- Gérardy, R., Emmanuel, N., Toupy, T., Kassin, V. E., Tshibalonza, N. N., Schmitz, M., & Monbaliu, J. C. M. (2018). Continuous Flow Organic Chemistry: Successes and pitfalls at the interface with current societal challenges, Eur J Org Chem, 2018(20–21), 2301–2351.
- Glasnov, T. (2016). Continuous-Flow Chemistry in the Research Laboratory. Springer New York, Dordrecht, Heidelberg, London.
- Baxendale, I. R. (2013). The integration of flow reactors into synthetic organic chemistry, J Chem Technol Biotechnol, 88, 519–552.
- Wegner, J., Ceylan, S., & Kirschning, A. (2011). Ten key issues in modern flow chemistry, Chem Commun, 47, 4583–4592.

- Hessel, V., Löwe, H., & Schönfeld, F. (2005). Micromixers – a review on passive and active mixing principles, Chem Eng Sci, 60(8–9), 2479–2501.
- Jähnisch, K., Hessel, V., Löwe, H., & Baerns, M. (2004). Chemistry in microstructured reactors, Angew Chem Int Ed, 43(4), 406–446

1.2 Introduction to the basics of microfluidics [14–16]

1.2.1 Basics

Fluid flow through microfluidic channels is characterized by low Reynolds numbers, a dimensionless parameter, which, when less than 2000, results in domination by laminar flow. In this flow regime, the mass transfer of solutes occurs transversely between the characteristic parallel flow profiles, and mixing occurs by diffusive forces.

The static nature of the fluid at the boundary produces a parabolic velocity profile within the channel.

The parabolic velocity profile has significant implications for the distribution of solutes transported within the channel, leading to nonuniformity of diffusion coefficients and greater dispersion of sample plugs [17]. Hydrodynamic (pressure-driven) flow is severely affected by the channel dimensions. A flow that is laminar, incompressible, and viscous, in a channel with its length much greater than its diameter is described by the Hagen–Poiseulle equation as follows:

$$\Delta P = \frac{12\eta LQ}{\pi d^4}$$

where ΔP is the pressure drop across length L of channel diameter, for a flow rate Q of a fluid with viscosity η. As the channel diameter decreases, the pressure needed to achieve the same flow increases dramatically.

Parabolic flow profile of hydrodynamic flow at different length and time is shown in Fig. 1.17. At the beginning of the channel, the velocity vectors are equal across the channel, but further down, fluid flows faster in the center of the channel than near the sides.

1.2.2 Laminar and turbulent flow regimes, the Reynolds number

The flow rate through the single channel determines flow velocity, flow regimes (characterized by the Reynolds number), and pressure loss in the system [18]. With information on the channel geometry and fluid properties, the mean flow velocity and Reynolds

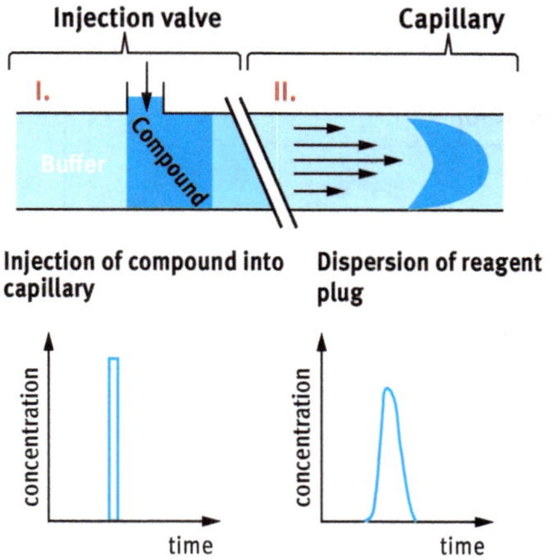

Fig. 1.17: Development of the parabolic profile after initiating the flow (adapted from Werner, M., Kuratli, C., Martin, R. E., Hochstrasser, R., Wechsler, D., Enderle, T., et al. (2014). Seamless integration of dose-response screening and flow chemistry: efficient generation of structure–activity relationship data of β-secretase (BACE1) inhibitors, Angew Chem Int Ed, 53, 1704–1708; and Golbig, K., Kursawe, A., Hohmann, M., Taghavi-Moghadam, S., & Schwalbe, T. (2005). Designing microreactors in chemical synthesis–residence-time distribution of microchannel devices, Chem Eng Commun, 192(5), 620–629).

number, Re, can be determined in the channel. Both give an indication of the flow regime, pressure drop, and energy dissipation rate in the mixing channel.

A substantial effect of reactor miniaturization is that fluid properties become increasingly controlled by viscous forces rather than inertial forces [19]. Reynolds number determines the flow regimes in the channels and describes the ratio between the inertial and viscous forces.

Re number increases with increasing the flow rate and channel diameter; this shifts Re number toward the turbulent flow regime (Fig. 1.18):

$$\mathrm{Re} = \frac{\text{inertial forces}}{\text{viscous forces}} = \frac{\rho v L}{\mu} = \frac{vL}{\nu}$$

where v is the mean velocity of the object relative to the fluid (SI units: m s^{-1}), L is a characteristic linear dimension (traveled length of the fluid; hydraulic diameter when dealing with river systems) (m), μ is the dynamic viscosity of the fluid (Pa·s or N·s m^{-2} or kg (m^{-1}·s^{-1})), ν is the kinematic viscosity ($\nu = \mu/\rho$) (m^2 s^{-1}), and ρ is the density of the fluid (kg m^{-3}).

The flow is

- truly laminar when Re < 100
- transient when 100 < Re < 2,500
- turbulent when 2,500 < Re [20, 21].

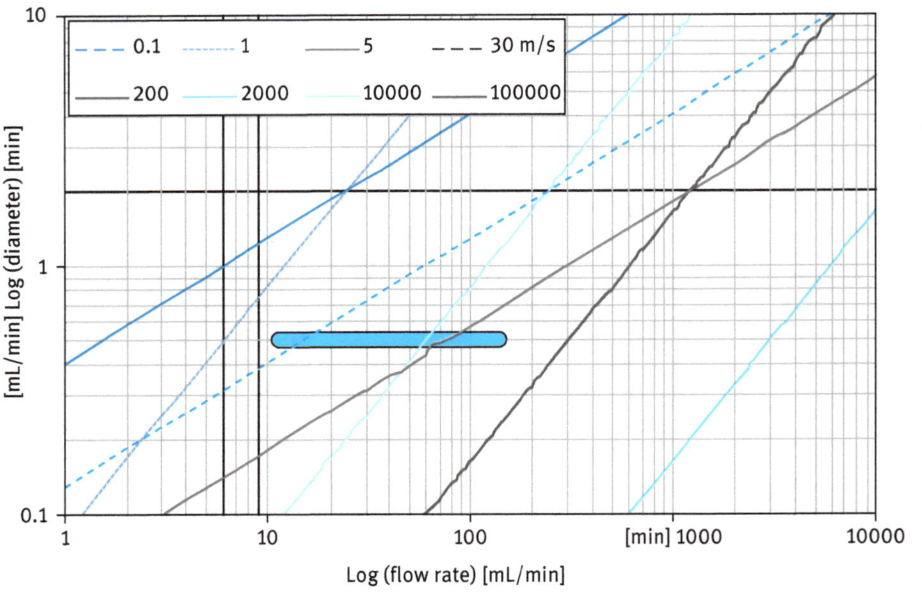

Fig. 1.18: Typical flow rates (mL min^{-1}) and Reynolds numbers (200, 2000, 10,000, and 100,000) in rectangular channels with water at 20 °C (adapted from Kockmann, N., Gottsponer, M., & Roberge, D. M. (2011). Scale-up concept of single-channel microreactors from process development to industrial production, Chem Eng J, 167, 718–726).

In practice, the flow is truly laminar if Re ≤ 10; however, most flow chemistry operates within the Re 100–1,000 range, which rather could be considered as an intermediate flow regime and not typically laminar [22]. At high Re (Re > 2,500), the flow regime is turbulent, and fluid elements exhibit a random motion, which facilitates convective mass transport. At macroscale, mixing is typically achieved by inducing a turbulent flow regime (e.g., generating mesoflow by disturbing the laminar flow; Fig. 1.19).

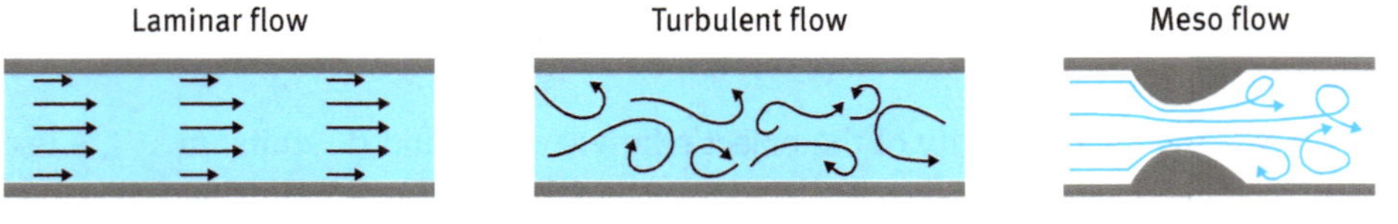

Fig. 1.19: The flow profile of the laminar, turbulent, and meso flow (adapted from John Goodel: http://ccc.chem.pitt.edu/wipf/Frontiers/John_G.pdf).

In packed-bed reactors, a modified Reynolds number (Re_{PB}) is applied. Re_{PB} is used to identify the boundaries of the different flow regimes in a packed bed as follows:

$$\mathrm{Re} = \frac{\rho d_{\mathrm{P}} U_0}{\mu(1-\varepsilon)}$$

where d_p is the spherical particle diameter, ρ is the density of the fluid, μ is the viscosity of the fluid, and ε is the porosity. U_0 is the superficial (empty tube) velocity. U_0 is calculated by dividing the volumetric flow rate (Q) by the cross-sectional area (A).

Porosity (ε) is the ratio of the volume of void space (such as fluids, V_V) and the total volume of material (V_T), including the solid and void components:

$$\text{Porosity} = \frac{V_v}{V_t}$$

The boundaries for the different flow regimes in a packed-bed microreactor include a laminar regime ($Re_{PB} < 10$), a transitional regime ($10 < Re_{PB} < 300$), and a turbulent regime ($Re_{PB} > 300$).

The existence of the particles within the bed will reduce the area available for fluid flow (Fig. 1.20); therefore, to preserve the fluid continuity with the entering superficial, the velocity within the bed (interstitial velocity – U) will be greater than the superficial. If the porosity (ε) is isotropic, U can be calculated as

$$U = \frac{U_0}{\varepsilon}$$

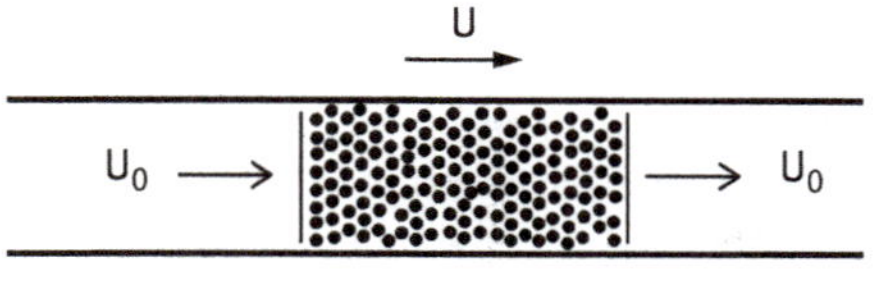

Fig. 1.20: Schematic representation of a packed-bed microreactor and the difference between superficial velocity and interstitial velocity (adapted from Noël, T., Su, Y., & Hessel, V. (2015). Beyond organometallic flow chemistry: the principles behind the use of continuous-flow reactors for synthesis, In: Organometallic Flow Chemistry (pp. 1–41). Springer, Cham).

1.2.3 Axial/radial dispersion versus convection (Bodenstein and Peclet numbers)

There are two dispersion effects along the flow line, due to axial and radial diffusion (Fig. 1.21). However, increasing convection has a more significant effect on the mixing and dispersion.

In flow reactors, convection dominates (axial or radial molecular diffusion is negligible), when the flow regime is in the transient or turbulent region (Reynolds > 1,000).

If axial dispersion is not measurable (in laminar flow), the microflow reactor is said to operate as an ideal PFR.

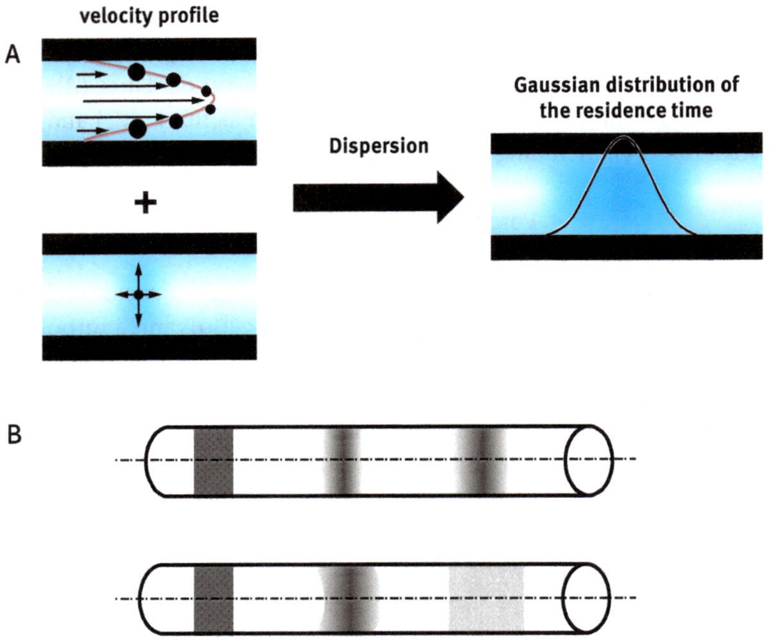

Fig. 1.21: (A) The Gaussian distribution of the residence time as a result of the axial and radial dispersion (Marre, S., & Aymonier, C. (2016). Preparation of nanomaterials in flow at supercritical conditions from coordination complexes. In: Organometallic Flow Chemistry (pp. 177–211). Springer, Cham); (B) Dispersion caused by axial diffusion (above) and convection plus axial and radial diffusion (Ekambara, K., & Joshi, J. B. (2004). Axial mixing in laminar pipe flows, Chem Eng Sci, 59(18), 3929–3944).

The degree of axial dispersion is evaluated by the Bodenstein number, Bo, which relates the axial convective forces from flow to the backmixing from axial dispersion:

$$\mathrm{Bo} = \frac{UL}{D^*}$$

where U is the average velocity of the reaction stream, L is the length of the reactor, and D^* is the dispersion coefficient. Small Bo values indicate that there is a large deviation from plug flow (Fig. 1.22).

Systems with a Bo > 100 have small deviations from plug flow, and systems with Bo < 100 display large deviations from plug flow. Systems with Bo > 1,000 can be approximated as having plug flow behavior.

The Peclet (Pe) number characterizes axial dispersion in another way by expressing the ratio of convection to diffusion. High axial dispersion is particularly due to convection (Pe > 500):

$$\mathrm{Pe} = \frac{N_{\mathrm{conv}}}{N_{\mathrm{diff}}} = \frac{UL}{D_{\mathrm{AB}}}$$

where the molecular diffusion coefficient, D_{AB}, is used to characterize the radial diffusion.

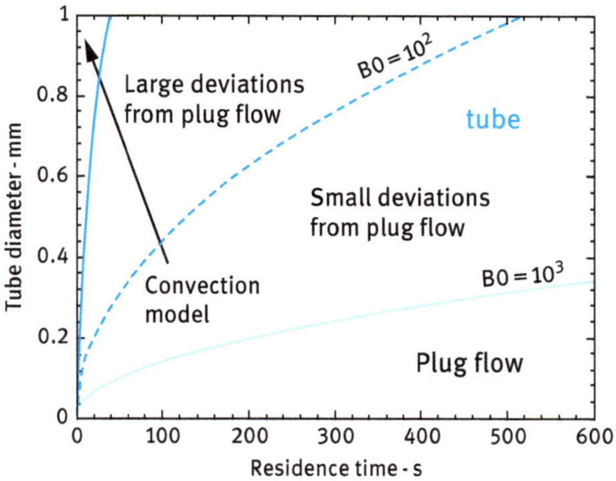

Fig. 1.22: Relationship between the tube diameter, residence time, plug (laminar) flow, and convection (Nagy, K. D., Shen, B., Jamison, T. F., & Jensen, K. F. (2012). Mixing and dispersion in small-scale flow systems, Org Process Res Dev, 16(5), 976–981).

It is obvious that high residence time (low flow rate) and small tube diameter leads to the ideal laminar (plug) flow, while in the opposite case, convection dominates. Similarly, for fast reactions and short tubing length, convection is the primary drive for mixing and mass transport.

1.2.3.1 Controlling the dispersion effect in multiflow systems with inline analytics

The effect of dispersion is important for multistep (flow) reactions. Inline infrared monitoring tool could determine the dispersion effect of the reaction stream. Reagent plugs are sequentially injected in the reaction stream, and this addition needs to be matched in terms of concentration and, thus, stoichiometry, to obtain high conversions and selectivities (Fig. 1.23).

1.2.4 Segmented and Taylor flow [23, 24]

Laminar flow occurs when parallel phases do not interrupt each other's longitudinal flow; slug (segmented) flow with T-mixer could enhance phase mixing. The segmented flow system delivers pulses of reactants that are segregated by an immiscible solvent into the flow reactor. Each segment may consist of a different combination of reactants for different reactions to occur (Fig. 1.24).

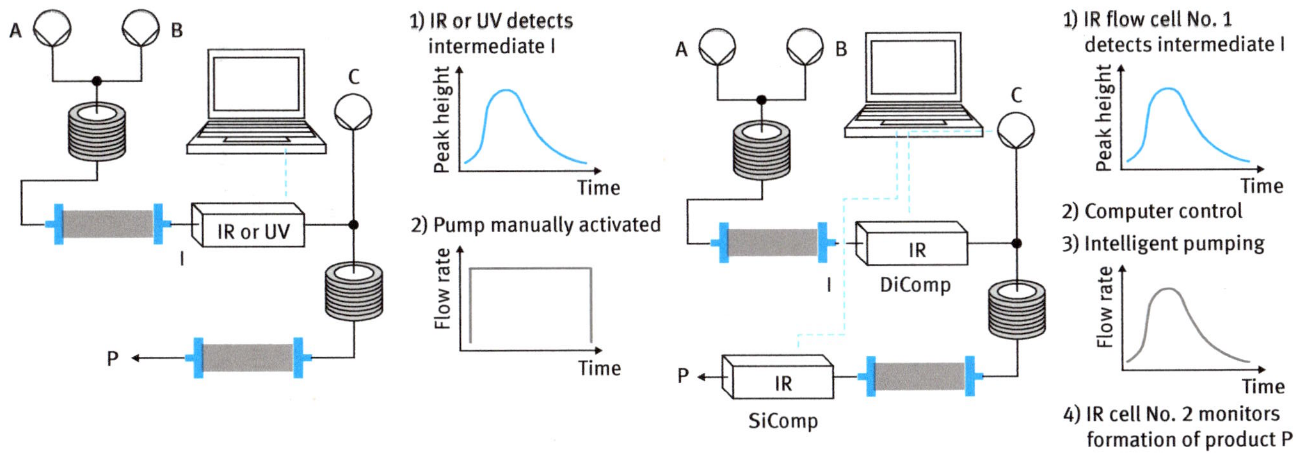

Fig. 1.23: Controlling the dispersion effect in multiflow systems with inline analytics (adapted from Fitzpatrick, D. E., & Ley, S. V. (2018). Engineering chemistry for the future of chemical synthesis, Tetrahedron, 74(25), 3087–3100).

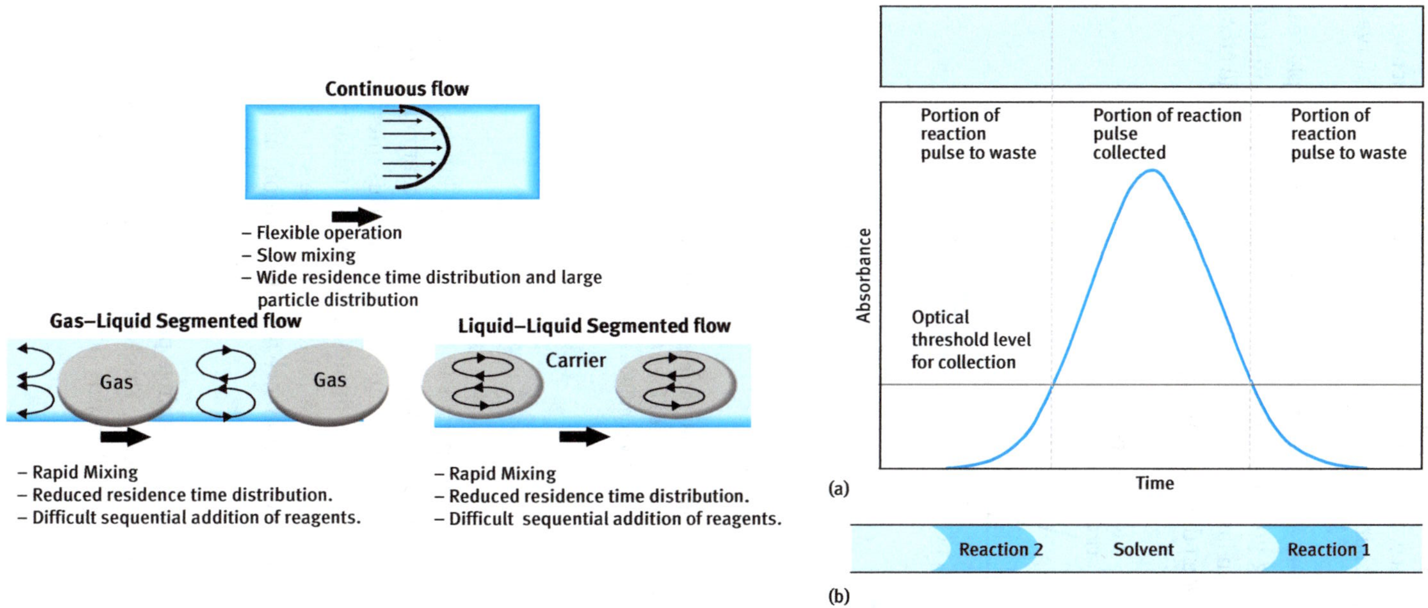

Fig. 1.24: Gas–liquid and liquid–liquid segmented flow, the typical distribution curve, and velocity profile in segmented flow (reaction pulse) (adapted from John Goodell presentation: http://ccc.chem.pitt.edu/wipf/Frontiers/John_G.pdf and Cabeza, V. S. (2016), In: Advances in Microfluidics – New Applications in Biology, Energy, and Materials Sciences. InTech, Rijeka).

A vortex circulation (axial dispersion) is generated within each fluid segment, continuously regenerating the interface. It is believed that the large interface generated, coupled with the internal vortex mixing, is responsible for the experimentally observed rate acceleration [25] (Fig. 1.25).

The slip velocity can be defined as the difference in velocity between two phases. The internal circulation in a slug greatly increases the mixing as compared to that achieved by molecular diffusion in laminar flow. In segmented flow systems, no cross contamination is observed between the reaction slugs.

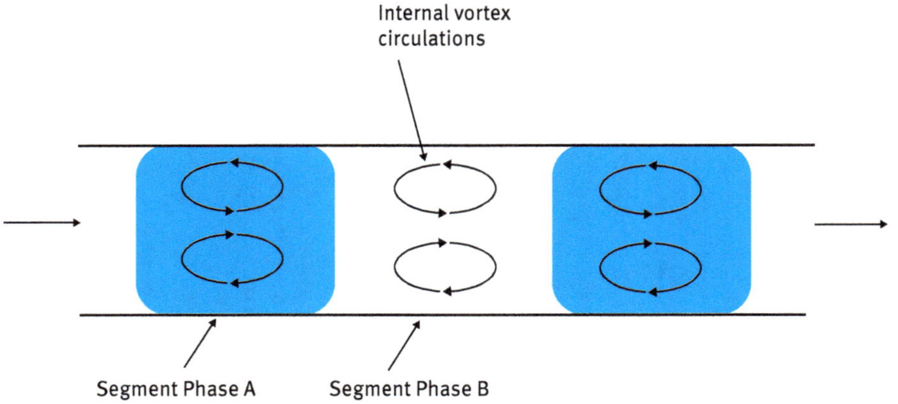

Fig. 1.25: The internal vortex circulation in a slug greatly increases the mixing compared to that achieved by molecular diffusion in laminar flow (adapted from Ahmed-Omer, B., Barrow, D. A., & Wirth, T. (2009) Heck reactions using segmented flow conditions, Tetrahedron Lett, 50, 3352–3355).

1.2.4.1 Exploration of Taylor's flow in gas–liquid segmented flow

Taylor flow is a very uniform flow pattern and is characterized by gas bubbles, which are separated by liquid slugs and a thin liquid film on the channel walls. In the liquid and gas bubble, internal circulations are established. This internal circulation results in homogeneous mixing and facilitates a fast mass transfer between the two phases. A microreactor operating in the Taylor flow regime shows minimal axial dispersion and provides an excellent residence time distribution, making it behave like an ideal PFR [26] (Fig. 1.26).

Example:

The liquid reagents were infused using a syringe pump and merged with an oxygen stream delivered by a mass flow controller. This setup resulted in the formation of a segmented flow regime, in which ideal mixing occurs through Taylor recirculation patterns [27] (Fig. 1.27).

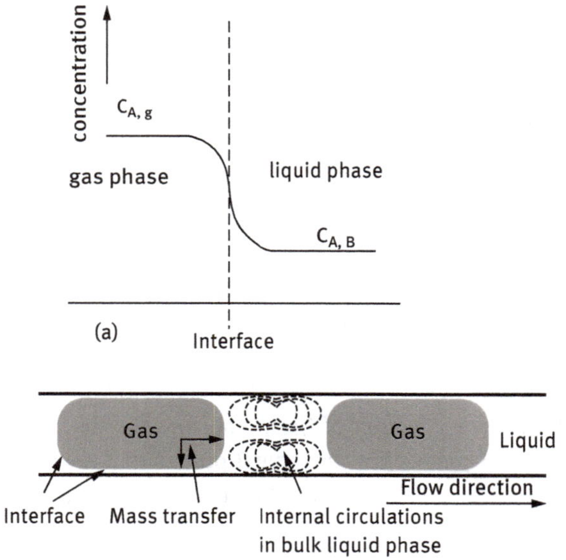

Fig. 1.26: Schematic representation of the Taylor's flow (adapted from Gemoets, H. P., Su, Y., Shang, M., Hessel, V., Luque, R., & Noel, T. (2016). Liquid phase oxidation chemistry in continuous-flow microreactors, Chem Soc Rev, 45(1), 83–117).

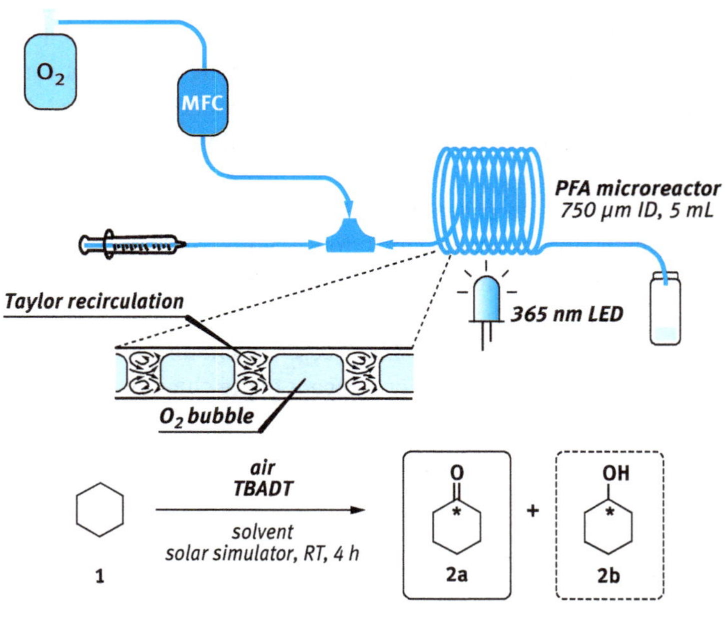

Fig. 1.27: Photochemical oxidation exploiting Taylor's flow (adapted from Laudadio, G., Govaerts, S., Wang, Y., Ravelli, D., Koolman, H. F., Fagnoni, M., et al. (2018). Selective C (sp^3) – H aerobic oxidation enabled by decatungstate photocatalysis in flow, Angew Chem Int Ed, 57(15), 4078–4082).

1.2.5 How to generate continuous flow?

There are three types of pumping/delivering [28] liquids in microchannel: hydrodynamic, electrokinetic, and centrifugal.

1.2.5.1 Hydrodynamic (pressure-driven) pumping

Hydrodynamic pumping exploits conventional or microscale pumps, notably syringe-type pumps, to deliver solutions around the channel network. In that case, flow allows fluid to be moved via positive or negative displacement, such as syringe or peristaltic pumps. Positive displacement pumps are pressure oriented and operate by forcing a fixed volume of fluid from the inlet pressure section of the pump into the discharge zone of the pump. Negative displacement pumps are centrifugal pumps that are flow-oriented pumps in which flow energy is first get converted into kinetic energy (velocity energy) and finally converted into pressure energy.

1.2.5.2 Electroosmotic (electrokinetic) flow (EOF)

Electroosmotic flow (EOF) induces fluid flow by the application of an electrical potential across a microchannel. Ions in solution migrate to the opposite charge, lining the channel wall and creating an electrical double layer of counter ions. The velocity profile of electroosmotic movement in an open channel is flat, exhibiting plug flow. An advantage of EOF is that the velocity profile reduces the diffusional nonuniformity (Fig. 1.28).

EOF is applicable only for aqueous or polar systems, low flow rates, and small channel dimensions.

Principles of EOF: At appropriate pH, a negative surface charge is present on the microreactor walls, which attracts positive ions from solution and forms an electrical double layer. When an electric field is applied along a microreactor's channel, the mobile cations move toward the negative electrode, dragging along the rest of the solution. The flow velocity profile is nearly flat across the channel, except for a thin (few nanometer) diffusive layer immediately adjacent to the channel wall.

The EOF fluid velocity [29] (v_{eof}) is

$$v_{eof} = -\frac{E\varepsilon\varepsilon_0\zeta}{\eta}$$

where E is the electric field (voltage divided by electrode separation), ε is the relative dielectric constant of the liquid, ε_0 is the permittivity of free space, ζ is the zeta potential of the channel wall–solution interface, and η is the liquid viscosity. Thus, solvents that possess a high dielectric constant (i.e., polar solvents) and low viscosity (η) will have higher flow rate. The solvent flow rate is directly proportional to the field strength applied.

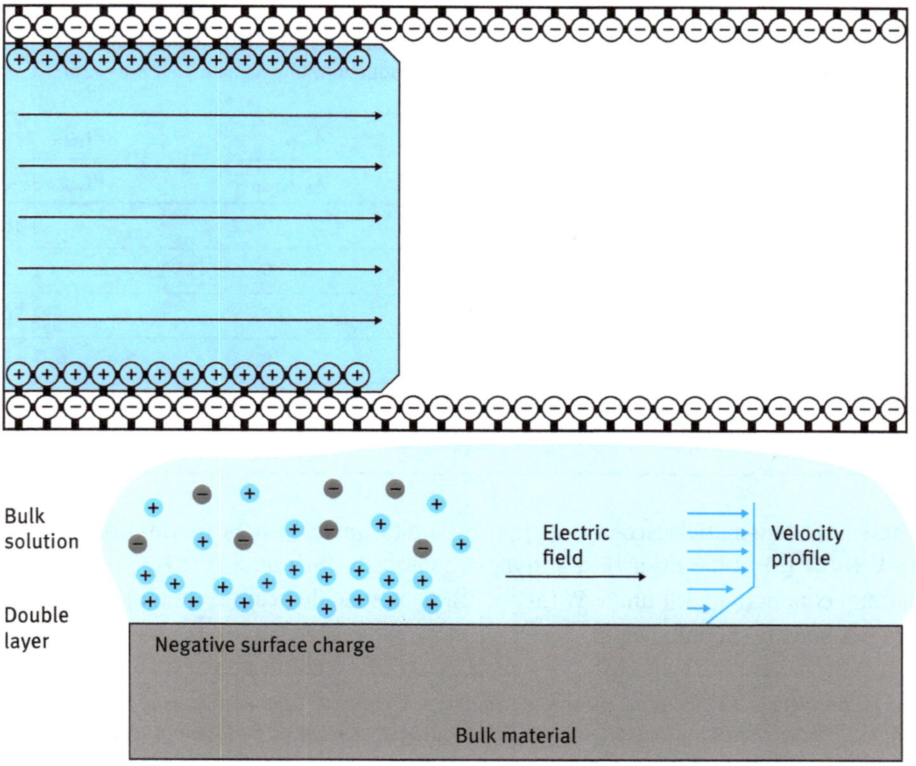

Fig. 1.28: The principles of electroosmotic flow and the flat velocity profile (adapted from Mason, B. P., Price, K. E., Steinbacher, J. L., Bogdan, A. R., & McQuade, D. T. (2007). Greener approaches to organic synthesis using microreactor technology, Chem Rev, 107, 2300–2318 and Webster, A., Greenman, J., & Haswell, S. J. (2011). Development of microfluidic devices for biomedical and clinical application, J Chem Technol Biotechnol, 86, 10–17).

1.2.6 Requirements for the solvent concerning flow regimes

General requirements include low viscosity, odorless (nonvolatile), anhydrous, noncorrosive, and biodegradable. Important characteristics regarding flow kinetics include kinematic viscosity, heat conductivity, and Prandtl number (Tab. 1.3).

The Prandtl number, Pr, is a dimensionless number, the ratio of momentum diffusivity (kinematic viscosity) to thermal diffusivity. It is defined as

$$\mathrm{Pr} = \frac{\nu}{\alpha} = \frac{\text{viscous diffusion rate}}{\text{thermal diffusion rate}} = \frac{c_p \mu}{k}$$

Tab. 1.3: Properties of organic solvents in relation to water properties at 22 C (adapted from Kockmann, N., Gottsponer, M., & Roberge, D. M. (2011). Scale-up concept of single-channel microreactors from process development to industrial production, Chem Eng J, 167, 718–726).

		$\frac{\nu_{solv}}{\nu_{water\ 20\ °C}}$	$\frac{\lambda_{solv}}{\lambda_{water\ 20\ °C}}$	$\frac{Pr_{solv}}{Pr_{water\ 20\ °C}}$
Toluene	295 K	0.24	0.22	0.38
	235 K	0.89	0.25	1.23
Ethanol	295 K	1.63	0.28	2.50
	235 K	6.56	0.31	7.44
THF	295 K	0.55	0.23	1.29
	235 K	1.44	0.45	1.35

where ν is kinematic viscosity, $\nu = \mu/\rho$ (SI units: $m^2\ s^{-1}$); α is thermal diffusivity, $\alpha = k/(\rho c_p)$ (SI units: $m^2\ s^{-1}$); μ is dynamic viscosity (SI units: Pa s = N s m^{-2}); k is thermal conductivity (SI units: W ($m^{-1}\ K^{-1}$)); C_p is specific heat (SI units: J ($kg^{-1}\ K^{-1}$)); and ρ is density (SI units: kg m^{-3}).

i *(For further details on this issue, please see Volume 1, Chapter 2, Title: Principles of controlling reactions in flow chemistry and Chapter 4, Title: Practical aspects of performing continuous flow chemistry)*

Definitions and relevant knowledge

- Reynolds number determines the flow regimes in the channels and is defined as the ratio of momentum (inertial) forces to viscous forces.
- The segmented flow system delivers pulses of reactants that are segregated by an immiscible solvent into the flow reactor. Each segment may consist of a different combination of reactants for different reactions to occur.
- Axial dispersion causes some backmixing in the flow reactors. The degree of axial dispersion is evaluated by the Bodenstein number (Bo). Small Bo values indicate that there is a large deviation from the ideal plug flow. During multistep (flow) reactions, inline infrared monitoring could determine the dispersion effect of the reaction stream allowing matching of the concentration profile by the sequential reagent stream.
- Taylor flow is a very uniform flow pattern and characterized by gas bubbles, which are separated by liquid slugs and a thin liquid film on the channel walls. In the liquid and gas bubble, internal circulations are established. This internal circulation results in homogeneous mixing and facilitates a fast mass transfer between the two phases.
- Hydrodynamic pumping exploits conventional or micro-scale pumps, notably syringe-type pumps, to deliver solutions around the channel network.
- EOF induces fluid flow by the application of an electrical potential across a microchannel. Ions in solution migrate to the opposite charge, lining the channel wall and creating an electrical double layer of counter ions.

Further readings

- Rossetti, I., & Compagnoni, M. (2016). Chemical reaction engineering, process design and scale-up issues at the frontier of synthesis: flow chemistry, Chem Eng J, 296, 56–70.
- Hone, C. A., & Kappe, C. O. (2020). The use of molecular oxygen for liquid phase aerobic oxidations in continuous flow, In: Accounts on Sustainable Flow Chemistry (pp. 67–110). Springer, Cham.
- Rossetti, I. (2018). Continuous flow (micro-)reactors for heterogeneously catalyzed reactions: main design and modelling issues, Catal Today, 308, 20–31.
- Jensen, K. F. (2017). Flow chemistry – microreaction technology comes of age, AIChE Journal, 63(3), 858–869.
- Fitzpatrick, D. E., & Ley, S. V. (2018). Engineering chemistry for the future of chemical synthesis, Tetrahedron, 74(25), 3087–3100.
- Hawbaker, N., Wittgrove, E., Christensen, B., Sach, N., & Blackmond, D. G. (2016). Dispersion in compartmentalized flow systems: influence of flow patterns on reactivity, Org Process Res Dev, 20, 465 – 473.
- Cabeza, V. S. (2016). In: Advances in Microfluidics – New Applications in Biology, Energy, and Materials Sciences. InTech, Rijeka.
- Noël, T., Su, Y., & Hessel, V. (2015). Beyond organometallic flow chemistry: The principles behind the use of continuous-flow reactors for synthesis, In: Organometallic Flow Chemistry (pp. 1–41). Springer, Cham.
- Kolb, G., & Hessel, V. (2004). Micro-structured reactors for gas phase reactions, Chem Eng J, 98(1–2), 1–38.

1.3 Flow arts: theory and practice

1.3.1 High-resolution reaction time control

Residence time inversely correlates with the flow rate as was shown in Section 1.1 (Fig. 1.4). The residence time between the addition of a reagent and that of a quenching agent or the next reagent in a flow microreactor is the reaction time, and the reaction time can be greatly reduced by adjusting the length of a reaction channel in a flow microreactor or by increasing the flow rate. If a reaction has consecutive or multiple elementary steps, the potential products or intermediates can be isolated as major products by simply adjusting or focusing the residence time ("reaction window") to that particular chemical step. The principle of high-resolution reaction time control is simply adjusting the residence time to a particular step in a multistep reaction sequence. Thus, the isolation of the intermediate products can be resolved by the residence time (partial reaction time), which is not possible in batch reactors [30, 31].

A typical example is the selective hydrogenation of 3-methyl-1-pentyn-3-ol over a Pd catalyst to the desired 3-methyl-1-penten-3-ol (P1); but overhydrogenation could also happen to 3-methyl-3-pentanol (P2) [32] (Fig. 1.29).

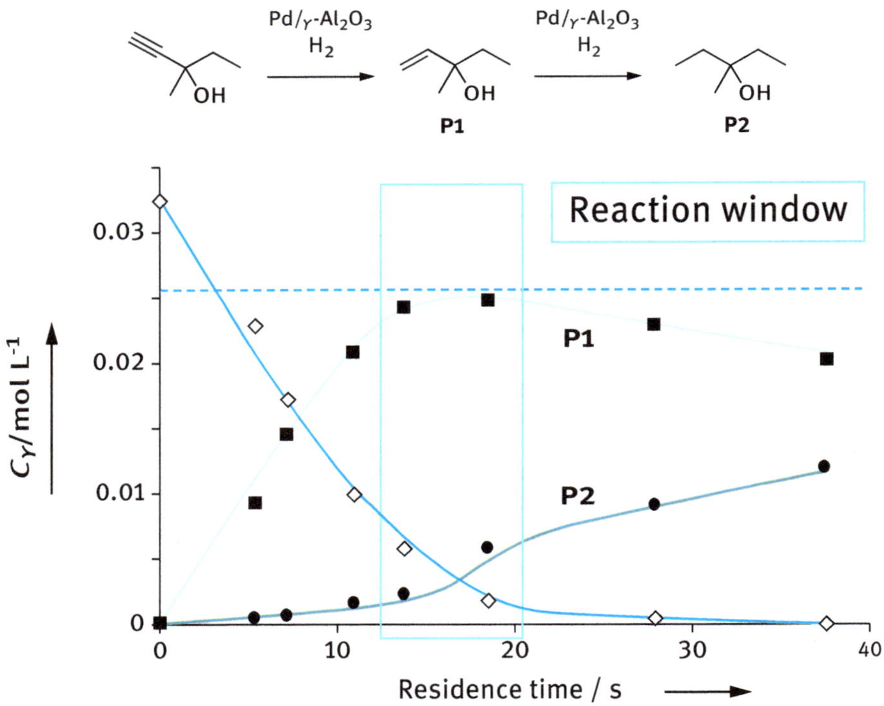

Fig. 1.29: Selective hydrogenation of 3-methyl-1-pentyn-3-ol (adapted from Bakker, J. J., Zieverink, M. M., Reintjens, R. W., Kapteijn, F., Moulijn, J. A., & Kreutzer, M. T. (2011). Heterogeneously catalyzed continuous-flow hydrogenation using segmented flow in capillary columns, ChemCatChem, 3(7), 1155–1157).

1.3.2 Mass transfer in continuous flow [33]

The small dimensions facilitate increased mass transfer in microflow reactors. Comparison of the typical liquid–liquid extraction methods in batch and flow: While in batch, rigorous shaking is necessary to improve the mass transfer between the aqueous and organic phases, in the extraction process in flow, simple laminar flow with high interfacial surface allows the rapid mass transfer. On the other hand, for fast reactions, the rate of molecular diffusion should be increased by convective mass transport and mixing (Fig. 1.30).

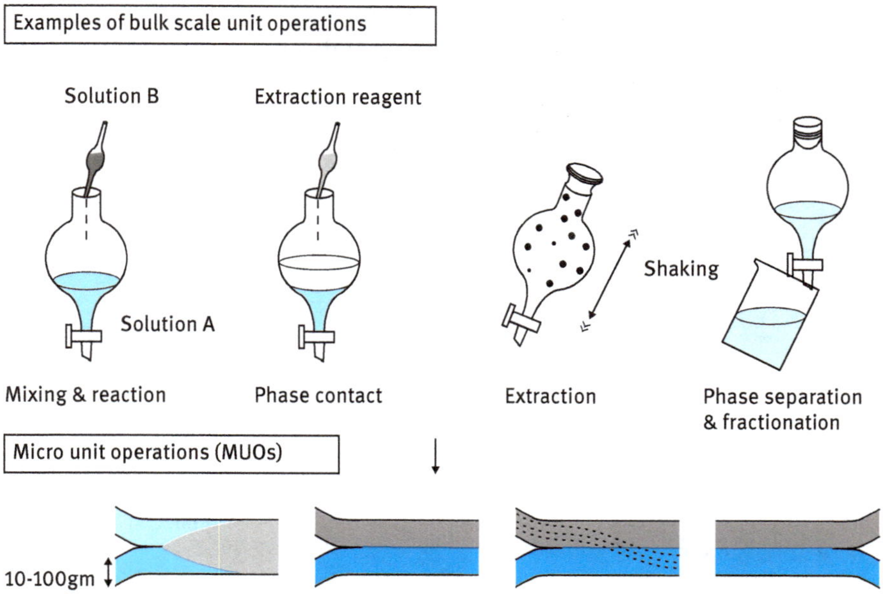

Fig. 1.30: Comparison of the mass transfer in batch and flow (adapted from Mawatari, K., Kazoe, Y., Aota, A., Tsukahara, T., Sato, K., & Kitamori, T. (2012). Microflow systems for chemical synthesis and analysis: approaches to full integration of chemical process, J Flow Chem, 1(1), 3–12).

1.3.3 Temperature control in continuous flow [34]

Heat is transferred between the interior and exterior of a reactor via the reactor surface, according to the theory of heat transfer. Therefore, area per unit volume of the reactor is a crucial factor for heat transfer (surface-to-volume ratio).

Example 1: Heat transfer

$$\frac{Q}{t} = -kS\frac{T1 - T2}{d}$$

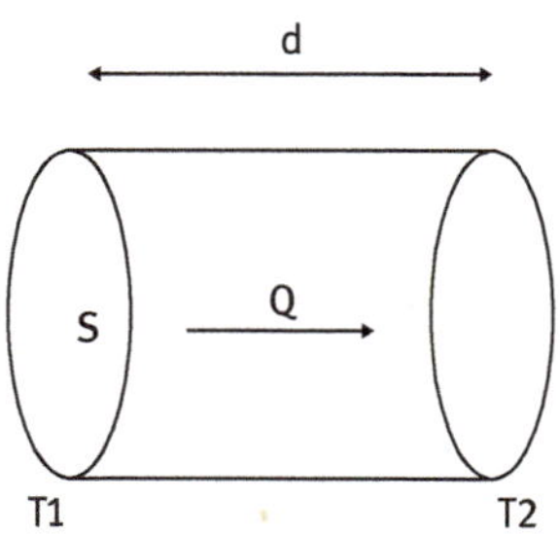

Heat transfer per unit time (Q amount of heat transferred; t time taken) depends on the following parameters: k is the conductivity of the material; S is surface area; d is the distance between the two ends; T1 is higher temperature end; and T2 is lower temperature end.

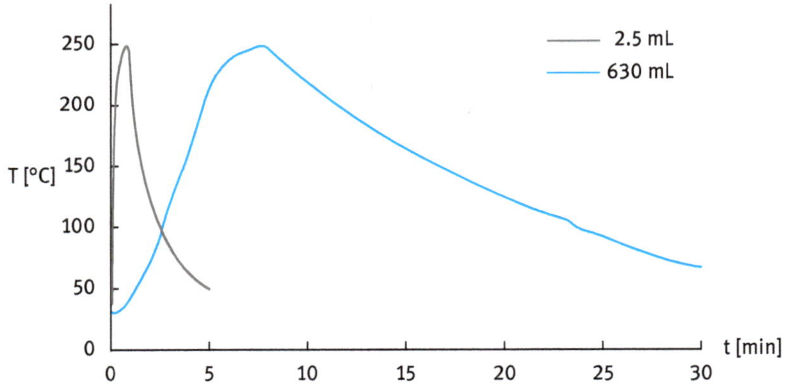

Fig. 1.31: Temperature profile of microreactor and batch reactor upon heating (adapted from Glasnov, T. N., & Kappe, C. O. (2011). The microwave-to-flow paradigm: translating high-temperature batch microwave chemistry to scalable continuous-flow processes, Chem Eur J, 17, 11,956–11,968).

A standard microreactor with channel or capillary dimensions of 1 mm is inherently very fast, owing to the high surface-to-volume ratio in these systems, especially when using steel- or silicon-based reactor materials that possess high thermal conductivity. Fig. 1.31 shows that a 2.5 mL volume microreactor could reach the targeted temperature (250 °C) in less than a minute, while a 630 mL flask requires 8 min to reach the same temperature.

Example 2: Heat removal in exothermic reactions

Similarly removing the excess heat in highly exothermic reactions is only feasible in so-called heat absorbing microreactors rather than in flasks. For exothermic reactions, the rate of heat transfer in a reaction system is dependent on several factors:

- rate of heat generated from the reaction, $-r\Delta H_{rxn}$
- rate of heat removed from reaction mixture through conduction or convection to the reactor wall
- rate of heat removed from the reactor wall through conduction to the surroundings

The heat of reaction (ΔH_{rxn}), adiabatic temperature rise (ΔT_{ad}), and reactor dimension (diameter, d_f) can help us make a useful comparison of heat generation and removal represented by a ratio. (Adiabatic temperature rise (ΔT_{ad}) is the increase in temperature of a reaction mixture, when there is no heat transfer to or from the environment). For flow systems, thermal conduction from the center of the channel to the reactor walls is the driving force for heat transfer. The ratio of heat generation and removal is then estimated as

$$\beta_B = \frac{\text{heat generated}}{\text{heat removed}}$$

$$\beta_B = \frac{-r\Delta H_{\text{rxn}} d_F^2}{4\Delta T_{\text{ad}} \kappa}$$

where d_F is the diameter of the flow reactor channel and k is the thermal conductivity of the reaction mixture. Heat removal can be achieved in microreactors when the respective β value is less than 1, unlike in batch reactors.

Fig. 1.32 outlines the temperature characteristic of the microreactor (inner volume 2 ml, surface-to-volume ratio 95) compared to a 100 ml flask (surface-to-volume ratio 1) in a cryogenic lithium/bromide exchange experiment. Since the inner temperature of the vessel rises close to the boiling point of the solvent (THF), the cooling system of the microreactor keeps it strictly to 0 °C, after a short period of equilibration.

In summary, microreactors achieve efficient input or removal of heat and nearly constant reaction temperatures because of their high surface-to-volume ratios.

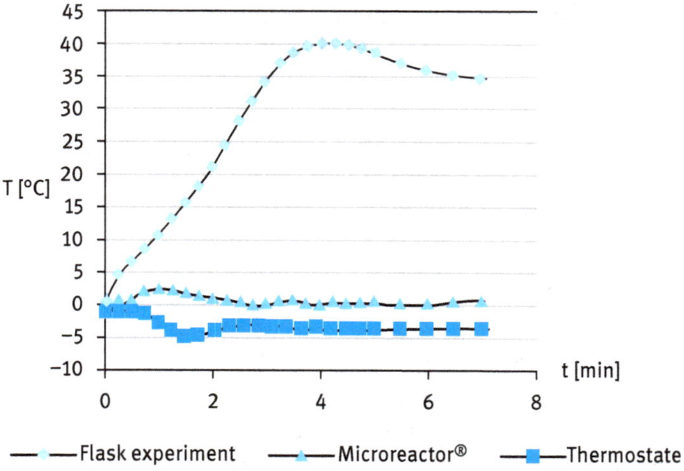

Fig. 1.32: Temperature increase in the microreactor and 100 ml flask (ice cooling) for the lithium/bromide exchange.

Example 3: Low-temperature control (rapid cooling)

Accelerated heat transfer allows rapid cooling in multistep flow synthesis. Jamison and coworkers reported and integrated a three-step continuous-flow system for the preparation of ortho-functionalized phenols [35]. In order to avoid side reactions, the reaction flow had to be cooled from 80–120 °C to –25 °C within the residence time of 1.2 min. Such rapid cooling can not be achieved in batch reactor systems (Fig. 1.33).

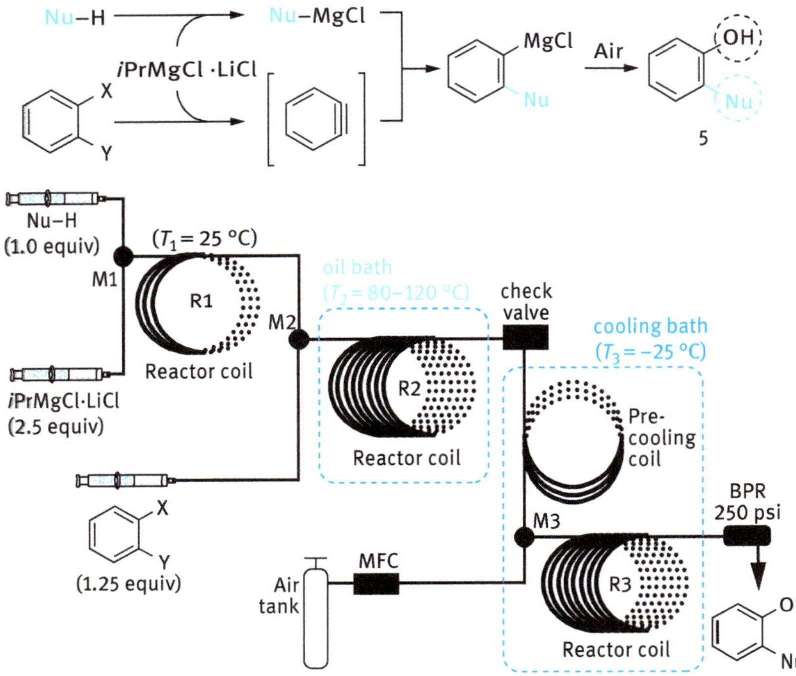

Fig. 1.33: Synthesis of ortho-functionalized phenols by an integrated flow system (adapted from He, Z., & Jamison, T. F. (2014). Continuous-flow synthesis of functionalized phenols by aerobic oxidation of Grignard reagents. Angew Chem Int Ed, 53(13), 3353–3357).

Highly reactive intermediates react with a substrate to generate the products within milliseconds to seconds.

These flash chemical reactions are highly exothermic and often diffusion-controlled; therefore, miniaturized flow devices are ideal for performing such transformations due to the excellent mixing and heat transfer properties. While such reactions require very low temperature in batch (typically –78 °C), they could be carried out at ambient or moderately low temperatures in these types of reactors (Fig. 1.34).

Example 4: Precise energy (heat) transfer to achieve selective activation

Reactions that offer two potential products from either kinetic or thermodynamic pathways are very sensitive to temperature. Batch reactors often provide broad temperature profiles that can allow access to multiple pathways, when only one pathway is desired. Fig. 1.35 compares the temperature distributions in batch and in a microreactor to the kinetic energy needed to access a by-product-forming pathway.

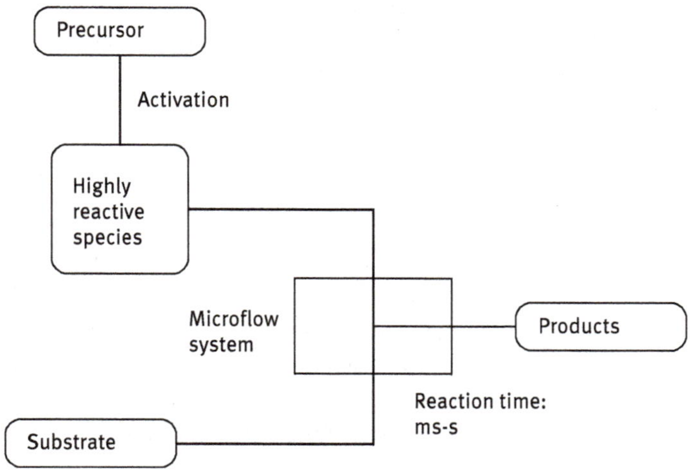

Fig. 1.34: Typical flow reactor connection to fast chemical reactions (adapted from Yoshida, J. I., Nagaki, A., & Yamada, T. (2008). Flash chemistry: fast chemical synthesis by using microreactors. Chem Eur J, 14, 7450–7459).

The batch reactor's broad temperature distribution allows the production of an undesired product, C, but the narrow temperature distribution in the microreactor restricts the reaction to the target product B [36].

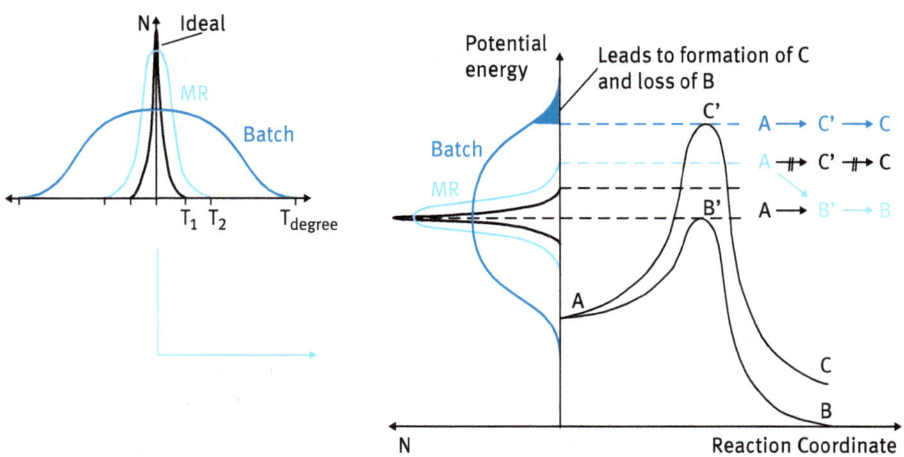

Fig. 1.35: Energy profile of a competing chemical reaction indicating the effect of a narrow (flow) and broad (batch) temperature distribution profile (adapted from Schwalbe, T., Autze, V., Hohmann, M., & Stirner, W. (2004). Novel innovation systems for a cellular approach to continuous process chemistry from discovery to market. Org Process Res Dev, 8(3), 440–454).

(For further details on this issue, please see Volume 1, Chapter 2, Title: Principles of Controlling Reactions in Flow Chemistry and Chapter 5, Title: When and How to Start Flow Chemistry)

1.3.4 Novel process windows

Slow reactions (reaction time >30 min) require significant rate acceleration. All these efforts are expressed in a systemic approach "novel process windows" [37, 38] (Fig. 1.36). NPW operates at process conditions that are beyond the usual condition – that considerably speed up conversion rates, while maintaining selectivity. This can be achieved by an increase in temperature, pressure, or concentration (solvent-free operation), by simplification of process protocols or by function integration [39]. Combined high-temperature and high-pressure flow regime, that is, >200 °C and >50 bar, is one of the main directions, with many applications focusing on the generation of high-temperature or supercritical water (scH_2O). NPW combines many different techniques including supercritical fluids, ionic liquids, fluorous media, microwaves, ultrasound, high pressure, physico- and electrochemical activation, microreactors, and ball-milling conditions [40].

High-temperature/high-pressure conditions shift routine process to harsher conditions that became an achievable reality with microreactors (Fig. 1.37).

Two emerging areas in NPW are the superheated and supercritical conditions.

Transformations in supercritical conditions have significant advantages, since all reactants (solvents and gases) can be brought into one phase. Supercritical fluids are gases, and they mix completely with other gases, such as hydrogen. In these cases, the access of the gaseous reactant to the catalyst is not limited by mass transfer resistance across a phase boundary (Fig. 1.38; Tab. 1.4).

1.3.5 Process intensification

Process intensification (PI) [41] is a term that is closely related to NPW and can be defined as the ability to obtain equivalent or better results in terms of purity, selectivity, and yield of the desired product in a reduced period of time, and therefore with an enhanced throughput, by increasing parameters such as temperature and pressure. A recent definition is provided by Baldea et al.: "any chemical engineering development that leads to substantially smaller, cleaner, safer and more energy efficient technology or that combine[s] multiple operations into fewer devices" [42].

Most chemical reactions are not processed under kinetically controlled conditions, but rather under mass- or heat transfer-controlled conditions. The class of such "slow" reactions is quite substantial, since it includes many major chemical transformation such as nucleophilic substitutions. Applying process intensification

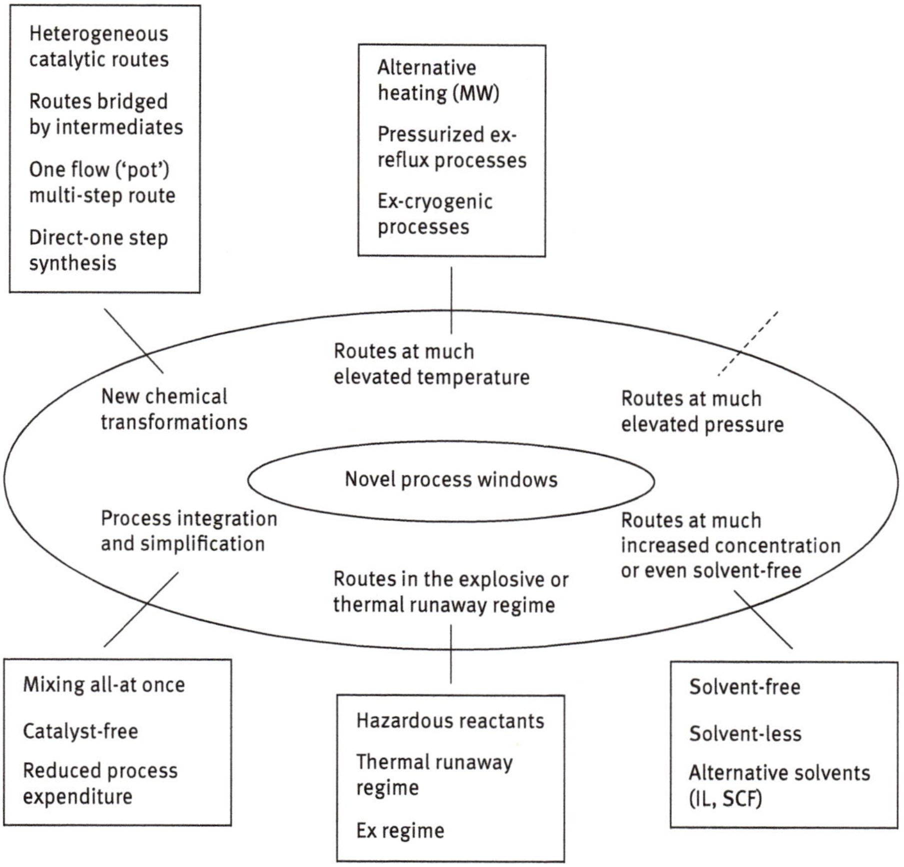

Fig. 1.36: Major path for novel process windows (adapted from Illg, T., Löb, P., & Hessel, V. (2010). Flow chemistry using milli- and microstructured reactors – from conventional to novel process windows. Bioorg Med Chem, 18, 3707–3719).

in microreactors, reaction times can be further reduced from minute to second level by exploiting much faster kinetics, which could be termed "intensified intrinsic kinetics." Similarly, reaction speed increases were found for the sealed microwave-operated vessels. That approach was then transferred to superheat processing in pressurized tubes, which bridge to the continuous-flow microreactor technology. In summary, various chemical reactions, including flow reactions, could be shifted to faster reaction rate category (intensified chemistry), by applying NPW and MCPT (microchemical processing technology) [43] (Fig. 1.39).

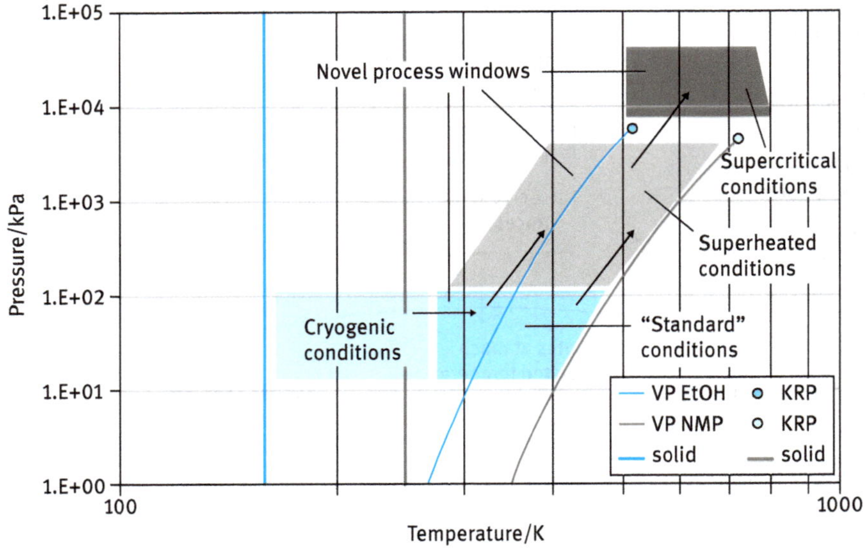

Fig. 1.37: Novel process windows shifts the reactions toward high temperature/pressure realization, while increasing the reaction rate (adapted from Hessel, V. (2011) Adding a Chemical and Process Intensification Field to Flow Chemistry Transport-Engineered Through Microreactors – Novel Process Windows, FROST3, Budapest).

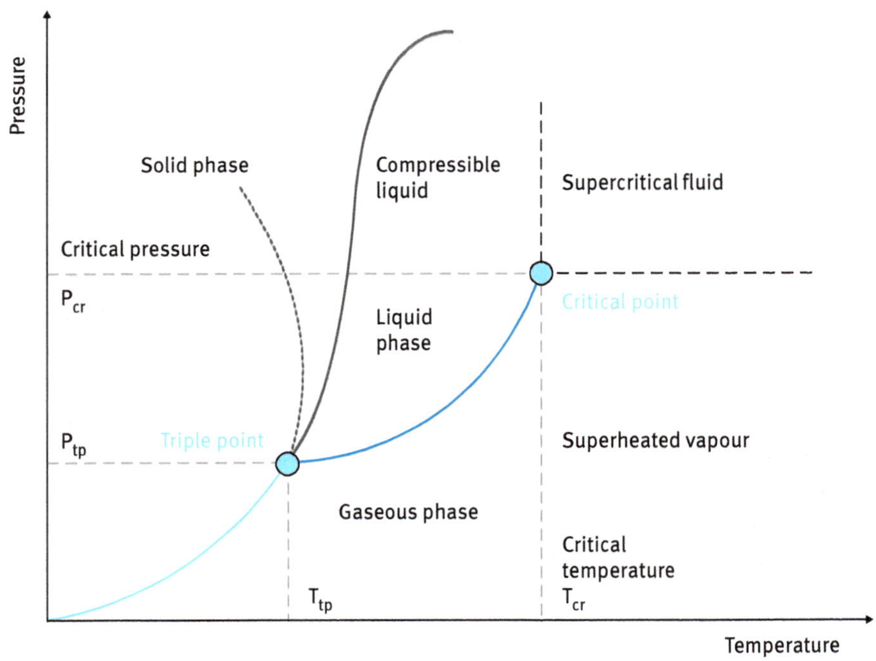

Fig. 1.38: A typical temperature/pressure plot displaying superheated and supercritical conditions.

Tab. 1.4: Common organic solvents and their critical temperatures and pressures (adapted from Razzaq, T., & Kappe, C. O. (2010). Continuous flow organic synthesis under high-temperature/pressure conditions, Chem–Asian J, 5, 1274–1289).

Solvent	Critical temperature T_c (°C)	Critical pressure p_c (bar)	Solvent	Critical temperature T_c (°C)	Critical pressure p_c (bar)
Acetic acid	322	58	Ethyl acetate	250	38
Acetone	236	48	Hexane	234	30
Acetonitrile	275	48	Methanol	240	80
Benzene	289	49	*n*-Octane	296	25
1-Butanol	290	44	2-Propanol	235	48
Chloroform	263	55	Pyridine	347	57
Cyclohexane	280	41	Toluene	321	42
DCM	237	61	Tetrachloromethane	283	46
DMF	376	44	THF	267	52
DME	263	39	Trifluoroacetic acid	218	33
Dioxane	315	52	Water	374	221
Ethanol	243	64	*p*-Xylene	345	34

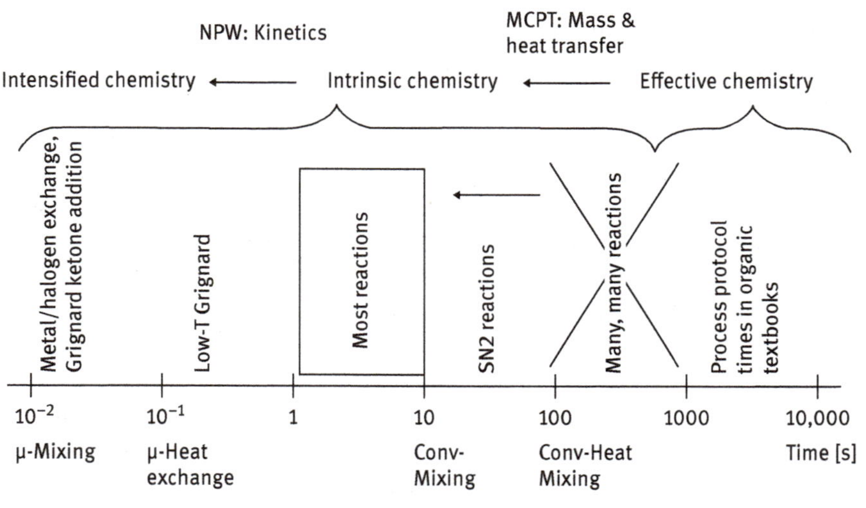

Fig. 1.39: Process intensification unites NPW and microchemical processing technology in order to accelerate reaction rates (adapted from Hessel, V., Cortese, B., & De Croon, M. H. J. M. (2011). Novel process windows – concept, proposition and evaluation methodology, and intensified superheated processing, Chem Eng Sci, 66, 1426–1448).

(For further details on this issue, please see Volume 1, Chapter 8, Title: Mitigation of chemical hazards under continuous flow conditions)

1.3.6 On-demand flow synthesis

Recent advances in flow chemistry, including closed (safe) environment, small-volume instrumentation, process intensification, multistep procedures with telescoping, in-line analytics, process control, and remote, AI-assisted operation, all lead to greatly improved synthesis platforms.

Such modules open up new opportunities and transform the present fine chemicals and pharmaceutical batch production practice into continuous manufacturing. Furthermore, the fine chemicals and pharmaceutical industry is taking numerous steps to improve the efficiency, safety, and sustainability of their manufacturing processes.

Integrated flow synthesis modules contributed to the emergence of a novel production paradigm – on-demand manufacturing. Thus, production could only be initiated if the products are needed or the precursors are available (JIT = just-in-time production). On-site, on-demand production of hazardous starting materials that are immediately used after generation saves transport, storage, and handling of these agents. On-demand production is also feasible at remote (e.g., in deserts or in space) or multiple locations.

1.3.6.1 On-demand continuous-flow manufacturing and remote control

For on-demand continuous-flow manufacturing, Adamo and Jensen [44] developed a compact, reconfigurable, on-demand manufacturing platform (Fig. 1.40). This flexible plug-and-play approach combines the advances of continuous-flow synthesis, complex multistep sequence telescoping, reaction engineering equipment, and real-time formulation.

A robotically controlled experimental flow platform [45] that combines artificial intelligence–driven synthesis planning relieves chemists from various routine tasks. Synthetic routes are proposed through generalization of millions of published chemical reactions and validated in silico to maximize their likelihood of success. The modular continuous-flow platform is automatically reconfigured by a robotic arm to set up the required unit operations and carry out the reaction. The platform is supported by computational solubility calculation and by predicting suitable purification procedures.

Poliakoff et al. [46] introduced a remotely controlled flow production line and the concept of cloud chemistry. The flow line contains a self-optimizing catalytic reactor, which feeds data from the in-line analysis into an algorithm to calculate new reactor parameters (flow rate, temperature, etc.) for iterative optimization. A single computer is used to control all components of the reactor and the in-line analysis system (pumps, heaters, gas chromatograph, and so on), as well as to implement the optimization algorithm.

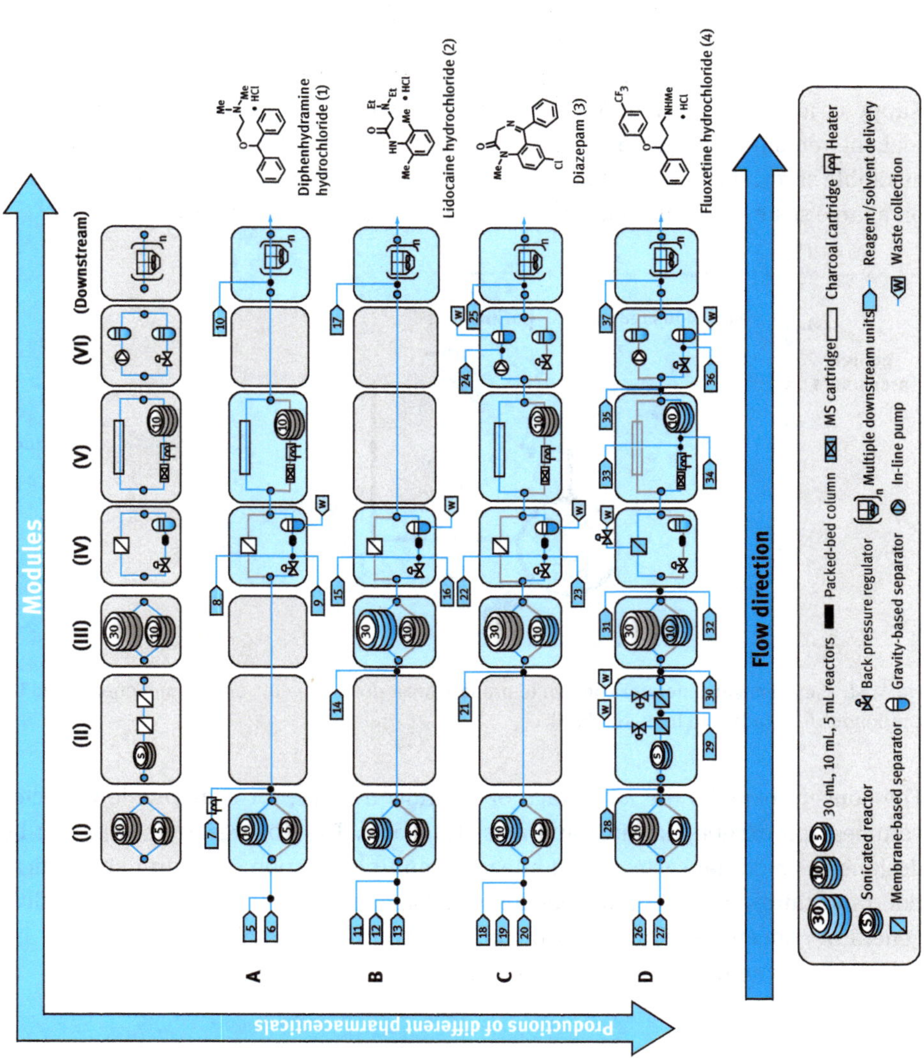

Fig. 1.40: A compact, reconfigurable, on-demand manufacturing platform (adapted from Ref. [44]).

Cloud chemistry requires remote operators located across the world. The remote operator has full control of the local computer, can activate any of the automated functions available on the reactor, and can remotely access all of the data generated during the experiment. In this way, experimental reactions can be controlled globally.

Ley et al. [47] describe state-of-the-art tools for the future of automated reaction control including remote control supported by various hardware (e.g., webcams), and remotely hosted desktop software tools. The automated reaction control system also implements many cloud-based tools to assist them with literature searching, electronic laboratory notebooks, and internet-based chemical inventory management.

1.3.6.2 On-demand synthesis of hazardous materials

Kappe et al. [48] developed the chemical generator concept to prepare hazardous reagents on-site and on-demand in continuous flow from inexpensive and readily available starting materials (Fig. 1.41). Such reagents improve safety risk, especially on a large scale, and particularly in transport, storage, and handling.

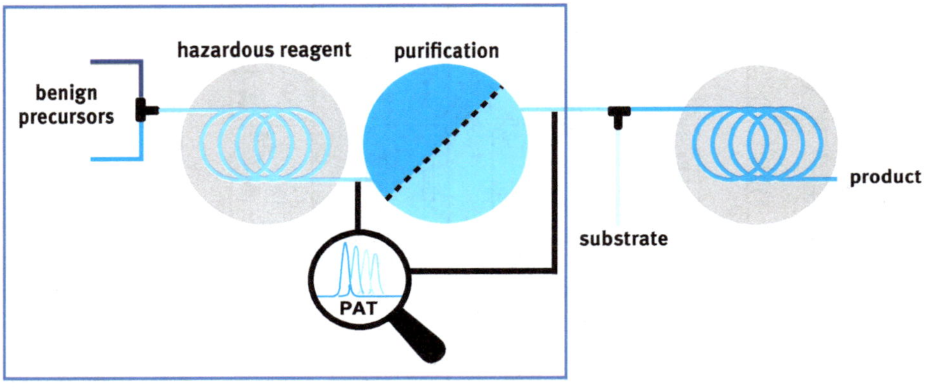

Fig. 1.41: The chemical generator concept to prepare hazardous reagents on-site and on-demand in continuous flow (adapted from Ref. [48]).

The concept exploits the advantage of the flow devices, for example, low reactor volumes and exceptional heat and mass transport. The flow reactor setups can be designed as modular systems that integrate multiple reaction steps, in-line purification procedures, and real-time reaction control, employing PAT tools. The chemical generator concept provides a reasonable solution for a chemical society that is increasingly committed to efficiency, sustainability, and safety.

1.3.6.3 On-demand space chemistry

One promising and fast developing area where flow chemistry has a decisive role is space chemistry.

Space chemistry refers, here, to chemical and certain physical transformations performed in the outer space. The physical transformations we refer here are normally associated with chemical reactions (like purification, crystallizations, or other morphological transformations [49].

There are a number of main differences between space and the Earth in terms of performing chemical reactions. The most important factor is the presence of microgravitation. As a consequence, mixing in the space is very poor, leading to serious difficulties in performing batch reactions. Strong and peculiar radiation also complicates the job of chemists.

The first chemical reactions in space were performed by Russian scientists in the 1960s. Systematic space chemistry started around 2015, when flow chemistry started to substitute batch reactions. Today, the large majority of chemical reactions and associated physical transformations are performed in flow reactors that are fully automated, remote-controlled, and suitable to provide very effective mixing.

A space chemistry laboratory is operated either under local or by remote control.

Local control can be exerted only at the International Space Station (ISS), where the EU, USA, Japanese, and Russian modules provide separate opportunities for experiments. Instruments and chemicals are delivered to the ISS by a rocket (e.g., SpaceX), or by a space shuttle (e.g., provided by Russian corporations). Return deliveries are organized in the same way.

Starting 2023, ESA's Space Rider (a returnable space shuttle) [50] will also provide opportunity for returnable experiments and instruments.

Experiments in microgravity produced by freefall (shorter than 8 min) can be organized onboard special aircrafts, which, for example, are provided by ESA for research projects.

Remote-controlled space chemical laboratories certainly will have a growing importance for space chemistry in the future [51]. The control can be directly human, automatic, or robot- operated. It is expected that the micro- and nanosatellites (that have growing importance in world telecommunication) will be fitted with small-size chemical laboratories.

While the chemical laboratories without human presence in space will certainly gain importance in the future, presently, only one pioneering realization of this brilliant concept is known: the DIDO satellites, developed by the SpacePharma corporation in Israel and Switzerland [52].

The DIDO satellites, also serving pharmaceutical and biological experimentations, however, cannot be returned to the Earth. Thus, they have to communicate their analytical results to their control station on the Earth.

Flow chemistry instruments enabling chemistry in space are, at present, adapted versions of instruments developed for producing flow reactions on the Earth. Examples are the microfluidic synthesis chip on DIDO's board or the flow reactor produced by Boston University Beeler Research Group and Space Tango [53]. It is expected that the fast progress in the field will lead to more advanced reactors, as in instruments performing photochemistry, liquid–liquid, gas–liquid, and heterogeneous catalytic reactions in space.

Flow instruments for purification and separations have been provided by Zaiput and were used already in experimentations on ISS, supported by NASA [54].

For crystallization, co-crystal formation and flow nanoformulation, special diffusion chambers (introduced otherwise by Steve Ley in flow chemistry) are applied, for example, by JAMSS (KIRARA) [55].

Space chemistry has an astonishingly large number of applications areas, already. First of all, oxygen recovery from CO_2 has been provided on the ISS (by batch reactor) from the start of its operation. In the field of drug discovery and drug development, drug formulation is the most important practical application area. Crystallization of proteins for producing pure materials for biologics is a favorite application area from which large pharmaceutical corporations have already profited.

The efforts in producing ultra-pure enzymes as biological targets in the space are very important. Enhanced resolution of the ultra-pure proteins on X-rays provides excellent opportunities for increasing the performance of docking experiments.

New areas that produce new polymorphs and co-crystals for enhancing the druggability of existing and new active principles for drugs, agrochemicals, and cosmetics are also being studied.

Chemical reactions performed in space are opening up a new world for chemists. It is absolutely important to have a deeper knowledge on the effect of space and on the theoretical chemistry foundation of chemical transformations. A good example of such research on reaction kinetics is provided by the Horvath Group at University of Szeged, Hungary, with ESA's support [56].

Supported by theoretical and practical knowledge and with new types of versatile flow reactors operating in space, new avenues that establish new synthetic routes to hitherto unsynthesized structures and for finding new reactions/reaction classes will open up. There is strong preference for green reactions in space due to the increased danger associated with explosible or inflammable solvents. NPW [57] are also needed in space.

New chemistry will significantly support the API and formulated drug production in space. Astronauts on long-term space travels or those staying on the Moon and Mars are facing the problem of the fast decomposition of APIs in space. This makes establishment of on-demand pharmaceutical production that provides the crew with necessary drugs important. For more details, see on-demand synthesis of APIs in this chapter.

Last but not the least, is the abundance of emerging important new application areas in space chemistry: on-site polymer synthesis of structural materials by flow reactors, synthesis of rocket propellants from human and other waste materials, and conversion of solar energy to useful chemicals, as well as space mining.

In summary, flow chemistry offers unforeseeable opportunities that create a novel research and manufacturing paradigm, including increased sustainability, safe and green operations, biocatalytic, electro-, and photochemical transformations, nanotechnology, 3D reactor engineering, more efficient scale-up and production with real time analytics, and self-optimization. All these opportunities, which could contribute to saving our planet will be described in the chapters of the textbook, and the realization that all these various fields exploit the same fundamental and unique features of continuous-flow chemistry is described in this introductory chapter.

(For further details on these issues, please see Volume 1, Chapter 8, Title: Mitigation of chemical hazards under continuous flow conditions, Volume 2, Chapter 6, Title: From green chemistry principles to sustainable flow chemistry, Chapter 7, Title: Flow chemistry in fine chemical production, and Chapter 11, Title: Outlook, future directions and emerging applications)

i

Definitions and relevant knowledge

- Damkoehler number (Da) shows if a reaction is under diffusion or kinetic control in laminar flow. It expresses the reaction rate relative to the mass transport rate (diffusion).
- The principle of high-resolution reaction time control rests simply on adjusting the residence time to a particular step in a multistep reaction sequence.
- NPW operates at process conditions that are beyond the usual condition – that considerably speed up conversion rates, while maintaining selectivity.
- Adiabatic temperature rise (ΔT_{ad}) is the increase in temperature of a reaction mixture, when there is no heat transfer to or from the environment.

Further readings

i

- Yoshida, J. I. (2010). Flash chemistry: flow microreactor synthesis based on high-resolution reaction time control, Chem Rec, 10, 332 – 341.
- Hessel, V., Cortese, B., & De Croon, M. H. J. M. (2011). Novel process windows–Concept, proposition and evaluation methodology, and intensified superheated processing, Chem Eng Sci, 66, 1426–1448.
- Razzaq, T., Glasnov, T. N., & Kappe, C. O. (2009). Accessing novel process windows in a high-temperature/pressure capillary flow reactor, Chem Eng Technol, 32, 1–16.
- Hessel, V., Kralisch, D., Kockmann, N., Noël, T., & Wang, Q. (2013). Novel process windows for enabling, accelerating, and uplifting flow chemistry, ChemSusChem, 6(5), 746–789.
- Coley, C.W., Imbrogno, J., Mo, Y., Thomas, D.A., and Jensen, K.F. Flow chemistry system design and automation, In: Jamison, T., & Koch, G. (Eds.). (2018). Science of Synthesis Flow Chemistry in Organic Synthesis. Flow Chemistry in Organic Synthesis, Ed: Jamison T. F., Koch G., Thieme Verlag,
- Yoshida, J. I. (2015). Flash chemistry, In: Basics of Flow Microreactor Synthesis. Springer, Tokyo.
- Gérardy, R., Emmanuel, N., Toupy, T., Kassin, V. E., Tshibalonza, N. N., Schmitz, M., & Monbaliu, J. C. M. (2018) Continuous flow organic chemistry: successes and pitfalls at the interface with current societal challenges. Eur J Org Chem, 2018(20–21), 2301–2351.
- Glasnov, T. (2016). Continuous-Flow Chemistry in the Research Laboratory. Springer, Amsterdam.

Study questions

?

1.1 What is a typical flow reactor setup and the main parts?
1.2 What are the advantages of the packed-bed flow reactors?
1.3 What is the difference between diffusion and convection?
1.4 What are the major types of mixing?
1.5 How will you characterize mixing versus reaction rate in laminar flow?
1.6 What are the major causes of dispersion along the flow line?
1.7 How do we handle stoichiometry in batch and flow?
1.8 How will you define the reaction time in batch and flow?

1.9 How does one characterize the reaction progress in batch and flow?
1.10 What is surface-to-volume ratio, and what is their value in typical macro- and micro-(flow) reactors?
1.11 Why is there an increased mass transfer in micro-(flow) reactors when compared to typical batch reactors?
1.12 Why is there an increased heat transfer in micro-(flow) reactors when compared to typical batch reactors?
1.13 What is hydrodynamic (pressure-driven) pumping and its velocity profile?
1.14 What is segmented flow?
1.15 What is the Reynolds number?
1.16 What Re numbers characterize the laminar, transient, and turbulent flows?
1.17 What are the typical flow rates in various flow reactors?
1.18 What is the typical dimension of continuous-flow (micro) reactors and the typical material they are constructed of?
1.19 What is STY and its dimension?
1.20 What is flash chemistry?
1.21 What is the principle of high-resolution reaction time control?
1.22 What is "novel process windows" (NPW)?
1.23 What is process intensification (PI)?
1.24 What are the typical flow configuration reactors forming multiple lines of reactors, and what are the typical in-line analytical tools connected?
1.25 What is on-demand flow synthesis?

References

[1] Baxendale, IR, The integration of flow reactors into synthetic organic chemistry, J Chem Technol Biotechnol, 2013, 88, 519–552.
[2] Wegner, J, Ceylan, S, Kirschning, A, Flow chemistry – a key enabling technology for (multistep) organic synthesis, Adv Synth Catal, 2012, 354, 17–57.
[3] Valera, FE, Quaranta, M, Moran, A, Blacker, J, Armstrong, A, Cabral, JT, Blackmond, DG, The Flow's the thing . . . or is it? assessing the merits of homogeneous reactions in flask and flow, Angew Chem Int Ed, 2010, 49, 2–10.
[4] Yoshida, JI, Nagaki, A, Yamada, T, Flash chemistry: fast chemical synthesis by using microreactors, Chem Eur J, 2008, 14, 7450–7459.
[5] Li, PH, Ting, H, Chen, YC, Urban, PL, Recording temporal characteristics of convection currents by continuous and segmented-flow sampling, RSC Adv, 2012, 2(32), 12431–12437.
[6] Zhang, J, Wang, K, Teixeira, AR, Jensen, KF, Luo, G, Design and scaling up of microchemical systems: A review, Annu Rev Chem Biomol Eng, 2017, 8, 285–305.
[7] Kuo, JS, Chiu, DT, Controlling mass transport in microfluidic devices, Annu Rev Anal Chem, 2011, 4, 275–296.
[8] Yoshida, JI, Flash chemistry: Flow microreactor synthesis based on high-resolution reaction time control, Chem Rec, 2010, 10, 332–341.
[9] Wegner, J, Ceylan, S, Kirschning, A, Ten key issues in modern flow chemistry, Chem Commun, 2011, 47, 4583–4592.
[10] Hessel, V, Cortese, B, De Croon, MHJM, Novel process windows–Concept, proposition and evaluation methodology, and intensified superheated processing, Chem Eng Sci, 2011, 66, 1426–1448.

[11] Irfan, M, Glasnov, TN, Kappe, CO, Heterogeneous Catalytic Hydrogenation Reactions in Continuous-Flow Reactors, ChemSusChem, 2011, 4, 300–316.
[12] Bogdan, AR, Poe, SL, Kubis, DC, Broadwater, SJ, McQuade, DT, Continuous flow synthesis of ibuprofen, Angew Chem Int Ed, 2009, 48, 8547–8550.
[13] Baumann, M, Baxendale, IR, Ley, SV, The flow synthesis of heterocycles for natural product and medicinal chemistry applications, Mol Divers, 2011, 15, 613–630.
[14] Webster, A, Greenman, J, Haswell, SJ, Development of microfluidic devices for biomedical and clinical application, J Chem Technol Biotechnol, 2011, 86, 10–17.
[15] Bogdan, AR, Mason, BP, Sylvester, KT, McQuade, DT, Improving Solid-Supported Catalyst Productivity by Using Simplified Packed-Bed Microreactors, Angew Chem Int Ed, 2007, 46, 1698–1701.
[16] Malet-Sanz, L, Susanne, F, Continuous flow synthesis. A pharma perspective, J Med Chem, 2012, 55(9), 4062–4098.
[17] Livak-Dahl, E, Sinn, I, Burns, M, Microfluidic chemical analysis systems, Annu Rev Chem Biomol Eng, 2011, 2, 325–353.
[18] Kockmann, N, Gottsponer, M, Roberge, DM, Scale-up concept of single-channel microreactors from process development to industrial production, Chem Eng J, 2011, 167, 718–726.
[19] deMello, AJ, Control and detection of chemical reactions in microfluidic systems, Nature, 2006, 442, 394–402.
[20] Noël, T, Su, Y, Hessel, V, Beyond organometallic flow chemistry: The principles behind the use of continuous-flow reactors for synthesis, In: Organometallic Flow Chemistry, Cham, Springer, 1–41.
[21] Nagy, KD, Shen, B, Jamison, TF, Jensen, KF, Mixing and dispersion in small-scale flow systems, Org Process Res Dev, 2012, 16(5), 976–981.
[22] Jähnisch, K, Hessel, V, Löwe, H, Baerns, M, Chemistry in microstructured reactors, Angew Chem Int Ed, 2004, 43(4), 406–446.
[23] Goodell, JR, McMullen, JP, Zaborenko, N, Maloney, JR, Ho, CX, Jensen, KF, et al., Development of an automated microfluidic reaction platform for multidimensional screening: reaction discovery employing bicyclo[3.2.1]octanoid scaffolds, J Org Chem, 2009, 74, 6169–6180.
[24] Chang, CH, Paul, BK, Remcho, VT, Atre, S, Hutchison, JE, Synthesis and post-processing of nanomaterials using microreaction technology, J Nanoparticle Res, 2008, 10(6), 965–980.
[25] Ahmed-Omer, B, Barrow, DA, Wirth, T, Heck reactions using segmented flow conditions, Tetrahedron Lett, 2009, 50, 3352–3355.
[26] Gemoets, HP, Su, Y, Shang, M, Hessel, V, Luque, R, Noel, T, Liquid phase oxidation chemistry in continuous-flow microreactors, Chem Soc Rev, 2016, 45(1), 83–117.
[27] Laudadio, G, Govaerts, S, Wang, Y, Ravelli, D, Koolman, HF, Fagnoni, M, et al., Selective C (sp3)–H aerobic oxidation enabled by decatungstate photocatalysis in flow, Angew Chem Int Ed, 2018, 57(15), 4078–4082.
[28] Watts, P, Haswell, SJ, Continuous flow reactors for drug discovery, Drug Discov Today, 2003, 8, 586–593.
[29] Watts, P, Haswell, SJ, The application of micro reactors for organic synthesis, Chem Soc Rev, 2005, 34, 235–246.
[30] Yoshida, JI, Flash chemistry: flow microreactor synthesis based on high-resolution reaction time control, Chem Rec, 2010, 10, 332–341.
[31] Mándity, IM, Ötvös, SB, Fülöp, F, Strategic application of residence-time control in continuous-flow reactors, ChemistryOpen, 2015, 4(3), 212.
[32] Bakker, JJ, Zieverink, MM, Reintjens, RW, Kapteijn, F, Moulijn, JA, Kreutzer, MT, Heterogeneously catalyzed continuous-flow hydrogenation using segmented flow in capillary columns, ChemCatChem, 2011, 3(7), 1155–1157.

[33] Mawatari, K, Kazoe, Y, Aota, A, Tsukahara, T, Sato, K, Kitamori, T, Microflow systems for chemical synthesis and analysis: Approaches to full integration of chemical process, J Flow Chem, 2012, 1, 3–12.
[34] Yoshida, JI, Kim, H, Nagaki, A, Green and sustainable chemical synthesis using flow microreactors, ChemSusChem, 2011, 4, 331–340.
[35] He, Z, Jamison, TF, Continuous-flow synthesis of functionalized phenols by aerobic oxidation of grignard reagents, Angew Chem Int Ed, 2014, 53(13), 3353–3357.
[36] Mason, BP, Price, KE, Steinbacher, JL, Bogdan, AR, McQuade, DT, Greener approaches to organic synthesis using microreactor technology, Chem Rev, 2007, 107, 2300–2318.
[37] Hessel, V, Cortese, B, De Croon, MHJM, Novel process windows–Concept, proposition and evaluation methodology, and intensified superheated processing, Chem Eng Sci, 2011, 66, 1426–1448.
[38] Hessel, V, Kralisch, D, Kockmann, N, Noël, T, Wang, Q, Novel process windows for enabling, accelerating, and uplifting flow chemistry, ChemSusChem, 2013, 6(5), 746–789.
[39] Razzaq, T, Glasnov, TN, Kappe, CO, Accessing novel process windows in a high-temperature/ pressure capillary flow reactor, Chem Eng Technol, 2009, 32, 1–16.
[40] Illg, T, Löb, P, Hessel, V, Flow chemistry using milli- and microstructured reactors – From conventional to novel process windows, Bioorg Med Chem, 2010, 3707–3719.
[41] Keil, FJ, Process intensification, Rev Chem Eng, 2018, 34(2), 135–200.
[42] Baldea, M, From process integration to process intensification, Comp Chem Eng, 2015, 81, 104–114.
[43] Hessel, V, Löwe, H, Hardt, S, Chemical Micro Process Engineering: Fundamentals, Modelling and Reactions, Weinheim, New York, VCH-Wiley, 2004.
[44] Adamo, A, Beingessner, RL, Behnam, M, Chen, J, Jamison, TF, Jensen, KF, et al., On-demand continuous-flow production of pharmaceuticals in a compact, reconfigurable system, Science, 2016, 352(6281), 61–67.
[45] Coley, CW, Thomas, DA, Lummiss, JA, Jaworski, JN, Breen, CP, Schultz, V, Jensen, KF, A robotic platform for flow synthesis of organic compounds informed by AI planning, Science, 2019, 365, 6453.
[46] Skilton, RA, Bourne, RA, Amara, Z, Horvath, R, Jin, J, Scully, MJ, Poliakoff, M, Remote-controlled experiments with cloud chemistry, Nat Chem, 2015, 7(1), 1–5.
[47] Fitzpatrick, DE, Ley, SV, Engineering chemistry for the future of chemical synthesis, Tetrahedron, 2018, 74(25), 3087–3100.
[48] Dallinger, D, Gutmann, B, Kappe, CO, The concept of chemical generators: on-site on-demand production of hazardous reagents in continuous flow, Acc Chem Res, 2020, 53(7), 1330–1341.
[49] Jones, R, Darvas, F, Janáky, C, New space for chemical discoveries, Nat Rev Chem, 2017, 1(7), 1–3.
[50] https://www.esa.int/Enabling_Support/Space_Transportation/Space_Rider/ESA_signs_contracts_for_reusable_Space_Rider_up_to_maiden_flight.
[51] Hessel, V, Tran, NN, (2018) Tiny spaces for the infinite space: flow chemistry mini-labs as assets of space manufacturing? 18th Australian Space Research Conference.
[52] https://en.globes.co.il/en/article-spacepharma-mini-lab-launched-in-italy-israel-collaboration-1001341647
[53] https://www.issnationallab.org/blog/boston-university-flow-chemistry-facilitation-spacexcrs20/
[54] https://www.zaiput.com/zaiput-in-space/
[55] https://www.jamss.co.jp/en/kirara/
[56] Horváth, D, Budroni, MA, Bába, P, Rongy, L, De Wit, A, Eckert, K, et al., Convective dynamics of traveling autocatalytic fronts in a modulated gravity field, Phys Chem Chem Phys, 2014, 16(47), 26279–26287.
[57] Hessel, V, Kralisch, D, Kockmann, N, Novel Process Windows: Innovative Gates to Intensified and Sustainable Chemical Processes, Hoboken, NJ, Wiley, 2014.

Jun-ichi Yoshida†

2 Principles of controlling reactions in flow microreactor chemistry

Edited and revised by Aiichiro Nagaki

2.1 Introduction

Flow chemistry has many advantages over conventional chemistry in both laboratory synthesis and industrial production of chemicals, pharmaceuticals, and agrochemicals [1–18]. Indeed, flow chemistry makes it possible to carry out various types of reactions that are either very difficult or quite impossible to carry out in practice in batch chemistry. Flow chemistry, therefore, opens up new possibilities; especially, extremely fast reactions involving highly reactive and unstable short-lived reactive intermediates. Such reactions are very difficult to control in laboratory flasks or industrial batch macro-scale reactors. They can be generated in flow microreactors within a short time and transferred elsewhere for subsequent reaction, before they decompose. Using flow microreactors to manage the fast reaction time is central to enabling chemical conversions that would otherwise be very difficult or practically impossible in batch macro-scale reactors [19]. This relatively modern technique, known as flash chemistry [20–22], is expected to complement flask chemistry, which is ubiquitous in laboratory synthesis. Flash chemistry is also likely to become important in industrial production. This chapter focuses on features of flow microreactors that are essential for controlling homogeneous reactions involving unstable intermediates. The principles behind flow microreactors are highlighted through their application to a selection of synthetic reactions and polymerization reactions.

2.2 Reactions in a flow microreactor

2.2.1 Reaction time in a batch reactor

As our introduction to reactions in flow microreactors will be compared with conventional batch chemistry, the fundamental features of reaction profiles of a batch reactor, such as a flask, are reviewed here. The reaction is typically initiated by the addition of a reagent or a catalyst to a solution containing a reactant. The concentration of the product increases from close to zero, as a function of time, whereas the concentration of the reactant decreases with time. Their concentrations will be homogenous throughout the reactor, provided the reaction mixture is well-stirred. Therefore, their concentrations can be measured by taking samples of the reaction

https://doi.org/10.1515/9783110693676-002

mixture, at periodic intervals, and concentrations can be determined by gas chromatography (GC), high performance liquid chromatography (HPLC), or some other method. A plot of the results will typically look like Fig. 2.1. When most of the reactant has been consumed, the product concentration will have reached a maximum if the product has not reacted further or decomposed. Around this point, the reaction is usually stopped and the product is isolated from the reaction mixture.

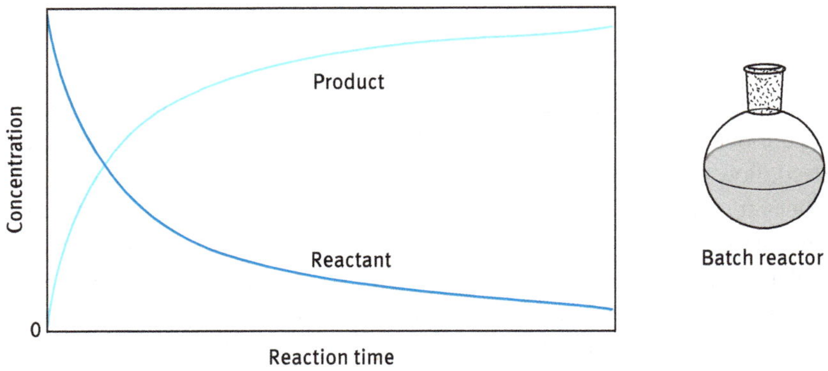

Fig. 2.1: Profile of a reaction in a batch reactor.

2.2.2 Residence time control in a flow reactor

In a flow reactor, under continuous flow conditions, the reaction takes place as the reactant travels along the reactor. Thus, the concentration of the reactant decreases and that of the product increases, with distance from the inlet of the reactor. Therefore, the reaction profile looks like Fig. 2.2, where the space position is simply the distance from a point, the inlet in this case. The product concentration reaches a maximum at the outlet. At any position along the reactor, the concentrations of the reactant and the product remain constant. In other words, under steady state, the concentrations do not change with operation time. It is important not to confuse operation time with reaction time; they are different in flow chemistry, but usually the same in batch chemistry. It is also important to note that the reaction time in a flow reactor corresponds to the space position in the reactor. Note also that the reaction time can be defined as the residence time, that is, the amount of time that the solution spends in the reactor. The residence time, and hence the reaction time, can be controlled by adjusting the length of the flow reactor or the flow speed. In fact, this is a major advantage of flow chemistry over batch chemistry.

The mean residence time in a flow reactor will be the flow speed (or flow rate) multiplied by the length of the reactor. So, for a length of 10 mm and a flow speed of 1 mm s^{-1} (linear velocity), it will be 10 s (Fig. 2.3a). However, the flow speed inside a

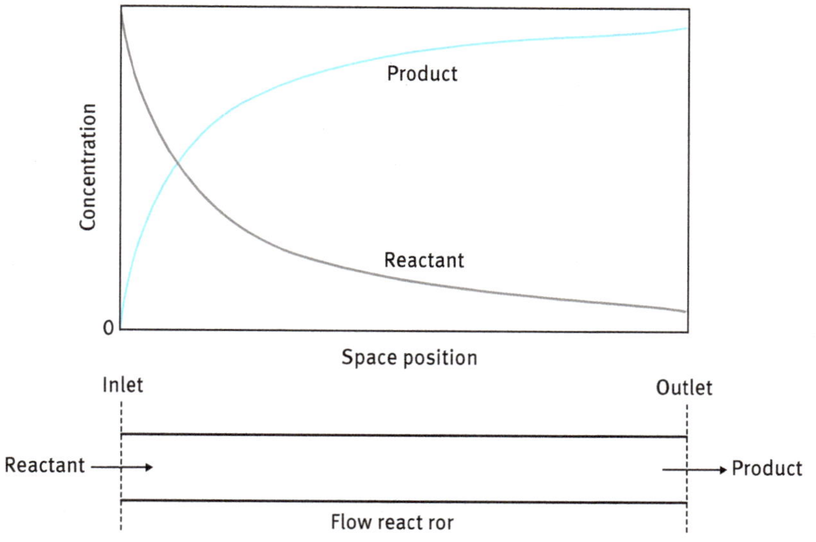

Fig. 2.2: Profile of a reaction in a flow reactor.

tube or a channel is not uniform; it is faster near the center than near the wall. Therefore, there would actually be a distribution of residence times but the distribution would be narrow if the diameter of the channel or tube is small. For convenience, only the mean residence time will be included in further discussions.

In the above scenario, the residence time could be reduced to 5 s by reducing the length of the reactor to 5 mm. Conversely, it could be increased to 20 s by increasing the reactor length to 20 mm. The crucial point is that the residence time can be controlled by adjusting the reactor length.

The residence time can also be adjusted by changing the flow speed. For example, increasing the flow speed to 1 m s^{-1} in a 10 mm pipe reduces the residence time to 10 ms (Fig. 2.3b). This flow speed and length are both well within the realms of human perception and manipulation. For example, 1 m s^{-1} is slightly less than walking pace, and reactors of 1 cm in length are easy to physically handle. However, controlling something without aid on a time scale of 10 ms is beyond our capacity, since humans take more than 200 ms to react to a visual stimulus. Therefore, it is beyond human capability to start a reaction, with control and precision, by manually adding some reagent to a flask containing a reactant and after 10 ms, add a quenching agent manually to stop the reaction. Note, therefore, that the features of flow chemistry allow us to easily control time lengths that we cannot control dexterously in batch chemistry. This means that flow chemistry enables us to carry out chemical reactions that we cannot accomplish manually in batch reactors.

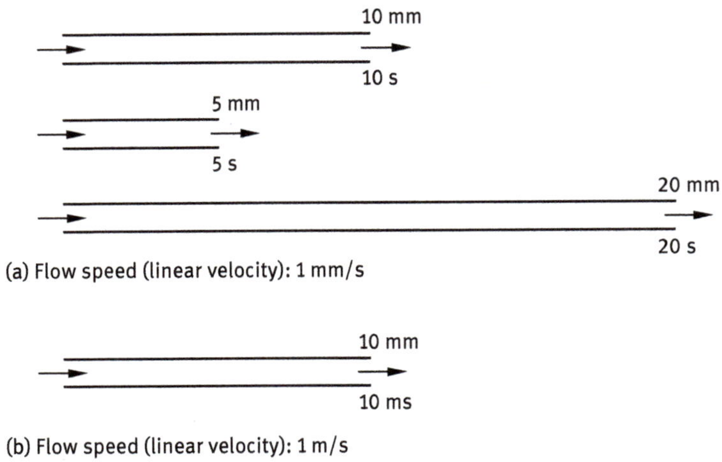

Fig. 2.3: Relationship between residence time, reactor length, and the flow speed in a flow reactor.

2.2.3 The importance of micromixing

The above discussion is applicable to both flow macroreactors and flow microreactors, except when dealing with extremely fast reactions. In those cases, only microreactors can offer suitable control. Why should that be? As mentioned above, in a flow reactor, the reaction time is the same as the residence time. The latter is the time for the fluid to flow between the reagent and quencher inputs, respectively, where the reaction is initiated by mixing of the reaction components, and ends when the reaction mixture mixes with the quenching reagent. The reaction time can be precisely controlled by adjusting the physical length between the inlets, which is typically about 1–10 mm in a microreactor, or by adjusting the flow speed (Fig. 2.4). Therefore, it is not the reactor length or reaction time that defines a reactor at the micro scale but the mixing times.

If the mixing time required to make a solution homogeneous is longer than the reaction time, the solution will not be homogeneous during the course of the reaction. Therefore, the concentrations of the reaction components cannot be adequately ascertained. Consequently, the kinetics of the reaction cannot be adequately controlled. Thus, the mixing time should be shorter than the reaction time and, preferably, much shorter. So, if the reaction time is 1 s, the mixing time for the reaction components should be much less than 1 s. The same is true for quenching. However, such rapid mixing cannot be achieved in conventional batch reactors by stirring, as will be discussed later. It is also difficult to achieve in flow macroreactors. A reactor with microstructures is typically necessary.

Let us now look at the mixing process in more detail. Mixing in a solution phase is defined as a phenomenon that increases homogeneity of all species in the solution. Figure 2.5 illustrates mixing in a batch reactor. When a solution of **2** is

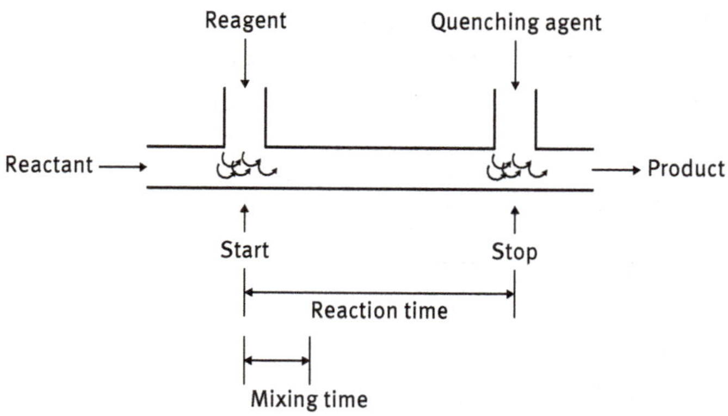

Fig. 2.4: Reaction time in a flow reactor.

added to a solution of **1**, eddies of solution **2** form in solution **1**. The eddies are dispersed into the solution upon stirring and their contents diffuse into the surrounding solution. During this step, the eddies are destroyed and homogeneity of the solution is achieved at a molecular level. Because the molecular diffusion is the slowest step in the mixing process, it is the key step to focus upon. According to the theory of molecular diffusion, the time needed for molecular diffusion is proportional to the square of the length of the diffusion path. The diffusion path is related to the size of the eddy. Small eddies mean shorter diffusion paths and fasting mixing. Hence, increasing the turbulence by more vigorous stirring results in faster mixing.

The minimum mean radius of eddies in optimal turbulence is between 10 and 100 μm. It would, therefore, take at least 0.05–5 s to obtain a homogeneous solution. Hence, the overall mixing time in a batch reactor, with conventional stirring, is likely to take several seconds; so reactions that finish in less than a second cannot be controlled.

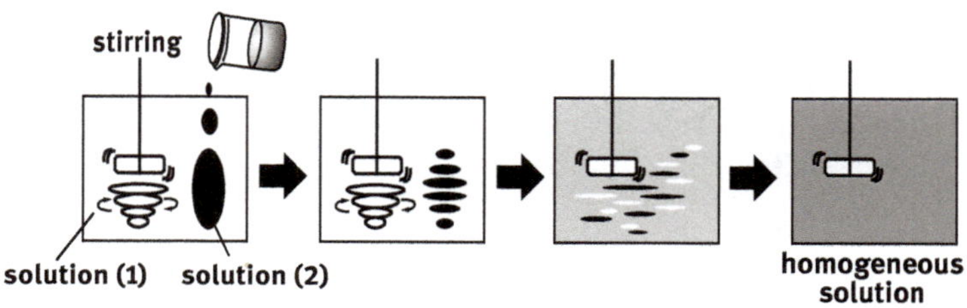

Fig. 2.5: Mixing by stirring.

Micromixers solve the problem by having very short diffusion paths in small segments of the solution. One method is multi-lamination or interdigital micromixing [23], shown in Fig. 2.6. The feed flow of solutions **1** and **2** is split into multiple channels by a microstructure. Diffusion occurs across the interfaces of the laminar streams. Since the diffusion path is short, mixing is rapid and a homogeneous solution can be obtained very quickly.

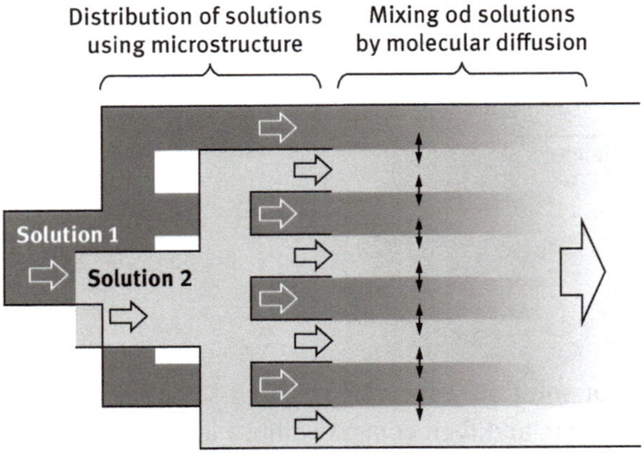

Fig. 2.6: Principle of multilamination-type micromixers.

Another method is the engulfment flow regime [24], shown in Fig. 2.7. Solutions **1** and **2** flowing at high speed meet at a T junction. Small segments of the solution are created, presumably, by shear stress. Their diffusion paths will also be very short. High flow speeds are required to avoid the flow being chiefly laminar, as this results in slower mixing, unless the channel is very small. Owing to rapid mixing, micromixers enable control of fast reactions that cannot be achieved in a macroreactor.

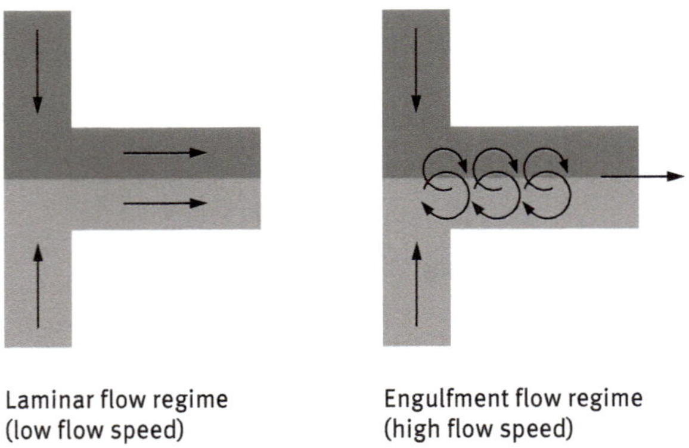

Fig. 2.7: Laminar flow regime and engulfment flow regime in a T-shaped mixer.

Thus, continuous flow systems need to be equipped with micromixers for precise control of reaction times of a second or less. A typical flow system comprising two micromixers, two microtube reactors, and three precooling units is shown in Fig. 2.8. The precise control of short reaction times in such flow systems enables good control of reactions involving unstable short-lived reactive intermediates or products that are unstable under the reaction conditions.

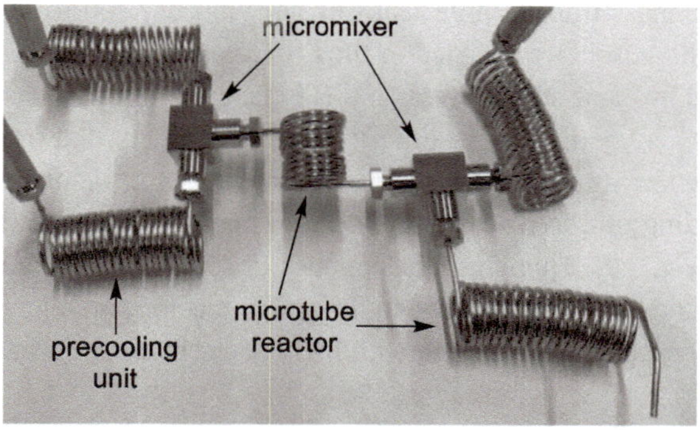

Fig. 2.8: A flow microreactor system comprising two micromixers and two microtube reactors.

(For related issues please see Volume 1, Chapter 1, Title: Fundamentals of flow chemistry)

i

2.2.4 The importance of heat transfer

Since fast reactions are usually highly exothermic, heat removal is also an important factor in controlling extremely fast reactions. Heat transfers through the reaction solution by convection and conduction. In microreactors, the short distance that heat needs to travel through the solution to the wall is an advantage for conduction through the solution. Conduction through the wall of the reactor is important if the system is being externally cooled.

The surface-to-volume ratio of a body is an important factor in the cooling of that body, as heat is lost through the surface; the larger the ratio, the faster the cooling. As the surface-to-volume ratio of microreactors is large when compared with that of macroreactors, heat conducts more rapidly, compared to a macroreactor, from the interior of a reactor to the exterior via the reactor wall. The efficient removal of heat that is generated by the highly exothermic reactions enables better control of the reaction temperature within microreactors.

A common way of achieving fast heat exchange in flow microreactor systems is by a shell and tube heat exchanger. As the name suggests, this typically consists of several microtubes, though there may be just one, located in a shell containing coolant to moderate the heat generated.

2.2.5 Reaction time and reaction regime

Often, kinetics are used to selectively control chemical reactions to deliver a particular product. However, the use of kinetics cannot be employed for extremely fast reactions due to insufficient homogeneity of the reaction environment. As mentioned above, in batch reactors and macroreactors, mass transfer and heat transfer limit our control of fast reactions. In these reactors, the boundary of reaction time between the kinetic regime region and the region limited by mass and heat transfer is typically in the order of minutes, making it difficult to choose pathways, leading to certain products that are based on kinetics (Fig. 2.9). However, in a flow microreactor, the boundary can be milliseconds to seconds, depending on the reactor size and design, due to the rapid mass transfer and heat transfer (Fig. 2.9). Therefore, the kinetics of reactions that are uncontrollable in batch macroreactors may be controlled in flow microreactors, enabling formation of the desired products based on the kinetics, with higher selectivity and with less unwanted by-products.

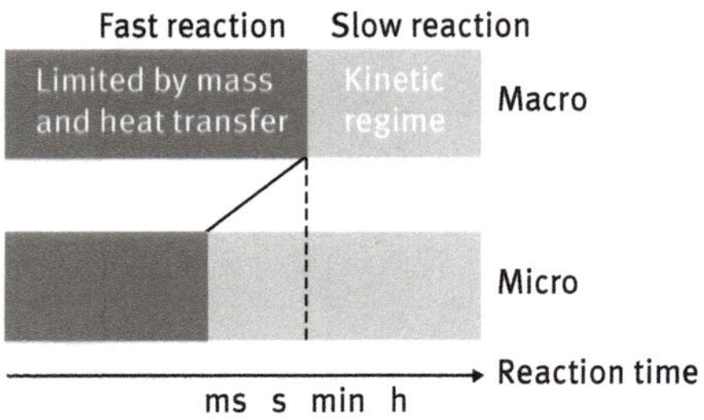

Fig. 2.9: Reaction time and reaction regime.

2.3 Very precise control of reaction times in flow systems

2.3.1 The principle

Problems in controlling fast reactions that involve highly unstable short-lived intermediates such as carbocations and carbanions is, unfortunately, quite common. Consider the simple model in Fig. 2.10, where **A** is a reactant and **I** is an unstable reactive intermediate that decomposes very quickly into the unwanted by-product **B**. For simplicity, assume first-order processes with rate constants k_1 and k_2. Also assume that $k_1 > k_2$, so **I** is generated faster than it decomposes, allowing it to accumulate. If a suitable quenching reagent is added before **I** decomposed, **I** could be transformed into the desired product **C**. Of course, the quenching reaction would have to be much faster than the decomposition reactions in order to obtain **C** at a quantitative yield.

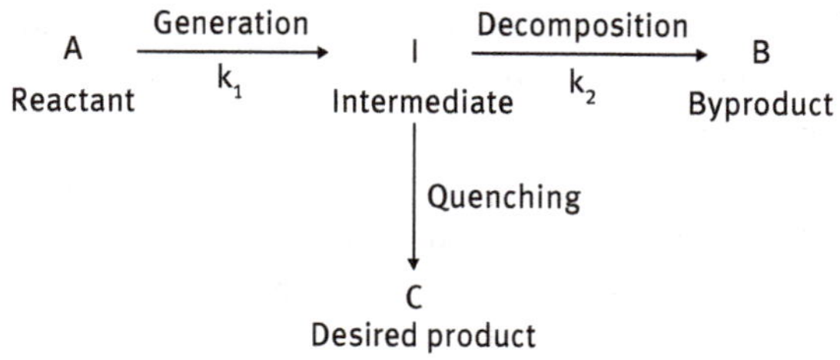

Fig. 2.10: A general scheme of reactions involving unstable intermediates.

When k_2 is much smaller than k_1, the generation of **I** goes nearly to completion, while only a little of **I** is decomposed into **B**. Addition of a quenching agent at this stage gives the desired product **C** at a good yield. In fact, the quenching agent can be added over a wide range of reaction times to obtain **C** at a good yield. A higher k_2 changes the situation even if k_2 remains smaller than k_1. No sooner has **A** produced **I**, **I** decomposes to **B**. Therefore, **I** can be accumulated in adequate amounts only over a limited time range.

Let us consider it from a more quantitative way. Let $k_1 = 10\ h^{-1}$ and $k_2/k_1 = 0.01$. As shown in Fig. 2.11, **A** is consumed within 0.5 h and the concentration of **I** increases to a maximum and then decreases gradually, because it is now decomposing faster than it is being produced. Therefore, the reaction time for the generation of **I** should be between ca. 0.3 h and 0.6 h to get the desired product **C** at a yield higher than 90%. The quenching agent needs to be added ca. 0.5 h after the start of the reaction. This time length is, of course, easy to control manually, enabling the reaction to be carried out in a batch reactor such as a flask.

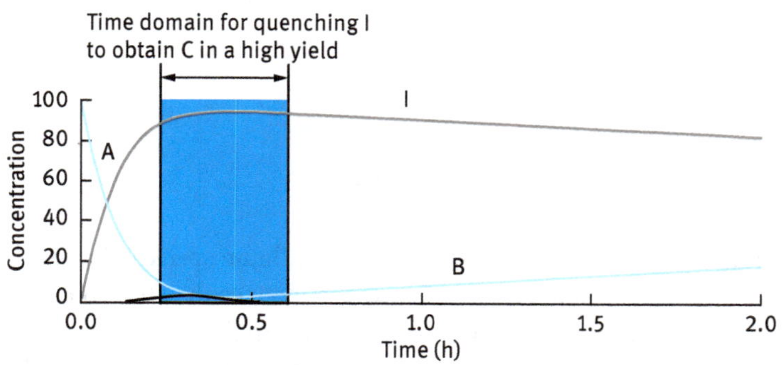

Fig. 2.11: Change in concentrations of reaction components A, I, and B with time when $k_1 = 10\ h^{-1}$ and $k_2/k_1 = 0.01$.

When the reaction is much faster, controlling it becomes much more difficult. For example, when $k_1 = 10\ s^{-1}$ and $k_2/k_1 = 0.01$, the reaction time for generation of **I** should be between about 0.3 and 0.6 s to get a yield higher than 90%, as shown in Fig. 2.12. This means that the quenching agent must be added and mixed within a very narrow window of time, otherwise a good yield cannot be obtained. As discussed above, a time length shorter than a second cannot be controlled manually and the reaction cannot be controlled in a batch microreactor. This is true even if k_2 and k_1 are the same. In other words, such precise reaction-time control cannot be achieved in flask chemistry; a flow microreactor system equipped with micromixers is essential for this reason.

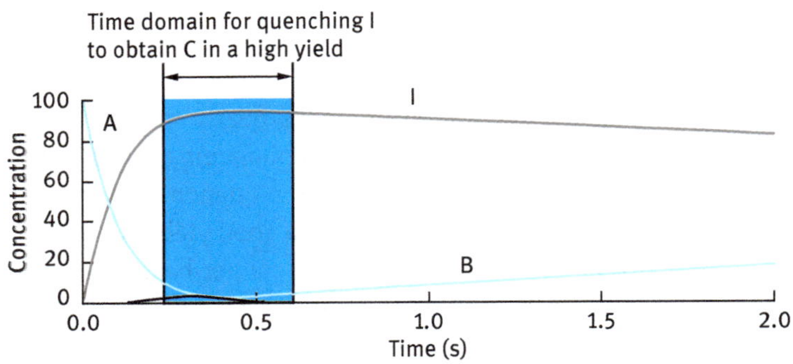

Fig. 2.12: Change in concentrations of reaction components A, I, and B with time when $k_1 = 10\ s^{-1}$ and $k_2/k_1 = 0.01$.

2.3.2 Example 1: phenyllithiums bearing carbonyl groups [26]

The following sections give examples of the use of flow microreactors to control reactions involving unstable reactive intermediates. The first process is the generation of a phenyllithium species with a pendant ester group, as, shown in Fig. 2.13.

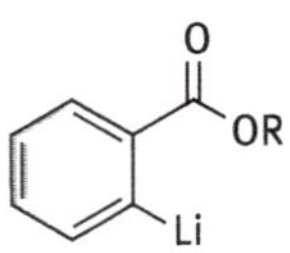

Fig. 2.13: Phenyllithium bearing a carbonyl group.

Carbonyl groups are known to react very rapidly with phenyllithium species. For example, the reaction between phenylithium and a carbonyl group gives the corresponding alcohol, as shown in Fig. 2.14. This example illustrates internal, functional group incompatibility between reactive organometallic species with a carbonyl group in the same molecule. This is a common problem in synthesis in batch mode, often requiring multiple steps and protecting groups to overcome the problem. Furthermore, the generation of, and reactions with, electrophiles are usually conducted at very low temperatures. Despite this, aryllithium compounds bearing a reactive pendant sidechain are therefore not considered viable or useful precursors under batch conditions.

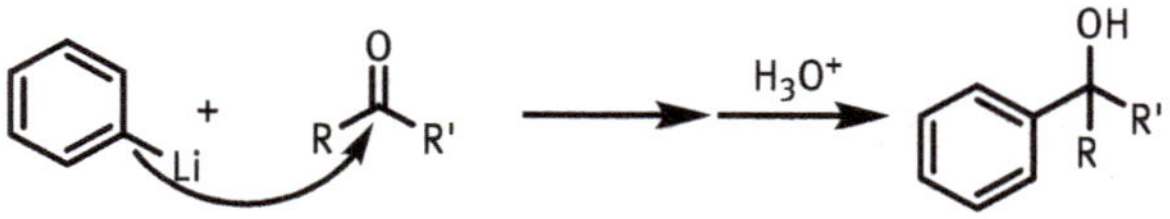

Fig. 2.14: Reaction of phenyllithium with a carbonyl group.

Consider the I/Li exchange reaction of *o*-pentanoyliodobenzene, using mesityllithium (MesLi) as a lithiating reagent to generate *o*-pentanoyl-substituted phenyllithium. *o*-Pentanoyl-substituted phenyllithium is very unstable and apt to dimerize to give undesired by-products, even at very low temperatures, as shown in Fig. 2.15. Rapid quenching with a suitable electrophile (in this case, a proton from methanol) before the intermediate aryllithium species decomposes can yield pentanoylbenzene as the primary product.

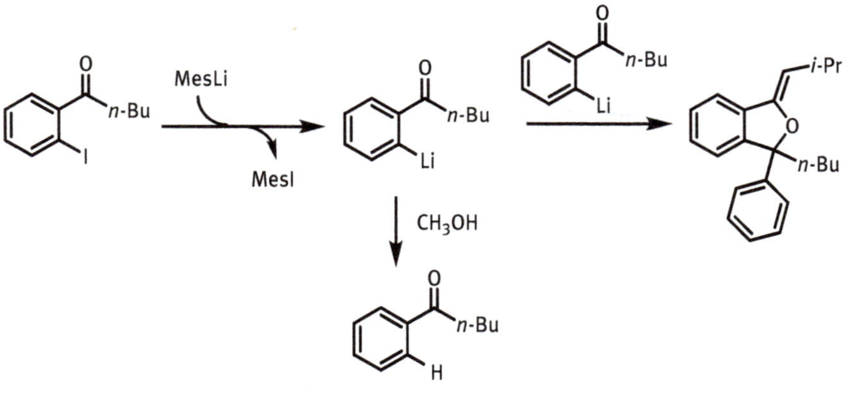

Fig. 2.15: Generation of *o*-pentanoyl-substituted phenyllithium.

This can be better achieved by using the flow microreactor system of two micromixers (**M1** and **M2**) and two microtube reactors (**R1** and **R2**), shown in Fig. 2.16. *o*-Pentanoyliodobenzene and mesityllithium are mixed in **M1** and the I/Li exchange reaction occurs in **R1**. Then, the resulting *o*-pentanoyl-substituted phenyllithium is reacted with methanol at **R2** after mixing at **M2**, giving the desired protonated product, pentanoylbenzene.

Figure 2.17 is a plot of the pentanoylbenzene yield against residence time in **R1** at −70 °C. All the *o*-pentanoyliodobenzene is consumed within 0.003 s. The pentanoylbenzene yield decreases with increase in the residence time, presumably due to dimerization of the *o*-pentanoyl-substituted phenyllithium intermediate. The plot indicates the appropriate time window to quench the intermediate in order to get the desired product in high yields.

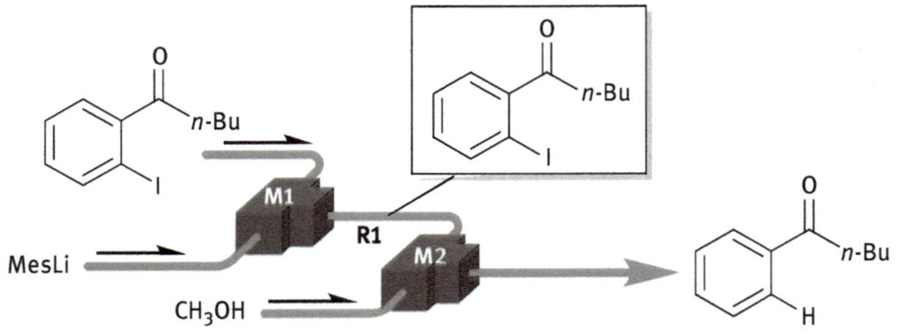

Fig. 2.16: The I/Li exchange reaction of *o*-pentanoyliodobenzene followed by quenching with methanol using a flow microreactor system.

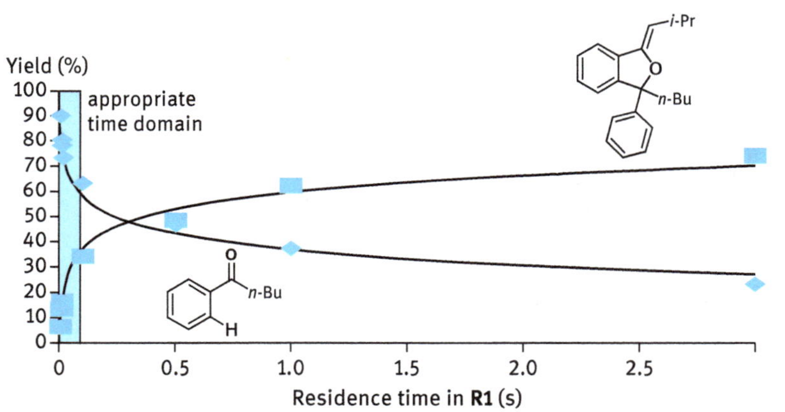

Fig. 2.17: Plot of the yield of the protonated product and the decomposed dimerization byproduct with the residence time in R1 at –70 °C.

2.3.3 Temperature-residence time map

Because temperature affects the rate of reaction, it is important to know how it affects the reaction profile. Therefore, reactions in flow microreactors are often carried out at various temperatures and residence times to produce a time-residence map.

Temperature–residence time maps are not restricted to plots of the product yield; other outcomes of the reaction such as the conversion of a reactant may be plotted. Generally, they serve as useful tools for analyzing reactions under continuous flow conditions.

In Fig. 2.18, at low temperatures and short residence times, the product yield is low because the starting material remains unchanged. The yield increases to a while maximum with increase in temperature and residence time, decreasing with further increases in the temperature and the residence time due to dimerization or some other undesired reaction of the ethyl *p*-lithiobenzoate intermediate [27, 28]. The map, therefore, indicates the appropriate temperature and residence times for obtaining the desired product at good yields.

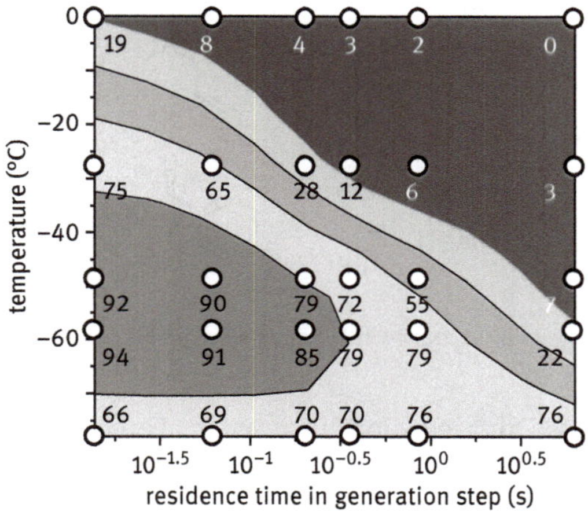

Fig. 2.18: A temperature–residence time map for the yield of the product in the I/Li exchange reaction of ethyl *p*-iodobenzoate, followed by quenching with ethanol using a flow microreactor system.

An important advantage of temperature–residence time maps is that they can reveal the stability and reactivity of highly reactive intermediates, such as aryllithium species bearing electrophilic functional groups. Therefore, they serve as powerful tools for analyzing the mechanisms of reactions involving highly unstable intermediates. Note that they are an important tool for optimizing the reaction conditions for synthesis [29], the information being revealed at a glance. A map consisting of 30–40 data points can be made in a day or two by collecting product solutions under various conditions and analyzing them by GC or HPLC.

Under optimized conditions, with ethanol as a quencher or an electrophile, ethyl *p*-lithiobenzoate intermediate can be used in reactions with various electrophiles (Fig. 2.19). For example, the intermediate can react with methyl trifluoromethanesulfonate (methyl triflate, $CH_3OSO_2CF_3$) to give the corresponding methylated product. Incidentally, it gives the same product when it reacts with methyl iodide (CH_3I), but at only 22% yield compared with 77% in the case of methyl triflate, because CH_3I is a much weaker electrophile; so the reaction is not fast enough to render insignificant the competition with dimerization of the intermediate. The ethyl *p*-lithiobenzoate intermediate reacts with trimethylsilyl trifluoromethanesulfonate (trimethylsilyl triflate, $(CH_3)_3SiOSO_2CF_3$)) to give the corresponding silylated product. To emphasize, the above transformations are impossible to carry out in conventional macro-batch reactors owing to the intermediate being too unstable. It is the precise control of the short residence times in flow microreactors that makes such transformations possible.

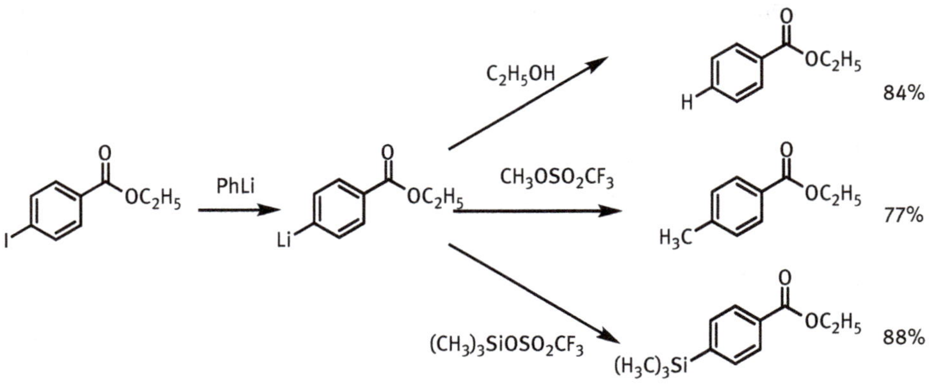

Fig. 2.19: Generation of ethyl *p*-lithiobenzoate and its reactions with various electrophiles.

The flow-microreactor can generally be used for generation and reaction of aryllithiums bearing an alkoxycarbonyl group at all three ring positions – para, meta, and ortho. Alkyl lithiobenzoates can be generated by a I/Li exchange reaction with phenyllithium (PhLi), while alkyl *o*-lithiobenzoates are successfully generated by a Br/Li exchange reaction with *s*-BuLi. The temperature-residence time maps for some lithiobenzoates, with an alkyl group in each of the three positions, are shown in Fig. 2.20. The stability of the reactive species is easy to see. For example, *t*-butyl lithiobenzoate products are obtained at high yields over a wide range of temperature and residence times, though low yields are also observed in the high-temperature long-residence–time region because of the dimerization of the intermediates. The steric bulkiness of the *t*-butyl group retards the attack on the carbonyl group, making the intermediates highly stable, relatively. A similar reaction profile is seen with isopropyl esters, though the high-yield region is smaller because the isopropyl group is less bulky. With ethyl esters, the high-yield regions are very small because they are sterically less hindered. These results clearly demonstrate that the stability of aryllithiums that bear alkoxycarbonyl groups, strongly depends on the size of the alkyl group attached to the carbonyl carbon and decreases in the order, *t*-butyl > isopropyl > ethyl.

The stability also depends on the position of the lithium. The high-yield region is bigger for *o*-lithiobenzoates than for *p*- and *m*-lithiobenzoates, regardless of the alkyl group. This seems to be due to the coordination of the carbonyl group to the Li, making the species more stable.

Precise control of reaction time and temperature-residence time maps can generally be applied to various reactions involving highly unstable reactive intermediates,

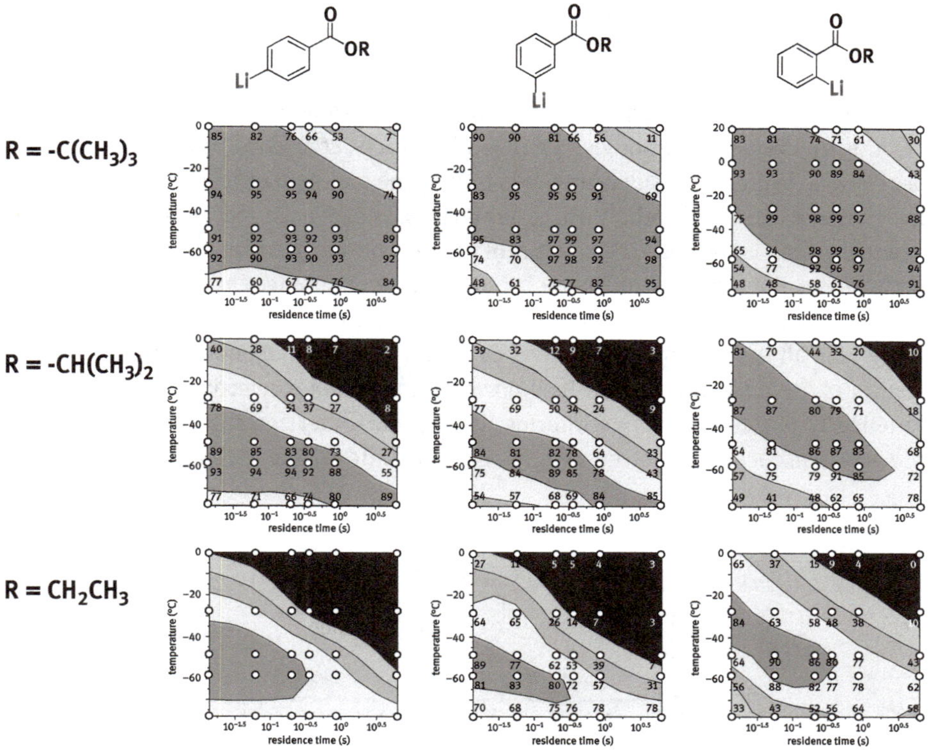

Fig. 2.20: Temperature–residence time maps for the yield of the protonated product in the halogen/lithium exchange reaction of alkyl *p*-, *m*-, and *o*-halobenzoates in a flow microreactor system.

such as aryllithiums bearing electrophilic functional groups [30], for example, cyano groups [31], benzyllithiums bearing electrophilic functional groups [32], pyridyllithiums [33, 34], oxyranyllithiums [35, 36], aziridinyllithiums [37], vinyllthiums [38], propargyllithiums [39], perfluroalkyllthiums [40], and alkoxysulfonium ions [41], and onium ions [42–44].

2.3.4 Example 2: control of isomerization of aryllithium species [45]

The nitro group is another fast reacting electrophilic group in organolithium reactions, but flow microreactor systems enable the generation and control of the reactions of nitro-substituted aryllithium species. Aryl iodides and phenyllithium (PhLi) are suitable for the generation and I/Li exchange reactions can be executed with *o*-, *m*-, and *p*-iodonitrobenzene, to give the corresponding aryllithium species. These species can be reacted with various electrophiles, such as methyl triflate, trimethylsilyl triflate, and benzaldehyde to give the corresponding *o*-, *m*-, and *p*-substituted nitrobenzenes. An interesting reaction is that of 1-bromo-2,5-dimethoxy-3-nitrobenzene, shown in Fig. 2.21. By treating it with PhLi at −48 °C, the corresponding aryllithiums can be generated by the exchange of Br with Li. Presumably, the exchange is facilitated by the coordination of methoxy oxygen with Li. Li can migrate to another position on the ring, creating an isomer. Interestingly, the migration can be controlled by changing the residence time. When the residence time is 0.06 s, the subsequent reaction with an aldehyde takes place at the carbon where the Br/Li exchange occurred. However, when the residence time is increased, a significant amount of the isomeric product is obtained because reaction with the aldehyde can take place at another carbon to which the Li has migrated. When the residence time is 63 s, the isomer is obtained exclusively, indicating that all the lithium has migrated by then. The reaction demonstrates that precise control of the residence time in flow microreactors enables selective use of either the kinetically preferred reactive intermediates (nonisomerized intermediates) or the thermodynamically preferred reactive intermediates (isomerized intermediates).

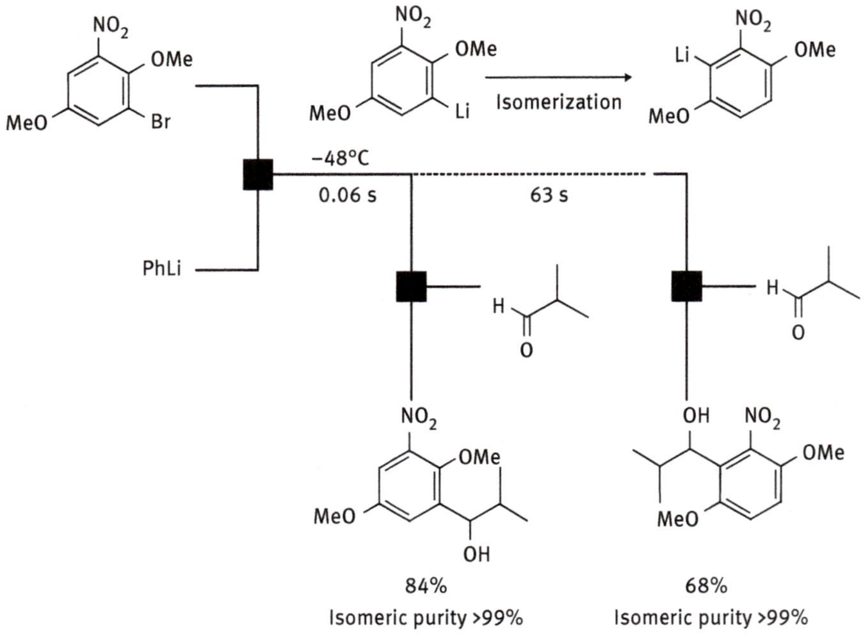

Fig. 2.21: Control of isomerization by adjusting the residence time.

The concept of controlling the isomerization process by the precise manipulation of reaction times by adjusting the residence time is also applicable to cis-trans isomerization of oxyranyllithiums [35, 35], epimerization of propargyllithiums in the synthesis of optically active allenes [46], ring opening of aziridinyllithiums [47], and ring opening of heteroaryllithiums [48].

2.4 Space integration of reactions [49]

2.4.1 The concept

Complex organic molecules are usually synthesized stepwise. This often involves performing a series of several reactions with intervening isolation of intermediate products. Ideally, the steps should be integrated into a single operation in a one-pot or one-flow system for enhanced power and speed. Integration of the reactions can be classified into three types, as depicted in Fig. 2.22 (a) time and space integration, where all the reaction components are mixed at once in one pot (domino, tandem, or cascade reactions) and where a sequence of reactions then takes place, (b) time integration, where a sequence of reactions is conducted in one pot by adding components at intervals (one-pot sequential synthesis), and (c) space integration, where components are added at different locations in a flow system, producing a sequence of reactions in a continuously flowing stream. Although time and space integration is very easy to carry out, time integration and space integration methods afford a wider choice of reagents and/or catalysts because they can be added stepwise. Flow chemistry has some of the benefits of space integration reactions [50–52] and flash chemistry enables space integration of reactions using short-lived unstable reactive intermediates. The following sections briefly describe space integration of reactions in integrated flow microreactor systems.

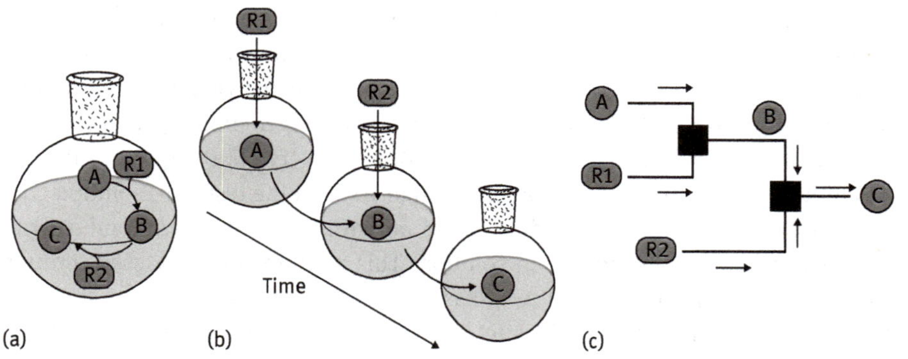

Fig. 2.22: Classification of reaction integration: (a) time and space integration, (b) time integration, and (c) space integration.

2.4.2 Example 3: synthesis of fluoro-substituted tetrasubstituted epoxides and aziridines from fluoroiodomethane [53]

The sequential introduction of two electrophiles using deprotonation can be carried out in an integrated flow microreactor system. This is one of the novel methods for synthesizing fluoro-substituted and tetrasubstituted epoxides and aziridines from commercially available fluoroiodomethane. Figure 2.23 depicts the setup. The flow microreactor system consists of four T-shaped micromixers (**M1**, **M2**, **M3**, and **M4**) and four microtube reactors (**R1**, **R2**, **R3**, and **R4**).

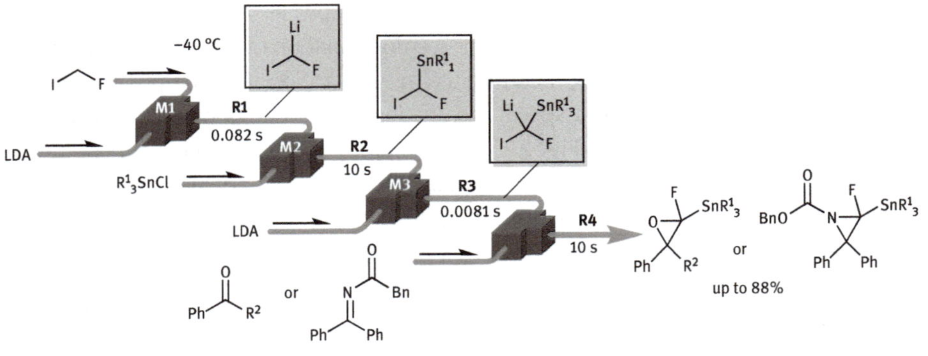

Fig. 2.23: The sequential introduction of two electrophiles from fluoroiodomethane, based on deprotonation, using an integrated flow microreactor system.

As an example, fluoroiodomethane in Et_2O and LDA (lithium diisopropylamide) in Et_2O are mixed in **M1**. The mixture is then passed through **R1**. The resultant solution containing *o*-fluoroiodomethyllithium is fed into **M2** and mixed with the first electrophile, chlorotrialkylstannanes in Et_2O. The mixture is passed through **R2** and the resultant solution containing the fluoroiodomethylstannane is fed into **M3**, where it is mixed with a solution of LDA. This generates an unusual and very unstable organolithium bearing two metal atoms (Li and Sn) and two halogen atoms (F and I) on the same carbon. The generated organolithium requires a very short residence time of 8.1 ms in R3. This mixture is then passed through **R3**, and the resultant solution containing the second lithium intermediate is fed into **M4** to mix with a second electrophile, such as ketones and imines in solution. This mixture is then passed through **R4**. The sequential introduction of two electrophiles via the first and the second unstable lithium intermediates provides tetrasubstituted epoxides and tetrasubstituted aziridines having a fluorine atom in good yields.

2.4.3 Example 4: switching of the reaction pathways [54]

The power of linear integration of two consecutive deprotonations of 1,2-dichloroethene is also applicable for switching reactions. By choosing an appropriate residence time and temperature in each reactor, the reaction pathway toward 1,2-dichlorovinyllithium, generated from *trans*-1,2-dichloroethene can be switched, and the corresponding alkenes and alkyne product can be selectively produced at will (Fig. 2.24). By using short residence times, 1,2-dichlorovinyllithium can be reacted with an electrophile to give the corresponding 1,2-dichloroethene substituted product. A second deprotonation, using *s*-BuLi, followed by a reaction of the resulting organolithium species with a second electrophile, gives the 1,2-disubstituted 1,2-dichloroethene product. By using longer residence times, 1,2-dichlorovinyllithium eliminates lithium chloride and the resulting chloroacetylene intermediate undergoes deprotonation. The resulting 2-chloroethynyllithium species reacts with an electrophile to provide the corresponding chloroacetylene substituted product. The subsequent chlorine–lithium exchange reaction using *s*-BuLi, followed by a reaction with a second electrophile, offers the corresponding 1,2-disubstituted acetylene product. Thus, alkenes and alkynes can be selectively produced as desired. A similar transformation can also be achieved via trichlorovinyllithium intermediates, generated by the deprotonation of trichloroethene, which is much cheaper than *trans*-1,2-dichloroethene.

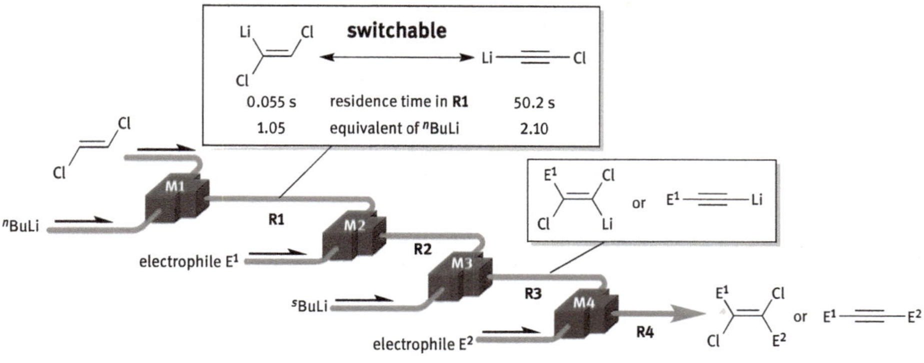

Fig. 2.24: Switching of the reaction pathways of two consecutive deprotonations of *trans*-1,2-dichloroethene in an integrated flow microreactor.

2.4.4 Example 5: synthesis of TAC-101 [55]

The sequential introduction of several electrophiles using halogen–lithium exchange reactions, together with deprotonation, can be carried out in an integrated flow microreactor system [55–57]. TAC-101 (4-[3,5-bis(trimethylsilyl)benzamido]benzoic acid) is a synthetic retinoid having wider importance, such as antitumor activity against liver cancer and inhibition of its metastasis into the colon. The methyl ester of TAC-101 is conventionally synthesized by batch chemistry in six steps. These steps, however, can be combined by space integration into a one-flow system. For that, a flow microreactor system comprising six micromixers (**M1 to M6**) and six microtube reactors (**R1 to R6**), as shown in Fig. 2.25, can be used. In short, 1,3,5-tribromobenzene undergoes three sequential, space-integrated Br/Li exchange reactions, each followed by a reaction with an electrophile. The total reaction time, that is, the residence time from M1 to R6, is ca. 13 s, and 100–200 mg min^{-1} of product can be obtained. Various

TAC-101 methyl ester analogs with two different silyl groups can also be synthesized from 1,3,5-tribromobenzene in one flow. Note that, using conventional batch methods, it is very difficult to synthesize compounds selectively with two different silyl groups, due to dilithiation occurring in the first step. The flow-microreactor method, however, is a selective, efficient, versatile, and practical method for the synthesis of TAC-101 and its analogues.

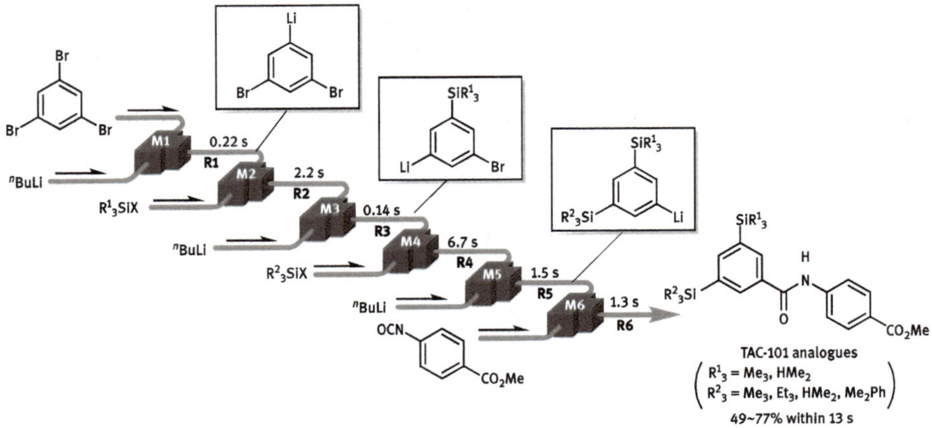

Fig. 2.25: Synthesis of TAC-101 in an integrated flow microreactor system.

2.4.5 Linear integration and convergent integration

An important application of flow chemistry is in multistep synthesis [58]. It can, in principle, be classified into linear and convergent synthesis, as illustrated in Fig. 2.26, though they are often used in combination for synthesizing complex molecules. Linear synthesis is suitable for iterative reactions, including repetition of similar reactions or even sets of reactions as well as repetition of the same reactions, and is frequently used for synthesizing peptides and oligosaccharides. Moreover, two components can be reacted together to generate a short-lived intermediate, which can be followed by subsequent generation of other short-lived intermediates upon adding more reaction components. This linear procedure can be repeated within a single-flow process to obtain the target molecule. This sequence of reactions can also be easily regulated by varying the combinations of the reaction component. However, in general, its drawback is that the overall yield quickly drops with an increase in the step number. In convergent synthesis, between two and several fragments are synthesized separately and then combined. This improves the efficiency of multistep synthesis. Moreover, several different short-lived intermediates can be generated separately and then combined to create the target molecule.

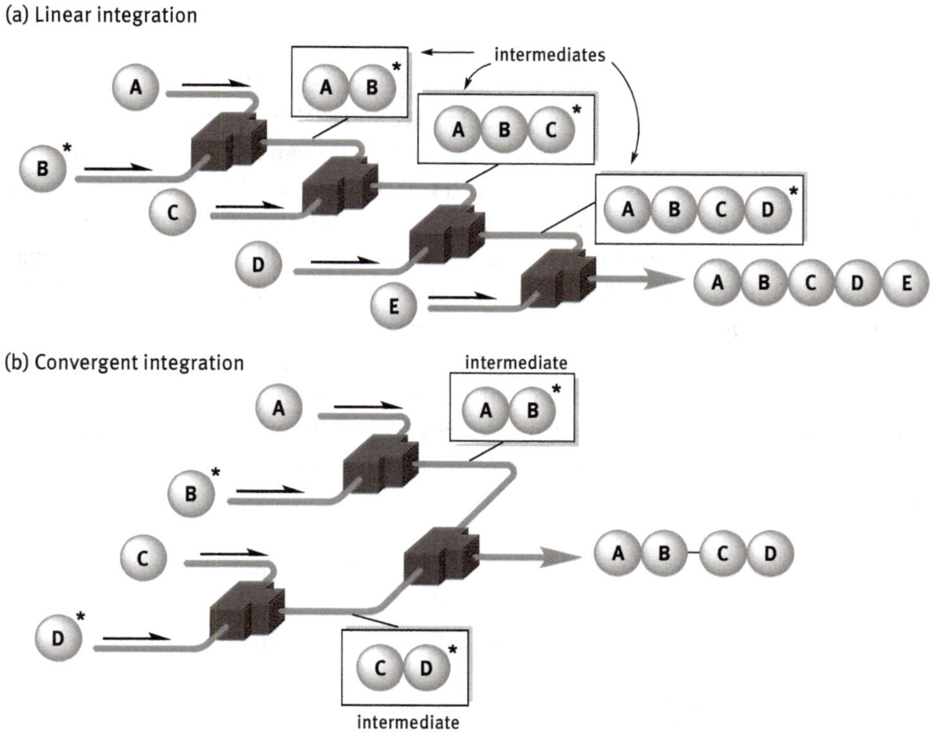

Fig. 2.26: Two types of space integration reactions involving several unstable intermediates: (a) linear and (b) convergent. Asterisk mark means reactive species.

The above-mentioned synthesis is an example of linear synthesis. Being an integration of reactions, the process can be called liner reaction integration. However, the concept of space integration of reactions can also be applied to convergent synthesis; so reaction integration for multistep synthesis can be classified into linear integration and convergent integration. Interestingly, the two types can be easily distinguished by the structure of the integrated flow system used for the synthesis [59, 60].

2.4.6 Example 6: three-component coupling by convergent integration [61]

Arynes, including benzyne, are highly reactive intermediates and extensive studies on their synthetic applications have been reported so far. In particular, carbometalation of benzyne serves as a powerful method for constructing organic structures containing an *o*-disubstituted benzene ring. In fact, carbolithiation of a benzyne, with functionalized aryllithiums, followed by reactions with various electrophiles is a nice example of convergent-integrated organolithium reactions involving several lithium intermediates that proceed in a flow-integrated microreactor system. In the example shown in Fig. 2.27, *o*-bromoiodobenzene is mixed with PhLi in **M1** and short-lived *o*-bromophenyllithium species are generated in **R1**. On the other hand, *p*-bromochlorobenzene is mixed with BuLi in **M2** and *p*-chlorophenyllithium species are generated in **R2**. Such two different organollithiums can be separately generated and then integrated at −70 °C in **M3** and **R3**. In **R3**, 2-bromophenyllithium decomposes at −30 °C to generate a benzyne without affecting the functional aryllithium. This process can be followed by the spontaneous carbolithiation of the benzyne with the aryllithium. The resulting functional biaryllithium can then be reacted with an electrophile, in reactor **R4**, to yield the corresponding three-component coupling product. The precise optimization of the reaction conditions using temperature–residence time mapping is responsible for the success of this three-component coupling. Furthermore, this method has been successfully used in the synthesis of Boscalid, an important fungicide belonging to a class of succinate dehydrogenase inhibitors.

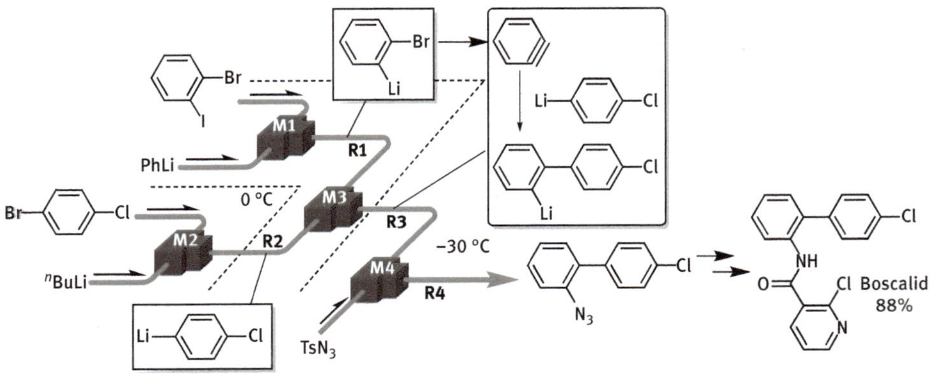

Fig. 2.27: Synthesis of an unsymmetrical diarylethene using an integrated flow microreactor system.

2.4.7 Example 7: three-component coupling by linear and convergent integration [62, 63]

Convergent and linear integration using the flow microreactor method can also be used in chemoselective three-component couplings. A central issue in organic synthesis is chemoselectivity, which refers to the preferential reaction of a chemical reagent or reactive species with one of two or more distinct functional groups. In general, chemoselective nucleophilic reactions of difunctional electrophiles are straightforward if the reactivity of one functional group is higher than that of the other. However, this is not the case for very rapid reactions. In fact, the reaction of 4-benzoylbenzaldehyde with an equivalent amount of phenyllithium in a batch reactor leads to the formation of a mixture of three products, even though aldehydes are generally more reactive than ketones. Conversely, remarkable chemoselectivity has been achieved by employing rapid micromixing, offering the desired compound with high selectivity at high flow rates. Reaction with *p*-cyanophenyllithium at the aldehyde carbonyl, followed by reaction with *p*-nitrophenyllithium at the ketone carbonyl, chemoselectively offers the three-component coupling product at 61% yield (Fig. 2.28).

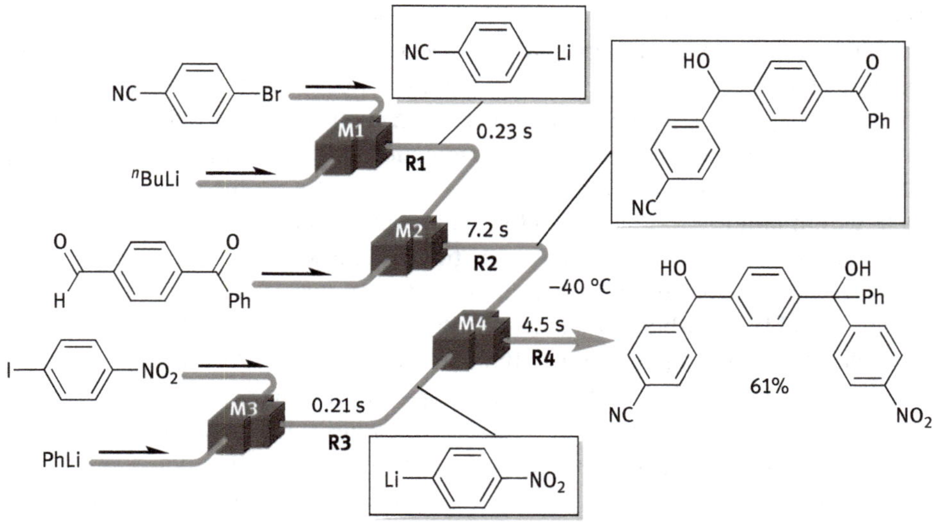

Fig. 2.28: Chemoselective three-component coupling reaction of difunctional electrophiles with functionalized aryllithiums in a linear and convergent integrated flow microreactor.

2.4.8 Example 8: integration of lithiation and cross-coupling [64, 65]

Different types of reactions can be combined by space integration. In the following example of space integration, halogen (X)/Li exchange of aryl halides (ArX) and Pd-catalyzed cross-coupling with aryl halides (Ar'X) are combined.

The palladium-catalyzed cross-coupling reactions of aryl–metal species with organic halides are widely used for carbon–carbon bond formation in the synthesis of a variety of biologically active compounds and functional materials. In particular, the Suzuki–Miyaura cross-coupling reaction involving aryl–boron compounds [66] is used extensively because boronic acids and their derivatives are usually air- and moisture-stable. In contrast, cross-coupling reactions using aryllithiums are rather limited, even though aryllithiums are often used to synthesize many aryl–metal compounds, including aryl–boron compounds. Therefore, using aryllithiums directly for the palladium-catalyzed cross-coupling reaction should be given serious consideration, especially because of its potential atom- [67] and step-economy [68].

The Pd-catalyzed cross-coupling of organolithium compounds with organic halides is called Murahashi coupling [69]. However, due to the serious side reactions, it is not popular in organic synthesis As an example, consider the cross-coupling of *p*-methoxyphenyllithium and bromobenzene. As shown in Fig. 2.29, *p*-methoxyphenyllithium can be obtained by Br/Li exchange of *p*-methoxybromobenzene using *n*-BuLi. The reaction also produces the by-product, 1-bromobutane (*n*-BuBr). The *p*-methoxyphenyllithium is next cross-coupled with bromobenzene in the presence of Pd catalyst, but as is often the case, if this reaction is slow, the *n*-BuBr reacts with the *p*-methoxyphenyllithium, to give *p*-methoxybutylbenzene. Actually, Pd-catalyzed cross-coupling reactions usually take hours to go to completion at room temperature or above, whereas reactions between aryllithiums and alkyl halides, such as *n*-BuBr, complete within minutes at 0 °C. Therefore, this problem needs to be solved in order to integrate the halogen/Li exchange reaction and the Murahashi coupling.

One way to avoid the problem might be to use two equivalents of *tert*-butyllithium (*t*-BuLi). The *t*-butyl bromide (*t*-BuBr), generated by the Br/Li exchange, reacts with the second equivalent of *t*-BuLi, to give isobutene, 2-methylpropane, and LiBr, none of which react with aryllithium species. However, this approach is impractical, especially for large-scale laboratory synthesis and industrial production, because *t*-BuLi ignites spontaneously in air, even at room temperature.

The reaction with *n*-BuBr is not the only undesired reaction. When the cross-coupling reaction is slow, the Br/Li exchange between *p*-methoxyphenyllithium and bromobenzene, generates *p*-methoxybromobenzene and phenyllithium. The *p*-methoxybromobenzene reacts with *p*-methoxyphenyllithium, giving the corresponding homo-coupling product. In addition, phenyllithium reacts with bromobenzene, giving the homo-coupling product, biphenyl.

The rate of the cross-coupling reactions depends upon the nature of the catalyst. Therefore, it is essential to find a catalyst that enables the cross-coupling to be faster than the side reactions. Conventional Pd catalysts, such as $Pd(PPh_3)_4$ and Pd $(Pt\text{-}Bu_3)_2$, are inadequate. Pd catalysts containing carbene ligands (PEPPSI-IPr and PEPPSI-SIPr) [70], on the other hand, make the cross-coupling reaction much faster than the side reactions, offering the desired products at good yields.

The desired integration of Br/Li exchange and the cross-coupling reactions can be successfully carried out in an integrated flow microreactor system that consists of three micromixers (**M1**, **M2**, and **M3**) and three microtube reactors (**R1**, **R2**, and **R3**), shown in Fig. 2.30. First, the *p*-methoxybromobenzene is mixed with *n*-BuLi in **M1** and passed into **R1** for the exchange reaction to take place. The temperature in **M1** and **R1** should be 0 °C and the residence time in **R1** should be 2.62 s. In the second step, the resulting *p*-methoxyphenyllithium species is mixed with a solution of bromobenzene and 0.05 equiv of PEPPSI-SIPr as a catalyst in **M2**. Because

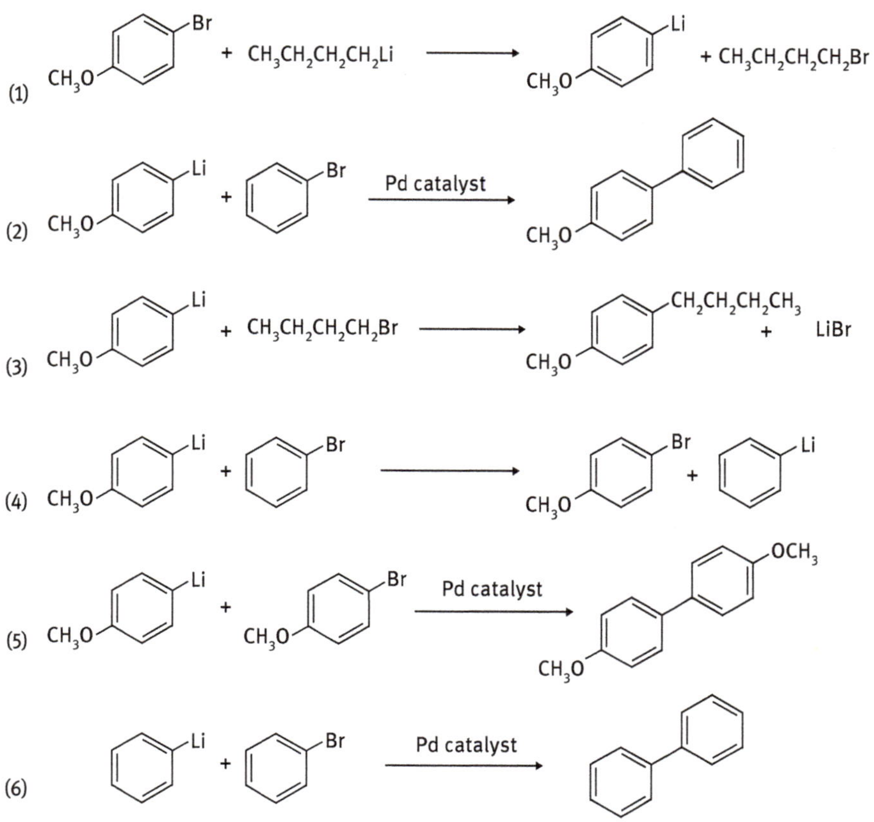

Fig. 2.29: Individual reactions, including unwanted side reactions, occurring in the generation of *p*-methoxyphenyllithium by a Br/Li exchange reaction of *p*-methoxybromobenzene, followed by a Pd-catalyzed cross-coupling reaction with bromobenzene.

the catalyst does not react with the bromobenzene, they are mixed together before being fed into **M2**. The cross-coupling reaction should be carried out in **R2** for 94 s at 50 °C. The resultant solution is then passed into **M3**, where it is mixed with methanol to destroy any unchanged *p*-methoxyphenyllithium species and *n*-BuLi. The overall transformation takes less than 2 min to complete. The reactions can be successfully carried out with various aryl bromides as both precursors of aryllithium species and coupling partners.

Halogen/Li exchange reactions can also be integrated with homo-coupling reactions promoted by $FeCl_3$ [71]. In this case, the homo-coupling is much faster than the side reactions. Space integration of halogen/Li exchange, borylation, and Pd-catalyzed Suzuki–Miyaura coupling are all possible [72–74] (Fig. 2.31). Functionalized aryl groups can be used, thanks to the very precise control of the reaction time [58].

These examples demonstrate that different types of reactions can be spatially integrated using integrated flow microreactor systems. However, it is important to keep

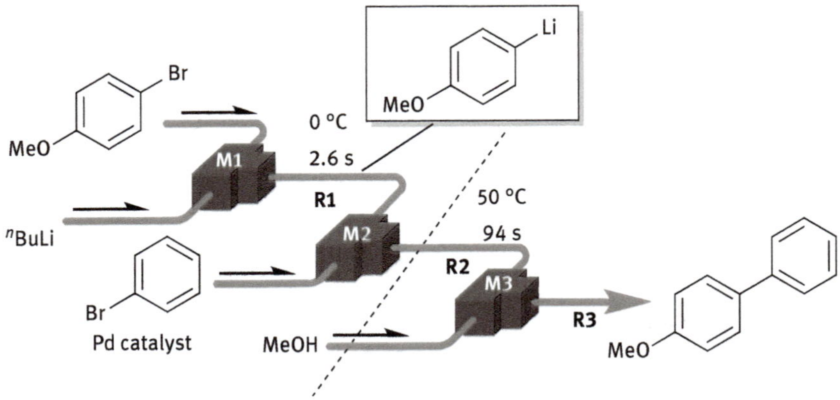

Fig. 2.30: An example of integration of lithiation and a cross-coupling reaction in a space-integrated flow microreactor system: generation of *p*-methoxyphenyllithium by Br/Li exchange of *p*-methoxybromobenzene, followed by Pd-catalyzed cross-coupling with bromobenzene.

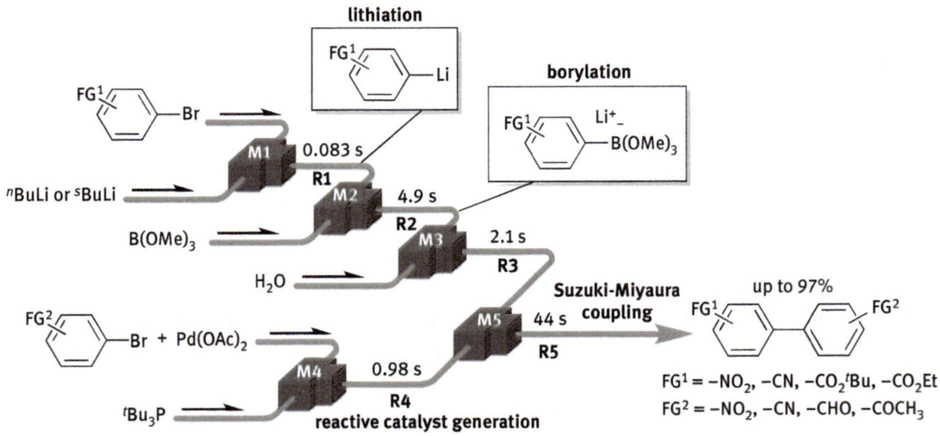

Fig. 2.31: An example of integration of lithiation, borylation, and a cross-coupling reaction in a space-integrated flow microreactor system.

in mind that the desired second reaction should be faster than the side reactions of the intermediate. Another important point is that the temperatures of the first and subsequent reactions can be quite different.

2.4.9 Example 9: anionic polymerization [75–77]

Polymerization reactions are very important in chemistry and widely used in industry for making macromolecules with a variety of functions. A powerful tool for studying polymerization processes is flow chemistry. There are various types and methods of polymerization. Particularly important types of polymerization are living and anionic polymerization. Anionic polymerization can be viewed as reactions involving unstable reactive intermediates. Anionic polymerization of vinyl monomers is an excellent method for synthesizing living polymers. They have well-defined end structures because the living ends of the anionic polymer are reactive organolithium species that can be used for end-functionalization reactions with various electrophiles and block copolymerization. Because the living polymer ends decompose readily in polar solvents such as THF, conventional anionic polymerization in macro-batch reactors using polar solvents should be carried out at around −78 °C or lower. This obviously places severe limitations on its use in industrial batch chemistry [79, 80]. In industrial production, maintaining such low temperatures can be a huge environmental and economic burden, limiting its useful polymerization. Living anionic polymerization of alkyl methacrylates, initiated by 1,1-diphenylhexyllithium, using a flow microreactor produces the corresponding poly(alkyl methacrylate)s with controlled molecular weight distribution under more accessible temperatures (methyl methacrylate (MMA): Mw/Mn = 1.16, −28 °C), (butyl methacrylate (BuMA): Mw/Mn = 1.24, 0 °C), (*tert*-butyl methacrylate (*t*-BuMA): Mw/Mn = 1.12, 24 °C) [81–83]. Precise control of the reaction temperature and the fast mixing of a monomer and an initiator seem to be responsible for successful polymerization (Fig. 2.32). In addition, one of the most important advantages of a flow microreactor is that a laboratory-scale reaction can be readily adapted to an industrial-scale production, simply by continuously running a reaction for a long period of time. In fact, in the case of flow anionic polymerization, about 1 kg of polymer was produced over a 3 h reaction time with high molecular weight and narrow molecular weight distribution (Mn = 8,000, Mw/Mn = 1.1). Moreover, polymethacrylates, synthesized by flow microreactor polymerization, might give a different structure compared to the conventional batch method, because optimized reaction temperatures would be different.

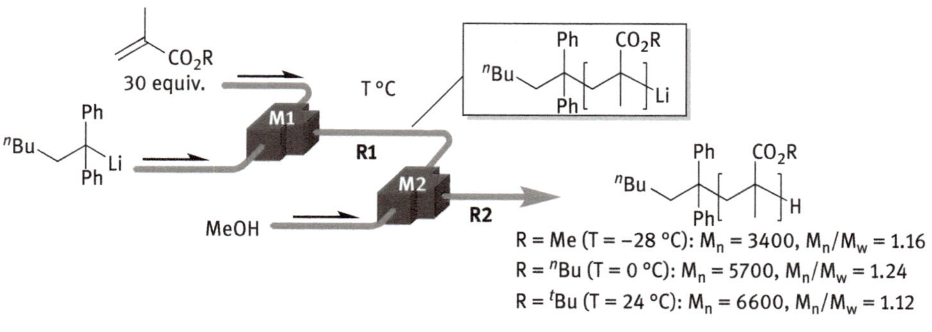

Fig. 2.32: A flow microreactor system for anionic polymerization of alkyl methacrylates initiated by 1,1-diphenylhexyllithium. M1, M2: T-shaped micromixer; R1, R2: microtube reactor.

For syntheses of end-functionalized polymers and block copolymers, it is crucial to control and maintain longevity of reactive carbanionic polymer ends. The lifetime of the polymer ends can be evaluated in a flow microreactor system, as shown in Fig. 2.33.

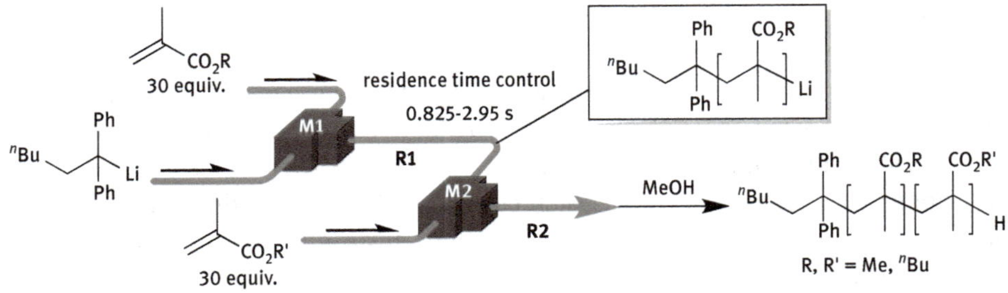

Fig. 2.33: An integrated flow microreactor system for the sequential anionic polymerization of alkyl methacrylates initiated by 1,1-diphenylhexyllithium. M1, M2: T-shaped micromixer; R1, R2: microtube reactor.

A solution of an alkyl methacrylate and that of 1,1-diphenylhexyllithium is mixed in micromixer **M1** and polymerization is carried out in the microtube reactor **R1**. Then, a solution of the same monomer is introduced into micromixer **M2**, which is connected to microtube reactor **R2**, where the sequential polymerization takes place. By changing the length of **R1** with the fixed flow rate, the influence of residence time in **R1** on the product can be analyzed. With any residence time, the addition of the second monomer solution resulted in an increase in molecular weight. However, longer the residence time in **R1**, the molecular weight distribution also increases, presumably because of decomposition of the polymer end (Fig. 2.34). By choosing an appropriate residence time in **R1** (MMA: 2.95 s, BuMA: 0.825 s), the sequential polymerization can be successfully carried out without significant decomposition of the living polymer end [84,85].

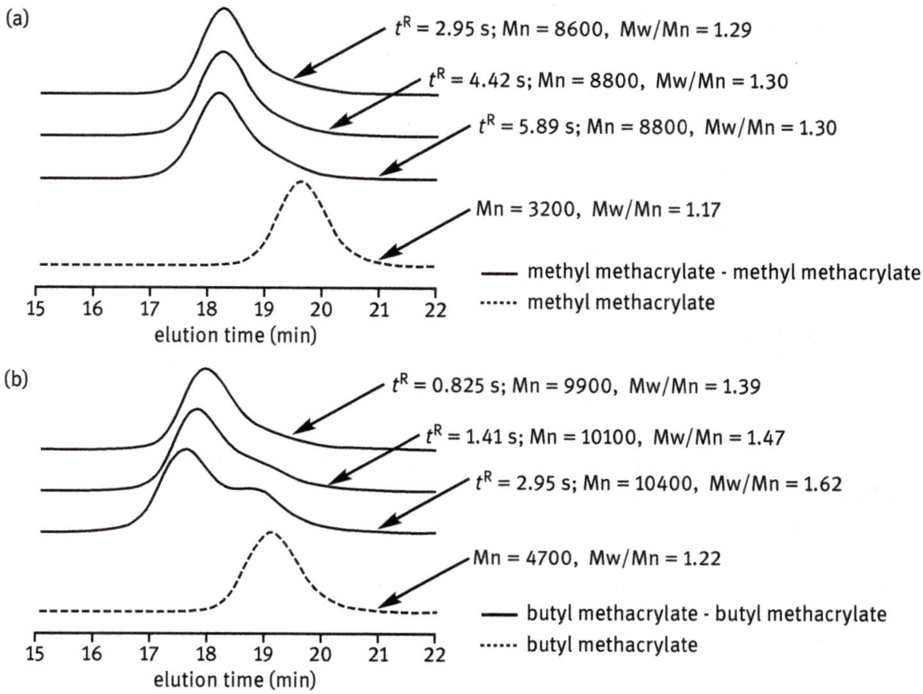

Fig. 2.34: Size-exclusion chromatography traces polymers obtained in the integrated flow microreactor system. Effect of residence time on the molecular weight distribution. (a) methyl methacrylate – methyl methacrylate, (b) butyl methacrylate – butyl methacrylate.

2.4.10 Example 10: heterotelechelic polymer synthesis by anionic polymerization initiated by a functional alkyllithium [85]

Structurally well-defined polymers are important functional materials. They can be synthesized by a highly controlled polymerization. Anionic polymerization is an excellent method as the anionic reactive polymer ends are living and can be used for subsequent polymerization with a different monomer. However, the synthesis of heterotelechelic polymers by anionic polymerization remains a major challenge because the initiators are usually nonfunctional alkyllithium compounds, such as *n*-butyllithium and *sec*-butyllithium. Our laboratory recently reported the generation and reaction of functional alkyllithiums using a flow microreactor [86]. This method was further used for the synthesis of heterotelechelic polymer, with two different functional end-groups, in an integrated flow microreactor system. As shown in Fig. 2.35, anionic block copolymerization of styrene, initiated by functional alkyllithiums and *p*-methoxy styrene, followed by borylation, successfully provided the structurally well-defined heterotelechelic block copolymer at a good yield with high molecular weight and narrow molecular weight distribution.

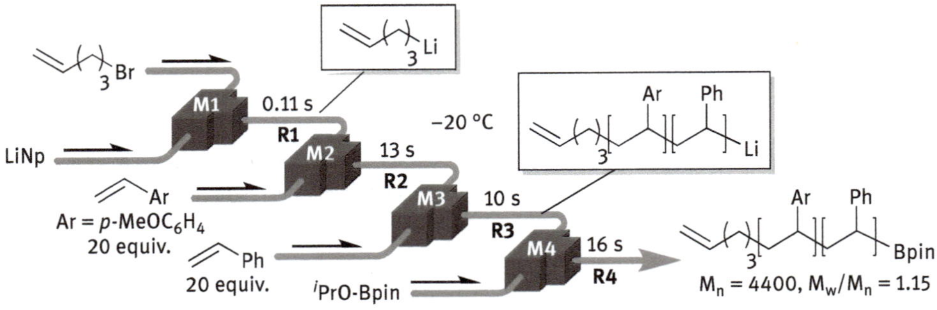

Fig. 2.35: Linear integration system for functional alkyllitnium-initiated anionic polymerization of a structurally well-defined heterotelechelic copolymer.

(For further details on this issue please see Volume 2, Chapter 4, Title: Polymer synthesis in continuous flow)

2.5 Looking forward

This chapter discussed the principles of controlling reactions involving highly unstable reactive intermediates in flow chemistry. Very precise control of the residence time is a key feature of flow microreactors. Rapid heat transfer, particularly out of the system, is also an important feature. Fast micromixing is crucial for the control of very fast reactions, which are complete within seconds. A selection of examples of such reactions, mainly focusing on organolithium reactions and anionic polymerization, are presented and discussed. It should be kept in mind that these reactions are studied not only for their own sake but also for generating potential ideas that may be useful for developing other types of reactions. The principles discussed are applicable to a wide range of reactions involving various unstable reactive intermediates.

Further readings

- Flowmicro synthesis: Yoshida J, Basics of Flow Microreactor Synthesis, Springer, Tokyo, 2015.
- Flash chemistry: Yoshida J, Flash Chemistry Fast Organic Synthesis in Microsystems. Wiley, Chichester, 2008.
- Organometallic chemistry: Noël T (Ed.), Organometallic Flow Chemistry, Springer International Publishing, Switzerland, 2016.
- Flowmicro synthesis: Reschetilowski E (Ed.), Microreactors in Preparative Chemistry Wiley-VCH, Weinheim, 2013.

- Flowmicro synthesis: Wirth T (Ed.), Microreactors in Organic Chemistry and Catalysis, 2nd ed., Wiley, Hoboken, 2013.
- Flow synthesis: Darvas F, Hessel V, Dorman G, Flow Chemistry, DeGruyter, Berlin, 2014.

Study questions

2.1 Design a flow system consisting of micromixers and microreactors suitable for the following transformations:

(a)

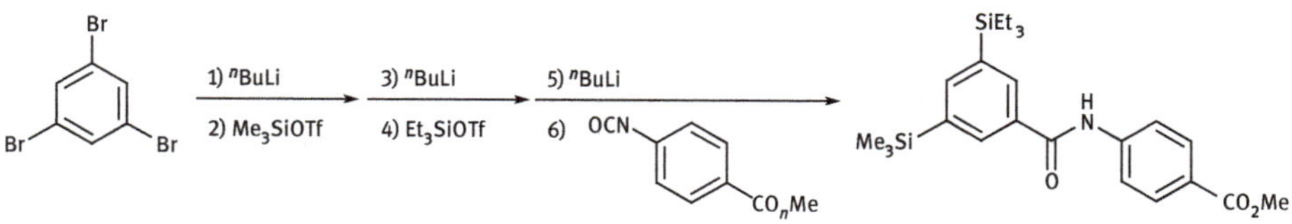

(b)

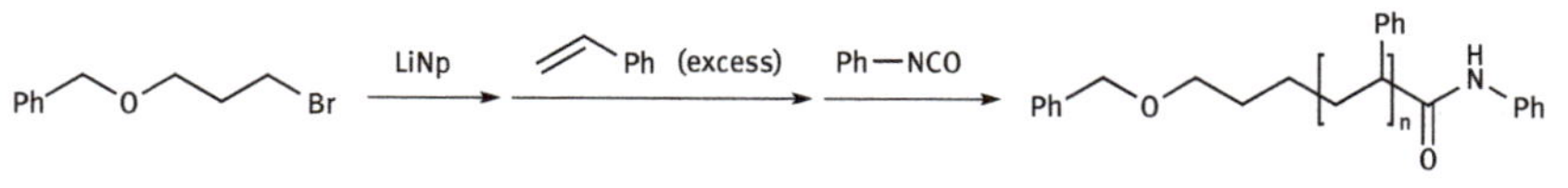

2.2 Draw the structures of aryllithium intermediates A to D, and design the flow system consisting of micromixers and microreactors suitable for the following transformations:

(a)

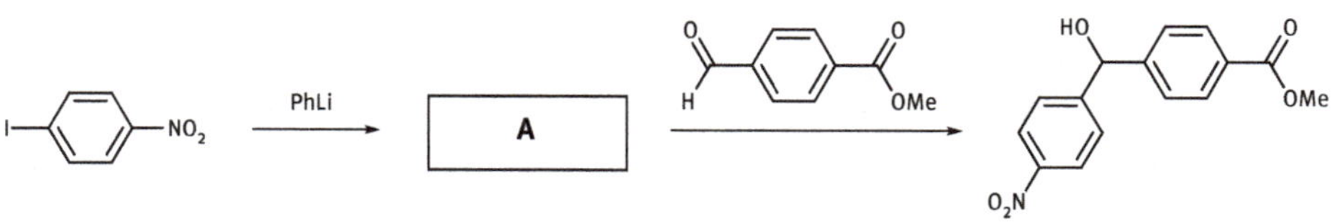

(b)

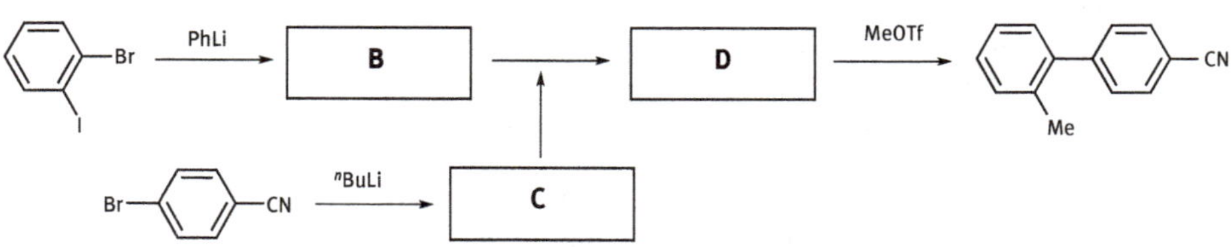

References

[1] Cantillo, D, Kappe, CO, Halogenation of organic compounds using continuous flow and microreactor technology, React Chem Eng, 2017, 2, 7–19.

[2] Plutschack, MB, Pieber, B, Gilmore, K, Seeberger, PH, The Hitchhiker's guide to flow chemistry, Chem Rev, 2017, 117, 11796–11893.

[3] Kobayashi, S, Flow "fine" synthesis: High yielding and selective organic synthesis by flow methods, Chem Asian J, 2016, 11, 425–426.

[4] Cambié, D, Bottecchia, C, Straathof, NJW, Hessel, V, Noël, T, Applications of continuous-flow photochemistry in organic synthesis, material science, and water treatment, Chem Rev, 2016, 116, 10276–10341.

[5] Fukuyama, T, Totoki, T, Ryu, I, Carbonylation in microflow: Close encounters of CO and reactive species, Green Chem, 2014, 16, 2042.

[6] Baxendale, IR, The integration of flow reactors into synthetic organic chemistry, J Chem Technol Biotechnol, 2013, 88, 519.

[7] Pastre, JC, Browne, DL, Ley, SV, Flow chemistry syntheses of natural products, Chem Soc Rev, 2013, 42, 8849–8869.

[8] McQuade, DT, Seeberger, PH, Applying flow chemistry: Methods, materials, and multistep synthesis, J Org Chem, 2013, 78, 6384–6389.

[9] Wiles, C, Watts, P, Continuous flow reactors: A perspective, Green Chem, 2012, 14, 38–54.

[10] Wegner, J, Ceylan, S, Kirschning, A, Ten key issues in modern flow chemistry, Chem Commun, 2011, 47, 4583–4592.

[11] Hartman, RL, McMullen, JP, Jensen, KF, Deciding whether to go with the flow: Evaluating the merits of flow reactors for synthesis, Angew Chem Int Ed, 2011, 50, 7502–7519.

[12] Ley, SV, The changing face of organic synthesis, Tetrahedron, 2010, 66, 6270–6292.

[13] Webb, D, Jamison, TF, Continuous flow multi-step organic synthesis, Chem Sci, 2010, 1, 675–680.

[14] Illg, T, Löb, P, Hessel, V, Flow chemistry using milli- and microstructured reactors – From conventional to novel process windows, Bioorg Med Chem, 2010, 18, 3707–3719.

[15] Tanaka, K, Fukase, K, Renaissance of traditional organic reactions under microfluidic conditions: A new paradigm for natural products synthesis, Org Process Res Dev, 2009, 13, 983–990.

[16] Fukuyama, T, Rahman, MT, Sato, M, Ryu, I, Adventures in inner space: Microflow systems for practical organic synthesis, Synlett, 2008, 151–163.

[17] Roberge, DM, Zimmermann, B, Rainone, F, Gottsponer, M, Eyholzer, M, Kockmann, N, Microreactor technology and continuous processes in the fine chemical and pharmaceutical industry: Is the revolution underway?, Org Process Res Dev, 2008, 12, 905–910.

[18] Ahmed-Omer, B, Brandt, JC, Wirth, T, Advanced organic synthesis using microreactor technology, Org Biomol Chem, 2007, 5, 733–740.

[19] Yoshida, J, Takahashi, Y, Nagaki, A, Flash chemistry: Flow chemistry that cannot be done in batch, Chem Commun, 2013, 49, 9896–9904.

[20] Yoshida, J, Flash Chemistry. Fast Organic Synthesis in Microsystems, Chichester, UK, Wiley, 2008.

[21] Yoshida, J, Nagaki, A, Yamada, T, Flash chemistry: Fast chemical synthesis by using microreactors, Chem Eur J, 2008, 14, 7450–7459.

[22] Yoshida, J, Flow microreactor synthesis based on high-resolution reaction time control, Chem Rec, 2010, 10, 332–341.

[23] Ehrfeld, W, Golbig, K, Hessel, V, Löwe, H, Richter, T, Characterization of mixing in micromixers by a test reaction: Single mixing units and mixer arrays, Ind Eng Chem Res, 1999, 38, 1075–1082.

[24] Soleymani, A, Yousefi, H, Turunen, I, Dimensionless number for identification of flow patterns inside a T-micromixer, Chem Eng Sci, 2008, 63, 5291–5297.
[25] Yoshida, J, Kim, H, Nagaki, A, Green and sustainable chemical synthesis using flow microreactors, ChemSusChem, 2011, 4, 331–340.
[26] Kim, H, Nagaki, A, Yoshida, J, A flow microreactor approach to protecting-group-free synthesis using organolithium compounds, Nat Commun, 2011, 2, 264.
[27] Nagaki, A, Kim, H, Yoshida, J, Aryllithium compounds bearing alkoxycarbonyl groups: Generation and reactions using a microflow system, Angew, Chem Int Ed, 2008, 47, 7833–7836.
[28] Nagaki, A, Kim, H, Moriwaki, Y, Matsuo, C, Yoshida, J, A flow microreactor system enables organolithium reactions without protecting alkoxycarbonyl groups, Chem Eur J, 2010, 16, 11167–11177.
[29] Rasheed, M, Wirth, T, Intelligent microflow: Development of self-optimizing reaction systems, Angew Chem Int Ed, 2011, 50, 357–358.
[30] Nagaki, A, Recent topics of functionalized organolithiums using flow microreactor chemistry, Tetrahedron Lett, 2019, 60, 150923.
[31] Nagaki, A, Kim, H, Usutani, H, Matsuo, C, Yoshida, J, Generation and reaction of cyano-substituted aryllithium compounds using microreactors, Org Biomol Chem, 2010, 8, 1212–1217.
[32] Nagaki, A, Tsuchihashi, Y, Haraki, S, Yoshida, J, Organolithiums bearing aldehyde carbonyl groups. a flash chemistry approach, Org Biomol Chem, 2015, 13, 7140–7145.
[33] Nagaki, A, Yamada, S, Doi, M, Tomida, Y, Takabayashi, N, Yoshida, J, Flow microreactor synthesis of disubstituted pyridines from dibromopyridines via Br/Li exchange without using cryogenic conditions, Green Chem, 2011, 13, 1110–1113.
[34] Nagaki, A, Yamada, D, Yamada, S, Doi, M, Ichinari, D, Tomida, Y, Takabayashi, N, Yoshida, J, Generation and reactions of pyridyllithiums via Br/Li exchange reactions using continuous flow microreactor systems, Aust J Chem, 2013, 66, 199–207.
[35] Nagaki, A, Takizawa, E, Yoshida, J, Generation and reactions of oxiranyllithiums by use of a flow microreactor system, Chem Eur J, 2010, 16, 14149–14158.
[36] Nagaki, A, Takizawa, E, Yoshida, J, Oxiranyl anion methodology using microflow systems, J Am Chem Soc, 2009, 131, 1654–1655 and 3787.
[37] Nagaki, A, Takizawa, E, Yoshida, J, Generations and reactions of N-t-butylsulfonylaziridinyllithiums using microreactors, Chem Lett, 2009, 38, 1060–1061.
[38] Nagaki, A, Takahashi, Y, Yamada, S, Matsuo, C, Haraki, S, Moriwaki, Y, Kim, S, Yoshida, J, Generation and reactions of vinyllithiums using flow microreactor systems, J Flow Chem, 2012, 2, 70–72.
[39] Tomida, Y, Nagaki, A, Yoshida, J, Carbolithiation of conjugated enynes with aryllithiums in microflow system and applications to synthesis of allenylsilanes, Org Lett, 2009, 11, 3614–3617.
[40] Nagaki, A, Tokuoka, S, Yamada, S, Tomida, Y, Oshiro, K, Amii, H, Yoshida, J, Perfluoroalkylation in flow microreactor: Generation of perfluoroalkyllithiums in the presence and absence of electrophiles, Org Biomol Chem, 2011, 9, 7559–7563.
[41] Kawaguchi, T, Miyata, H, Ataka, K, Mae, K, Yoshida, J, Room temperature swern oxidation using micro flow system, Angew Chem Int Ed, 2005, 44, 2413–2416.
[42] Saito, K, Ueoka, K, Matsumoto, K, Suga, S, Nokami, T, Yoshida, J, Indirect cation flow method. Flash generation of alkoxycarbenium ions and studies on stability of glycosyl cations, Angew Chem Int Ed, 2011, 50, 5153–5156.
[43] Suga, S, Yamada, D, Yoshida, J, Cationic Three-component Coupling Involving an Optically Active Enamine Derivative. From Time Integration to Space Integration of Reactions, Chem Lett, 2010, 39, 404–406.
[44] Ashikari, Y, Saito, K, Nokami, T, Yoshida, J, Nagaki, A, Oxo-thiolation of cationically polymerizable alkenes using flow microreactors, Chem Eur J, 2019, 25, 15239–15243.

[45] Nagaki, A, Kim, H, Yoshida, J, Nitro-substituted aryl lithium compounds in microreactor synthesis: Switch between kinetic and thermodynamic control, Angew Chem Int Ed, 2009, 48, 8063–8065.
[46] Tomida, Y, Nagaki, A, Yoshida, J, Asymmetric carbolithiation of conjugated enynes: A flow microreactor enables the use of configurationally unstable intermediates before they epimerize, J Am Chem Soc, 2011, 133, 3744–3747.
[47] Giovine, A, Musio, B, Degennaro, L, Falcicchio, A, Nagaki, A, Yoshida, J, Luisi, R, Synthesis of 1,2,3,4-Tetrahydroisoquinolines by Microreactor-Mediated Thermal Isomerization of Laterally Lithiated Arylaziridines, Chem Eur J, 2013, 19, 1872–1876.
[48] Asai, A, Takata, A, Ushiogi, Y, Iinuma, Y, Nagaki, A, Yoshida, J, Switching reaction pathways of Benzo[b]thiophen-3-yllithium and Benzo[b]furan-3-yllithium based on high-resolution residence-time and temperature control in a flow microreactor, Chem Lett, 2011, 40, 393–395.
[49] Yoshida, J, Saito, K, Nokami, T, Nagaki, A, Space integration of reactions: An approach to increase the capability of organic synthesis, Synlett, 2011, 1189–1194.
[50] Pastre, JC, Browne, DL, Ley, SV, Chem Soc Rev, 2013, 42, 8849–8869.
[51] Baxendale, IR, J Chem Technol Biotechnol, 2013, 88, 519–552.
[52] Mascia, S, Heider, PL, Zhang, H, et. al., End-to-end continuous manufacturing of pharmaceuticals: Integrated synthesis, purification, and final dosage formation, Angew Chem Int Ed, 2013, 52, 12359–12363.
[53] Colella, M, Tota, A, Takahashi, Y, Luisi, R, Nagaki, A, Fluoro-substituted methyllithium chemistry based on a novel external quenching method using flow microreactors, Angew Chem Int Ed, 2020, 59, 11924–11928.
[54] Nagaki, A, Matsuo, C, Kim, S, Saito, K, Miyazaki, A, Yoshida, J, Lithiation of 1,2-dichloroethene in flow microreactors: Versatile synthesis of alkenes and Alkynes by precise residence-time control, Angew Chem Int Ed, 2012, 51, 3245–3248.
[55] Usutani, H, Tomida, T, Nagaki, A, Okamoto, H, Nokami, T, Yoshida, J, Generation and reactions of *o*-bromophenyllithium without benzyne formation using a microreactor, J Am Chem Soc, 2007, 129, 3046–3047.
[56] Nagaki, A, Tomida, Y, Usutani, H, Kim, H, Takabayashi, N, Nokami, T, Okamoto, H, Yoshida, J, Integrated micro flow synthesis based on sequential Br-Li exchange reactions of *p*-, *m*-, and *o*-Dibromobenzenes Chem, Asian J, 2007, 2, 1513–1523.
[57] Nagaki, A, Imai, K, Kim, H, Yoshida, J, Flash Synthesis of TAC-101 and Its Analogues from 1,3,5-Tribromobenzene Using Integrated Flow Microreactor Systems, RSC Adv, 2011, 1, 758–760.
[58] Webb, D, Jamison, TF, Continuous flow multi-step organic synthesis, Chem Sci, 2010, 1, 675–680.
[59] Ushiogi, Y, Hase, T, Iinuma, Y, Takata, A, Yoshida, J, Synthesis of photochromic diarylethenes using microflow system, Chem Commun, 2007, 2947–2949.
[60] Asai, T, Takata, A, Nagaki, A, Yoshida, J, Practical synthesis of photochromic diarylethenes in integrated flow microreactor systems, ChemSusChem, 2012, 5, 339–350.
[61] Nagaki, A, Ichinari, D, Yoshida, J, Three-Component coupling based on flash chemistry. Carbolithiation of benzyne with functionalized aryllithiums followed by reactions with electrophiles, J Am Chem Soc, 2014, 136, 12245–12248.
[62] Nagaki, A, Imai, K, Ishiuchi, S, Yoshida, J, Remarkable chemoselectivity by flash chemistry. Reactions of difunctional electrophiles with functionalized aryllithiums, Angew Chem, Int Ed, 2015, 54, 1914–1918.
[63] Nagaki, A, Ishiuchi, S, Imai, K, Sasatsuki, K, Nakahara, Y, Yoshida, J, Micromixing enables protecting group-free-synthesis in organolithium chemistry, React Chem Eng, 2017, 2, 862–870.
[64] Nagaki, A, Kenmoku, A, Moriwaki, Y, Hayashi, A, Yoshida, J, Cross-coupling in a flow microreactor: Space integration of lithiation and murahashi coupling, Angew Chem Int Ed, 2010, 49, 7543–7547.

[65] Nagaki, A, Moriwaki, Y, Haraki, S, Kenmoku, A, Takabayashi, N, Hayashi, A, Yoshida, J, Cross-coupling of aryllithiums with aryl and vinyl halides in flow microreactors, Chem Asian J, 2012, 7, 1061–1068.
[66] Miyaura, N, Suzuki, A, Palladium-catalyzed cross-coupling reactions of organoboron compounds, Chem Rev, 1995, 95, 2457–2483.
[67] Trost, BM, Atom economy – a challenge for organic synthesis: Homogeneous catalysis leads the way, Angew Chem Int Ed Engl, 34, 259–281.
[68] Wender, PA, Verma, VA, Paxton, TJ, Pillow, TH, Function-Oriented synthesis, step economy, and drug design, Acc Chem Res, 2008, 41, 40–49.
[69] Murahashi, S, Yamamura, M, Yanagisawa, K, Mita, N, Kondo, K, Stereoselective synthesis of alkenes and alkenyl sulfides from alkenyl halides using palladium and ruthenium catalysts, J Org Chem, 1979, 44, 2408–2417.
[70] Organ, MG, Avola, S, Dubovyk, I, Hadei, N, Kantchev, EAB, O'Brien, CJ, Valente, C, A User-Friendly, All-Purpose Pd–NHC (NHC=N-Heterocyclic Carbene) Precatalyst for the Negishi Reaction: A Step Towards a Universal Cross-Coupling Catalyst, Chem Eur J, 2006, 12, 4749–4755.
[71] Nagaki, A, Uesugi, Y, Tomida, Y, Yoshida, J, Homocoupling of aryl halides in flow: Space integration of lithiation and $FeCl_3$ promoted homocoupling, Beilstein J Org Chem, 2011, 7, 1064–1069.
[72] Nagaki, A, Moriwaki, Y, Yoshida, J, Flow synthesis of arylboronic esters bearing electrophilic functional groups and space integration with Suzuki–Miyaura coupling without intentionally added base, Chem Commun, 2012, 48, 11211–11213.
[73] Noël, T, Kuhn, S, Musacchio, AJ, Jensen, KF, Buchwald, SL, Suzuki–miyaura cross-coupling reactions in flow: Multistep synthesis enabled by a microfluidic extraction, Angew Chem Int Ed, 2011, 50, 5943–5946.
[74] Takahashi, Y, Ashikari, Y, Takumi, M, Shimizu, Y, Jiang, Y, Higuma, R, Ishikawa, S, Sakaue, H, Shite, I, Maekawa, K, Aizawa, Y, Yamashita, H, Yonekura, Y, Colella, M, Luisi, R, Takegawa, T, Fujita, C, Nagaki, A, Synthesis of biaryls having a piperidylmethyl group based on space integration of lithiation, borylation and Suzuki-Miyaura coupling, Eur J Org Chem, 2020, 618–622.
[75] Takahashi, Y, Nagaki, A, Anionic polymerizations using flow microreactors, Molecules, 2019, 24, 1532.
[76] Nagaki, A, Yoshida, J, Controlled polymerization in flow microreactor systems, Advances Poly Sci, 2013, 259, 1–50.
[77] Wilms, D, Klos, J, Frey, H, Microstructured reactors for polymer synthesis: A renaissance of continuous flow processes for tailor-made macromolecules? macromol, Chem Phys, 2008, 209, 343–356.
[78] Nakahara, Y, Furusawa, M, Endo, Y, Shimazaki, T, Takahashi, Y, Jiang, Y, Nagaki, A, Practical continuous flow controlled/living anionic polymerization, Chem Eng Tech, 2019, 42, 2154–2163.
[79] Nagaki, A, Nakahara, Y, Furusawa, M, Sawaki, T, Yamamoto, T, Toukairin, H, Tadokoro, S, Shimazaki, T, Ito, T, Otake, M, Arai, H, Higashida, N, Takahashi, Y, Moriwaki, Y, Tsuchihashi, Y, Hirose, K, Yoshida, J, Feasibility study on continuous flow controlled/living anionic polymerization processes, Org Process Res Dev, 2016, 20, 1377–1382.
[80] Nagaki, A, Tomida, Y, Miyazaki, A, Yoshida, J, Microflow system controlled anionic polymerization of alkyl methacrylates, Macromolecules, 2009, 42, 4384–4387.
[81] Nagaki, A, Tomida, Y, Yoshida, J, Microflow system controlled anionic polymerization of styrenes, Macromolecules, 2008, 17, 6322–6330.
[82] Wurm, F, Wilms, D, Klos, J, Löwe, H, Frey, H, Carbanions on tap – living anionic polymerization in a microstructured reactor, Macromol Chem Phys, 2008, 209, 1106–1114.

[83] Nagaki, A, Takahashi, Y, Akahori, K, Yoshida, J, Living Anionic polymerization of tert-butyl acrylate in a flow microreactor system and its applications to the synthesis of block copolymers, Macromol React Eng, 2012, 6, 467–472.
[84] Nagaki, A, Miyazaki, A, Tomida, Y, Yoshida, J, Anionic polymerization of alkyl methacrylates using flow microreactor systems, Chem Eng J, 2011, 167, 548–555.
[85] Nagaki, A, Yamashita, H, Hirose, K, Tsuchihashi, Y, Yoshida, J, Generation and reaction of functional alkyllithiums using flow microreactors and its application to heterotelechelic polymer synthesis, Chem Eur J, 2019, 25, 13719–13727.
[86] Nagaki, A, Yamashita, H, Hirose, K, Tsuchihashi, Y, Yoshida, J, Alkyllithiums Bearing Electrophilic Functional Groups: A Flash Chemistry Approach, Angew Chem Int Ed, 2019, 58, 4027–4030.

Melinda Fekete and Toma Glasnov

3 Technology overview/overview of the devices

Edited and revised by King Kuok (Mimi) Hii and Benjamin J. Deadman

The design and assembly of a flow reactor system is no different from that for a batch reactor; that is, it requires an understanding of the role of each individual component, which can be combined to deliver the required reaction conditions. The choice of each apparatus is also dictated by the nature of the experiment, including physical and chemical characteristics, operating regime, and desired productivity (Tab. 3.1); for example, different mixers (flow) or stirrers (batch) may be deployed, depending on the miscibility and viscosity of the reacting fluids (liquid or gas).

The major difference between batch and flow operations is how the various components can be integrated, and their spatial arrangement. Typically, mixing, heating, cooling, and quenching are integrated in the same "reaction volume" in a batch reactor. In contrast, these different operations can be uncoupled and spatially separated in a continuous-flow process, thus affording the chemist greater control and flexibility to potentially fine-tune each individual step, toward a desired outcome. Thus, it is critical for the operator to understand the capability and limits of each component (just as they do for the various components of a batch reactor), in order to operate the process safely and effectively.

With the rapid expansion of interest in the area of continuous-flow chemistry in the last few years, particularly by industrial R&D synthetic chemistry laboratories, several commercial instruments have emerged on the market to satisfy the needs of the laboratory chemist to access the new technology quickly. These systems are compact and usually include all that is needed to perform a continuous-flow synthesis in the early stages of a research and development project. However, given sufficient knowledge of how to construct a flow reactor, home-built systems can also be very effective, offering flexible and affordable alternatives to the commercial instruments.

This chapter begins with a brief description of the essential components of a continuous-flow system, followed by some information on how to assemble a flow reactor system and successfully carry out experiments. For the less-experienced researchers, some recommendations on the operation and maintenance of the individual units will also be highlighted. Last but not least, a list of commercially available flow reactors, suitable for laboratory-scale R&D, is provided at the end of the chapter.

https://doi.org/10.1515/9783110693676-003

Tab. 3.1: Fundamental considerations for the construction of a flow reactor.

Questions	Devices/decisions
Phases – how many and what types, including reactants, reagents, catalysts, products, by-products, and side-products	Most important factor. Determines number and type(s) of pumps, mixer, reactor, fittings, valves, and downstream operations. Certain solids (e.g., catalyst particles) can be handled as slurry/in a packed bed but clogging of reaction channels can be an issue. For gas–liquid reactions, (simulations of) vapor–liquid–pressure diagrams are needed to control the phases of a reaction.
What is the nature of the chemicals involved?	Chemical compatibility of materials for the wettable parts. Solvents are especially important.
Do the reactions involve gases	Decide whether the gas will be dissolved in the solvent (pre-saturate) or remain in a different phase (segmented flow, bubble column, etc.).
What is the reaction rate –fast or super-fast? Is the reaction kinetically or diffusion controlled?	Note: Slow reactions do not generally benefit from flow. The choice of mixers is dependent on the reaction kinetics.
Is the reaction exothermic or does it involve the application of a light source (photochemistry)?	Maximize surface-to-volume ratio of the reactor for better heat management/photon flux.
Does the reaction require temperature control?	Energy source, temperature control (thermocouple)
Is pressure needed (e.g., to handle superheated solvent, gaseous reactants/products)?	Type of pressure regulator, type of tubing and connectors
Is phase separation required?	Choice of phase separators (membrane separators, off-gassing, settling stage, etc.)
Is reaction monitoring required?	Choice of suitable online, inline, or at-line analytical techniques.
What productivity is required of the process? The quantity (mg, g, or kg) per time unit (s, min, h) per volume (L) is also known as the "space–time yield."	Choice of pumps, flow rates, and reactor volumes.

3.1 General aspects

The simplest assembly of a continuous-flow reactor includes a *pump* to process the reaction mixture throughout the setup and a *reactor*, where the synthetic transformation happens. Further additions to the setup depend on the requirements of the reaction itself (Tab. 3.1). For example, carrying out reactions with two or more reagents in separate streams will require additional pumps and *mixers* to ensure good contact between the reagents, particularly if they are of different viscosities, or immiscible phases.

Similarly, a reaction may be an exo- or endothermic process. This will require the reactor usually to be either cooled or heated in order to dissipate the excess heat, or to provide the energy to initiate and maintain the reaction. On an industrial scale it is common to use the released heat from one process for heating another one, thus reducing the operational cost and making the overall synthesis greener (in terms of energy efficiency). In a research laboratory *heating or cooling of the reaction zone* may be simply achieved using a traditional oil bath/oven for heating, or an ice-bath for subambient temperatures.

In all the cases where elevated temperatures and/or pressure are involved, an essential piece of the continuous-flow reactor is the *back-pressure regulator* (BPR), to maintain a steady flow (critical for controlling residence times). In some special processes, where a gas is needed – for example, gas–liquid reactions, reactions in supercritical CO_2, and reactions under inert atmosphere – the use of a BPR is obligatory to control and maintain the requisite vapor–liquid equilibrium (Fig. 3.1). This is also the case, when the reaction is operated in a temperature regime above the boiling point of the solvents or reagents. The choice of pumps may also influence the selection of BPR. While all pumps will have an effective upper limit to the backpressure they can work against, some pumping technologies (e.g., many piston pumps) are designed to work against back pressure and will deliver inaccurate or variable flow rates when insufficient back-pressure is applied to the system.

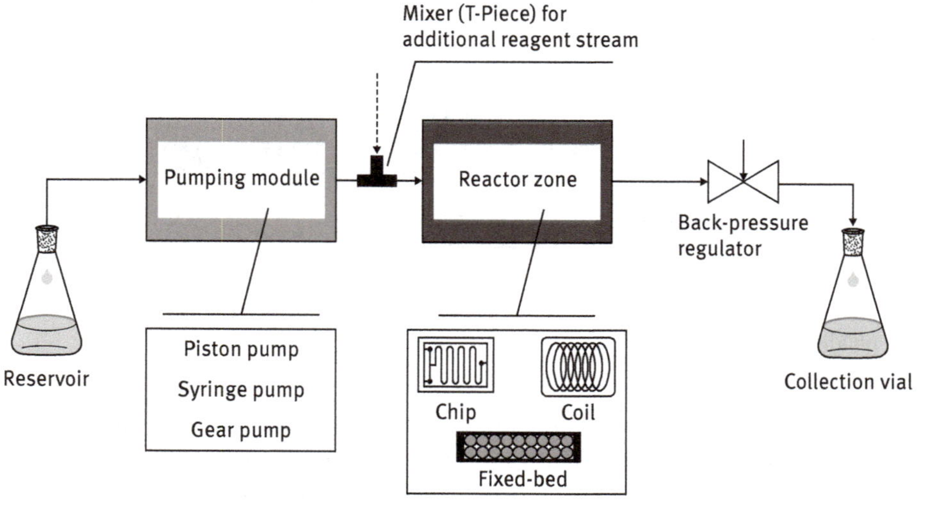

Fig. 3.1: A general scheme of a continuous-flow setup.

3.1.1 Pumps for liquid handling

The invention of the pump can be dated back to 2000 BC, when Egyptians invented the shadoof to draw water from wells. Ever since then, the development of pump technology has largely been driven by the need to redistribute water for irrigation and sanitation purposes. The dawn of the industrial era brought about more challenging problems, which were met with ever more creative technological solutions. Nowadays, it is possible to transport not only water but also gases, crude oil, and slurries, in a large scale across several continents, as well as a great many applications in industry.

Pumps can be classified into three main groups, by the method used to move the fluid: direct lift, displacement, or gravity [1]. Most pumping technologies employed in the synthetic laboratory for performing continuous-flow experiments are based on the displacement method, and their basic modes of operation are described below.

3.1.1.1 Syringe pump

A syringe pump (sometimes also known as an infusion pump) works on the principle of a reciprocating positive displacement of a fluid by moving a plunger in a cylinder (Fig. 3.2). In the simplest design, the actuation of the plunger is driven by a screw connected to stepper motor, providing a constant linear flow of the syringe content into the reactor zone. This type of syringe pump is widely deployed in medical care for the continuous administration of drugs to patients at low speed (typically few microliters to milliliter per minute), which may be achieved with high accuracy.

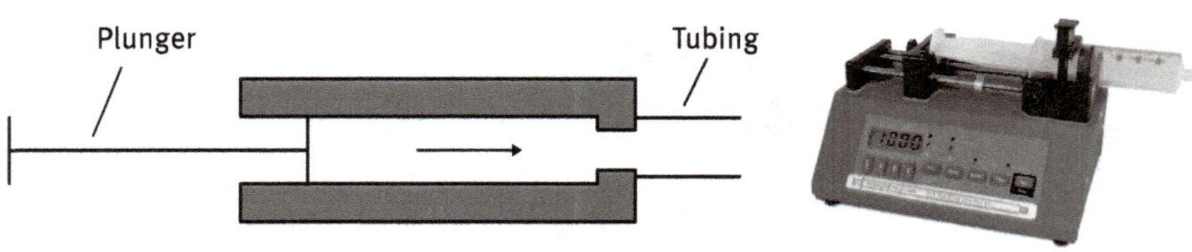

Fig. 3.2: Schematic of a syringe pump (left) and a syringe pump for laboratory use (right, NE-1000, Microfluidics).

For synthetic chemistry, many variations of the syringe pump are available commercially. To overcome the shortcomings of the single syringe pump system: namely, limited reservoir capacity and lack of mixing possibilities, a dual syringe pump can be deployed. Consisting of two electronically driven syringes that can be programmed to work synchronously, one of the syringes can be (re)charging fluid from a reservoir, while the other is delivering the fluid into the reaction zone. Alternately, syringe pumps can be combined with switch valves to deliver reagents or reactants at multiple points in a continuous-flow system.

A syringe pump can provide constant and pulsation-free flow at very low flow rates (microliter range) with high precision. Depending on the size of the syringes, different flow speeds are accessible. The syringe can be made from different materials – polymer, glass, stainless steel, and so on. While many laboratory syringe pumps are only designed for low-pressure applications, more specialized syringe pumps are available which can deliver liquids at high pressures of up to a few hundred bar.

3.1.1.2 Piston pump

The piston pump is a type of positive displacement pump used in nearly every modern liquid chromatography (LC) and high-performance liquid chromatography (HPLC) systems (Fig. 3.3). Piston pumps use a reciprocating piston, connected to a crank mechanism, force-moving a fluid through a cylindrical chamber consisting of non-return check valves to direct fluid flow. Piston pumps work at high-pressure regimes (up to ~140 bar) and are a popular choice for a continuous-flow system.

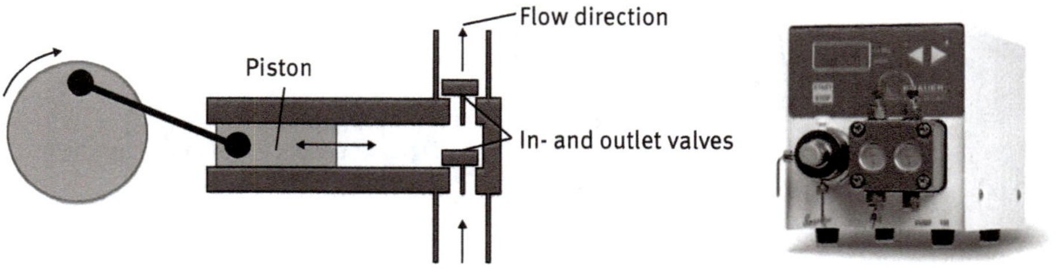

Fig. 3.3: Piston pump – working principle and a commercial double piston pump (SmartLine 100, Knauer) for continuous-flow applications.

A variation of the pump, where the piston is replaced by a plunger, can achieve even higher pressures of up to 2,070 bar. The pump can be made out of different materials: steel, stainless steel, nickel–molybdenum alloys (Hastelloy®), or ceramics. The Hastelloy® and ceramic pumps provide good resistance against many corrosive chemical reagents. Generally, a single pump is sufficient to provide an uninterrupted continuous-flow process without the need of additional valves if connected to a reservoir with a fluid.

There are few limitations to be taken into consideration when working with a piston or plunger pump:

- The mechanism generates a pulsating flow. Many piston pumps will employ two pistons to smooth the flow, but pulsation may still be problematic in some applications, particularly at low flow rates. For very low flow rates, syringe pumps are generally preferred.
- Processing of highly viscous materials is difficult.
- Processing of slurries or suspensions is difficult, if not impossible due to particulates impeding the operation of most nonreturn check valves.
- Airlocks in a piston/plunger pump can disrupt the flow and should be avoided at all costs.

For gas–liquid mixtures, a *diaphragm pump* (sometimes also known as membrane pump) may be deployed, where a reciprocating (flexing) membrane is used to displace a fluid through the check valves (Fig. 3.4). For aggressive chemicals, Teflon membranes may be deployed. However, as the diaphragm is generally driven by compressed air, the working pressure is typically lower than what can be achieved using a piston/plunger.

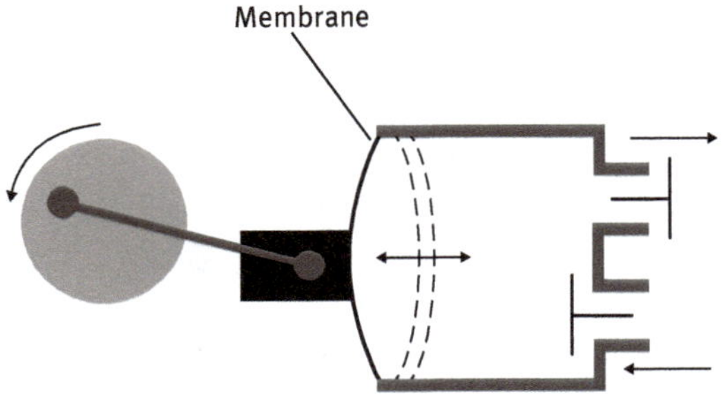

Fig. 3.4: Basic design of a diaphragm pump.

3.1.1.3 Gear pump

Rotary pumps form another class of positive-displacement pumps commonly used in the chemical laboratory in continuous-flow processes. A gear pump drives fluid flow by the meshing of two rotating gears, via an external (Fig. 3.5) or an internal mechanism. Such a pump is particularly suitable for the transfer of viscous fluids

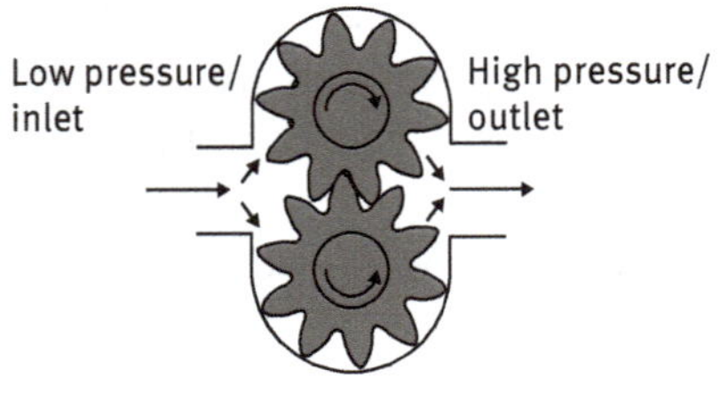

Fig. 3.5: Mechanism of an (external) gear pump.

and can work at pressures of up to ~200 bar. Furthermore, it can be designed to withstand highly corrosive liquids.

3.1.1.4 Peristaltic pump

Another popular rotary pump type is the peristaltic pump (Fig. 3.6). Here, the fluid is processed through a flexible tube fitted in a pump casing, which is compressed by a rotating mechanism so that the fluid is brought into motion. This pump closely resembles the gastrointestinal movement in the human body, where the movement of the intestinal content is realized by the muscle contraction. As such, the peristaltic pump is particularly suitable for handling slurries and viscous liquids.

The flow rate of a peristaltic pump is determined by the internal diameter (ID) of the tubing, the outer diameter (OD) of the pump head, as well as the rotation speed. As the only wettable part of the pump is the tubing, which is available in a wide range of materials and dimensions, the peristaltic pump is perhaps the most versatile in terms its chemical compatibility. Changing the tubing between applications is trivial, making it very easy and relatively cheap to maintain.

On the other hand, the pressure that can be delivered by the peristaltic pump is quite low (<10 bar) in comparison with the other pumps described above.

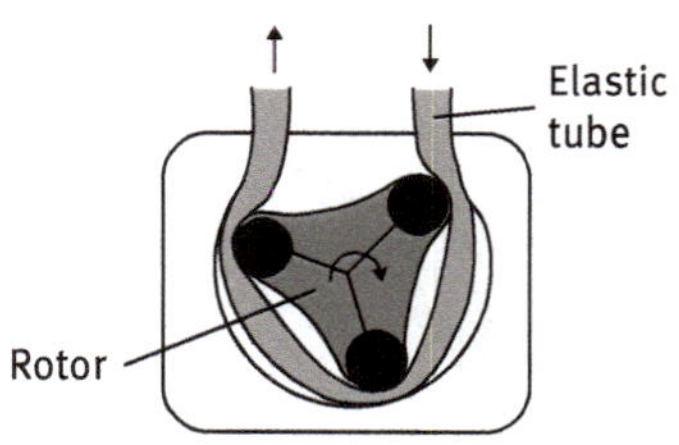

Fig. 3.6: Mechanism of a peristaltic pump.

3.1.2 Mass-flow controllers

A mass-flow controller (MFC) is a device that measures and controls the flow of liquids and gases. It is usually employed in flow chemistry to control the fluid flow to a set point. Several mechanisms exist for sensing mass flow, but the MFC devices typically employed for flow chemistry utilize a thermal sensor, consisting of two resistance thermometers wrapped around a small tube. During the operation, the two resistances heat up the tube to a given temperature [2–4]. When gas flows through the sensor tube, the gas flow cools down the tube and creates a temperature difference between the two elements, and the resultant change in resistances is detected. The size of the difference signal is dependent on the mass flow, the density, and the specific heat of the applied gas, which controls the opening/closing of a valve to regulate the gas flow to the desired set point. MFCs are not to be confused with BPRs (see Section 3.1.3).

3.1.3 Back-pressure regulators

BPRs are nonreturn valves that allow the precise maintenance of a set pressure (upstream) under controlled fluid flow conditions. The use of BPRs is essential for gas–liquid reactions, or when gas evolution occurs during the reaction, to preserve a steady flow of the fluid. The application of pressure is also necessary to maintain the volatile reactants in the liquid phase when the reaction temperature exceeds their boiling point.

There are many different types of BPRs that are manufactured from different materials, including different types of polymers and stainless steel. For continuous-flow operations in a laboratory scale, the most common types of BPRs are as follows:

"Fixed" or static BPRs (Fig. 3.7a): Being the simplest and the cheapest, this type of BPR contains a nonreturn check valve with a (fixed) set point, beyond which it will open and release the excess pressure, so that the upstream pressure will be maintained just below the limit. Widely deployed in HPLC pumps, they can be procured from various suppliers at a reasonable cost (e.g., Upchurch, Optimize Technologies, Scientific Systems Inc.).

(a) (b) (c)

Fig. 3.7: (a) Fixed/static pressure valve for different pressures, with a holder; (b) manual back-pressure regulator; and (c) adjustable back-pressure valve containing a PTFE membrane.

"Variable" or adjustable BPRs: These BPRs allow manual or electronic control of the desired working pressures. In its simplest form, an adjustable spring-loaded pressure relief valve is used to vary the pressure at which the valve will open [5]. They are reasonably low cost and various models are available on the market (Swagelok, Tescom, etc.) offering working pressures of up to ~1,034 bar (Fig. 3.7b). Another type of adjustable BPR is a dome-loaded device where a flexible diaphragm is sandwiched between the main body and the reference cap (Fig. 3.7c). To operate, gas or air is fed into the top to the desired pressure. As fluids flow through the main body of the unit, the diaphragm pushes up from, or down onto the orifice to release or apply the pressure, respectively. This type of BPR can be operated at high pressures up to 689 bar. It can also be constructed entirely using polytetrafluoroethylene (PTFE) or Hastelloy, thus offering excellent chemical compatibility.

Cracking pressure The pressure at which a relief valve opens to allow a fluid to flow through.

Full-flow pressure The pressure at which a valve is opened fully at its full flow.

Pressure differential (or pressure override) The difference between cracking and full-flow pressure.

3.1.4 Heating/cooling

In terms of heating the reactor or reaction zone there are many options. A common approach is to immerse the reactor (tube, coil, chip, etc.) in a water bath (<95 °C) or oil bath (from ambient, up to ~250 °C). Precise temperature control can be achieved by using a thermostat. Alternately, heating elements, in the form of a cartridge, a heating block, a mat, or a tape, can also be used to provide contact heating.

Conversely, cooling baths can be assembled using ice–salt mixtures (down to −20 °C) or, for even lower temperatures (down to −78 °C), using liquid N_2 or dry ice (CO_2) in combination with various solvents. For longer experiments lasting more than several hours, electronically controlled refrigerated circulators (e.g., from manufacturers such as Julabo, Huber, or Heidolph) can provide consistent cooling at cryostatic conditions (down to −95 °C). These are bulky in size and require a circulatory cooling liquid; nevertheless, they can be a good option for in-house reactor setups.

Another way of heating and/or cooling is by repurposing gas chromatography (GC) or HPLC column thermostats/cryostats (column ovens/chillers), which allow very precise maintenance of the set temperature but also temperature programming (ramping and holding times). Using Peltier elements, a temperature range from −10 to 200 °C can be achieved using a HPLC column thermostat without using cooling fluids. On the other hand, GC ovens can offer a much wider temperature range from −80 to 450 °C (depending on manufacturer), making it the heating/cooling device of choice when assembling an in-house continuous-flow reactor. However, they tend to have a sizeable footprint and will occupy significant bench space in a synthetic laboratory.

In recent years, other "unconventional" energy sources include the use of inductive heating (combined with a metallic reactor, or one that is packed with magnetic nanoparticles) [6]; ultrasound [7] and microwaves (Section 3.2.6) have also been deployed for certain reactions that require very fast heating. The existence of nonthermal effect of sonication or microwaves on enhancing the reaction rate is still a hotly debated topic, however [8].

Regardless of the choice of the heating/cooling device, it is important to consider the thermal resistance of the materials used for the fabrication of the reactor, connections, and tubing, in order to minimize temperature gradients across, or excessive heat loss through, the reactor walls.

3.1.5 Mixers

Good mixing between reacting fluids, either homogeneous or heterogeneous, is of utmost importance in a continuous-flow process, as the efficiency of heat and mass transfer directly affects the reaction rate and selectivity. The mixing processes strictly obey the laws of fluid mechanics and rheology [9], and can be modelled with a great level of sophistication in the field of computational fluid dynamics, which is beyond the scope of this book chapter. Nevertheless, it is important to have a basic understanding of the different types of flow regimes that can result from mixing, which is critical in the selection of an appropriate mixer and reactor, not only to achieve the desired reaction outcome but often also to ensure a safe operating regime.

In microreactors (≤1 mm, typically μm scale), the high surface-to-volume ratio enables mixing to be achieved through rapid diffusion (see Section 3.2.1). In larger reaction channels, however, mixing by diffusion is generally less efficient, particularly for multiphasic reaction mixtures. In these cases, a number of "static" (or "passive") or "dynamic" (or "active") mixers should be deployed. Static mixing is achieved by changing the internal structure of a reaction chamber or the reaction channel that disrupts the laminar flow path to enhance mixing. With no electrical components, static mixers are more energy efficient, and the absence of moving parts also make them cheaper and safer to implement and maintain, although the design of a mixing element can be quite complex.

In contrast, a dynamic mixer is usually a device that contains a moving part, for example, a stir bar, a propeller, or an impeller that folds or shears the fluid mechanically. As it imparts significant kinetic energy to the fluid flow, highly turbulent flow (high Reynold's numbers) can be achieved. This may be preferred for reactions that require longer residence times, or for processing highly immiscible liquids or suspensions, where high shear interactions between the phases may be particularly beneficial.

The plethora of mixers available may be initially daunting for a researcher new to the field of flow chemistry. The following subsections provide a description of the most common type of mixers.

3.1.5.1 Tees and crosses

The primary function of a union is to bring two (or more) fluids together to form a reaction mixture. Inevitably, mixing occurs at the confluence and the extent will depend on the dimension, speed (flow rate), and direction of the merging fluids. For example, using a Y-piece, a T-piece,, or a cross-piece (Fig. 3.8), solutions can be mixed at an angle (60° or 90°, "cross-flow") or from opposing directions (180°, "opposing flow") [10]. These "union pieces" can be made out of different polymers or stainless steel, depending on the chemical compatibility and pressure requirements, and a series of these pieces can be joined sequentially via tubings to perform a series of "telescoped" reactions (see Section 3.3.4).

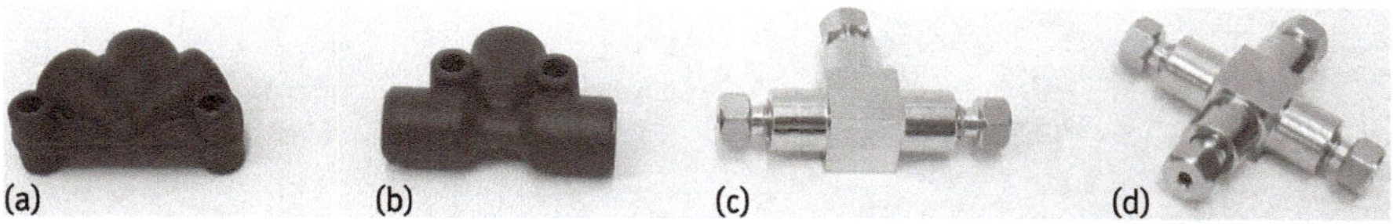

Fig. 3.8: (a) Y-piece; (b) T-piece (polymer); (c) T-piece (stainless steel); and (d) cross-mixer.

For relatively undemanding applications, for example, involving miscible fluids with similar viscosities, these union pieces can create effective mixing. Turbulence can be improved by increasing flow rates or by heat evolved during exothermic reactions.

For more demanding reactions, for example, when the reaction is biphasic/multiphasic, or only slow flow rates can be used, additional mixers will be required. Mixers can be either be deployed before the reactor, to ensure good mixing between reactants prior to exposure to reaction conditions (e.g., high temperatures, contact with catalyst), or can be incorporated into the reaction chamber or channel to provide good/even mixing between immiscible phases during the reaction.

3.1.5.2 Inert packings (packed-bed reactors)

The cheapest and often easiest solution of enhancing mixing is to pack the internal volume of a flow reactor with inert particles, such as sand, silica gel, silica carbide, or glass beads [11, 12], to form, effectively, a "packed-bed" reactor. Packing reduces the reactor volume, leading to a shorter residence time. Thus, it is important to determine the "void volume" of such packed-bed reactors for residence time calculations. [13] This may be achieved experimentally by weighing the reactor containing the dry bed, and again when it is saturated with a liquid of a known density, for example, water. The difference will allow the void volume to be determined.

Packed bed reactors are frequently deployed in heterogeneous catalysis, where the immobilized/supported catalyst particles act as a static mixer, with the reactions occurring at the active sites on the surface, as the mobile phase (containing the reactant) flows through the bed.

In general, smaller particles are better for mixing, but very fine particles can cause clogging or unacceptable pressure buildup in the system. To avoid this, the material can be passed through a sieve to remove the fine particles.

For reactions that are particularly exothermic or proceed at fast flow rates, metal (titanium, nickel, Fig. 3.9), plastic (PTFE), or even ceramic foams are alternatives to a pack bed reactor. The porosity and rigidity of these foams can be optimized to afford a high surface area to volume for efficient heat transfer, while also preventing unacceptable pressure drops.

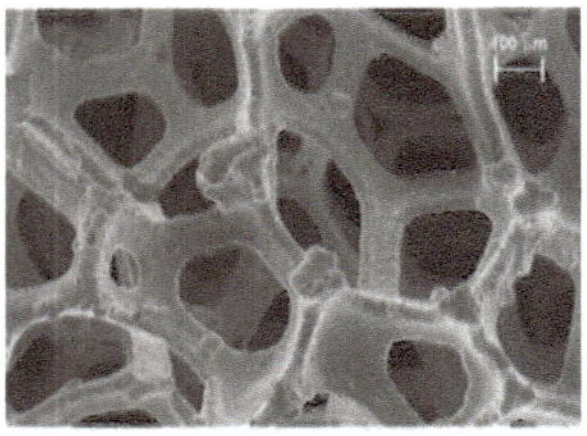

Fig. 3.9: Magnified image of a Ni foam produced by Recemat BV (Netherlands).

3.1.5.3 Structured mixers

Particles and foams induce mixing of fluids by pushing them through microchannels in a highly disordered network. For more precise control of flow patterns, a "structured" static mixer of a well-defined shape and geometry (e.g., wave, spiral, and baffles), can be placed directly inside the reactor tube (Fig. 3.10), to achieve excellent radial mixing while maintaining a nearly plug-like profile of the axial flow, for example. These static mixers can be obtained from various suppliers in different sizes, shapes, and materials (for a review on static mixers, see [14]). The availability of 3D printers in recent years has also greatly enabled researchers to design, produce, and test custom-made mixers for specific applications with ease.

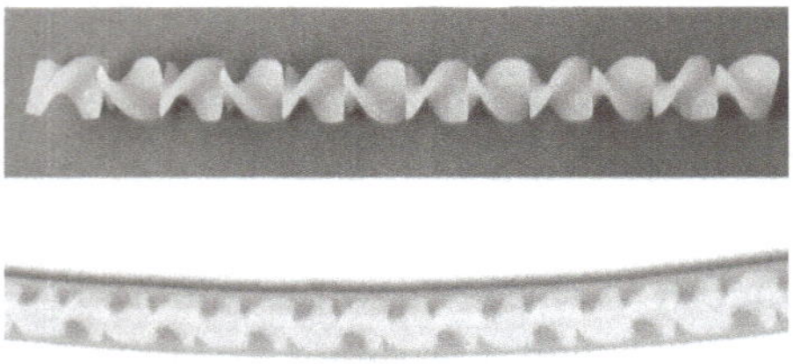

Fig. 3.10: A Teflon static mixer (top) inserted into a tube (bottom).

3.1.5.4 Gas–liquid contactors (bubble column or membrane reactors)

A porous media, such as a frit or microfiltration membrane (made of titanium, stainless steel, or PTFE), are deployed as a "contactor" in bubble column reactors, to generate very small gas bubbles in the reacting fluid (Fig. 3.11), creating a large interfacial contact between the two phases, providing good mixing and heat transfer. The size of the bubbles and droplets can be very precisely controlled by the pore size and shape of the membrane, as well as the gas flow rate. The tube-in-tube reactor (Section 3.2.4) also utilizes a membrane to control the delivery of gas in a flow reactor in a less turbulent manner.

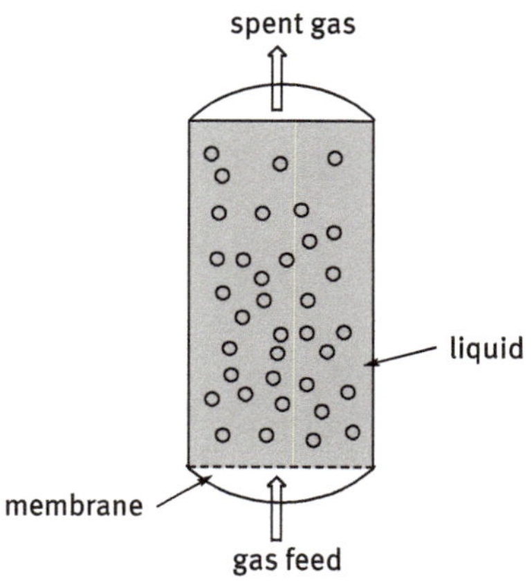

Fig. 3.11: A bubble column reactor.

3.1.5.5 Twisted pipes

It is essentially impossible to achieve turbulent flows (signified by high Reynold's numbers) in microchannels. In these cases, good mixing in a plug flow can be achieved by modifying the structure of the reaction channel themselves into so-called twisted pipes. As a fluid flows round a bend in the reaction channel, a pressure differential induced by the centrifugal forces causes molecules in the fluid to move toward the outside of the bend, causing radial mixing in the process.

In a "split-and-recombine" (SAR) design (Fig. 3.12), the reaction mixture is continually separated and merged to achieve so-called chaotic flow. This approach is adopted by Corning in the design of its advanced-flow reactor (see Section 3.5.6).

The concept was demonstrated on an industrial scale, in the design of the Sulzer mixer reactor for the manufacturing of polymers or fibers (Fig. 3.13). The reactor contains a series of tube bundles perpendicularly to enhance radial mixing of viscous liquids, while also allowing a high surface area for effective heat transfer [16].

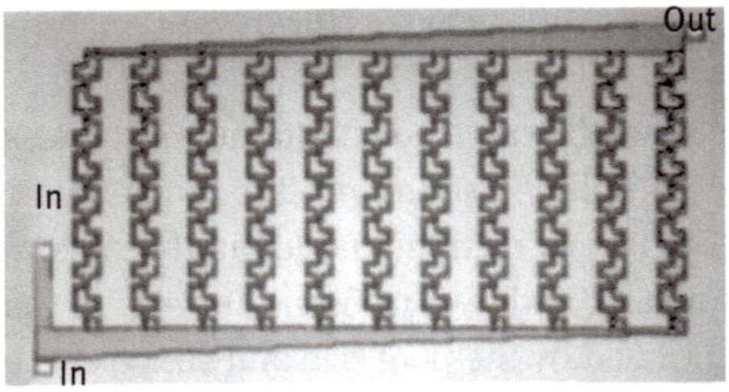

Fig. 3.12: An example of a split-and-recombined micromixer [15].

Fig. 3.13: The Sulzer mixer reactor (SMR) [16].

i *(For basic knowledge on mixing please see Volume 1, Chapter 1, Title: Fundamentals of flow chemistry)*

3.1.5.6 Continuous stirred tank reactors

Most synthetic chemists would be familiar with using magnetic or mechanical ("overhead") stirring as a means of agitating a reaction mixture in a flask by creating a vortex. Very similar mixers can also be incorporated in continuous stirred tank reactors (CSTRs).

An example of a lab-scale CSTR is a fReactor^{TM} (Fig. 3.14a), where mixing is achieved by using a magnetic stir bar, controlled by varying revolutions per minute (RPM). The small internal volume of the reaction chamber (2 mL) allows efficient mixing to be achieved. The individual units can be "daisy chained" (Fig. 3.14b) to increase the mean residence time, giving a more uniform, plug flow–like, processing environment [17]. A larger number of lower volume reactors will give a better residence time distribution than a single CSTR of the same combined volume.

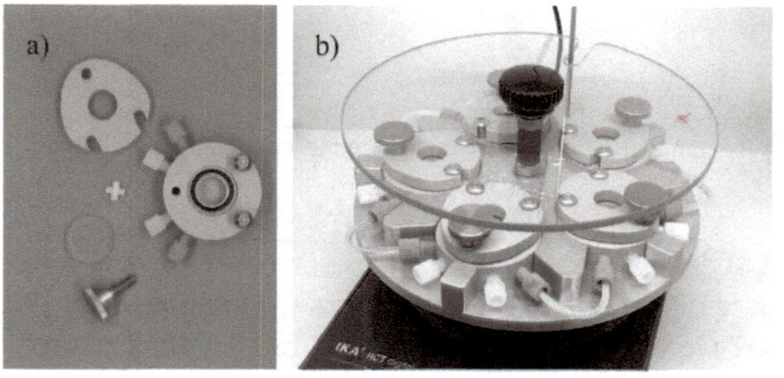

Fig. 3.14: The fReactor^{TM}: a lab-scale CSTR that utilizes magnetic stirring to achieve active mixing: (a) a disassembled unit; (b) deployed in a "daisy chain" arrangement on a magnetic stirrer. Reproduced with permission from Asynt Ltd.

3.1.5.7 Oscillatory flow

Dynamic mixing can also be achieved by changing pressure gradients to achieve back-mixing in the reactor. This is exemplified by a time-pulsating ("oscillatory") flow, where mixing is further enhanced by introducing constrictions along the tube, for example, baffle disks with orifices (Fig. 3.15). The upstroke of an oscillating pump pushes the fluid through the orifices of the baffles, creating eddies to mix the fluid axially, while a downstroke creates similar eddies on the opposite side of the baffle, thus creating a highly turbulent flow [18].

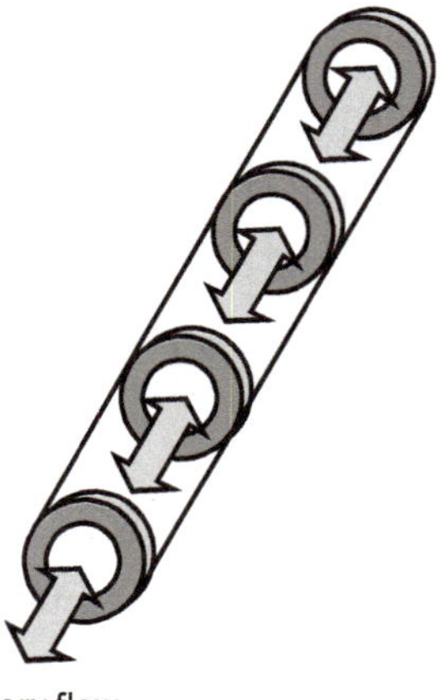

Fig. 3.15: Schematic of an oscillatory baffled reactor.

3.2 Reactors

"Where the magic happens."
Reactor: a device in which reactive materials can be brought to undergo a reaction under controlled conditions.

The term "reactor" is used to describe the region in the continuous-flow setup where the actual reaction takes place [19]. In certain cases, reactions occur during the mixing process, for example, in the T-piece. More often than not, however, the reaction is initiated when the reaction mixture comes into contact with a catalyst, and/or is exposed to an external energy source, such as heating, sonication/UV light/microwave radiation.

3.2.1 Microfluidic (chip) reactors

A microfluidic reactor contains micro-structured channels that are typically etched or molded into a transparent material such as acrylic, glass, silicon, or polydimethylsiloxane (PDMS). Through holes in the chip serving as inlet (or outlet) ports, reactants are pumped into the microchannels, which can be interconnected to integrate

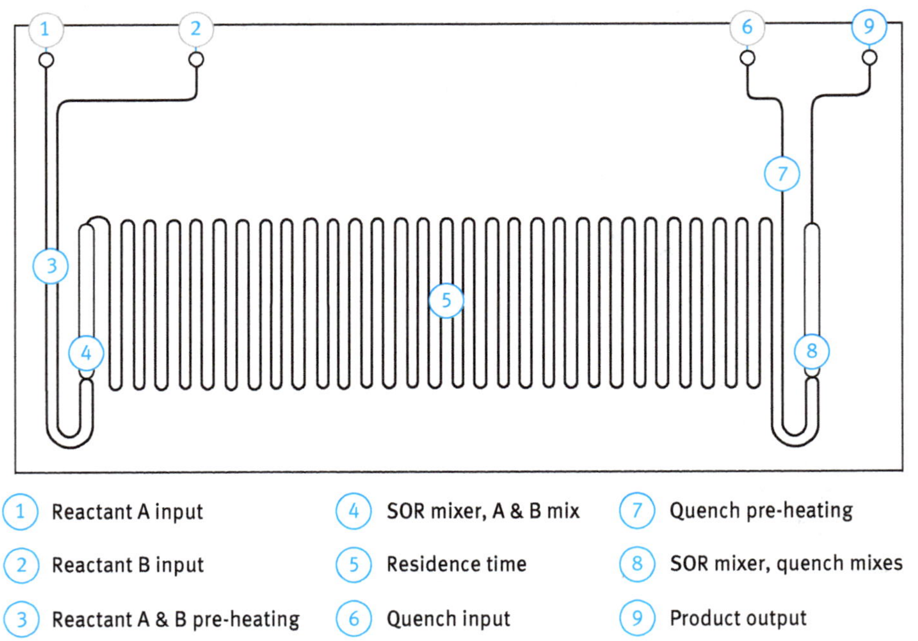

Fig. 3.16: A microfluidic reactor that incorporates (split-and-recombine) mixing, heating, and quenching.

several "units of operation" (mixing, heating/cooling, quenching, and product extraction) in one uninterrupted continuous flow (Fig. 3.16), a technique that is known as "lab on a chip."

The narrow channels allow efficient mixing to occur through diffusion in laminar flow, which cannot be achieved in larger (meso-scale) channels. Consequently, microfluidic chips are often also deployed as micromixers, to premix or pre-heat a reaction mixture, prior to its introduction into a different type of reactor.

The internal volume of microreactors can be as small as picoliters (pL). The miniaturized feature is highly desirable for applications where there are very limited quantities of material to be processed; typically in the bioanalytical sciences, for preparing bioassays for (rapid) analysis and diagnosis of diseases. In synthetic chemistry, the small reactor volume allows reactions to be optimized using a small amount of material. The large inner surface area-to-volume ratio (several tens of thousands $m^2\ m^{-3}$) or microreactors affords very high heat-transfer coefficients of up to 20–30,000 $W\ m^{-2}\ K^{-1}$, providing very efficient heating/cooling. This allows highly exothermic reactions to be performed safely, even with hazardous and reactive intermediates [20, 21].

3.2.1.1 Biphasic flow regimes in microchannels

In a microchannel, the contact interface between two immiscible phases (liquid–liquid or gas–liquid mixture) can follow various flow patterns (Fig. 3.17), depending on volumetric flow rates, physical properties of the liquid/gas (viscosity, surface tension), the inlet geometry, the size, shape, and length of the microchannel, as well as its inner surface (hydrophobicity and roughness).

For gas–liquid biphasic reactions in bubble flow, a gas is dispersed in the form of small bubbles in a continuous liquid phase. In segmented/slug flow, the bubbles coalesce to form bigger, elongated bubbles that divide the continuous liquid phase into slugs. In churn flow, the larger gas bubbles collapse in the liquid to form a disordered flow pattern. In annular flow, the gas becomes the continuous phase, with the liquid flowing in a continuous film along the surface of the channel, forming an annulus.

In principle, an annular flow regime affords a high surface area contact between the gaseous and liquid phases within a microchannel. Annular flow can be very beneficial for reactions that are mass-transfer limited, for example in ozonolysis reactions [23], or to remove the accumulation of an unwanted by-product, as demonstrated by Kappe and co-workers in a rhodium-catalyzed decarbonylation of aldehydes [24]. However, a high gas flow rate is required to maintain annular flow. This places a limit on the maximum length of the reaction channel, which may not be practical for reactions that require a long residence time.

For these reasons, segmented flow is often favored for biphasic reactions conducted in microreactors. Contact between the two phases can be controlled by adjusting the lengths of the gas and liquid slugs [25]. Another key advantage of conducting reactions

Fig. 3.17: Biphasic flow regimes in microchannels (a) bubbly flow; (b) slug/Taylor or segmented flow, (c) churn flow, (d) slug/annular flow, and (e) annular flow [22].

in segmented flow is that the reaction volume is no longer dependent upon the volume of the reactor. This allows multiple reaction mixtures to be generated in each slug (e.g., with different stoichiometries, or combinations of reactants) and subjected to the same residence time and temperature. This is particularly beneficial for the reaction optimization and the generation of chemical libraries, with small quantities of material.

The same principles apply to liquid–liquid biphasic reactions conducted in segmented flow. In a microchannel, discrete volumes of fluids can be generated in droplets that are separated from each other by an immiscible phase, to form serial trains of fluid packets. Within each segment (e.g., phase A), rapid mixing is achieved by an internal fluid vortex, which also simultaneously refreshes the diffusion interface to facilitate mass transfer between the segments (phases A and B, Fig. 3.18).

For liquid–liquid segmented flow, the cross section of each segment must be smaller than its length, to avoid the formation of emulsions. It is also important to take into account the material used in the construction of the microchannel, which affects interfacial tension and surface energies between the liquid and the reactor wall.

To expand the application of microfluidic reactors to more challenging chemistry, new production techniques, including sintering and 3D printing, have been used to produce microreactors from a variety of materials, including metal alloys and

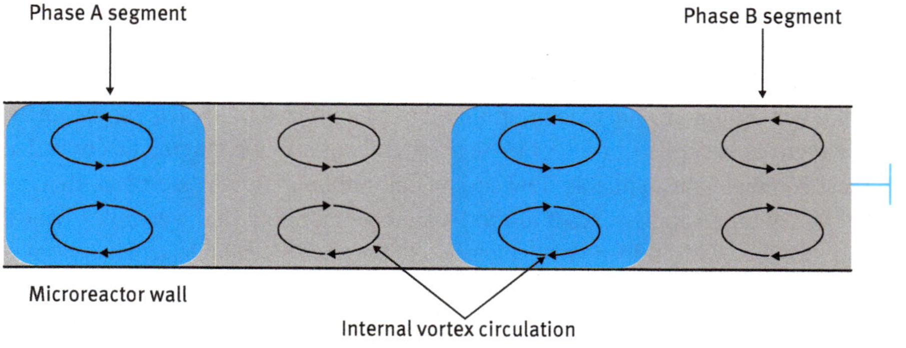

Fig. 3.18: Schematic representation of segmented flow in a microchannel.

ceramics – but the cost of producing microchannels using these materials can escalate quickly. As such, glass reactors are still the cheapest and most popular choice.

3.2.1.2 Pressure drop and scalability issues

Pressure drop is the pressure difference between two points of the reactor, usually between inlet and outlet of the reactor. This is generally undesirable in a tubular flow reactor because: (1) it consumes more energy (greater pressure is required to move the fluid); (2) pressure buildup can exert strain on the reactor wall and fittings, which can be potentially dangerous; and (3) affects the solubility of gases in the solvent.

Pressure drop in a microreactor is dependent on the material of its construction, the length and inner diameter of the channels, the viscosity of the liquid, and its flow rate.

The narrow channel of a microfluidic reactor is also its Achilles' heel, in that it is very prone to fouling and blockages caused by micro-sediments or precipitates, which can lead eventually to the cracking of the glass chip. For the same reasons, scaling up of chemical processes using reactors with such small volumes is not trivial. In principle, the reactors can be stacked (scaling-out) or numbered up (parallelization of microreactors). However, achieving even or equal flow distributions between the reactors can be a considerable challenge, if not a costly exercise.

At this juncture, it is important to address the problem of pressure drop in microreactors (text box). For a chip reactor fabricated from PDMS, it is generally recommended that the pressure drop should not be more than 2–3 bar. Due to the narrow channels, reactions involving highly viscous liquids and a high flow rate can be problematic.

3.2.2 Coil (tubular) reactors

In many aspects, a coiled tubular reactor may be considered as a larger version (mesoscale) of the microfluidic chip reactor, in that it can be utilized to implement the different flow regimes described in Fig. 3.17. Reactor coils of specific lengths are provided with most commercially available flow equipment, although it can also be easily constructed by winding capillary tubes around a suitable manifold into a compact helical coil (Fig. 3.19). A wide selection of tubes is commercially available (stainless steel, Hastelloy® or different polymer tubes: PTFE, polyether ether ketone (PEEK), perfluoroalkoxy (PFA), ethylene tetrafluoroethylene (ETFE) etc.); the most common ODs – 1/32", 1/16", or 1/8" – are usually supplied as HPLC consumables. IDs of ≤2 mm are generally sufficient to assure a good mixing in a homogeneous reaction mixture without additional mixers.

Fig. 3.19: Coiled (tubular) reactors containing polymer and stainless steel tubes, wrapped round a manifold.

Unlike the chip reactor, the internal volume of a coil reactor can be changed by varying the length of a tube. This allows different residence times to be attained without changing the flow rate(s), thus preserving the desired flow characteristics/regime. As the ID of the tubing is generally larger than that found in chip reactors, it can also tolerate faster flow rates, longer residence times and more viscous liquids before pressure drop becomes an issue. That said, the narrow ID of the tubing is still susceptible to clogging by fouling or precipitation of the insoluble reaction components. While it is generally easy and cheap to replace the tubing, clogging can cause dangerous pressure buildups in the system. Monitoring of the reactor pressure should be employed where there is potential for blockage events, and ideally the system will be configured to automatically shut down in the event of a blockage being detected.

Metal coils allow reactions to be performed at elevated temperatures (up to 500 °C) and pressures (e.g., a 1/16" stainless steel tubing can withstand pressures up to 12,000 psi, or 827 bar, depending on wall thickness); with the high surface-to-volume ratio providing good heat transfer. Although metal tubings made from titanium and tantalum can withstand strongly acidic and basic conditions, polymer tubings are generally preferred for these chemically aggressive reagents – not only are they cheaper, this can also eliminate the problem of leaching of catalytic amounts of iron, nickel or other metals from the surface of the metal under such extreme pH values, which may influence the outcome of a reaction. However, while polymer tubings can generally withstand high pressures, they are not so robust toward high temperatures, for example, a 1/16" tubing made of PEEK offers good chemical and pressure tolerance (up to 6,000 psi, 413 bar), but the material has a glass transition temperature of 143 °C.

3.2.3 Packed-bed reactors

In chemical engineering, a packed bed is a hollow tube, pipe, or vessel that is filled with a packing material. The packing can be small particles such as sand, small objects with defined shapes (e.g., catalyst pellets), or have specifically designed structures (e.g., in a catalytic converter) [26] (Fig. 3.20). Packed-bed reactors are widely deployed in the chemical industry, particularly for catalytic processes.

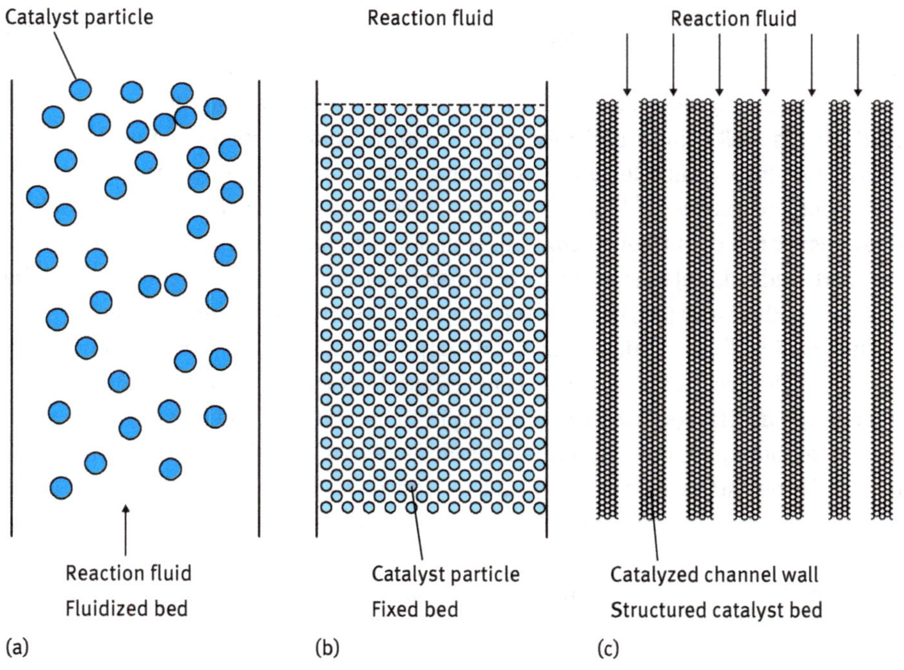

Fig. 3.20: Three main types of packed-bed reactors [26].

In the laboratory, a fixed-bed reactor can be simply prepared by filling a chromatographic column or stainless steel tube with solid catalyst particles. Both ends of the tube can be capped with sinters or filters to retain the catalyst material, before fitting with an inlet and outlet (Fig. 3.21). The catalyst bed can be mounted in a heating block, or heated in an oven. In continuous-flow operation, a mobile phase of liquid, gas, or a mixture of both is pumped through the catalytic material, typically in an upward direction to expel dead volumes in the reactor.

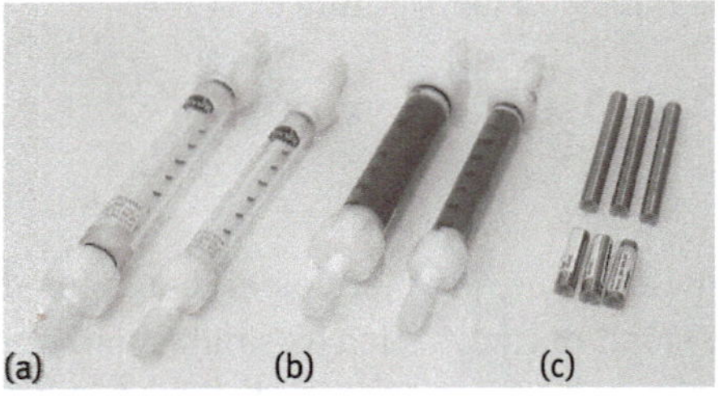

Fig. 3.21: Laboratory-scale fixed-bed reactors: (a) empty chromatography columns; (b) columns filled with steel beads; and (c) stainless steel cartridges filled with catalysts (pictures provided by ThalesNano).

3.2.3.1 Heterogeneous catalysis

A wide range of heterogeneous catalysts can be deployed in the fixed-bed reactor for liquid–solid, gas–solid or gas–liquid–solid reactions, when the appropriate gas is introduced into the system via a dosing unit or MFC. The pressure in the system is controlled by the BPR. These typically comprise metal nanoparticles embedded onto/ encapsulated into a suitable support (alumina, silica, zeolites), although catalytically active metal complexes, or biocatalysts (enzymes), can also be similarly immobilized for deployment in continuous flow. The support material does not need to be chemically inert. Indeed, catalyst stability and reactivity are very often enhanced by the redox and acidic/basic properties of the support ("metal-support interactions").

The physical form of support is clearly important for mixing (see Section 3.1.5.2). For optimal catalyst performance, it is necessary to consider the rates of diffusion of the reacting substrate(s) from the bulk liquid to catalytically active sites located on the surface (internal mass transfer), and from there to within the pores (internal mass transfer). If the rates of these mass transfer processes are slow, it could lead to an under-estimation of the catalytic activity, as the system operates under a "mass transfer," rather than a "kinetic," regime. In general, smaller particles facilitate better mass transfer (due to increase in the surface area); however, fine particles can also cause unfavorable pressure drops across the fixed-bed reactor, which also compromises the efficiency of the system. Generally, the optimal particle size d_p is dependent on the length (L) and diameter (d_r) of the fixed bed, with recommended ratios of $d_p < L/25$ and $d_p < d_r/10$.

The key advantages of conducting a heterogeneous catalytic reaction in a fixed-bed reactor include:
- Within each reactor volume, a high catalyst-to-substrate ratio ensures a high conversion per unit mass of catalyst, which is beneficial for space–time yields required for process intensification.
- Precise residence time control suppresses competitive reactions that lead to side products, improving the selectivity of the process.
- As the product and by-product(s) do not accumulate in the reactor, catalyst poisoning by these species can be greatly reduced, if not eliminated. This improves catalyst longevity (increasing turnover numbers).
- In principle, the catalyst can be recovered and reused easily. In certain cases, deactivated catalysts may be regenerated (e.g., catalyst poisoned by coking can be treated by calcinating under an oxidizing atmosphere).

Other potential issues that can arise with the use of a fixed-bed reactor are:
- The development of undesirable temperature gradients either radially between the wall of the reactor and the center of the catalyst bed (due to its thickness and inefficient heat transfer), or axially along the length of the fixed bed (due to exothermic reactions causing an adiabatic temperature rise). This will affect the reaction outcome or accelerate catalyst aging/deactivation.
- Channeling within the catalyst material may occur, particularly for gas–liquid–solid reactions where there is insufficient control of pressure.
- Depending on the physical properties of the catalyst, high pressure drops may occur.
- Erosion of active catalytic species to the mobile phase ("catalyst leaching") is an irreversible process that not only reduces the overall efficiency, but the metal residues also contaminate the product.

3.2.3.2 Solid-phase synthesis and in-line purification

Fixed-bed reactors are also widely employed in solid-phase synthesis (SPS), where a reagent is immobilized onto an insoluble material (typically silica gel or polymer support) in a fixed-bed. The reactant is dissolved in a suitable solvent and is transformed as it passes through the fixed bed. As the by-product(s) of the reaction remains bound to the solid support, workup procedures can be eliminated. This is particularly beneficial for multistep synthesis, where consecutive reactions can be performed by passing the synthetic precursor through several fixed beds containing solid-supported reagents, without the need for intermediate purification steps. Some examples of (commercially available) immobilized reagents and their applications:

- polymer-supported bromine for electrophilic and radical addition and substitution reactions [27]
- sodium borohydride on silica for reduction of ketones and aldehydes [28]
- azide exchange resin, azide on Amberlite resin [29]
- polymer-bound triphenylphosphine, for Wittig and Mitsunobu reactions [30]

By the same token, a wide range of scavenger resins can also be implemented as fixed-beds in a continuous-flow process, to purify the newly made compounds by removing excess reagents, unwanted by/side products, and impurities [31].

The volume of the fixed-bed reactor required for solid-phase-supported reagents or inline scavengers – both of which are employed in a stoichiometric amount, often in excess – will need to be quite large. For this reason, liquid chromatographic columns, such as those produced by Omnifit® (Fig. 3.21a), are often employed for these applications.

3.2.4 Tube-in-tube reactor

A tube-in-tube reactor is a type of membrane reactor (see Section 3.1.5.4), whereby a Teflon tube made of semipermeable AF-2400 membrane is used to introduce a gaseous reactant through the walls, into a concentric tube containing the reactant in a liquid phase (Fig. 3.22), thus achieving rapid gas/liquid contact along the direction of the flow.

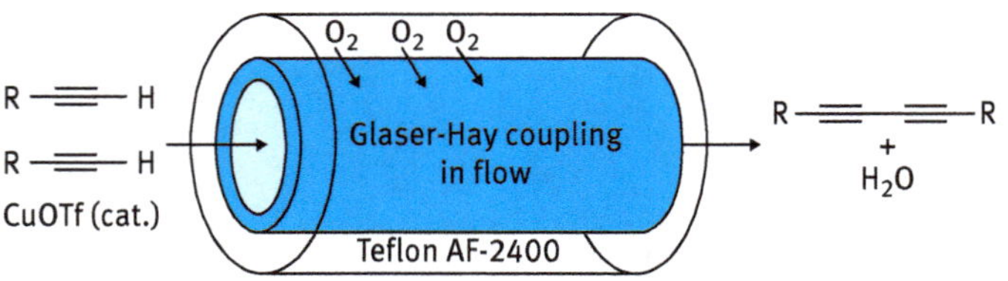

Fig. 3.22: Tube-in-tube reactor for copper-catalyzed aerobic alkyne coupling reactions [32].

In this example, the gas flows from the outer tube to the inner tube, but a reverse arrangement is also possible [33].

The delivery of gas into the liquid phase is determined by the pressure difference across the membrane, rather than by metered flow delivered by a MFC. Provided that the rate of gas permeation across the membrane is matched by the rate of its uptake by the liquid (through dissolution and consumption), formation of gas bubbles can be avoided – necessary to maintain steady state in the reactor. This means of delivery also allows a more even distribution of gaseous concentration along the reaction volume, compared to, for example, pre-saturating the liquid with gas before the reactor.

3.2.5 Falling film reactors

Falling film flow is widely found in applications in modern process industries dealing with heat and mass transfer; for example, in heat exchangers. In a falling film reactor, the mobile phase consists of thin films of liquid that flow between two layers of glass sheet by gravitational force. The large interfacial area also makes it suitable for gas–liquid contacting processes. The surface area-to-volume may be further increased by introducing grooves to one of the glass layers (Fig. 3.23).

The geometry of a falling film reactor is characterized by its small thickness and large interfacial area, which inherently places an upper limit on its scalability. Although reactors with dimensions of up to 1 m exist, it is usually more practical to "scale out," by stacking the reactor layers, rather than increase the surface area.

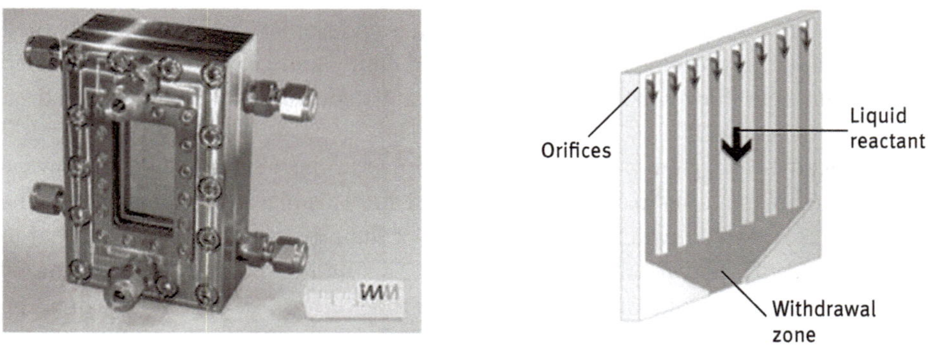

(a) (b)

Fig. 3.23: (a) A falling film microreactor and (b) schematic of the grooved reaction plate (picture provided by IMM).

The key features of the falling-film reactor are fully demonstrated in the direct α-fluorination of ethyl acetate with F_2 (Fig. 3.24) [34]. The biphasic gas–liquid reaction is highly exothermic, thus requiring good mass and heat transfer to achieve good selectivity for the mono-α-fluorinated product safely.

The large surface area of the falling film reactor is also highly beneficial for photochemical reactions [35], allowing good transmission of photons across the glass and into the thin film.

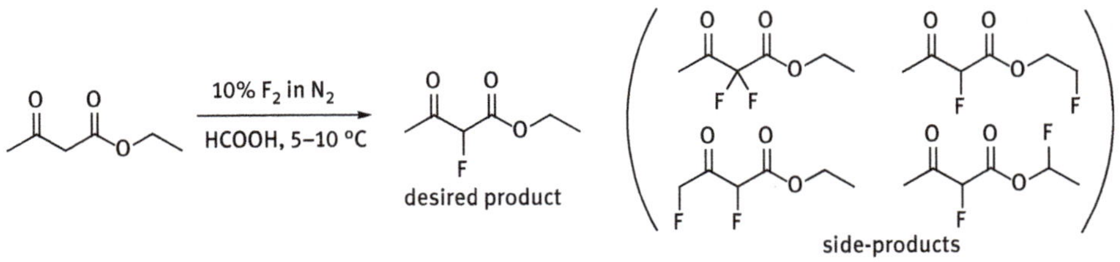

Fig. 3.24: Selective α-fluorination of ethyl acetate enabled by a falling film reactor.

3.2.6 Microwave flow reactors

Microwave chemistry is based on the fast and efficient heating of materials by electromagnetic radiation in the microwave range. Molecules containing electric charges heat up very fast, while apolar molecules heat up much slower. This selectivity allows some objects to heat more quickly or more slowly than others (for example the reaction vessel). As the heating is achieved by the radiation and not by thermal convection, a more uniform heating of the bulk mixture can be achieved.

Although the use of microwave irradiation is widespread in the laboratory, their use in industrial processes is limited. In 2012, Kappe and coworkers demonstrated how continuous flow can enable microwave-assisted processes to be achieved on an industrial scale [36].

Typically, a frequency of 2.45 GHz is adopted in most single-mode microwave reactors, corresponding to a wavelength of 12.2 cm. This places an upper limit to the diameter of a reactor, which has to be narrower than the standing wave in order to achieve effective and homogeneous heating. This limitation may be overcome by aligning several modes along the flow reactor tube (Fig. 3.25), effectively extending the range of the microwave irradiation.

To date, the largest continuous-flow microwave reactor available can be operated up to 250 °C and 30 bar, with a productivity of up to 2 ton h^{-1} [37].

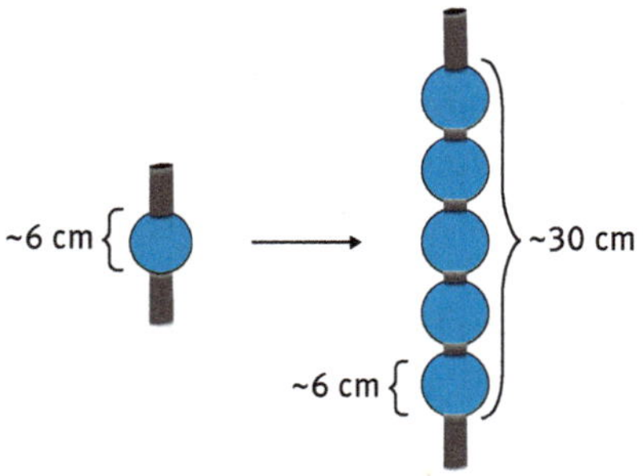

Fig. 3.25: The arrangement of a series of monomode resonators to apply microwave irradiation in continuous flow (an idealized cylindrical reaction tube is shown in yellow) [36].

3.2.7 Photoreactors (UV or visible light)

Photochemistry is driven by the delivery of photons from a light source to a reaction mixture, which will clearly benefit from the large surface-area-to-volume ratio afforded by flow reactors. Compared to the falling film reactor (Section 3.2.5), a simple continuous-flow photoreactor can be easily constructed by wrapping a transparent tubing around a light source [38], which was deployed by Seeberger and coworkers to achieve the synthesis of the antimalaria drug Artemisinin, by a photooxygenation reaction in slug flow [39] (Figs. 3.26 and 3.27).

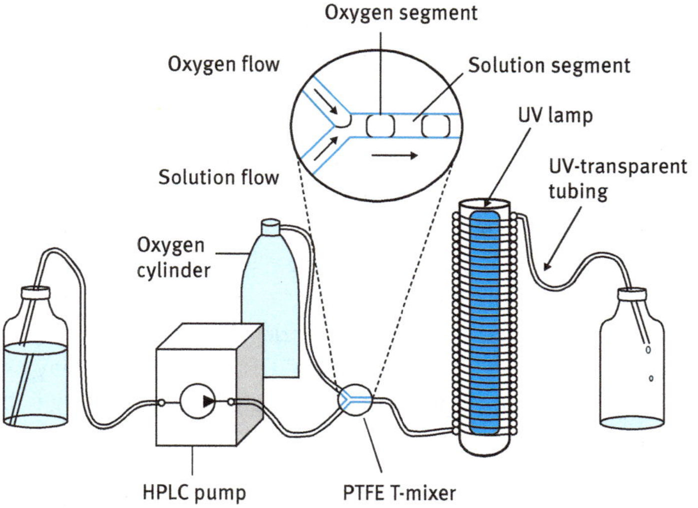

Fig. 3.26: Schematic of the photoreactor system.

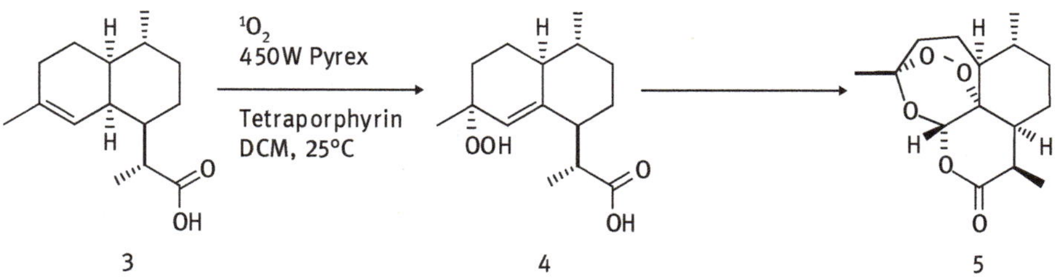

Fig. 3.27: Photooxygenation of dihydroartemisinic acid.

3.3 Assembling and using a flow reactor

In this section, the assembly and safe operation of a flow reactor, using the individual components mentioned above, will be described.

3.3.1 Fittings and tubings

Appropriate pressure fittings are needed to connect tubings between the separate units, into a working flow system (Fig. 3.28). Known as compression fittings, these are assembled using three basic elements: A nut, one or more ferrule(s), and a compression fitting body (e.g., end of a T-piece). The general assembly procedure for compression fittings is to position the nut and ferrule(s) on the tubing, before inserting the tubing fully into the fitting body and tightening the nut, which forms a seal by compressing the end of the ferrule onto the OD of the tubing. For these reasons, the use of rigid tubes, such as that made of metal or hard polymers (e.g., PEEK, nylon, Teflon®, Kynar®, or polyethylene), is recommended, which can resist the compression applied by the ferrule to form a tight seal. The choice of tubes and fittings is dictated by the temperature and pressure working regimes of the system, as well as the chemical compatibility required.

The fittings are available as standard HPLC accessories, supplied in different sizes and materials to fit tubes with 1/8", 1/16", and 1/32" OD. Fitting types may be broadly classified as having flat-bottomed ports, or coned ports, and the fitting will also be defined by port thread. Many commercial systems use ¼-28 UNF threaded flat-bottomed connections for 1/8" and 1/16" tubing where medium pressures (e.g., up to 20–30 bar) are expected. Where flow systems are designed to work at higher pressures (>30 bar) it is more common to encounter coned ports as they are generally rated to higher pressures than similarly sized flat-bottomed connections. The 10–32 UNF thread is a common standard employed by many HPLC manufacturers.

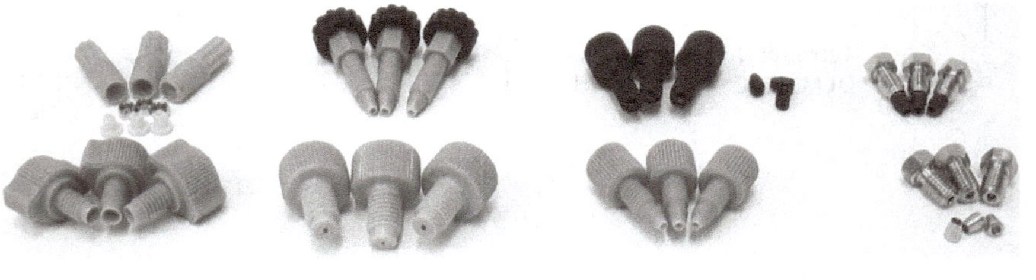

Fig. 3.28: Commercially available nuts and ferrules used in compression fittings.

As compression nuts are threaded, it is important not to under- or over-tighten a fitting, which can result in leaks. Installation of compression fittings also differs between manufacturers and designs, so it is important to follow the recommended procedures provided by the manufacturer. Extreme caution should be exercised before intermixing and interchanging tube fittings from different manufacturers particularly for high-pressure work. Provided that the ferrule and nut are matched, most ¼-28 flat-bottomed connections are generally interchangeable in low to medium pressure (<30 bar) flow reactor applications.

Researchers should however be aware that coned ports can differ between manufacturers and using the incorrect ferrule(s) for a port could reduce the fittings ability to contain pressure. It has been experimentally observed that geometrically similar twin ferrule metal fittings designed for high-pressure applications (such as those manufactured by Swagelok®) exhibited different swaging behavior due to dimensional and metallurgical differences of the fittings, and therefore intermixing such fittings is an *unsafe practice, to be avoided at all costs* [40].

Last but not least, while it is possible to disconnect and reconnect most fittings many times, the user should always inspect the fitting components for signs of damage and wear before reconnection. If there are any doubts about the condition of the fittings, it is recommended that either the ferrules and/or the fitting body should be replaced to ensure a leak-tight seal. For the same reasons, polymer tubings are generally mechanically less robust than SS tubes and can become brittle over time from exposure to chemical reagents, heat or (UV) light, and should be replaced periodically to avoid catastrophic failures.

Appropriate selection of fitting components can help prolong the lifetime of fittings which need to be disassembled and reassembled frequently. For example, when selecting parts for a ¼-28 flat bottomed connection it is advisable to select two-piece ferrules instead of the lower cost single-piece ferrules. The single-piece ¼-28 fitting ferrules have a tendency to become stuck to the nut during installation, and any subsequent disassembly can result in twisting of the tubing, with potential to damage the tubing wall. The stainless steel locking ring on the two-piece ¼-28 fitting ferrules allows free rotation of the nut against the ferrule, and allows the connection to be disassembled and reassembled multiple times without applying rotational stress to the tubing.

3.3.2 Pressure control and monitoring

The valve in a ***pressure regulator*** is normally ***opened*** – this is installed at the start of a system (between the pump and the reactor) to regulate ("inlet") pressure.
The valve in a ***BPR*** *(*Section 3.1.3*)* is normally ***closed*** – this is installed after the reactor to provide an obstruction to flow and thereby regulate upstream ("back") pressure.

BPRs may be installed after pumps (particularly piston pumps) as working against pressure can help the mechanical action of pump valves, and will prevent gravity feeding the mobile phase through the pump at a faster flow rate than intended. Ideally each pump will also have a pressure sensor installed between the pump and the BPR. Monitoring the pressure at the pump outlet can be used to troubleshoot and diagnose pump faults such as air locks and valve failures.

A BPR is usually installed at the end of the flow process, usually before the collection vessel, to keep the process under pressure and prevent outgassing and solvent evaporation. Inclusion of a pressure sensor before the end BPR (but after any pump BPRs) allows the pressure drop of the system to be monitored, which is important for the detection of system blockages and/or leakages.

3.3.3 Pumps

All pumps involve moving parts, and regular maintenance (e.g., cleaning of check valves, replacement of gaskets and O-rings) is required to ensure smooth operation as pump failure can be an expensive exercise, and potentially dangerous, for example, where accurate stoichiometric control is required for chemical safety reasons. Most pumps work via a positive displacement mechanism, which dispenses a given volume for each cycle of operation. Periodic recalibration is necessary to ensure reliable and accurate dispensing of the required volumetric rate for each experiment.

The type and size of the pump will govern the ultimate productivity of the process (in terms of mass per unit time). Good chemical resistance of the wettable parts of the pump is also important, not least to avoid introducing unwanted impurities to the reaction mixture (Tab. 3.2). The choice of the type of pump(s) to be used in a flow process will affect the rest of the reactor design – including the size and type of pipes,

Tab. 3.2: Direct comparison between three most commonly employed pumps for flow chemistry.

Pumps	Syringe	Peristaltic	Piston
Pressure	Typically low but specialized syringe pumps can go to very high pressure	Low to medium (typically <10 bar)	Low to very high
Flow rate	Very low to medium	Medium to very high	Low to high
Dispensing volume	Good for small volumes (down to picoliters)	Good for moving high volumes of liquid	Mid-range

Tab. 3.2 (continued)

Pumps	Syringe	Peristaltic	Piston
Chemical compatibility	Good tolerance for most chemicals	Dependent on tubing used. Chemically resistant tubing formulations are available but can be expensive.	Have more wetted parts which all need to be checked for chemical compatibility. Specialized piston pumps with chemically resistant check valves, pump heads, and pistons are available.
Cost of maintenance	Low	Low	High
Fluid characteristics	Low-to-high viscosity liquids can be dispensed	Good for moving undissolved solids	Viscous fluids can be challenging

fitting, and valves, to make sure that they can withstand the process conditions. Conversely, the type of reactor selected will also influence the choice of the pump used to achieve the desired flow regime (laminar/turbulent, segmented/annular flow).

3.3.4 Telescoped processes

Very often a reaction has two or more chemical transformation steps, including quenching and scavenging, that can be "telescoped"; that is, the reaction mixture emerging from a reactor unit can be directly introduced into another uninterrupted without the need for intermediate purification. In some cases, this can be achieved within one microfluidic device (lab on the chip, Section 3.2.1). In other cases, this can be achieved by connecting different reactors sequentially (Fig. 3.29), illustrated using coil reactors), particularly when different temperatures are required for each step of the sequence.

Additional pumps will be required to introduce reagents or solvents at the intermediate stages. In these cases, it is important to consider that the volumetric flow is an additive quantity; that is, when two (or more) reactants are combined, the resultant flow rate will be the sum of the flow rates of the single pumps. This will not only result in a shortened residence time, the reaction mixture will also be diluted by the added solution, which must be compensated, for example, by lengthening the second reactor (increase residence time) or increasing the reaction temperature (to accelerate reaction rate). Without suitable intervention, for example, introducing a means of removing solvents, the solutions will ultimately become too dilute, which places an upper limit on the number of reactions that can be telescoped in this way.

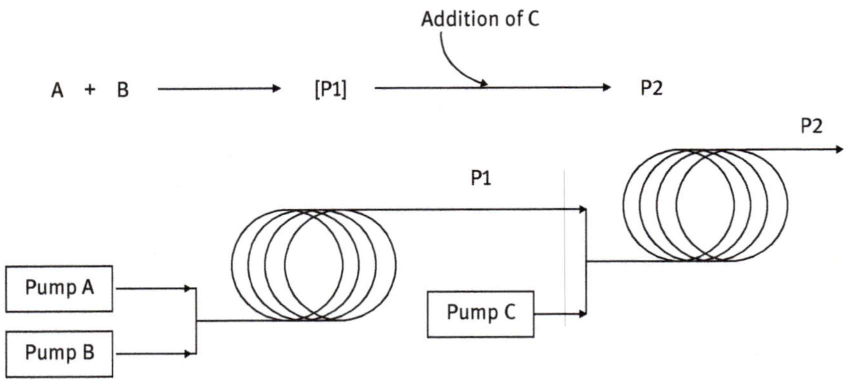

Fig. 3.29: Schematics of a reaction setup for a telescoped process: A + B → P1 + C → P2.

3.4 Operation of a flow reactor

3.4.1 Preparation

Before initiating any flow experiment, it is important to take into account not only the volume of the reactor ("void" volume in case of a packed bed) but also the associated dead volumes in the fittings and connecting tubings. The dead volume can be determined by the difference between the total volume (end to end) of the reactor system, and the reactor volume itself. These volumes can be measured by using an accurately calibrated pump and the relationship between flow rate, volume, and residence time (duration of time between the initiation of the pump, and collection of the liquid):

$$\text{flow rate (volume per unit time)} = \frac{\text{reactor volume}}{\text{residence time}}$$

Knowing the residence time of the reactor is important to determine when the output of the reactor has reached steady state, whereupon product collection can commence. The residence time of the reactor is also important for calculating the productivity of the process. Referred to as space velocity, this refers to how many reactor volumes of the feed can be treated per unit time, for example, a reactor with a space velocity of 5 h^{-1} is able to process a total volume of five times the reactor volume per hour. Particularly for custom-made reactors, it is also necessary to measure the residence time distribution, in order to understand the flow characteristics of the device. This can be achieved by introducing a known tracer at the reactor inlet (at $t = 0$) and monitoring its concentration at the outlet over time [19].

To prevent the formation of solid precipitates that can cause blockages, it is important to test the solubility limits of the reactant, reagent, product(s), and side-product(s) in the reaction solvent(s), before attempting the reaction in flow. Where it is challenging to test the solubility of all reaction species, it can sometimes be helpful to perform a small scale batch reaction using the feed stock solutions proposed for the flow system to check for any unexpected solid formation. A batch microwave reactor can be an effective tool for testing potential superheated flow conditions.

From a safety perspective, it is important to test for loose fittings and connections in the reaction assembly, by pumping a nontoxic liquid, such as water or 2-propanol, through the system. Fittings that deliver gaseous reactants should also be leak tested using soapy water. Before a reaction, it is advisable to fill the reactor system with the reaction solvent, as the presence of gas pockets (which are compressible) can prevent the system from reaching a stable back-pressure. During this "priming" period, the temperature and pressure of the reactor can be adjusted to the set values and allowed to equilibrate, before the introduction of the reactants and reagents, in a synchronized manner.

3.4.2 Initiation

In some cases, the reaction initiates with the mixing of the reactants in the T-piece, and proceeds as the reaction mixture flows through the reaction channel. In other cases, the reaction can only be initiated under certain conditions within a reactor device, such as heating/cooling, ultrasonication, or microwave radiation, or where the mixture comes into contact with the catalyst. Depending on the residence time profile of the reactor, it may require a few reactor volumes of the reaction mixture to pass through the system before a consistent output can be attained ("steady state"). The exact point for collection of the product can be estimated from the total volume of the system, or more precisely determined by using an in-line analytical technique. Alternatively, a fraction collector could be used to collect the output from the reaction, and the conversion in the collected fractions can be quantified by GC or HPLC.

(For further details on this issue please see Volume 1, Chapter 8, Title: Fundamentals of continuous manufacturing PAT: Sampling, Analysis, and Automation) i

In-line monitoring systems have been also developed to gather real-time data about the reaction which is particularly important for optimizing the reaction as the output responds to changes in temperature, pressure, gas, and liquid flow rates, concentrations of the reagents and stoichiometries. One of the emerging areas of research is "self-optimizing" reactor systems, which utilize a series of feedback loops, in combination with machine-learning algorithms to achieve the desired reaction outcome with minimum intervention from the operator (see Chapter 6).

3.4.3 Post reaction

Reactions can be quenched by physical or chemical means: the former usually involves cooling or diluting; the latter involves the adjustment of pH, removal of catalyst/reagent by using scavengers or venting the (gaseous) reactant. Product purification (workup) procedures can also be performed automatically in continuous flow by implementing in-line purification techniques, including pre-packed columns with scavengers, phase separation (for extraction), crystallization, or even an automated column chromatography that allows continuous purification of the crude product (see Chapter 6).

3.4.4 Shutting down

Following the reaction, the system should be flushed through with the reaction solvent as it returns to ambient temperature before the pressure (if used) can be released. If the system is not required any further, it can be rinsed with an organic solvent such as ethanol or 2-propanol, before it is disassembled.

3.5 Commercially available systems for the laboratory use

Due to the growing interest in performing continuous-flow synthesis on laboratory scale during the recent years, many companies emerged on the market, offering sophisticated instrumentation to the synthetic chemists.

3.5.1 Chemtrix (Netherlands)

Labtrix® is a flow reactor platform that is either manually operated (Labtrix® Start) or automated (Labtrix®-S1). The system is based on the use of glass micro reactors for the research and optimization of liquid phase reactions with reagents dosed using syringe pumps and thermally controlled using a Peltier with a thermal range of −20 to 195 °C. By varying the number of pumps employed, the platform is used for the evaluation of reactions of the type $A + B + Q \rightarrow P$ up to $A + B = P1 + C = P2 + Q \rightarrow P$ (Fig. 3.30). By integrating efficient thermal control and mixing elements into small volume microreactors (1–19.5 µL), rapid reaction screening of high value materials can be performed to generate robust process data suitable for scaling to kg scale and beyond.

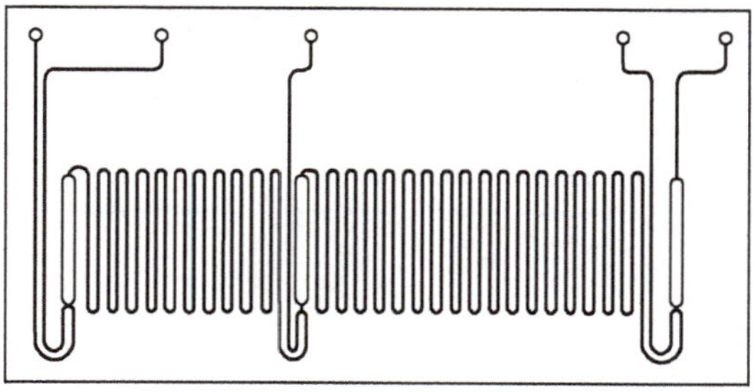

DESIGN 3224
Reactor with four inlets and one outlet:
A+B=P1+C=P2+Q=P

- Width channel: 300 µm
- Depth channel: 120 µm
- Reactor volume: 15 µl (=5+10 µl)

Fig. 3.30: **Labtrix®** chip microreactor.

KiloFlow® is a scalable, turnkey continuous-flow system capable of processing up to 140 L of reaction mixture per day (Fig. 3.31). Again utilizing glass reactors (0. 8–19 mL), efficient mixing and thermal control enables the scaling of reaction conditions developed at the lab to provide g to kg material production. The KiloFlow® basic system comprises two dual syringe pumps and a reactor holder, which contains a series of glass devices, in which the thermostating and reaction of two reagent feeds (A + B → P) can be performed; however, it can be flexibly expanded to incorporate additional pumplines and reaction steps. Positioned within a standard laboratory fume hood, researchers can readily transfer processes from Labtrix® where milligram quantities are produced to KiloFlow® where tens of gram per hour can be realized.

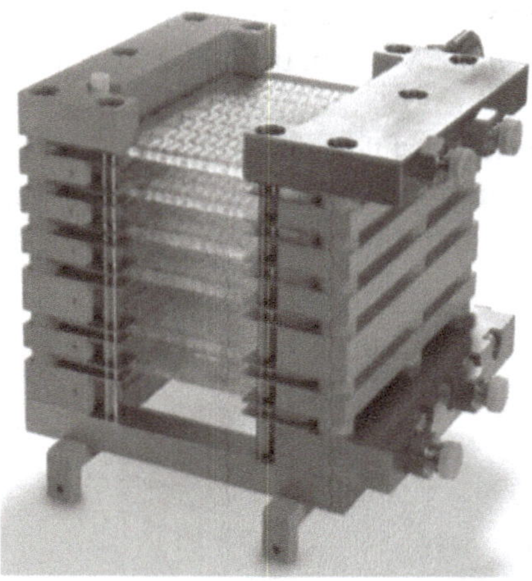

Fig. 3.31: **KiloFlow®** chip microreactor set for large-scale production.

Plantrix® is an industrial flow platform that uses EKasic® silicon carbide (SiC) reactors for the performance of flow reactions at the 10's (MR260) to 100's (MR500) liter per hour scale. Employing integrated heat exchange and mixing, Plantrix® reactors allow the performance of controlled flow reactions at a scale suitable for chemical manufacture. Exploiting the thermal performance and high chemical resistance of

SiC, the flow reactor affords access to processing windows otherwise not accessible in glass, steel, or Hastelloy reactors.

Further information is available on the Chemtrix webpage: www.chemtrix.com.

3.5.2 FutureChemistry (the Netherlands)

FutureChemistry has developed several modules for carrying out flow chemistry at small scales. The philosophy behind these instruments is low usage of chemical compounds, combined with easy operation and ease of understanding the actual chemical processes.

The base module (**FlowStart Evo**) is built for carrying out reactions in the liquid state (Fig. 3.32). Up to three different stock solutions are placed in syringes (with a typical volume of 5 mL each), operated by syringe pumps. The stock solutions flow through flexible inlet modules (tubing with internal volume ca. 100 µL) to a microreactor. The microreactor has integrated mixing channels (with folding flow mixing units) and residence time channels, with a typical combined volume of 100 µL (for normal flow chemistry) or 1 µL (for ultra-fast reactions, also called "flash chemistry"). The microreactor is contacted with a heating/cooling plate which enables a temperature range of –10 to 200 °C. The microreactor is connected to an outlet, which can be complemented with a BPR to increase the pressure inside the microreactor up to 500 kPa, sufficient to keep, for example, water in the liquid state up to 155 °C and prevent outgassing of the liquids.

Because the syringe pumps can reach a wide range of flow rates (below 1 µL min^{-1} up to 2.9 mL min^{-1}), residence times of 0.05 s up to an hour can be theoretically reached. In practical situations, times between 0.1 s and 15 min are normally used.

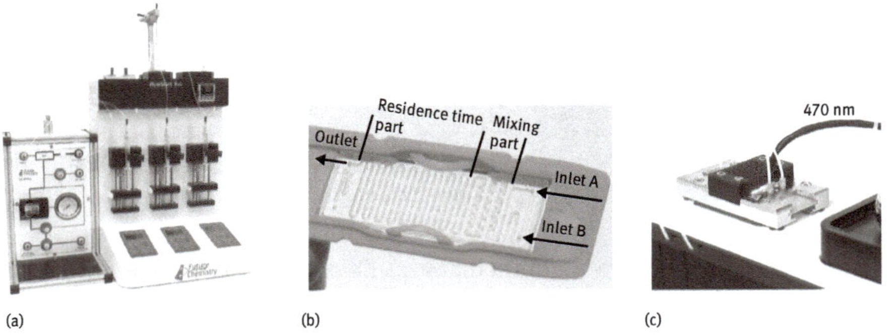

Fig. 3.32: From left to right: **FlowStart Evo** with gas module, close-up of microreactor, photochemistry module.

Examples of liquid-phase reactions performed on the FutureChemistry platform include Prilezhaev dihydroxylations [41], Vilsmeier–Haack formylations [42], and a Grignard reaction [43].

FutureChemistry has a photochemical reactor module for irradiating microreactors at a 100 µL scale (Fig. 3.32). LED light sources are available from 365 nm to the visible range, while a mercury lamp module can reach low UV (210 nm) wavelengths.

Further information is available on the FutureChemistry webpage: www.futurechemistry.com.

3.5.3 Uniqsis (United Kingdom)

The **Uniqsis FlowSyn™** is a compact and flexible, dual-channel *meso*-scale continuous-flow reactor system designed for bench-top use (Fig. 3.33). It utilizes two high-pressure pumps to deliver a maximum combined flow rate of 20 mL min^{-1} at pressures of up to 200 bar. A larger version of the instrument, the **FlowSyn Maxi**, is also available with a maximum flow rate of 100 mL min^{-1} and a 100 bar max pressure rating.

FlowSyn can be specified with stainless steel or Hastelloy C flow-paths suitable for high-pressure or microwave-like chemistry [44, 45], or with a full perfluoropolymer flow path for optimal chemical resistance [46]. An embedded user control interface permits individual experiments to be run either automatically or with full manual control. The system is fitted with three pressure transducers and BPRs which enable priming to be performed "on the fly" without needing to first cool/depressurize the reactors.

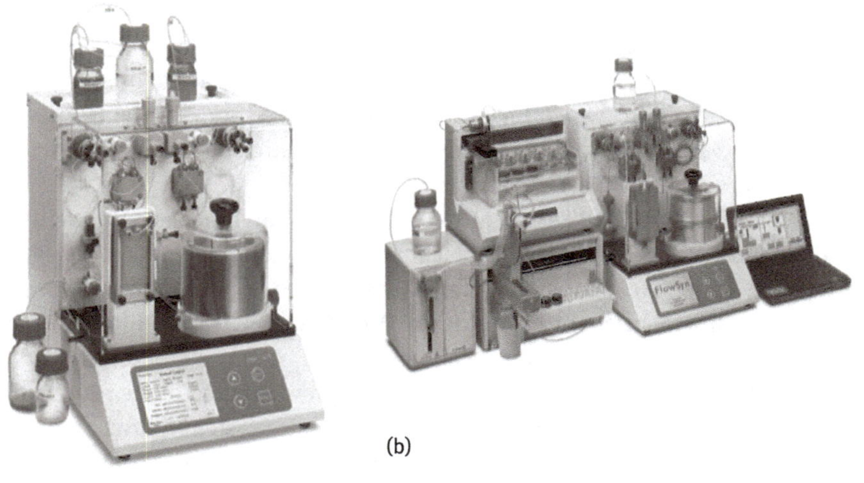

(a) (b)

Fig. 3.33: **Uniqsis FlowSyn™** stand-alone and with automated loop filling (**FlowSyn™ Auto-LF**).

Each channel has a low-pressure inlet valve to allow switching between solvent and reagent solution feeds. Reagent solutions may be delivered from stock bottles (passing through the pump heads) or alternatively from sample loops fitted to high-pressure injection valves positioned inline after the pump heads.

The unit is fitted with two independent heated reactor modules, a coil reactor heater (T_{max} 260 °C) and a module which can accommodate either column reactors or glass static mixer/reactor chips (T_{max} 150 °C). The latter may be used as mixers which then flow into a coil reactor to extend residence time [47], or as short residence time reactors affording excellent temperature control for rapid exothermic reactions [48].

A wide range of coil reactor sizes fabricated from materials including Hastelloy® C, 316L stainless steel, Inconel®, PTFE, PFA, and copper are available. In addition gas–liquid reactions where the gas is fully dissolved in the reaction mixture can be performed using "tube-in-tube" modules utilizing the semi-permeable fluoropolymer AF2400 [49, 50].

A number of system upgrades are available. The **Multi-X** package adds a simple fraction collector that allows a sequence of experiments to be run automatically using the same starting reagent solutions [51]. For combinatorial library generation and reaction optimization requiring a variety of starting materials or reagents, the **Auto-LF** upgrade package adds an independent liquid handler that automatically fills the sample loops according to a preprogrammed list of experiments (Fig. 3.33) [52].

For more complex flow chemistry, the number of flow channels can be increased up to a maximum of four either by adding stand-alone pumps or by the addition of the **Uniqsis Binary Pumping Module** (BPM) and additional reactor modules such as the **ArrheniusOne Flow™** microwave or the innovative **Polar Bear *Plus* Flow™** compact heater/chiller module [53, 54]. The **Polar Bear *Plus*** is an advanced reactor system which combines heating and cooling (–40 to 150 °C) in a compact system, and is available in versions which take coil or chip reactors. Stand-alone **HotCoil™** and **HotChip™** reactor modules are also available for heating coil and chip reactors (ambient to 260 °C and to 300 °C respectively).

Uniqsis have recently released the **PhotoSyn™** high-power LED photoreactor. Available in UV-A, blue, green and custom light options, this reactor is capable of outputting up to 700 W of photons for intensification of photoflow processes. Combining the **PhotoSyn™** with a **ColdCoil™** or **Polar Bear *Plus* Flow** allows independent temperature control of the coil reactor.

The BPM itself can also be used in combination with stand-alone reactor modules as a hub around which to build a modular flow reactor system (**FlowLab Plus™**). In this more complex configuration, full system control is achieved using **FlowControl™** software running on a separate computer.

While developed primarily for homogeneous flow chemistry applications, FlowSyn has also been used for the preparation of nanoparticles [55, 56] and engineered silicas [57].

Further information is available on the Uniqsis webpage: www.uniqsis.com.

i

3.5.4 ThalesNano (Hungary)

ThalesNano received an R&D 100 award for their **H-Cube® Flow Reactor** which has been utilized worldwide both in academia and in industry since its release in 2005. Since then they have developed further systems, such as the modular **H-Cube® Pro, Phoenix Flow Reactor™**, and **Gas Module™** (Fig. 3.34). ThalesNano systems provide solutions to many safety and hazardous chemistry-related problems. For instance, with the H-Cube® reactor family, hydrogen cylinders can be eliminated from laboratory environment. This is due to the built-in hydrogen generation system which requires only water for the generation of high-pressure (up to 100 bar) hydrogen, allowing the process to be greener, safer, faster, and an easy alternative compared to conventional batch methods. Furthermore, the cartridge technology (**CatCart®**), a patented ThalesNano product, provides a safe handling of even the most pyrophoric catalysts.

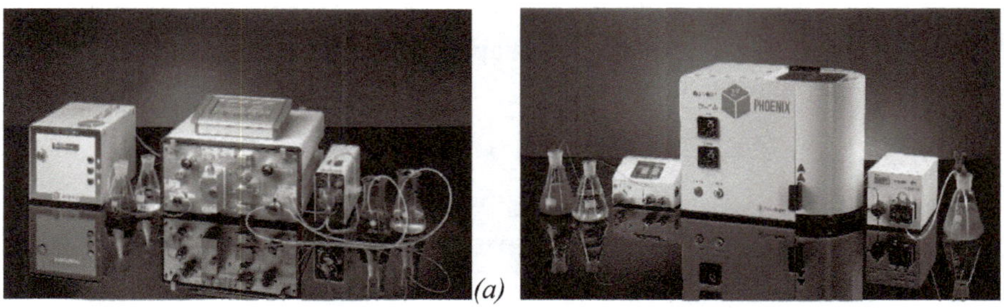

(a) (b)

Fig. 3.34: **H-Cube® Pro** hydrogenation reactor with **Gas Module™** (a) and **Phoenix Flow Reactor™** (b).

While the H-Cube® reactors are mainly used for hydrogenation reactions, they are also capable of performing heterogeneous reactions in flow on different scales. The **Phoenix Flow Reactor™** extended the temperature range to 450 °C both for homogeneous and heterogeneous reactions, and up to 200 bar allowing the use of different gases in a complete automated manner.

The **IceCube™** allows the user to conduct low temperature reactions such as diazotation, Swern oxidation, and even multistep, homogeneous reactions in one reactor system (Fig. 3.35). Modularity also allows the chemist to perform ozonolysis within the confines of a small footprint with the highly efficient ozone generator, the **Ozone Module™**.

Fig. 3.35: **IceCube™** reactor.

In 2016 ThalesNano launched the **H-Cube® Mini Plus** reactor as an entry-level flow system based on their H-Cube hydrogenation technology. The H-Cube® Mini Plus aims to provide universities with access to an affordable instrument for hydrogenations and other reactions in flow.

In 2020, ThalesNano launched its newest product, called the PhotoCube™ (Fig. 3.36). **The PhotoCube™** is a multiwavelength instrument for advanced photochemical applications. Various configurations can be applied to a diverse set of photochemical

Fig. 3.36: PhotoCube™.

reactions using batch, flow, stop-flow, and CSTR methods. The multicolor option ranges from UVA to red. With the opportunity to select the color of the LEDs, the system can be fine-tuned for specific wavelengths and applications; furthermore, multiple wavelengths can be utilized at the same time.

Further information is available on the ThalesNano webpage: www.thalesnano.com.

3.5.5 Vapourtec (United Kingdom)

Vapourtec offers two modular systems for flow chemistry applications. The **R-Series** flow chemistry system enables several modules to be combined to create the exact system required (Fig. 3.37). It offers a range of pumping options for feeding reagents (from strong acids, organometalic, gases to slurries) with precision, inline detection systems and liquid handlers to run library synthesis with an unlimited number of reactions.

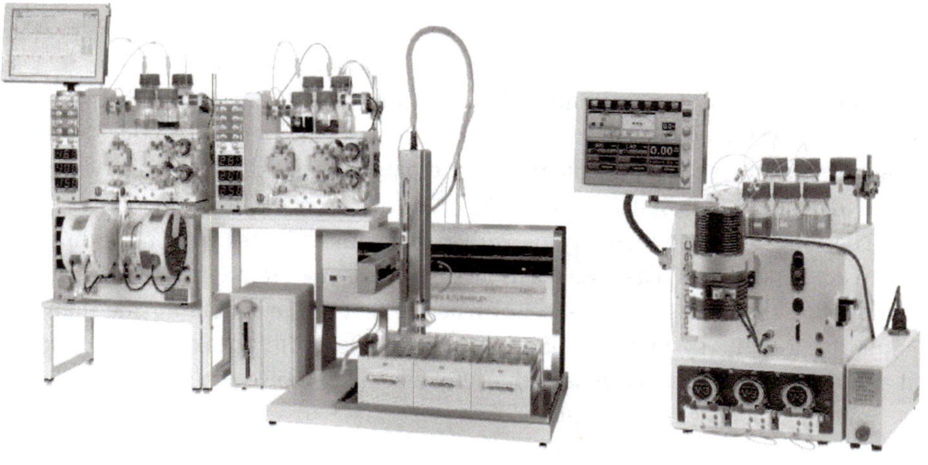

Fig. 3.37: Left: **R-Series** reactor with four reactor positions, two pump modules, and a liquid handler. Right: **E-Series** reactor block with two reactor positions and featuring **V3 pumps** and the photochemical reactor **UV-150.**

Vapourtec offers a range of options for feeding reagents into the reaction. For example, by combining two pumping modules, the system can deliver up to four reagent channels. All pumping modules use a unique continuous automatic monitoring system to monitor performance of pumps to ensure accurate reporting of flow rates and shutdown safely in the event of a leak or blockage.

The R-Series system allows separate temperature control of up to eight reactors with a broad temperature range (from −70 to 250 °C). These reactors can be combined for multistep synthesis or used to allow increased total reactor volume, for higher throughput or longer residence times. With the use of **Flow Commander** software, all the data is captured in a single file, which accurately predicts steady state conditions and allows for full automation.

The available reactors are suitable for both thermally mediated and nonthermally mediated chemical reactions.

The photochemical reactor **UV-150** eliminated the problems of traditional batch photochemistry, allowing for a safe, precise, efficient, consistent, and scalable process. It covers the UV region with a variety of exchangeable UV sources. The most recent high-power LEDs provide a radiant power of 60 W.

The **Ion** electrochemical reactor enables continuous-flow electrochemistry with a rapid assemble/disassemble operation that allows evaluation of >20 available electrode materials. The Ion can work up to 5 bar pressure, which allows the use of gaseous reagents and working at high temperatures.

Vapourtec launched in 2020 the **Variable Bed Flow Reactor**, which can control and adjust the packing density of a solid media, making it ideal for solid phase synthesis (e.g., solid phase peptide or oligonucleotide synthesis) and for reactions that consume a solid reagent (i.e., making Grignard reagents in flow).

The **E-Series** system continues the innovative trend that the R-Series started. With the same forced convection reactor heating system, and equipped with the robust and easy to use **V-3 pumps**, the E-Series simply makes flow chemistry accessible to a broader range of users (Fig. 3.37).

The **E-Series** instruments come with a maximum of three reagent pumps and two reactor positions. The two reaction steps are separately temperature controllable. Using the same extensive set of reactors that are available for the high-end Vapourtec R-Series system, the **E-Series** offers the flexibility to take on a wide range of applications. All models come with useful chemistry tools such as:
- Solvent vapor pressure calculator
- Arrhenius reaction rate calculator
- Residence time calculator
- Dispersion and residence time distribution tool

i *Further information is available on the Vapourtec webpage: www.vapourtec.com.*

3.5.6 Dow Corning (USA)

Since 2002, Corning has provided the chemical process industry with their **Corning® Advanced-Flow™ Reactors** process intensification platform (Fig. 3.38). Corning provides a portfolio of continuous-flow reactor products that are designed and optimized to quickly and seamlessly scale-up from academia and lab feasibility to process development to multiton industrial production of chemicals.

Corning® Advanced-Flow™ reactors are also well suited for reactions where conditions related to process exothermicity, concentrations, mixing, and so on make batch processing difficult or impossible; where mass transfer is limiting (e.g., nonhomogeneous liquid–liquid, gas–liquid, gas–liquid–solid); and/or where residence time distribution has a direct impact on final quality (nanoparticles, microparticles, high-grade and technical polymers, etc).

Corning's patented fluidic module design provides vastly improved mixing (× 10–100), very narrow residence time distribution, and superior heat transfer (×1,000), which in turn enables faster, inherently safer reactions that produce better yields of higher purity output. Corning Advanced-Flow Reactors Technology is designed to enable rapid scale-up to quickly and easily transition work from the lab to production scale without requiring new development work.

Corning's flow chemistry education platform, Corning **Nebula™ Education Kits**, was designed to specifically enhance the academic curricula of chemistry-related majors and provide students with a hands-on learning experience to become proficient in flow chemistry and flow reactor operation.

Corning also provides a full range of services to enhance customers' projects from development through to implementation and operation, including:
- workshops and trainings
- basic and detailed auxiliary systems engineering (Fig. 3.39)
- customized turnkey solutions
- technical support
- FAT/SAT and industrial startup
- assistance for equipment qualification following cGMP/FDA requirements
- pre- and post-purchase technical support
- compliance with international standards (ATEX, ASME, PED, SELO, KGS, etc.)

Further information is available on the Corning Advanced-Flow Reactors website: www.corning.com/reactors.

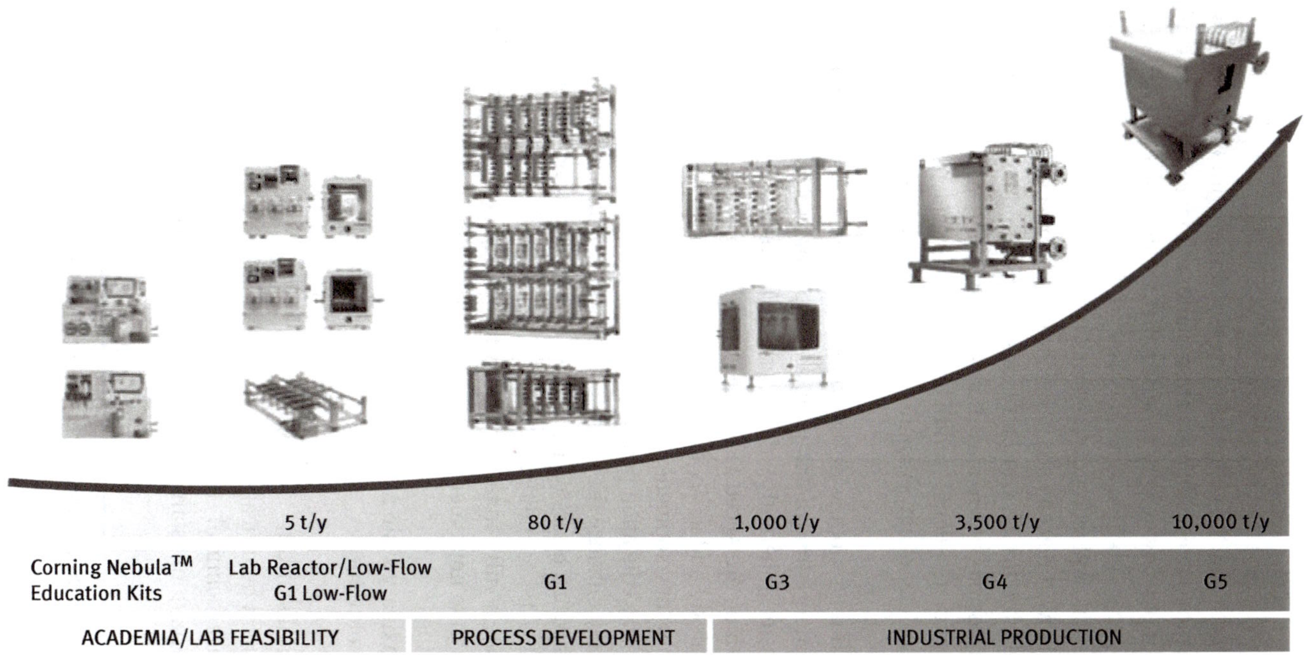

Fig. 3.38: Corning® **Advanced-Flow™ reactors** product portfolio ranges from academia and lab feasibility to large-scale industrial production, in glass or silicon carbide (SiC) and includes photochemistry capabilities.

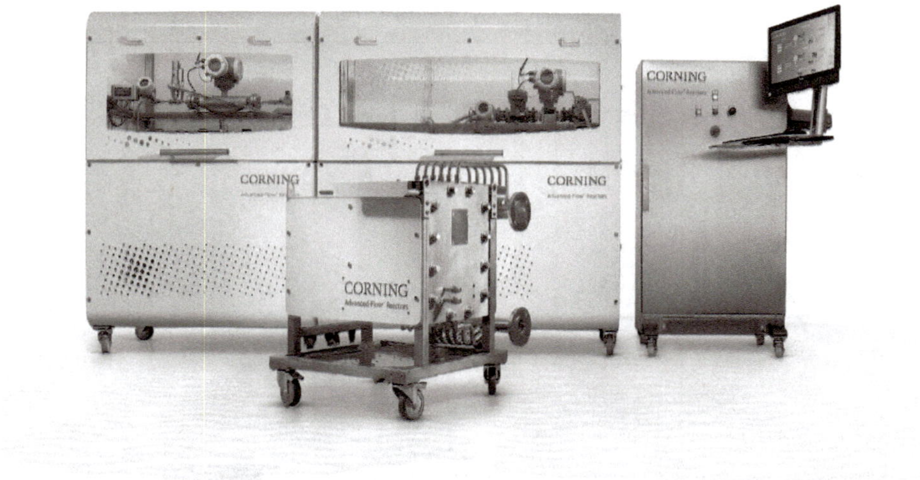

Fig. 3.39: Corning's **Advanced-Flow™ reactors** auxiliaries for the G4 industrial-scale reactor provide customers with complete turnkey solutions for chemical production.

3.5.7 Syrris (United Kingdom)

The current Asia product range is the fourth-generation system specifically designed for flow chemistry. Previous flow chemistry systems include Africa (commercialized in 2004) and FRX (commercialized in 2006).

The **Asia** product range, winner of a prestigious R&D100 award, features a variety of modules that can be used in stand-alone mode or arranged in any desired system to enable the widest variety of chemical processes (Fig. 3.40). All wetted parts are made from either glass or PTFE, thereby offering maximum chemical compatibility. Each module can be controlled either manually or in an automated way using a computer. Key modules of the Asia range include:

- The **Asia Syringe Pump**: the first pump designed specifically for flow chemistry. Its patented pumping system enables smooth pumping of solvents or reagents and can be used with low boiling point solvents such as dichloromethane, and ether. The Asia pump design enables a very low pulsation over a wide range of flow rates (1–10 mL min^{-1} for each channel).
- The **Asia Flow Liquid–Liquid Extractor** (FLLEX): this popular module is the flow equivalent of a separation funnel and is used for continuous workup or extraction. By using an advanced membrane technology, the FLLEX can separate two phase mixtures in-line.
- The **Asia FLUX** electrochemistry module enables a wide range of electrode materials to be changed in seconds, tool-free to explore continuous electrochemistry applications.

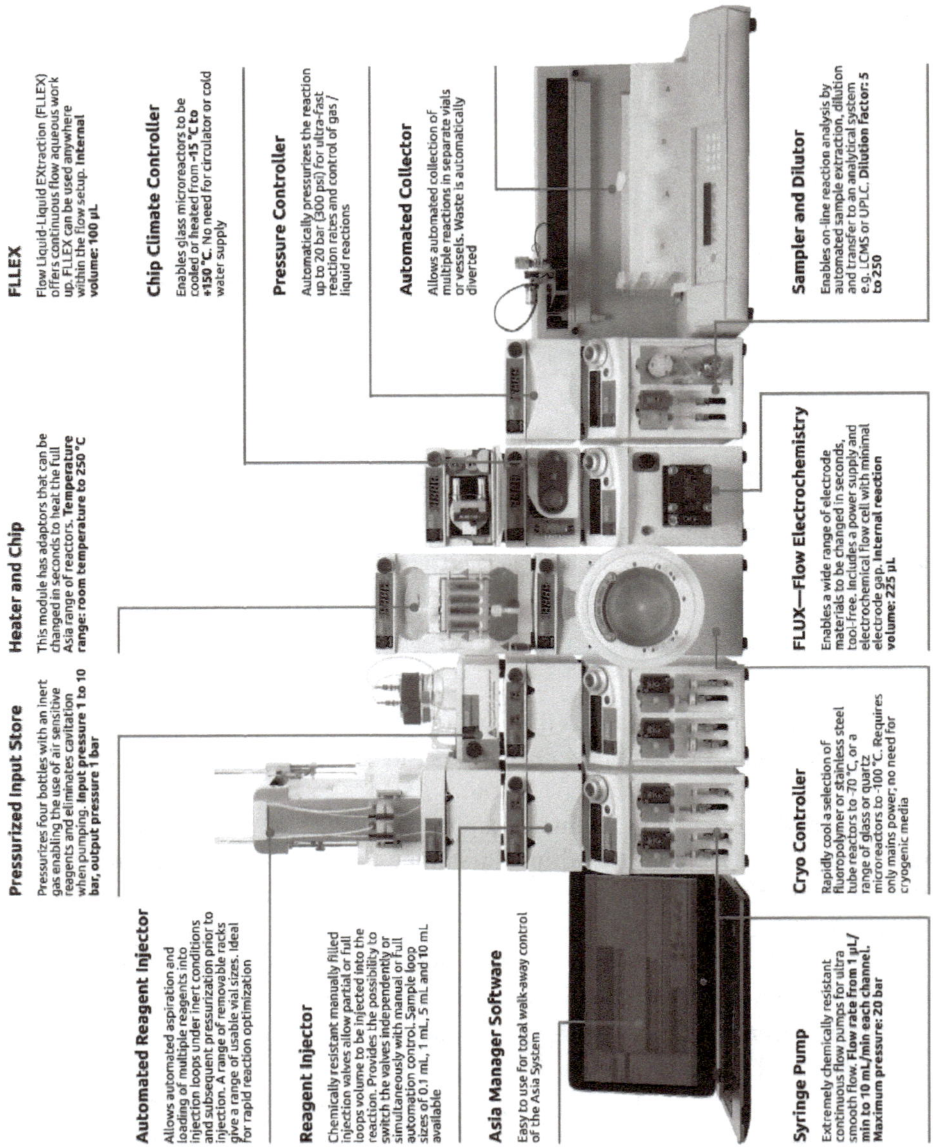

Fig. 3.40: The Asia flow chemistry system modules. The system can be configured with different modules to suit the application or budget.

- The **Asia Photochemistry** module (launched in 2020) offers a range of single wavelength LED light sources that can be combined for increased power or multiple wavelengths, variable light intensity, and interlocked safety features.
- The **Asia Automated Reagent Injector** is a small footprint autosampler, enabling discontinuous variables such as reagents, catalysts, and enzymes to be explored. The dual-core needles allow aspiration from sealed vials for air and moisture sensitive reagents and pre-pressurize the sample loops to give precise control of small aliquot additions. This ensures the Asia system is ideal for library synthesis and reaction optimizing.
- Microreactors: the Asia range enables the use of all types of reactors: glass microreactors and micromixers, fluoropolymer or metal tube reactors for larger volumes and glass solid phase reactors. The wide range of volumes and materials gives much needed flexibility to the flow chemist.
- The **Asia Sampler and Dilutor**: this device is used for the sampling in-line of a flow experiment, then dilution and transfer to an analytical device such as a HPLC, GC-MS, or LCMS.
- The **Asia Cryo Controller**: this device enables microreactors to be controlled to −100 °C and tube reactors to −70 °C. The system does not require any thermal fluid and only requires standard mains power supply for all low temperature applications.

The modularity and wide range of available modules enable the system to be used for different applications such as organic synthesis, process optimization, library synthesis, nanoparticle synthesis, enzymatic catalysis, biofuel synthesis, and transfer of microwave reactions. The automated control of the system, via Asia Manager software, enables complex multistep chemistries, rapid reaction optimization, and library generation with precise segment tracking to be performed. Asia systems can be modified and expanded as the requirements evolve and new modules (e.g., electrochemistry or photochemistry in flow) can be added afterward with full compatibility.

Further information is available on the Syrris website: www.syrris.com. i

3.5.8 MICROINNOVA (Austria)

MICROINNOVA specializes in flow chemistry, process intensification, and continuous manufacturing. Combining chemistry knowledge and plant competence, MICROINNOVA provides plants from laboratory scale through to pilot scale for continuous-flow chemistry technology as well as chemical development services.

Their **Flow Miniplant** bridges the gaps between benchtop, process development, and manufacturing. A setup for a one-step reaction is shown in Fig. 3.41. The scope of application ranges from easy processes to difficult areas (complex chemical reactions, high complexity of the process, new process windows with high temperature and high pressure). Chemical synthesis, workup, and liquid formulations can be covered with their flexible approach.

The Flow Miniplant combines modules of various kinds (e.g., feed module, reaction module, product module) to create the desired process. Exchanging modules can be easily done by the lab chemist. For multiple-step reactions, more modules can be combined as well as for workup or other unit operations. Different unit operations or reactor types can be used within the same system. This flexibility enables an adjustment of the plant according to the needs of the process.

Fig. 3.41: Example installation with two feed modules, one reaction module and one product module.

Feed module: The central part of the feed module is the annular gear pump. The flowrate of the module is 0.5–10 L h^{-1}. A filter protects the pump from particles. A burst disk protects the plant from overpressure. The flow is set via a mass flow meter by controlling the pump speed. The pressure is measured and the system will be stopped by the automation once the maximal pressure is reached, so as not to cause the opening of the burst disk.

Reaction module: The reaction takes place in a micro reactor. The reactor consists of ten glass plates submerged in a heating/cooling bath. The first two plates function as heat exchangers; they bring the two reactants to the required reaction temperature. The third plate contains a mixing structure for intensive mixing of the two reactants. The remaining plates are plug flow reactors that provide the residence time for the reaction. The number of plates can be modified freely. Temperature sensors are installed between the plates.

Product module: It contains a tube-in-tube heat exchanger where the product is cooled down or warmed up to room temperature. The flow of the heating/cooling medium is set by a manual control valve. After the heat exchanger, the temperature and the pressure are measured. The pressure is set by a manual pressure regulator.

A manual three-way valve enables the user to switch between two outlets, where one is usually used to take samples.

Automation system: The automation hardware is partly located on the modules. Each module carries the necessary input modules and signal converters. The main control is done on the personal computer by a control software, which allows the full control of the connected modules, logging of the measured values and reporting. Further options such as PAT tools, MTP (new Protocol for Modular Plant Systems), Digital Twin, and Model Predictive Control can be integrated upon request.

Further information is available on the MICROINNOVA website: www.microinnova.com. i

3.5.9 Little Things Factory GmbH (Germany)

Little Things Factory offers the **Flow Chemistry Starter Set** (Fig. 3.42) to introduce trainees in flow chemistry applications and to provide further education for those who already work in this field. The starter set includes a unique manual with detailed instructions for 13 reactions to introduce trainees in flow chemistry applications and it permits the simple translation from batch process into a continuous one by using standard laboratory equipment.

Fig. 3.42: Little Things Factory Flow Chemistry Starter Set.

The Flow Chemistry Starter Set consists of 4 MR-LAB microreactors, 2 syringe pumps, a wide range of connection equipment as well as a manual with detailed instructions and analysis for 13 reactions from the field of organic chemistry.

Little Things Factory has modified microreactors of the MR-LAB Series in two different ways to enable the introduction of a heterogeneous catalyst into the microreactor.

The **LTF-MS-Kat-G** microreactor has two reaction channels, a catalyst-powder-containing channel and a plurality of intermediate channels, which are connecting the reaction channels (Fig. 3.43). The channel containing the catalyst is arranged between the reaction channels.

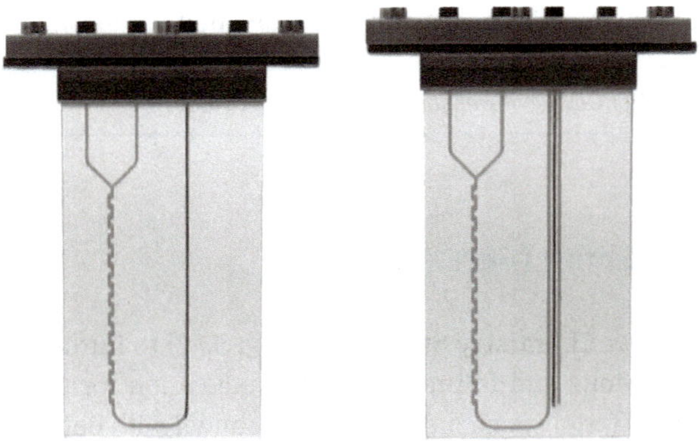

Fig. 3.43: Modified microreactors for using heterogeneous catalysts. The **LTF-MS-Kat-G** (left) includes a channel which can be filled with powder, and the **LTF-MS-Kat-R** (right) includes catalyst supported on a removable rod.

The **LTF-MS-Kat-R** catalyst-containing channel comprises a catalyst rod made of a metal wire (e.g., copper, nickel, or platinum), an alloy, or a glass (Fig. 3.43). In this case, the catalyst rod can be pushed in and out of the channel.

i *Further information can be found on the Little Things Factory webpage: www.ltf-gmbh.com.*

3.5.10 AM Technology (United Kingdom)

Coflore® flow reactors from AM Technology are actively mixed and multi-stage, a combination that decouples mixing and plug flow from residence time (Fig. 3.44). Residence times are controlled by the flow rate of the pump without compromising mixing or plug flow performance. Coflore® flow reactors are scalable and compatible with many different reaction types from liquid–liquid, solid–liquid, liquid–gas to solid–liquid–gas, and can produce up to thousands of tons of material per year in a single reactor.

Coflore® flow reactors combine attributes from batch and flow technologies, to afford a reactor which is as versatile and productive as batch, while maintaining the efficiencies of flow. Residence times from seconds to hours are interchangeable, while

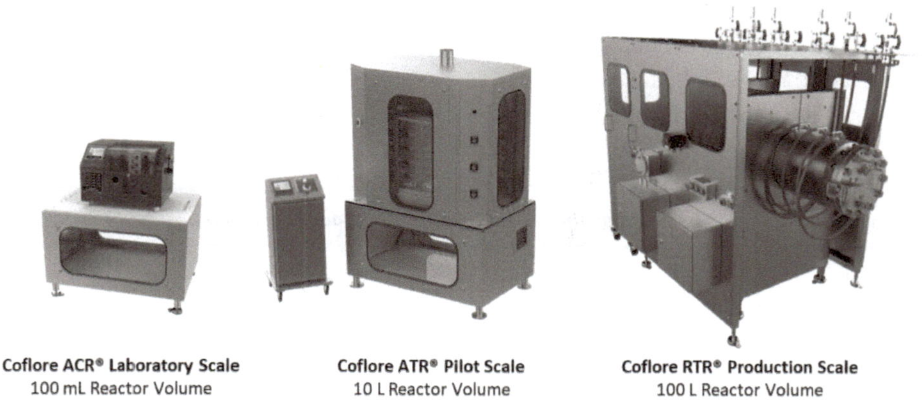

Fig. 3.44: Range of **Coflore®** flow reactors.

maintaining orderly flow, and without the need for configurational changes or add-ons to the reactor.

The **Coflore ACR** is an actively mixed, bench-top flow reactor, used within a laboratory environment for developing and optimizing scalable flow chemistry solutions, or for when limited quantities of reactants or material are available or required. The Coflore ACR reactor process channel is divided into ten reactor cells, equivalent to ten CSTRs in series for orderly flow.

The **Coflore ATR** is an actively mixed, tubular flow reactor of up to eight reactor tubes in series for orderly flow, operating up to 10 L in reactor volume to produce material at the pilot scale. Each of the tubes can be heated or cooled independently, and reagents can be added, or samples removed, at each end of each reactor tube. For a 5-minute residence time, the Coflore ATR produces 120 L of material per hour.

The **Coflore RTR** has a 100 L capacity that is capable of processing countless reactor volumes without interruption. Such throughputs have the potential to produce many thousands of tons of material per year in a single reactor. The Coflore RTR operates as a 10-stage, actively mixed, self-baffling flow reactor with orderly flow characteristics. For a 5-min residence time, the Coflore RTR produces 1,200 L of material per hour.

Reactor specifications:

- Temperature range:
  - ACR: −40 to 140 °C
  - ATR: −40 to 200 °C
  - RTR: −40 to 150 °C
- Operating pressure: up to 10 bar (50 bar for the ATR)
- Wetted materials: 316 stainless steel, Hastelloy C276 (and PTFE for the ACR)

- Throughput:
  - ACR: 0.1–500 mL min^{-1}
  - ATR: up to 30 L min^{-1}
  - RTR: up to 300 L min^{-1}

i *Further information can be found on the AM Technology webpage: www.amt.uk.*

3.5.11 fReactor (United Kingdom)

The **fReactor™** is an entry level cascaded CSTR system suitable for bench-scale flow chemistry (Fig. 3.45). The system comprises five reactor modules or stages, with each module forming a separate reaction zone which can be flexibly configured. Heating and stirring is achieved by placing the fReactor on a stirrer hotplate. Each stage has 4 × ¼-28 flat-bottomed ports which can be used for connecting fReactor stages, introducing reagent feeds, feeding out product streams, or introducing additional sensors. The wetted materials on the standard configurations are PEEK, glass, and Viton, but other materials are available on request. The stirrer modules are rated to withstand pressures of up to 100 psi (6.89 bar).

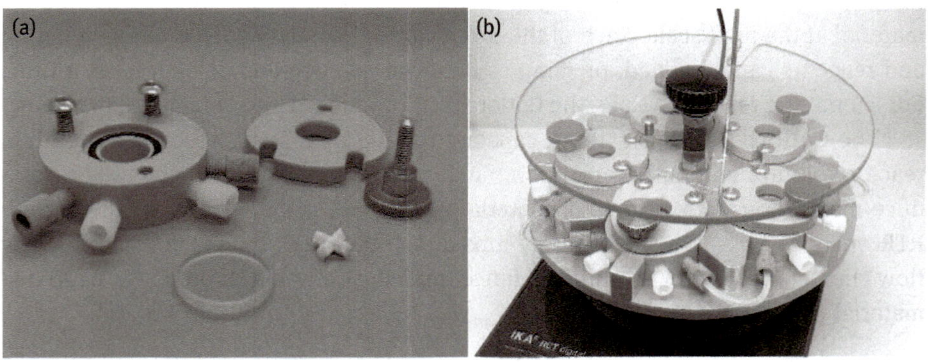

Fig. 3.45: The **fReactor™**: a lab-scale CSTR that utilizes magnetic stirring to achieve active mixing: (a) a disassembled unit; (b) deployed in a "daisy-chain" arrangement on a magnetic stirrer. Reproduced with permission from Asynt Ltd.

The **fReactor™** stirrer modules are currently available in two sizes. The **fReactor™ – Classic** has 1.75 mL stage volumes for a total volume of 8.8 mL in the five-stage system, while the **fReactor™ – Maxi** has 7.5 mL stage volumes for a total volume of 37.5 mL over five stages.

The **fReactor**^{TM} is sold and distributed by Asynt.

*Further information can be found on the **fReactor**^{TM} webpage: www.freactor.com.*

i

3.5.12 Zaiput Flow Technologies (USA)

Zaiput have phase separator and back-pressure regulating modules which are designed for use in flow chemistry systems. The Zaiput separator modules allow researchers to perform in-line liquid–liquid or liquid–gas separations. A variety of membranes are available to suit different applications, and the modules come in three sizes to suit a wide range of flow rates. The **SEP-100** is optimized for laboratory systems operating at 0–10 mL min^{-1} total flow rate, while the **SEP-200** and **SEP-3000** are for pilot (20–200 mL min^{-1}) and production (200–3,000 L min^{-1}) scales.

The Zaiput adjustable BPRs use a compressed gas and an elastic membrane to provide resistance to the flow path. The wetted parts are chemically resistant ETFE and PFA, and the regulators can provide back pressures of up to 2 MPa (290 psi or 20 bar). Two configurations are available: The **BPR-10** is for total flow rates of 0.05–20 mL min^{-1}, while the **BPR-1000** is for total flow rates of 20–1,000 mL min^{-1}.

Further information can be found on the Zaiput webpage: www.zaiput.com.

i

3.5.13 H.E.L Group (United Kingdom)

H.E.L has the **FlowCAT** reactor, a benchtop unit for high-pressure flow chemistry such as hydrogenations. The standard FlowCAT is capable of heating to 300 °C but customized versions are available for up to 550 °C. Pressures of up to 100 bar can be safely handled in the standard FlowCAT. The reactors available for the FlowCAT includes a range of column reactors in diameters from 6 to 24 mm, with a working length of 15 cm (4–60 mL internal volume range), and tubular reactors. The FlowCAT includes full software control of pressure, temperature, and feed rates of liquids and gasses. The FlowCAT platform is highly customizable with options for multiple reactors, integration of PAT and analytics, larger-scale units, and specialized materials.

The **FlowCAT SA** is a lower cost version of the FlowCAT platform. It has the same operating range (up to 300 °C and 100 bar) as the FlowCAT, but it only has local digital controllers for temperature, pressure, and flow rate.

Further information can be found on the H.E.L Group webpage: www.helgroup.com

i

Further readings

- Hessel, V, Schouten, JC, Renken, A, Wang, Y, Yoshida, JI: Handbook of Micro Reactors. Chemistry and Engineering, Weinheim, Wiley-VCH, 2008.
- Wirth, T, Ed., Microreactors in Organic Synthesis and Catalysis, Weinheim, Wiley-VCH, 2008.
- Hessel, V, Renken, A, Schouten, JC, Yoshida, JI, Eds., Micro Process Engineering: A Comprehensive Handbook, Weinheim, Wiley-VCH, 2009.
- Yoshida, JI, Flash Chemistry: Fast Organic Synthesis in Microsystems, Chichester, UK, John Wiley & Sons, 2008.
- Luis, SV, Garcia-Verdugo, E, Eds., Chemical Reactions and Processes under Flow Conditions, Cambridge, RSC Publishing, 2010.
- Antonio de la Hoz, André Loupy, Microwaves in Organic Synthesis (Third Edition), 2 Volume Set; Weinheim, Wiley-VCH, 2013.

Databases:

- SciFinder – https://www.cas.org/products/scifinder
- Reaxys – http://www.elsevier.com/online-tools/reaxys
- Search for keywords and phrases: microreactor, continuous-flow synthesis, flow reactor, continuous processing.

Dedicated journals:

- *Journal of Flow Chemistry* – http://www.jflowchemistry.com
- *Beilstein Journal of Organic Synthesis* (*Thematic Series in Flow and Process Chemistry*):
  - *Chemistry in Flow Systems* – https://www.beilstein-journals.org/bjoc/series/4
  - *Chemistry in Flow Systems II* – https://www.beilstein-journals.org/bjoc/series/17
  - *Chemistry in Flow Systems III* – https://www.beilstein-journals.org/bjoc/series/33
  - Automated chemical synthesis – https://www.beilstein-journals.org/bjoc/series/65
  - Integrated multistep flow synthesis – https://www.beilstein-journals.org/bjoc/series/76
- *Organic Process Research & Development* (Special Issues)
  - Continuous processing, microreactors and flow chemistry – https://pubs.acs.org/toc/oprdfk/20/2
  - Flow chemistry enabling efficient synthesis – https://pubs.acs.org/toc/oprdfk/24/10
- *European Journal of Organic Chemistry* (Special Collection in Flow Chemistry – https://chemistry-europe.onlinelibrary.wiley.com/doi/toc/10.1002/(ISSN)1099-0690.Flow-Chemistry)
- *ChemSusChem* (Special Issue in Flow Chemistry – https://chemistry-europe.onlinelibrary.wiley.com/toc/1864564x/5/2)

Study questions

3.1 What are the main parts of a continuous-flow reactor setup?

3.2 What should be taken into account when assembling a continuous-flow rector setup for a specific synthetic process?

3.3 You plan to run a reaction similar to the one in Fig. 3.29. Components A and B (both 0.1 M) are both pumped at a rate of 1 mL min^{-1} into the reactor which has a volume of 8 mL.

(a) How long is the residence time in the reactor?

(b) You decided to use an excess of the B reactant. For this, you increased the flow rate of B to 1.5 mL min^{-1} and left the flow rate of A at 1 mL min^{-1}. What is the stoichiometric ratio of A:B when the two reagent streams are mixed?

(c) What is the new residence time in the reactor at this flow rate?

(d) Having performed the reaction, you realized that with the reduced residence time the reaction is incomplete. Suggest what you can do to achieve full conversion while retaining the new stoichiometry ratio.

(e) You make another attempt to perform the same reaction, but this time you also would like to quench it. The initial flow rates of A and B (both at 0.1 M) are 1 mL min^{-1}. Your quenching reactor coil has a volume of 10 mL and you would like to have a residence time of 2 min in this coil. Exactly 3 equivalents of the quenching reagent are required. What should be the concentration and flow rate of the quenching solution?

References

[1] Karassik, IJ, Messina, JP, Cooper, P, Heald, CC, Pump Handbook, 4th edn., NY, USA, McGraw-Hill Education, 2007.

[2] Thermal mass flow meter / controller (principle of operation). Bronkhorst, 2014. (Accessed December 2, 2020, at https://www.youtube.com/watch?v=G62ma2IFh9o&feature=youtu.be.)

[3] How thermal mass flow meter technology works. Sierra Instruments, 2014. (Accessed December 2, 2020, at https://www.youtube.com/watch?v=-VUL0xWfUeY&feature=youtu.be.)

[4] Mass flow meter. OMEGA Engineering, (Accessed December 2, 2020, at https://www.omega.co.uk/prodinfo/mass-flow-meter.html.)

[5] Swagelok® back pressure regulators made simple. Swagelok, (Accessed December 2, 2020, at https://youtu.be/HxiZGnZmsQM.)

[6] Ceylan, S, Coutable, L, Wegner, J, Kirschning, A, Inductive heating with magnetic materials inside flow reactors, Chem-Eur J, 2011, 17, 1884–1893.

[7] Tu, NP, Hochlowski, JE, Djuric, SW, Ultrasound-assisted click chemistry in continuous flow, Mol Divers, 2012, 16, 53–58.

[8] Kappe, CO, Pieber, B, Dallinger, D, Microwave effects in organic synthesis: myth or reality?, Angew Chem Int Ed, 2013, 52, 1088–1094.

[9] Paul, EL, Atiemo-Obeng, V, Kresta, SM, Handbook of Industrial Mixing: Science and Practice, Hoboken, John Wiley & Sons, 2004.

[10] Zhao, Y, Chen, G, Yuan, Q, Liquid–liquid two-phase mass transfer in the T-junction microchannels, AIChE J, 2007, 53, 3042–3053.

[11] Parrott, AJ, Bourne, RA, Akien, GR, Irvine, DJ, Poliakoff, M, Self-optimizing continuous reactions in supercritical carbon dioxide, Angew Chem Int Ed, 2011, 50, 3788–3792.

[12] Löwe, H, Axinte, RD, Breuch, D, et al., Flow chemistry: Imidazole-based ionic liquid syntheses in micro-scale, Chem Eng J, 2010, 163, 429–437.

[13] Breuch, D, Löwe, H, Heterogenously catalyzed one-step alkylation and carboxylation of N-methylimidazole with dimethylcarbonate in continuous flow, Green Process Synth, 2012, 1, 261–267.

[14] Ghanem, A, Lemenand, T, Della Valle, D, Peerhossaini, H, Static mixers: Mechanisms, applications, and characterization methods – A review, Chem Eng Res Des, 2014, 92, 205–228.

[15] Löbbecke, S, Ferstl, W, Panić, S, Türcke, T, Concepts for modularization and automation of microreaction technology, Chem Eng Technol, 2005, 28, 484–493.

[16] Sulzer heat exchanger SMR. (Accessed December 4, 2020, at https://www.sulzer.com/en/shared/products/heat-exchanger-smr.)

[17] Chapman, MR, Kwan, MHT, King, G, et al., Simple and versatile laboratory scale CSTR for multiphasic continuous-flow chemistry and long residence times, Org Process Res Dev, 2017, 21, 1294–1301.

[18] Avila, M, Fletcher, DF, Poux, M, Xuereb, C, Aubin, J, Mixing performance in continuous oscillatory baffled reactors, Chem Eng Sci, 2020, 219.

[19] Fogler, H, Elements of Chemical Reaction Engineering, 4th edn., Prentice Hall PTR, 2005.

[20] Wiles, C, Watts, P, Micro Reaction Technology in Organic Synthesis, Boca Ratton, CRC Press-Taylor & Francis, 2011.

[21] Ehrfehld, W, Hessel, V, Löwe, H, Microreactors: New Technology for Modern Chemistry, Weinheim, Wiley-VCH, 2000.

[22] De Loos, SRA, Van Der Schaaf, J, Tiggelaa, RM, Nijhuis, TA, De Croon, MHJM, Schouten, JC, Gas–liquid dynamics at low Reynolds numbers in pillared rectangular micro channels, Microfluid Nanofluid, 2010, 9, 131–144.

[23] Roydhouse, MD, Ghaini, A, Constantinou, A, Cantu-Perez, A, Motherwell, WB, Gavriilidis, A, Ozonolysis in flow using capillary reactors, Org Process Res Dev, 2011, 15, 989–996.

[24] Gutmann, B, Elsner, P, Glasnov, T, Roberge, DM, Kappe, CO, Shifting chemical equilibria in flow-efficient decarbonylation driven by annular flow regimes, Angew Chem Int Ed, 2014, 53, 11557–11561.

[25] Gong, X, Miller, PW, Gee, AD, Long, NJ, De Mello, AJ, Vilar, R, Gas–liquid segmented flow microfluidics for screening Pd-catalyzed carbonylation reactions, Chem Eur J, 2012, 18, 2768–2772.

[26] Liu, W, Multi-scale catalyst design, Chem Eng Sci, 2007, 62, 3502–3512.

[27] Zupan, M, Šegatin, N, Bromination of organic molecules with polymer-supported bromine complexes, Synthetic Commun, 1994, 24, 2617–2626.

[28] Ranu, BC, Das, AR, Silica gel-supported zinc borohydride. 2. Regioselective 1,2-reduction of conjugated ketones and aldehydes to the corresponding allylic alcohols, J Org Chem, 1991, 56, 4796–4798.

[29] Hassner, A, Stern, M, Synthesis of alkyl azides with a polymeric reagent, Angew Chem Int Ed, 1986, 25, 478–479.

[30] Moussa, Z, Judeh, ZMA, Ahmed, SA, Polymer-supported triphenylphosphine: application in organic synthesis and organometallic reactions, RSC Adv, 2019, 9, 35217–35272.

[31] Seeberger, PH, Organic synthesis: Scavengers in full flow, Nat Chem, 2009, 1, 258–260.

[32] Petersen, TP, Polyzos, A, O'Brien, M, Ulven, T, Baxendale, IR, Ley, SV, The oxygen-mediated synthesis of 1,3-butadiynes in continuous flow: Using Teflon AF-2400 to effect gas/liquid contact, ChemSusChem, 2012, 5, 274–277.

[33] Cranwell, PB, O'Brien, M, Browne, DL, et al., Flow synthesis using gaseous ammonia in a Teflon AF-2400 tube-in-tube reactor: Paal-Knorr pyrrole formation and gas concentration measurement by inline flow titration, Org Biomol Chem, 2012, 10, 5774–5779.

[34] Chambers, RD, Fox, MA, Holling, D, Nakano, T, Okazoeb, T, Sandford, G, Elemental fluorine. Part 16. Versatile thin-film gas–liquid multi-channel microreactors for effective scale-out, Lab Chip, 2005, 5, 191–198.

[35] Coyle, EE, Oelgemöller, M, Micro-photochemistry: photochemistry in microstructured reactors. The new photochemistry of the future?, Photochem Photobio S, 2008, 7, 1313–1322.

[36] Morschhäuser, R, Krull, M, Kayser, C, et al., Microwave-assisted continuous flow synthesis on industrial scale, Green Process Synth, 2012, 1, 281–290.

[37] C-Wave, PRD. (Accessed December 4, 2020, at https://www.ctechinnovation.com/product/c-wave-prd/.)

[38] Oelgemöller, M, Shvydkiv, O, Recent advances in microflow photochemistry, Molecules, 2011, 16, 7522–7550.

[39] Lévesque, F, Seeberger, PH, Highly efficient continuous flow reactions using singlet oxygen as a "green" reagent, Org Lett, 2011, 13, 5008–5011.

[40] Mihsein, M, Seymour, MD, Intermixing twin ferrule tube fittings – Is it safe?, J Eng Design, 1999, 10, 391–416.

[41] Samw, VDB, Becker, R, Kössl, F, et al., Prilezhaev dihydroxylation of olefins in a continuous flow process, ChemSusChem, 2011, 5, 289–292.

[42] Van Den Broek, SAMW, Leliveld, JR, Becker, R, et al., Continuous flow production of thermally unstableintermediates in a microreactor with line IR-analysis: Controlled Vilsmeier-Haack formylation of electron-rich arenes, Org Process Res Dev, 2012, 16, 934–938.

[43] Van Gool, JJF, Van Den Broek, SAMW, Ripken, RM, Nieuwland, PJ, Koch, K, Rutjes, FPJT, Highly controlled gas/liquid processes in a continuous lab-scale device, Chem Eng Technol, 2013, 36, 1042–1046.

[44] Bagley, MC, Fusillo, V, Jenkins, RL, Lubinu, MC, Mason, C, Continuous flow processing from microreactors to mesoscale: the Bohlmann–Rahtz cyclodehydration reaction, Org Biomol Chem, 2010, 8, 2245–2251.

[45]. Hodgkinson, JT, Galloway, WRJD, Saraf, S, et al., Microwave and flow syntheses of Pseudomonas quinolone signal (PQS) and analogues, Org Biomol Chem, 2011, 9, 57–61.

[46] Battilocchio, C, Baxendale, IR, Biava, M, Kitching, MO, Ley, SV, A flow-based synthesis of 2-aminoadamantane-2-carboxylic acid, Org Process Res Dev, 2012, 16, 798–810.

[47] Pieber, B, Martinez, ST, Cantillo, D, Kappe, CO, In situ generation of diimide from hydrazine and oxygen: Continuous-flow transfer hydrogenation of olefins, Angew Chem Int Ed, 2013, 52, 10241–10244.

[48] Browne, DL, Baxendale, IR, Ley, SV, Piecing together the puzzle: understanding a mild, metal free reduction method for the large scale synthesis of hydrazines, Tetrahedron, 2011, 67, 10296–10303.

[49] O'Brien, M, Taylor, N, Polyzos, AS, Baxendale, IR, Ley, SV, Hydrogenation in flow: Homogeneous and heterogeneous catalysis using Teflon AF-2400 to effect gas–liquid contact at elevated pressure, Chem Sci, 2011, 1250–1257.

[50] Mastronardi, F, Gutmann, B, Kappe, CO, Continuous flow generation and reactions of anhydrous diazomethane using a Teflon AF-2400 tube-in-tube reactor, Org Process Res Dev, 2013, 15, 5590–5593.

[51] Thompson, CM, Poole, JL, Cross, JL, Akritopoulou-Zanze, I, Djuric, SW, Small molecule library synthesis using segmented flow, Molecules, 2011, 16, 9161–9177.

[52] Guetzoyan, L, Nikbin, N, Baxendale, IR, Ley, SV, Flow chemistry synthesis of zolpidem, alpidem and other GABAA agonists and their biological evaluation through the use of in-line frontal affinity chromatography, Chem Sci, 2013, 4, 764–769.

[53] Pastre, JC, Browne, DL, O'Brien, M, Ley, SV, Scaling up of continuous flow processes with gases using a tube-in-tube reactor: Inline titrations and fanetizole synthesis with ammonia, Org Process Res Dev, 2013, 17, 1183–1191.

[54] Newby, JA, Huck, L, Blaylock, DW, Witt, PM, Ley, SV, Browne, DL, Investigation of a lithium–halogen exchange flow process for the preparation of boronates by using a cryo-flow reactor, Chem Eur J, 2014, 20, 263–271.

[55] Mehenni, H, Sinatra, L, Mahfouz, R, Katsiev, K, Bakr, OM, Rapid continuous flow synthesis of high-quality silver nanocubes and nanospheres, RSC Adv, 2013, 3, 22397–22403.

[56] Pan, J, El-Ballouli, AO, Rollny, L, et al., Automated synthesis of photovoltaic-quality colloidal quantum dots using separate nucleation and growth stages, ACS Nano, 2013, 7, 10158–10166.

[57] Patwardhan, SV, Perry, CC, Synthesis of enzyme and quantum dot in silica by combining continuous flow and bioinspired routes, Silicon, 2010, 2, 33–39.

Kian Donnelly, Mara Di Filippo, Cormac Bracken
and Marcus Baumann

4 Practical aspects of performing continuous flow chemistry

4.1 Preface and general outline including learning objectives

Over the last two decades the advent of new enabling technologies such as continuous flow synthesis has revolutionized the way chemists make molecules of interest. While initially seen as an academic curiosity, the many advantages offered by running a chemical reaction continuously in a miniaturized flow setup rather than traditional glassware (round-bottomed flask, beaker, condenser, etc.) soon attracted the attention of the entire chemical community. This has led to the development of numerous flow-based processes from both academic and industrial laboratories. The value of flow chemistry in industrial settings has been a key driver within the last ten years resulting not only in numerous case studies describing continuous approaches on kilogram scale but moreover the manufacture of several active pharmaceutical ingredients (APIs) in a continuous manner [1–3]. Besides the intrinsic advantages of flow chemistry over batch processing, such as improved heat and mass transfer, additional benefits such as in-line analysis and purification techniques embedded in telescoped multistep sequences, the safe generation and use of hazardous reagents and intermediates as well as variable temperatures and pressures that in combination with automated protocols impart superb reaction control are commonly cited. Of interest for industrial applications are furthermore the readily scalable nature of continuous processes, the small footprint of reactors, the comparatively low investment cost in equipment, and its compatibility with existing infrastructure.

While there is no doubt that there are numerous advantages of using and developing continuous processes, the training of future chemists and chemical engineers to use and advance this technology has become the focus point within recent years [4]. To aid in this process, this chapter will specifically discuss the practical considerations needed to understand and subsequently apply continuous flow processing to the synthesis of organic entities. The key learning objectives of this chapter are therefore:

- to understand the key parameters for performing continuous flow reactions and their effect on optimizing these,
- to appreciate the diversity and applicability of commercially available flow reactors as well as basic home-built equivalents,
- to gain an understanding on which transformations benefit from continuous processing and how to develop an effective flow process from start to finish,

https://doi.org/10.1515/9783110693676-004

- to learn specific features and approaches from multiple published case studies showcasing the breath of chemistries that are precedented in continuous flow mode.

4.2 On the choice of flow equipment: commercial, home-made

While flow chemistry is often associated with modular components that are arranged in a manner so as to render a setup most suitable for a given process, at first it may be difficult to appreciate the reasons for choosing a specific setup. In general, there are two options: (a) using commercial flow reactors, or (b) building a bespoke flow setup.

A commercial flow reactor allows using a purpose-built unit that enables the operator to directly start working with this system. Today's widespread availability of such small foot-print reactors from various vendors provides compatibility, user-friendliness, and competitive pricing. These ready-to-use systems furthermore include all key components (pumps, mixers, reactors, and appropriate software control) that are easily interconnected and complemented with additional devices (back-pressure regulators, in-line analysis tools, etc.). In addition to quickly starting on running a flow reaction, these commercial systems also provide higher reproducibility and purchase often involves training and service contracts through the vendor.

In contrast, many first-time users may prefer building a flow setup from individual components in accordance with a specific application in mind. While this approach offers great flexibility and oftentimes at lower cost, the operator must choose from various types of pumps, connectors, mixers, reactor materials and so on to realize a setup suitable for the intended application. In recent years this has been complemented by the advent of 3D-printing that can produce specific components made of various polymers and more recently metals, that can be accessed quickly and often at low material cost. Consequently, an increasing number of reports discuss 3D-printed components including their limitations such as chemical compatibility and robustness toward high pressures and temperatures [5, 6]. As such, this approach is attractive especially in early stages of development where disposable components are acceptable to provide quick insight into the feasibility of performing a given transformation in flow mode.

For both options (commercial and home-build systems), researchers can select from specific components, for instance pumps (piston, peristaltic or syringe pumps). Piston pumps are very commonly used in high-performance liquid chromatography (HPLC) equipment and while they tolerate high pressures and offer a good flow rate range (typically up to 10 mL min^{-1}), they do not tolerate insoluble materials as these lead to inaccurate flow rates and eventually blockages. Peristaltic pumps can handle light slurries and offer a good range of flow rates, but often at limited maximum pressures and with significant pulsing. Lastly, syringe pumps are often available in research labs and provide very accurate flow rates especially at lower flow rates (μL min^{-1}); however, they are less tolerant toward high pressures and have limited capacity by volume, unless used in a dual fashion (e.g., one syringe is recharged while the other one delivers the solution). In a similar fashion, different tubing materials are available (typical diameters 1/16–1/4 inch) for both the reactor unit as well as connecting segments. These include fluorinated polymers (PFA, FEP, polytetrafluoroethylene (PTFE)), polyether ether ketone (PEEK), metals (stainless steel, copper, etc.) or glass, and the selection will be determined based on application, compatibility (corrosive chemicals, temperature, and pressure), and cost.

While the availability of different flow reactor components as indicated in the previous section offers great flexibility and modularity, the operator thus will need to develop an understanding on when to choose a particular option which commonly will be informed on precedent from previous reports as well as training via more experienced researchers.

(For further details on this issue please see Volume 1, Chapter 3, Title: Technology overview/overview of the devices.) i

---

4.3 Important variables in flow processing and their impact

Besides the nature of equipment to choose from, researchers must have a good grasp on several important concepts and their implications on continuous flow processes when embarking on applying flow chemistry.

A central term is **"residence time"'** (t_{Res}), which in general refers to the time the reaction mixture spends in the flow reactor (coil reactor, chip reactor, etc.). The residence time can therefore be calculated by dividing the volume of the reactor (in mL) by the flow rate (in mL min^{-1}). In case of flow reactions in which individual streams of solutions are mixed before entering a reactor, the total reactor volume must be divided by the combined flow rate, which is the sum of all individual flow rates (see example in Fig. 4.1). Individual flow rates are often chosen to be similar to each other for practical reasons, for example, processing similar volumes of solutions and mitigating pump inaccuracies.

Furthermore, for telescoped multistep flow sequences, the residence time in the last reactor (e.g., step 2) will be affected by the combined flow rates of the prior process (e.g., step 1). This means that a slow second step is not easily accommodated after a rapid first step (high flow rate) unless the reactor volumes are chosen appropriately (Fig. 4.2; e.g., small volume for first reactor and large volume for second reactor).

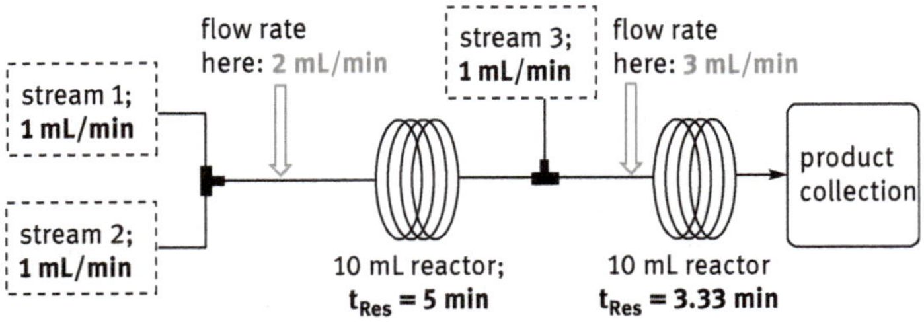

Fig. 4.1: Calculating residence times (t_{Res}) for simple flow sequences.

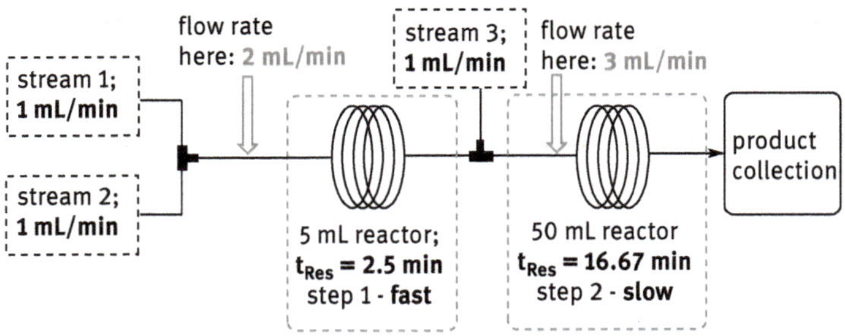

Fig. 4.2: Interdependence of flow rates, reactor volumes as t_{Res} in telescoped sequences.

Flow reactions are often characterized by relatively short residence times (<1 h) when compared to equivalent batch transformations (several hours) due to faster kinetics resulting from better **mass transfer** (e.g., mixing of reactants and reagents/catalysts). However, the total processing time needed to pass a larger volume of substrate solution through the reactor can still be in the order of hours, especially during longer runs for scale-up. The overall time for running a flow reaction therefore depends on both the residence time (often short) and the scale of the reaction (variable), whereas in batch the reaction time is equal to the residence time (e.g., time within the reaction vessel) and thus independent of scale. In cases of flow reactions occurring within packed-bed reactors (e.g., glass columns) filled with heterogeneous catalysts or reagents, the term "contact time" is sometimes used highlighting that the reaction only takes place upon contact of the substrate with the reagent/catalyst in the column. In these cases, faster rates are often observed as the substrate molecules experience an excess of reagent/catalyst while passing through a packed-bed column.

To benefit from faster reaction kinetics, flow reactions make use of different **mixers** that are usually small-diameter T- or Y-pieces or static mixers placed within the flow path. The aim of these devices is to generate significant turbulence to homogenize different streams of reactants and reagents in favor of faster reactions.

In addition, the diameter of the tubing will determine whether a laminar or turbulent flow regime is prevalent, which can be characterized by the Reynold's number of the system. Further mixing can be achieved through serpentine channels as well as patterned channel surfaces as often exploited in microchips. Consequently, better mixing in flow regimes is independent of reaction scale, whereas batch processes suffer from ineffective mixing as the dimension of the reaction vessel (and thus the scale) increases.

Once mixing of different flow streams is achieved, the combined reaction mixture enters the actual reactor in which the chemical reaction occurs under various stimuli, e.g., heat, irradiation, or sonication. Here, the superb **"heat transfer"** of flow reactions results in manifold benefits. For instance, in case of **cryogenic reactions** whose exothermic nature requires cooling, the large surface-to-volume ratio of flow reactors (e.g., reactor coil or microchip) allows for effective dissipation of the heat released. This does not only increase safety and control over the flow process, but moreover allows "higher" temperatures in many reported cases (e.g., −78 °C in batch lithiations, but −30 °C in flow), which affects the energy consumption of the process. Conversely, in case of endergonic reactions that require heating, flow processes are advantageous as only as small reactor volume must be heated. Effective heat input is again achieved because of the high surface-to-volume ratio. In addition, the use of back-pressure regulators allows to use and superheat "low-boiling" solvents (MeCN, THF, etc.) at temperatures far exceeding their atmospheric boiling point yet remaining in the liquid phase (e.g., >150 °C) (Tab. 4.1). This prevents the use of unfavorable high-boiling solvents (xylene, dichlorobenzene, NMP, etc.) in favor of "greener" solvents that are easier to separate after the reaction.

Tab. 4.1: Examples of calculated boiling points (bp) for pressurized solvents.

Solvent	bp at 1 bar	bp at 10 bar	bp at 25 bar
MeCN	82 °C	149 °C	183 °C
THF	66 °C	126 °C	157 °C

Upon assembly of a suitable reactor configuration and setting of all parameters, a continuous flow process is started by the experimentalist. Due to the various variables (flow rate, temperature, pressure) that affect the flow process it is common to allow the flow system to reach **steady state** before collecting the reaction product (e.g., by pumping 1–2 reactor volumes of substrate solutions through the system). This allows the experimentalist to ensure that the system has stabilized and is likely to run continuously without fouling. **In-line analysis tools** in combination with suitable software are available to capture real-time data and report any issues to the experimentalist who can quickly intervene if necessary.

While all the above principles and concepts demonstrate how flow reactions are potentially superior to analogous batch processes, experimentalists must always pay close attention to **solubility issues in flow processes**. While heterogeneous materials are not a major problem in batch reactions, insoluble starting materials or precipitation of products will lead to fouling within flow reactors due to blockages in microchannels, connectors, and back-pressure regulators. Inappropriately high concentration of solutions is thus a common challenge in flow reactions; however, solutions that avoid clogging and fouling are available to experienced experimentalists (e.g., exploiting peristaltic pumps, agitation, optimized solvents, and temperatures).

i *(For further details on this issue please see Volume 1, Chapter 1, Title: Fundamentals of flow chemistry.)*

4.4 How to approach running flow reactions – from inception to execution

Flow chemistry is recognized as an attractive and potentially superior alternative to traditional batch processing, and most researchers nowadays are well versed in the associated benefits. However, most practitioners will agree that a number of pitfalls are commonly encountered by first-time users and as this chapter aims to highlight successful case studies that cover a wide range of potential scenarios, we wish to also point out common challenges and sources of mistakes. Crucially, to realize a robust flow process, a solid knowledge on the published literature must meet an open-minded experimentalist. Creativity in combination with suitable experimentation can often lead to the identification of gaps and subsequently the generation of new tools that enable a better continuous process. Once initial considerations have resulted in a provisional approach, experimentation will guide the further steps. Oftentimes, researchers are conducting initial experiments in batch to evaluate solubility, reactivity, and stoichiometry for a given reaction before attempting this in continuous flow mode. It is also advisable to break down multistep sequences and integrate in-line purification aids as the process develops. Crucially, the chemist must keep in mind all parameters that will affect downstream operations (quenching, additional transformations, purification, etc.) in order to ascertain compatibility of the intermediate material (solvents, spent reagents, side-products) with the subsequent reagents as well as increasing dilution and flow rates. This multivariable scenario leads to rapidly expanding complexity if an innovative multistep sequence is targeted. The following selection of case studies presents the reader with the thinking process by highlighting the path to successful flow processes.

4.5 Case study – general solution phase chemistry in flow

Background: Homogeneous flow reactions represent the most common and simple scenario; however, it is crucial to assess the solubility of all components (substrates, intermediates, products [incl. side products]) within the relevant temperature window. Consequently, optimization of solvent choice, concentration, and temperature are important first steps. A recent report from the Charette group [7] demonstrates this through the in situ generation of difluorocarbene from the commercial TMS-CF_3 reagent to prepare a set of valuable difluorocyclopropanes and difluorocyclopropenes through flow-based cycloadditions [8].

Flow process: In this application the difluorocyclopropanation of 4-*tert*-butyl styrene was initially investigated as a model reaction in batch mode, revealing that full conversion can be obtained within 2 h by heating the substrate solution (THF) in the presence of TMS-CF_3 and the initiator (NaI) under reflux.

Flow design and optimization: Subsequently, the reaction was optimized in continuous flow in view of decreasing reaction times and improving productivity. The commercially available Vapourtec R-series reactor was used. The reactor setup consisted of a HPLC pump, a six-way valve with injection loop (5 mL), a 10 mL PFA heated coil reactor, and a back-pressure regulator (8 bar). This closed system allowed for dosing of reagents and solvents without exposing them to potential moisture which could lead to hydrolysis of the difluorocarbene source, minimizing the need for excessive quantities of TMS-CF_3. Solvent optimization in flow indicated that, as was the case in batch, THF was the most suitable. It emerged that the choice of initiator was crucial, with only NaI initiating the formation of difluorocarbene.

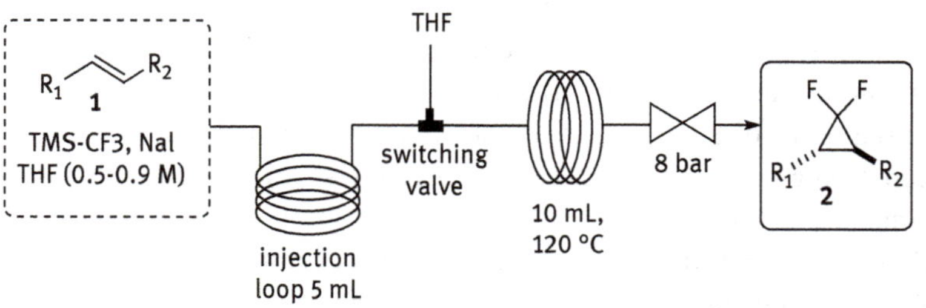

Fig. 4.3: Continuous flow setup for difluorocyclopropane synthesis.

Optimization of the reaction temperature showed that superheating of the THF solution to 120 °C significantly accelerated the reaction, whereas further increase (to 150–200 °C) favored decomposition. The final optimized conditions involved using a premixed solution consisting of substrate, TMS-CF_3 (2.0 equiv.) and NaI (0.1 equiv.) in THF (0.9 M). This solution was pumped through the reactor at 120 °C, at 8 bar back-pressure (allowing for superheating) and a residence time of 10 min, yielding the desired product in an isolated yield of 94%.

Substrate scope: Using these reaction conditions, a variety of alkene substrates were tested. It was observed that various electron-rich and electron-poor styrene derivatives were well tolerated, with yields greater 80%. Simpler alkenes, such as cyclooctene and allenes, were also tolerated, but a reduction in yield was observed with less nucleophilic alkenes. The substrate scope was then successfully expanded to alkynes which are known to undergo the analogous cyclopropenation reaction (Fig. 4.4).

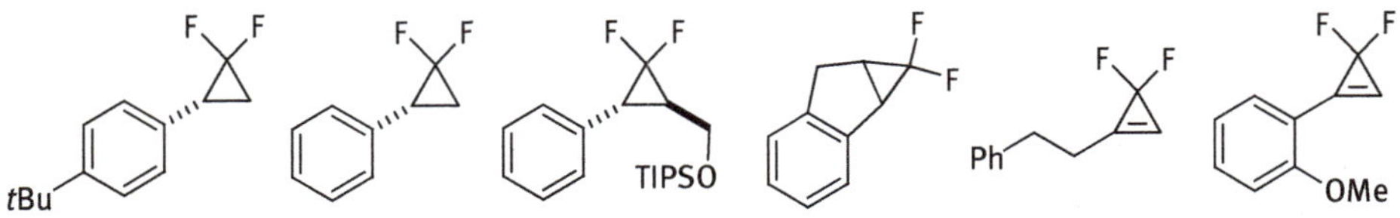

2a, 94% **2b**, 91% **2c**, 76% **2d**, 62% **2e**, 83% **2f**, 52%

Fig. 4.4: Selected substrate scope and scale-up example (**2c**).

A significant advantage of carrying out a reaction in continuous flow is the ease at which scale up can be carried out. The authors demonstrated the scalability of this process by synthesizing difluorocyclopropane **2c** on multigram scale. The conditions were modified slightly for scale-up with a larger quantity of TMS-CF_3 (2.5 equiv.) and a smaller quantity of NaI (0.06 equiv.) being used to reach a productivity of 27.4 mmol h^{-1}.

Experimental procedure: (2,2-Difluorocyclopropyl)benzene **2b** – in an oven-dried flask under argon styrene (3.0 mmol, 1.0 equiv.), sodium iodide (0.3 mmol, 0.1 equiv.) and TMS-CF_3 (6.0 mmol, 2.0 equiv.) were premixed in freshly distilled THF (3.5 mL, 0.9 M). Using the Vapourtec R-Series flow system, the reaction mixture was injected into a 5 mL loop and pumped into a 10 mL reactor preheated at 120 °C followed by a back-pressure regulator of 8 bar. The reaction was run with a residence time of 10 min (1 mL min^{-1} flow rate) and freshly distilled THF was used as the system solvent. The crude mixture was collected in an opened flask. Water was added and then the crude was diluted with hexanes.The biphasic mixture was transferred to a separation funnel. After extraction the combined organic layers were dried over anhydrous Na_2SO_4, filtered and evaporated to yield the crude product that was purified by flash chromatography over silica gel using hexanes (100%) as eluent, affording (2,2-difluorocyclopropyl)benzene (**2b**, 422 mg, 91% yield) as a colorless liquid.

4.6 Case study – exploiting gases in flow – aerobic oxidation reactions

Background: The chemoselective oxidation of alcohols to aldehydes and ketones is an important strategy to access these versatile building blocks. Specifically, the exploitation of benign oxidants such as oxygen gas in combination with suitable catalysts is industrially relevant as it provides a scalable means due to high atom economy and low process mass intensity (PMI). However, the resulting biphasic (gas/liquid) system presents safety concerns in batch mode due to the involvement of flammable solvents and the necessity to pressurize the vessel. Therefore, a recent report from the Stahl group [9] outlines the benefits of safely exploiting gases in continuous flow mode [10, 11]. This study therefore highlights an advantageous flow approach for safe and scalable aerobic oxidations.

Flow process: Although the authors had previously demonstrated a successful Pd-based approach for aerobic continuous flow oxidations of alcohols [12], this new protocol favored copper as catalyst due to potentially faster reactions, lower cost, lower toxicity, and, importantly, higher substrate tolerance (incl. heterocycles bearing nitrogen and sulfur atoms). The chosen catalyst system was utilizing Cu(OTf) (5 mol%), TEMPO (5 mol%), *N*-methylimidazole (NMI, 10 mol%) and bipyridine (bpy, 5 mol%) in combination with acetonitrile as solvent and a dilute oxygen-nitrogen mixture (9% O_2 in N_2). Initial batch tests focused on activated (benzylic, propargylic) and non-activated (aliphatic) alcohols exploiting different temperatures and pressures of oxygen gas and revealed suitable conditions based on 35 bar pressure and temperatures of 60–100 °C to give high conversion of substrates.

The resulting flow process was exploiting a slug flow regime in which interspersing gas and liquid segments pass through a heated tubular reactor (stainless steel or PTFE) prior to a set of vapor–liquid separator and expansion chamber units. The gas was vented into the fume hood while the solution was collected in a vessel after passing a backpressure regulator (Fig. 4.5).

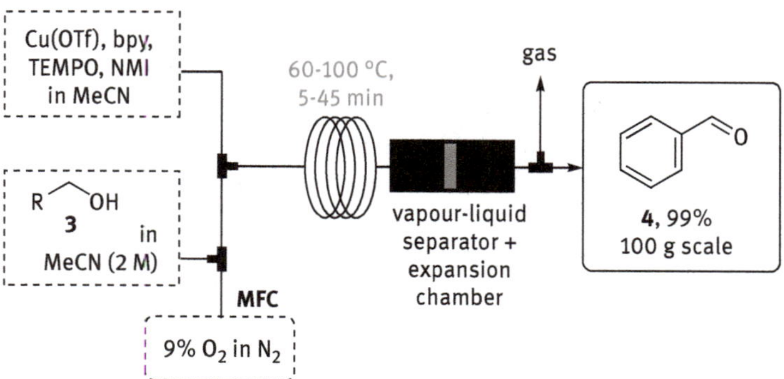

Fig. 4.5: Flow setup for aerobic oxidation of alcohols.

Further attention was given to preparing the stock solutions where initial tests revealed best outcomes when the alcohol substrate (in MeCN, stream A) was kept separate from the homogeneous catalyst mixture (in MeCN, stream B). Furthermore, appreciable deactivation of the Cu(OTf) catalyst was observed when using stainless steel syringes. This was avoided by using an HPLC pump in combination with inert PTFE tubing for this stream while keeping the second stream (substrate in MeCN) in a stainless-steel syringe as before. Upon completing these prior studies, the flow process was optimized for activated and nonactivated substrates. A short residence time of 5 min in combination with elevated temperature (100 °C) and a pressure of 35 bar provided high yields for several activated substrates (Fig. 4.6), whereas non-activated substrates showed significant amounts of decomposition (e.g., overoxidation) under these conditions. This was resolved by lowering the temperature (60 °C) and extending the residence time (30–45 min).

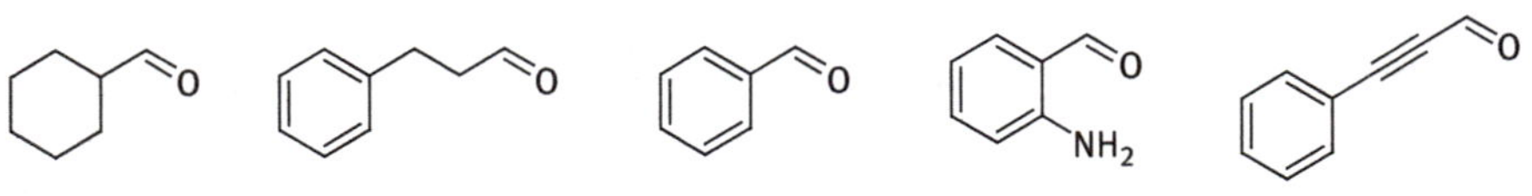

4a, 95% (45 min) | **4b**, 99% (30 min) | **4c**, 99% (5 min) | **4d**, 99% (5 min) | **4e**, 99% (5 min)

Fig. 4.6: Product scope and yields of flow process.

Lastly, reaction scale-up was successfully performed on 100 g of benzyl alcohol after re-evaluation of the catalyst loading (to 2.5 mol% Cu(OTf) and bpy, 5 mol% NMI and 0.25 mol% TEMPO in combination with 4 M substrate in MeCN). This resulted in a robust process operated for over 24 h in which a short residence time of 5 min ascertained near quantitative yield of the benzaldehyde product.

Experimental procedure: A stock solution of the alcohol substrate (8.4 mmol) was prepared in MeCN (20 mL, containing mesitylene as internal standard) and transferred to a syringe pump (260 mL, Teledyne ISCO 260D). The catalyst solution comprised [Cu(CH_3CN)$_4$]OTf (158 mg, 0.42 mmol, 5 mol%), bpy (66 mg, 0.42 mmol, 5 mol%), TEMPO (66 mg, 0.42 mmol, 5 mol%), and NMI (67 µL, 0.84 mmol, 10 mol%) dissolved in MeCN (23 mL) and was pumped via a HPLC pump (Hitachi L6000). A stainless-steel gas reservoir was filled with O_2 (8.7 bar) then by N_2 (79.5 bar) to afford an 88.2 bar mixture of 9% O_2 in N_2. The gas was delivered via a mass flow controller (MFC) that allowed for an O_2 to substrate molar ratio of 2:1. The feeds for the biphasic system mixed in a 1/16 inch T-piece mixer before entering the 1/8 inch tube reactor (60 or 100 °C, submerged in a heat transfer fluid). Aliquots of product solution (100–500 µL) were repeatedly removed from a small tee for analysis, and remaining liquid and gas were separated, with liquids aggregating at the bottom of the reactor, and gas being vented from the top using a pressure relief valve.

4.7 Case study – controlled precipitation via Coflore ATR flow reactor

Background: Most commonly flow synthesis is applied to homogeneous reactions as precipitation of solids can lead to fouling and clogging of reactors. Consequently, suitable precautions (solvent choice, concentration, temperature) are paramount to retain a homogeneous system [13]; however, this will impact the productivity of a flow process due to higher dilutions and thus slower kinetics. Being able to tolerate and exploit slurry-based flow processes can thus have a profound effect on reaction throughput, while minimizing solvent usage and downstream processing as the desired product can be isolated by simple filtration. A recent example by the Baxendale group [14] demonstrates the use of a Coflore ATR flow system to produce kg quantities of pyranone **6** from dehydroacetic acid **5**, an important biorenewable chemical building block (Fig. 4.7).

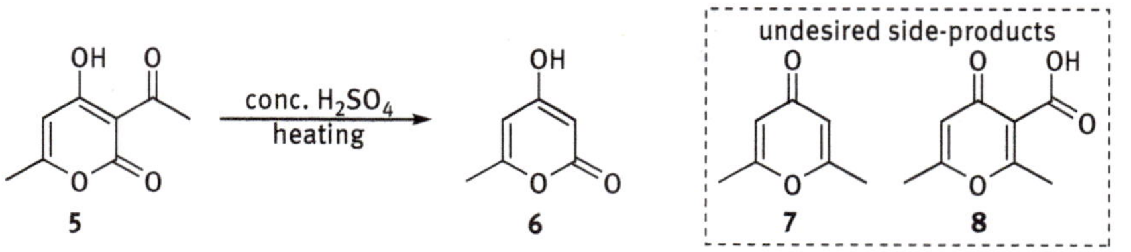

Fig. 4.7: Synthesis of pyranone **6** from dehydroacetic acid **5**.

Flow process: Initial batch reactions had revealed that the conversion of dehydroacetic acid to pyranone **6** can be accomplished by treatment with hot concentrated sulfuric acid (>110 °C) to promote the deacetylation process. While this worked well at small scale (<10 g, 85–88% yield), issues arose when scaling this process. These included the exothermicity of the aqueous quenching step during which the desired product would precipitate. Consequently, lower yields were observed along with side products (**7** and **8**). Additionally, discoloration of the solid product and an unfavorable physical appearance (sticky solid) were noted during batch scale-up. To overcome these limitations a flow synthesis was developed to affect the deacetylation and integrate the product precipitation through the aqueous quench. Flow was thus expected to provide for superior heat transfer, safe handling of hot conc. sulfuric acid and reproducible performance upon scale-up.

Specifically, the flow process was realized using a Vapourtec E-series reactor module for delivering two viscous solutions of substrate **5** (2 M, in concentrated H_2SO_4) into separate heated flow coils (Polar Bear Plus, 52 mL each, 130 °C, 13 min residence time) to affect the deacetylation process (Fig. 4.8). The two exiting streams (4 mL min^{-1} each) were combined in a T-piece and then united with a quench stream of water (60 mL min^{-1}, approximately 10 °C) at the entrance of a Coflore ATR reactor (internal volume 0.1–1.0 L). The Coflore ATR reactor [15] exploits internal static mixers to provide excellent mixing aided by constant transverse agitation at a specific frequency (here 4 Hz) to prevent sedimentation of the solid product formed. An external cooling system assured a constant temperature of 10 °C to facilitate heat transfer and realize excellent product yield (92%). The exiting product suspension was then collected on a sintered funnel under suction rendering the target product as an off-white free-flowing solid after drying. Further optimization toward increased throughput establish a process in which 250 g of **6** were produced per hour by doubling the flow rates and halving the residence time within the Coflore ATR reactor (1.5 min to 44 s) while maintaining high yield (94%) and excellent product purity.

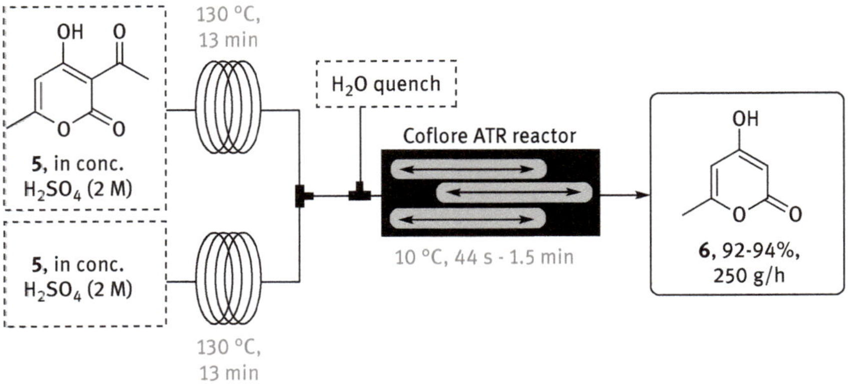

Fig. 4.8: Continuous flow synthesis of **6** via a Coflore ATR reactor module.

Experimental procedure: For the flow procedure, two 2 M solutions of 3-acetyl-4-hydroxy-6-methyl-2 H-pyran-2-one (**5**) were pumped at a flow rate of 4 mL min^{-1} into two 52 mL parallel coil reactors (13 min residence time) maintained at 130 °C. The two outflows were combined and mixed with a stream of water (60 mL min^{-1}, 10 °C) prior to entering one of the 10 channels of the 1 L Coflore ATR reactor chamber. The chamber was agitated at a frequency of 4 Hz (1.5 min residence time). The ATR chamber was cooled to 10 °C by fitting it with a Jubalo recirculating chiller. The emerging suspension from the ATR reactor was then collected onto a sintered filter (under suction) rendering a pale-yellow solid after drying in a vacuum oven to provide the target **6** in 92% yield.

4.8 Case study – high-temperature Gould–Jacobs quinolone synthesis in flow

Background: Heteroaromatic rings are a common feature of modern drugs, whose synthesis allows for variation of both the core and the peripheral substituents. As the cyclization process typically requires overcoming a high activation barrier, forcing conditions are necessary. In batch mode this leads to long reaction times that can favor side-product formation. Additionally, unattractive high-boiling solvents (DMF, NMP, dichlorobenzene) are needed. A recent report by researchers from Abbvie [16] highlights the benefits of continuous flow processing in safely accessing high temperatures (>200 °C) to affect the synthesis of various quinolones and pyrimidones via the Gould–Jacobs reaction [17]. This reaction involves the condensation of anilines with alkoxymethylene compounds (**10**) proceeding via various intermediates, that then yield the desired heterocyclic product (**16**, Fig. 4.9).

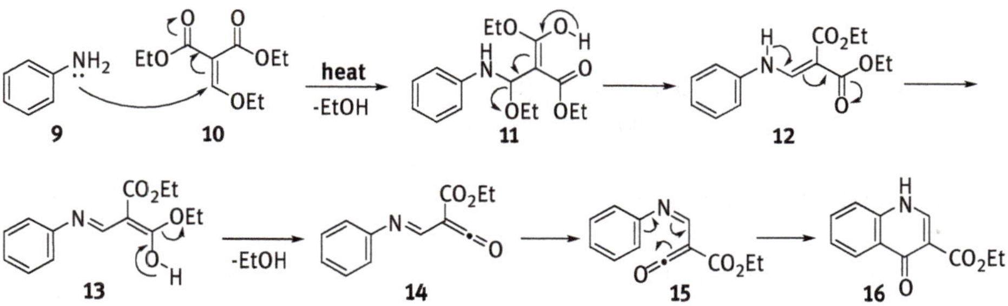

Fig. 4.9: Plausible Gould–Jacobs cyclization mechanism.

Flow process: Continuous flow technology is routinely applied to reactions which require high temperatures and pressures as solvents need to be safely heated to near- or supercritical temperatures [18, 19]. This case study exploits automation and fraction collection systems in combination with a Phoenix™ flow reactor that is dedicated to operating under high temperature and pressure. The continuous flow setup consisted of a HPLC pump (JASCO PU-2085 plus), a backpressure regulator (JASCO BP-2080 Plus) a Tecan MiniPrep Automated Sample Processor with a 10-port Valco Injector valve (with a pair of 5 mL sample loops) and a stainless steel tubing reactor coil (2 mL volume, 1.0 mm inner diameter) housed within the modified Phoenix reactor (Fig. 4.10).

Flow design and optimization: Aided by Design-of-Experiments (DoE) software (Stat-Ease Design Expert 7) a series of reactions based on alkylidene β-ester **10** were designed with key parameters such as reaction temperature (250–400 °C), concentration (0.05–0.4 M in THF), and flow rate (0.5–4.0 mL min^{-1}) being screened. In addition to the desired Gould–Jacobs-type cyclization, the quantity of decarboxylated side-

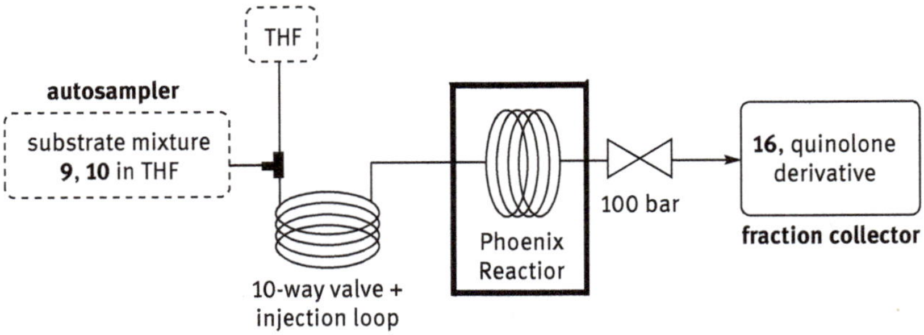

Fig. 4.10: Schematic setup of the continuous flow setup.

products was also monitored. From the DoE contour plots it was noted that elevated temperatures, higher concentrations and shorter residence times were required for high yields, whereas the formation of decarboxylated side products was only favored at longer residence times. Based on the data collected, the optimal conditions included a temperature of 390 °C with a flow rate of 4.0 mL min^{-1} (t_{Res} = 0.5 min) and a reaction concentration of 0.4 M.

Substrate scope: The optimized conditions were applied to the synthesis of a range of both quinolone and pyrimidone containing products, with the latter giving higher yields with higher substrate tolerance. Interestingly the alkoxy methylene-malonic ester (**10**) could be replaced with condensation products of β-keto esters and cyanoacetic acid esters while maintaining high yields. The synthesis of quinolones by comparison gave lower yields, but still tolerated a range of substitution patterns. Based on what was observed during reaction optimization, the selective synthesis of decarboxylated products could be achieved by decreasing the flow rate (4.0–0.5 mL min^{-1}) and the reaction concentration (0.4–0.05 M) while increasing the pressure (100–120 bar). The addition of a small quantity of water (ca. 1 vol%) was found to further favor the decarboxylated product, which combined with the lower concentration suggests a bimolecular reaction.

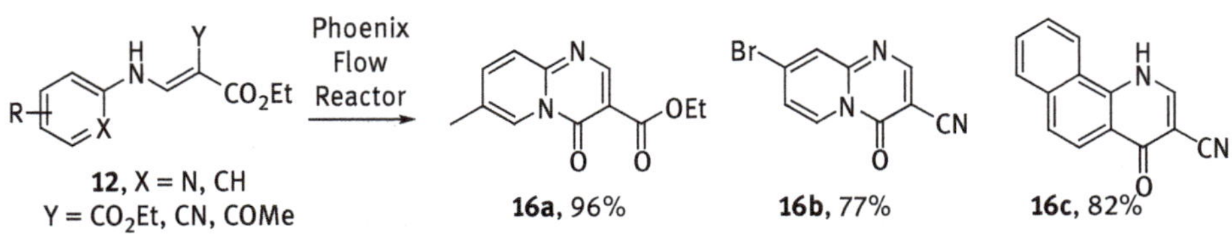

Fig. 4.11: Selected examples of Gould–Jacobs cyclization reaction.

One of the primary advantages of carrying out reactions in continuous flow is the ease at which reactions can be scaled up. In contrast to batch mode, processes can be run for longer periods of time while maintaining constant reaction conditions. To demonstrate the scalability of this reaction, 18 mL of stock solution were prepared and after 4.5 min of collection isolated to give 1.36 g of product in 82% yield equating to a productivity of 18.2 g h^{-1}.

Experimental procedure: Ethyl 7-methyl-4-oxo-4 *H*-pyrido[1,2-a]pyrimidine-3-carboxylate – a stock solution of diethyl 2-(((5-methylpyridin-2-yl)amino)methylene)malonate (56 mg, 0.2 mmol) was prepared in THF (500 µL, 0.4 M) and injected into the Phoenix reactor using an autosampler. A flow rate of 4.0 mL min^{-1} was used in combination with a set temperature of 390 °C. The crude reaction mixture was collected after passing BPR (100 bar), concentrated, and purified by trituration with ethyl acetate. The product was isolated as an off-white powder in 96% yield (45 mg).

4.9 Case study – low temperature in flow: terminal epoxides from ketones

Background: The improved heat transfer offered in flow processes results from high surface-to-volume ratio and allows to effectively and safely perform highly exothermic reactions that in batch require cryogenic conditions (below −78 °C). This opened the field of flash chemistry [20] in which miniaturization of flow mixers provides superb reaction control that enables fast transformations to be exploited safely and in a scalable manner. The Kappe group recently exploited a flash chemistry approach for the synthesis of terminal epoxides from ketones using bromomethyl lithium (Fig. 4.12) [21]. Bromomethyl lithium is a useful C-1 source that can be synthesized via lithium halogen exchange of readily available dibromomethane; however, it is a highly unstable reagent even at temperatures as low as −120 °C and therefore must be generated in situ [22]. Batch scale-up can prove extremely challenging due to the requirement to add the lithium base very slowly and the potential for various side-reactions.

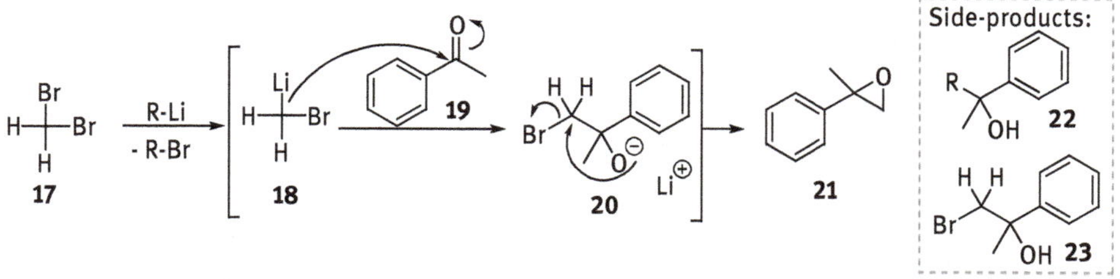

Fig. 4.12: Epoxidation of acetophenone.

Flow process: Initial batch experiments studied the epoxidation of acetophenone (**19**) as a model reaction for the screening of organolithium reagents. As expected, the main side-products observed in this case were based on addition of the organolithium species to the substrate carbonyl (**22**), and the non-cyclized product (**23**). While bases like LDA and LiHMDS favored deprotonation of dibromomethane over lithium halogen exchange, alternatives such as ^{n}BuLi and MeLi proved more promising, with MeLi being more selective than the more nucleophilic ^{n}BuLi. Interestingly, the use of MeLi–LiBr complex resulted in even higher selectivity. This was attributed to the stabilization of CH_2BrLi offered by the LiBr salt, increasing its half-life in solution.

With the optimal base selected, the reaction was transferred to flow. The initial flow reactor, constructed using PFA tubing (0.8 mm i.d.) and T-piece mixers, consisted of four streams: substrate (THF, 0.5 M) and dibromomethane (THF, 0.6 M) which were combined before being mixed with MeLi–LiBr (1.5 M in Et_2O) which was diluted using a feed of THF. As upon cooling the MeLi–LiBr complex was found to precipitate at this concentration (1.5 M) dilution with further THF was necessary and a suitable concentration of 0.2 M (at −80 °C) was found. With this knowledge in hand, the system was simplified to two separate streams: (a) MeLi–LiBr (1.5 equiv., 0.2 M in THF/Et_2O) and (b) a mixture of substrate (1.0 equiv.) and dibromomethane (1.1 equiv.). The optimized system incorporated "pre-cooling" loops before the T-mixer to ensure reagents were mixed at the desired temperature. The setup consisted of two temperature zones (−80 °C and room temperature), where lithium halogen exchange would occur at low temperature, followed by cyclization when allowed to warm to room temperature (Fig. 4.13). The reaction mixture was then quenched upon exiting using water with the product being isolated following a simple liquid–liquid extraction with diethyl ether.

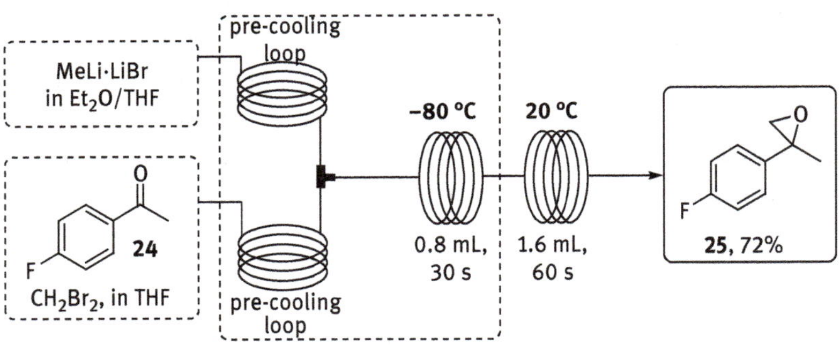

Fig. 4.13: Optimized continuous flow setup.

Using these optimized conditions, the substrate scope was investigated and indicated that various acetophenones as well as cyclic, aliphatic, and unsaturated ketones furnished the desired epoxides in good to excellent yields. Furthermore, it was found that α-chloroketones were tolerated, which enabled the exploitation of this methodology toward a scalable synthesis of the antifungal agent fluconazole (Fig. 4.14, **29**).

In this extension a metal plate microreactor (Lonza FlowPlate) replaced the setup shown in Fig. 4.14 in order to avail of the increased heat transfer and mixing of such systems which was vital in view of realizing a scalable process for this mass-transfer limited lithiation process. To this end a stream of MeLi.LiBr (0.2 M, in Et_2O/THF) was mixed within the plate reactor with a solution containing and dibromomethane and substrate **26** (in THF). Due to the improved heat transfer of this setup a shorter residence time of 10 s was viable at higher temperature (−20 °C) for the initial reaction which went to completion in a subsequent reactor coil (1 min) at ambient temperature. After an aqueous quench, the desired product **27** was obtained in high yield (throughput: 34 mmol in 3 h) allowing for its subsequent batch reaction with 1,2,4-triazole (**28**) to complete this effective synthesis of fluconazole (**29**).

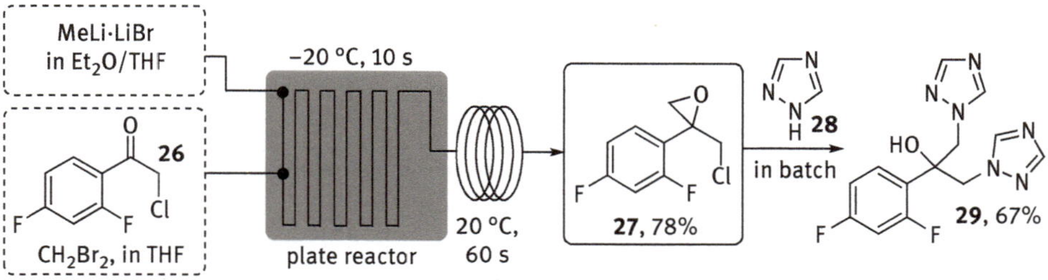

Fig. 4.14: Application to the synthesis of **27** using a plate reactor.

Experimental procedure: *2-Methyl-2-phenyloxirane* – equipment was setup as depicted in Fig. 4.14. A precooled (−80 °C) solution of acetophenone (0.5 M in THF, 0.26 mL min^{-1}, 1.0 equiv.) and dibromomethane (0.6 M in THF, 0.24 mL min^{-1}, 1.1 equiv.) was mixed with a precooled (−80 °C) solution of MeLi • LiBr (0.2 M in Et_2O/THF) using a T-connector (PEEK, 0.5 mm i.d.). The resulting stream entered a PFA coil reactor (0.8 mL, 0.8 mm i.d.) held at −80 °C (t_{Res} 30 s). After warming up to 20 °C in an additional PFA reactor coil (1.6 mL, 0.8 mm i.d.) for 1.0 min, the reaction output was quenched with water, extracted with Et_2O and the solvents were removed under reduced pressure. The crude product was purified by flash column chromatography (gradient cyclohexane/EtOAc 100:0 to 90:10) to obtain 2-methyl-2-phenyloxirane as a colorless liquid (79 mg, 0.59 mmol, 88%).

4.10 Case study – biocatalysis in flow using immobilized enzymes

Background: The use of enzymes as mild and selective catalysts that enable a multitude of important chemical reactions has garnered significant interest in academia and industry over the last decades. This development is supported by a desire to achieve greener chemical synthesis, and the ability to engineer enzymes to widen their substrate scope is a key factor in this process. Crucially, merging continuous flow processing with biocatalysis provides for further advances such as scalability and overall efficiency toward single-step and multistep processes [23]. Recently, Paradisi and coworkers demonstrated the latter approach through an enzyme cascade process that converts amines via a transaminase (HEWT) into aldehydes that are then directly reduced to alcohols using a reductase (HLADH, Fig. 4.15) [24].

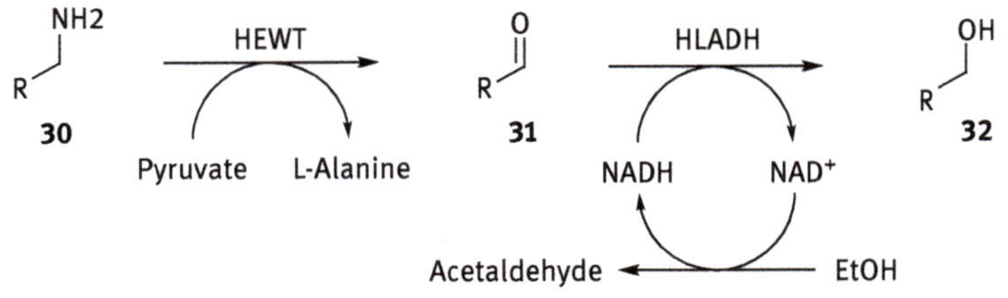

Fig. 4.15: Two-step biocatalyzed process converting amines into alcohols.

Flow process – To achieve an efficient flow process, both enzymes were immobilized onto epoxy resins prior to packing into Omnifit glass columns (6.6 or 10 mm i.d.; length 100 or 150 mm). Initial experiments aimed at optimizing the individual steps separately, before combining these into the enzyme cascade process including in-line extraction.

Flow design: Using a Vapourtec R-series flow reactor, the primary amine substrates were pumped (aq. EtOH containing NADH and phosphate buffer) and mixed via a T-piece with a second stream containing buffered pyruvate and pyridoxal phosphate (PLP, Fig. 4.16). Upon pumping this mixture through the first column containing immobilized transaminase HEWT at 37 °C, the transamination process rendered the corresponding aldehyde together with L-alanine as the by-product generated from pyruvate. Subsequently the intermediate aldehyde solution then passed through a second column containing immobilized HLADH (held at 28 °C), where the reduction to the final alcohol occurred. Crucially, this flow process provided increased mass transfer and thus intensified the desired transformation toward alcohol products in ca. 45 min residence time. Inline purification was achieved by mixing a stream of ethyl acetate with the aqueous reaction mixture and directing the resultant organic phase through a cartridge containing an amine scavenger (QP-BZA) to remove traces of aldehyde.

Substrate scope: The resulting flow cascade process was successfully applied to a variety of primary benzylic amines to produce the corresponding alcohol products in high isolated yields (Fig. 4.17; 86–97%). Furthermore, by initiating the sequence using chiral benzylic amines a series of acetophenones were obtained as intermediates that were stereoselectively reduced by exploiting a DKR process based on a ketoreductase enzyme (KRED1-Pglu) mixed in tandem with glucose-mediated co-factor recycling (glucose to glucuronic acid via immobilized BmGDH).

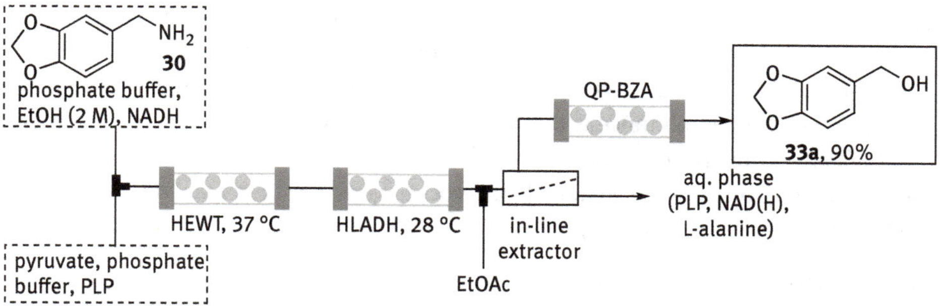

Fig. 4.16: Continuous biocatalysis cascade.

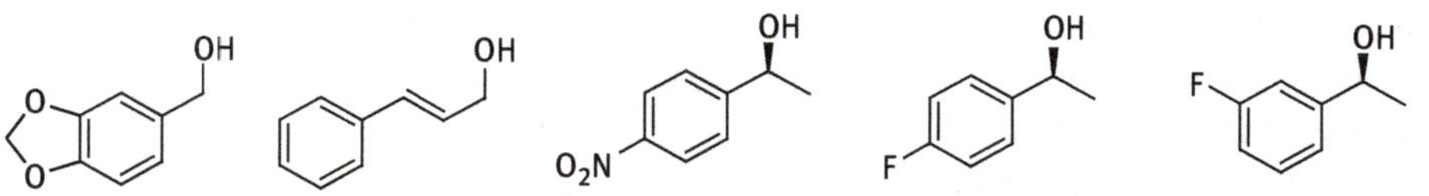

Fig. 4.17: Selection of substrates accessed via enzyme cascade process.

Experimental procedure: Continuous tandem flow biotransformations were performed using two Omnifit glass columns containing the biocatalyst. The first biocatalyst, imm-HEWT (5 mg g^{-1}), was employed for the transformation of aromatic amines into carbonyls, and imm-HLADH (2–0.1 mg g^{-1}) or premixed imm-KRED1-Pglu/imm-BmGDH (ratio of 1:25) was used for the preparation of primary or secondary (*S*)-alcohols, respectively. Sodium pyruvate (4, 6, 10, and 20 mM) in phosphate, HEPES, or Tris-HCl buffer (pH 8.0, 7.5 or 9) that contained catalytic amounts of cofactors (0.2 mM PLP, 0.02 mM NADH, or 0.2 mM NADP+) and 4, 6, and 10 mM amino donor solution with different concentrations of cosolvents (2 M EtOH or 20% DMSO) were prepared. The two solutions were mixed in a T-piece, and the resulting flow stream was directed into the columns packed with the biocatalysts. The flow rate was varied and optimized accordingly. The flow stream was extracted in-line using EtOAc, and the resulting organic phase subsequently purified using the QP-BZA scavenger. The reactions were analyzed by off-line HPLC, and the solvent evaporated to obtain the final pure alcohols.

4.11 Case study – photochemical synthesis of oxazoles in flow

Background: Recent years have witnessed a steady increase in photochemical applications that exploit continuous flow processing [25]. Crucially, flow offers uniform irradiation, avoids intrinsic limitations regarding the permeation of light into the reaction vessel and minimizes side-reactions arising from over-irradiation. This led to the discovery of new photochemical transformations as well as the exploitation of classical reactions that suffered from poor reproducibility or scalability issues in batch mode. Of industrial importance is that numerous bioactive molecules can be generated more effectively using continuous flow photochemistry [26]. In a recent example the Baumann group reported on a valuable heterocyclic transposition reaction creating various oxazole targets in an atom-economic and scalable manner [27].

Flow process: In this study a Vapourtec E-series flow reactor including its UV-150 flow module was exploited. This is based on a medium-pressure Hg lamp, a set of low-pass filters as well as a cooling mechanism that adjusts the reactor temperature by passing chilled air through the unit. Together with the temperature, the power of the Hg lamp can be regulated (75–150 W) providing effective means for identifying suitable reaction conditions. The use of low-pass filters helps in blocking undesired wavelengths of light (here >400 nm) that can lead to higher temperatures and increased decomposition. The reactor coil (10 mL) is made of UV-transparent FEP polymer and is housed within a fully enclosed reactor unit to avoid leaking of harmful UV light into the environment.

Optimization: Acetonitrile was identified as a suitable solvent based on providing high solubility for both substrates and products. Through variation of key parameters suitable conditions were identified and included a concentration of 10–20 mM, a residence time of 20 min and a temperature range of 25–50 °C. This allowed to process a series of isoxazole substrates (**34**) through the flow setup and realize their conversion into the desired oxazoles (**36**) in a process that is likely based on homolytic cleavage of the O-N bond followed by formation and subsequent collapse of an acyl azirine intermediate (**35**).

Several features of this flow process are noteworthy. Most importantly, this study demonstrates for the first time a practical means to exploit this intriguing transformation that provides new drug-like oxazole structures that are decorated with useful functionalities including further heterocycles. Additionally, a key observation was that high yields were realized if the absorption of the isoxazole substrates sufficiently matched the emission of the Hg lamp. While this was the case for aryl substituted systems (preferably electron-rich), alkyl substituents (e.g., ^{t}Bu – **36 c**) provided a mismatch which translated into lower yields; however, this could be increased by extending the residence time. Furthermore, the use of ester appendages was preferable as analogous amides resulted in lower yields along with previously unknown rearrangement products that enforce the operation of radical pathways and at the same time provide reasoning for further studies on such flow processes exploiting UV light in favor of milder visible light.

Experimental procedure: Using a Vapourtec E-series module a solution of the corresponding isoxazole precursor (**34**, MeCN, 10–20 mM) was pumped via a peristaltic pump at a flow rate of 0.5 mL min^{-1} (20 min residence time) through the UV150

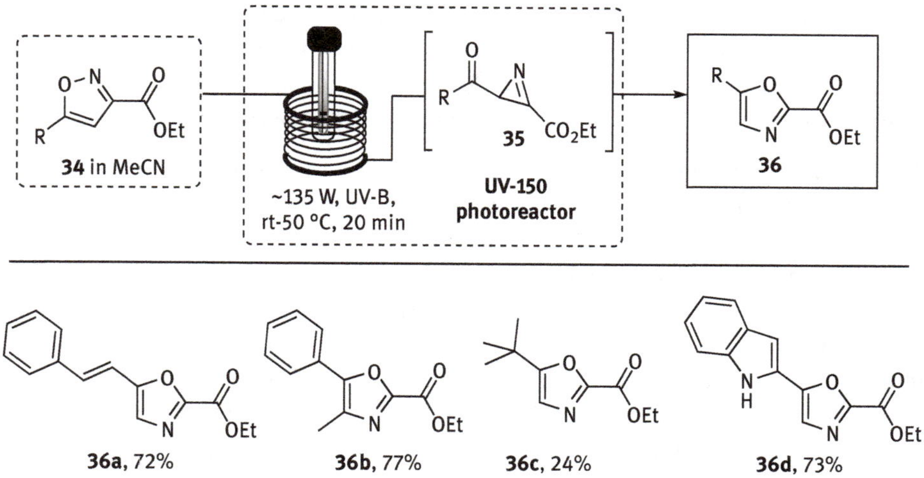

Fig. 4.18: Flow setup for photochemical synthesis of oxazoles from isoxazoles.

photoreactor. The medium-pressure Hg lamp was used at 90–100% power setting (equivalent to 135–150 W) in combination with a suitable low-pass filter. The temperature within the reactor was controlled via a stream of chilled air, resulting in a temperature of 25–50 °C. The crude reaction products were evaporated under reduced pressure and purified by silica column chromatography, yielding the desired oxazole products **36**.

(For further details on flow photochemistry please see Volume 2, Chapter 1, Title: Photochemical transformations in continuous-flow reactors.) i

4.12 Case study – electrochemistry in flow: oxidative electrosynthesis of amides

Background: Electrochemical synthesis has experienced a revival in recent years [28] due to the advent of improved processing technologies including flow chemistry [29, 30]. Exploiting electrons for various redox processes enables electrochemistry to achieve a greener, milder, and potentially more efficient means than the use of external chemical oxidants and reducing agents. Amides are pivotal structural motifs for whose preparation there are various synthetic methods; however, these typically require various functional group interconversions and multiple steps to achieve the desired result. Alternative methods have been proposed which involve oxidative conversions of aldehydes or alcohols to yield amides; however, these usually require the presence of an external chemical oxidant. The Brown group recently developed a platform for the synthesis of this important motif catalyzed by *N*-heterocyclic carbenes (NHCs), using a microfluidic electrolysis cell in place of chemical oxidants [31]. The reaction of NHCs with aldehydes forms a Breslow intermediate which can then be oxidized either with an external chemical oxidant, or in this case via anodic oxidation.

Flow process: In order to optimize the reaction, 3,4,5-trimethoxybenzaldehyde (**37**) and benzyl amine were chosen as the test substrates. Direct application of the previously developed NHC-mediated process, replacing the alcohol nucleophile with $BnNH_2$ failed to yield any desired product. This was attributed to competing imine formation of the amine with NHC in place of the desired Breslow intermediate (**39**) generation. A significant advantage of carrying out reactions in continuous flow is the modular nature of the setup. This allowed for a simple redesign of the reactor, which was modified to involve mixing of thiazolium salt **38** with the desired aldehyde **37** to preform intermediate **39**, prior to reacting with the amine nucleophile. The modified setup consisted of three streams. Firstly, a solution containing aldehyde substrate and thiazolium salt was combined with a stream of DBU in a T-piece mixer (Fig. 4.19).

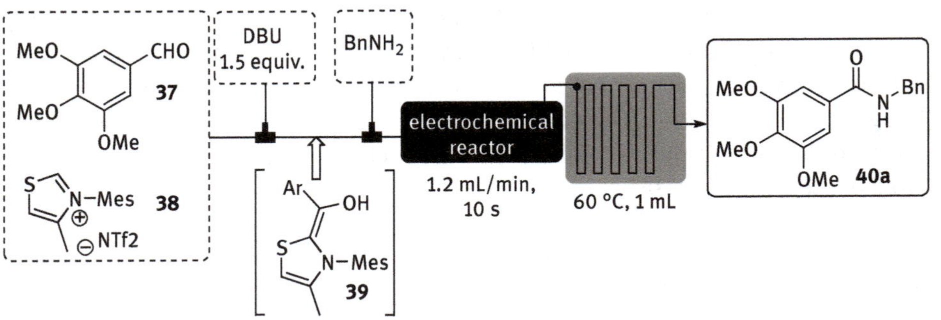

Fig. 4.19: Continuous flow setup for electrochemical amide synthesis.

This mixture containing the Breslow intermediate **39** was then combined in a second T-piece mixer, with a third stream consisting of the desired amine nucleophile. This final mixture was then passed through a microfluidic electrochemical reactor with the output material being collected in a flask for workup and isolation.

With the above setup utilizing a THF/DMSO solvent system, a total flow rate of 0.12 mL min^{-1} and a final concentration of 25 mM yielded the desired amide product (**40a**) in 30–40% yield. Switching the solvent to DMF, which is regarded as a favorable solvent for electrochemical reactions due to its large dielectric constant while being effective in solubilizing the thiazolium salt, resulted in a significant yield increase to 80%. To increase the productivity of the flow platform, the total flow rate was increased to 1.2 mL min^{-1} with a similar adjustment made to the cell current (10–105 mA). Unexpectedly a decrease in yield was observed which was found to be due to incomplete conversion of acyl thiazolium intermediate to the desired amide product. However, complete conversion was observed when the mixture was held for a further 2 h. To address this, the output of the electrochemical reactor was passed through a heating chip (60 °C) to decrease reaction times. The optimized conditions resulted in a quantitative reaction yield at residence times of 10 s in the electrochemical cell, demonstrating the high efficiencies flow electrolysis can offer.

With the optimized conditions in hand, the authors began to investigate the substrate scope of this reaction. A range of aromatic and heteroaromatic aldehydes were tolerated with the desired amides being obtained in excellent yields (Fig. 4.20). Interestingly, primary amines containing chemically oxidizable functionalities such as furan and indole groups were well tolerated. Attempts to utilize bulkier secondary amines proved unsuccessful which was attributed to their low reactivity toward acyl thiazoliums.

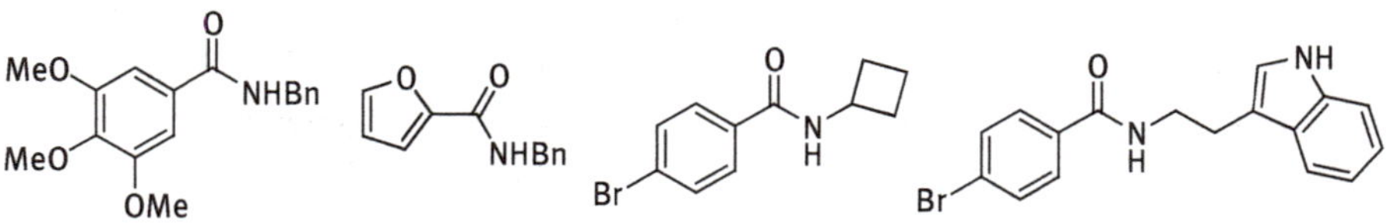

40a, 94% up to 2.5 g/h **40b**, 94% **40c**, 83% **40d**, 71%

Fig. 4.20: Selected amide products.

To demonstrate the scalability of the continuous flow setup, the authors again utilized the amidation of 3,4,5-trimethoxybenzaldehyde with benzylamine. To improve productivity increased substrate concentration (to 0.5 M) and the cell current (510 mA) in tandem with an increased chip temperature of 110 °C were necessary. An extended experiment over 8.3 h produced 21.5 g of product **40a**, equivalent to a productivity of ~2.5 g h^{-1} which would be difficult to match using a batch procedure.

Experimental procedure: *N*-Benzyl-3,4,5-trimethoxybenzamide **40a**. Three individual solutions A, B, and C in DMF (5 mL each) were sonicated: A – 3,4,5-trimethoxybenzaldehyde (**37**, 98.1 mg, 0.50 mmol, 1.0 equiv.) and thiazolium salt (**38**, 498 mg, 0.50 mmol, 1.0 equiv.), B – DBU (112 μL, 0.75 mmol, 1.5 equiv.) and C – $BnNH_2$ (55 μL, 0.50 mmol, 1.0 equiv., 10 mL DMF). The two 5 mL solutions were injected into separate sample loops and flowed at 0.30 mL min^{-1} into a T-piece mixer, to give a total flow rate of 0.6 mL min^{-1}. After mixing with the amine solution (0.6 mL min^{-1}) in a further T-piece the reaction mixture entered the electrochemical flow reactor at a constant current of 104 mA. The reaction mixture was finally passed through a glass chip reactor (1 mL, 60 °C), and the effluent was collected in a flask allowing removal of the H_2 gas by-product. The solvent was removed under reduced pressure to give the crude product that was purified chromatographically to give *N*-benzyl-3,4,5-trimethoxybenzamide **40a** (150 mg, 0.49 mmol, 99%) as colorless crystals.

(For further details on flow electrochemistry please see Volume 2, Chapter 2, Title: Electrochemical processes in flow)

4.13 Case study – telescoped multistep synthesis in flow

Background: The ability to link several distinct chemical reactions into a single telescoped reaction sequence is commonly highlighted amongst the attractive features of flow processing [32]. Although simple on paper, the realization of powerful telescoped flow sequences that render advanced intermediates, or indeed final products, is by no means trivial as many variables (concentration, solvents, reagents, residence time, etc.) need to be balanced to achieve the final goal. In a recent study Roper, Jamison, and co-workers report on multistep flow sequences to generate the complex structure of the important HIV integrase inhibitor dolutegravir [33]. In this case study the telescoping of the first three steps to generate a highly functionalized dihydropyridone core structure (**47**) will be discussed.

Flow process: The overarching aim of this study was the creation of a robust flow synthesis of dolutegravir based on cheap and readily available starting materials. Further emphasis was placed on achieving a high reaction throughput to evaluate the potential toward future development and scale-up. As can be seen in the following discussion the telescoped process was achieved by breaking down the desired sequence into individual steps that were initially evaluated in flow mode and subsequently linked.

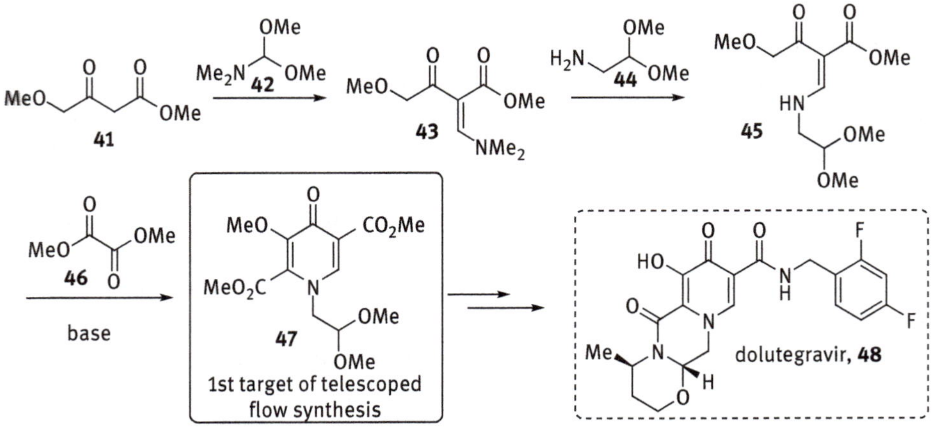

Fig. 4.21: Route toward building block **47**.

Flow design: In the first step, methyl 4-methoxyacetoacetate **41** was subjected to a condensation process with dimethylformamide dimethylacetal (**42**, DMF-DMA). As both substrates are liquid a solvent-free process was viable, which provides faster kinetics while improving the process mass intensity (PMI). It was quickly established that mixing equimolar amounts of both reagents in a T-mixer followed by reaction in a small flow coil (10 mL, PFA) generated the desired condensation product in only 10 min residence time. Increasing the reactor temperature from 30 to 85 °C and using an excess of 1.6 equivalents of DMF-DMA led to quantitative yield of **43** and allowed to integrate the second step via telescoping.

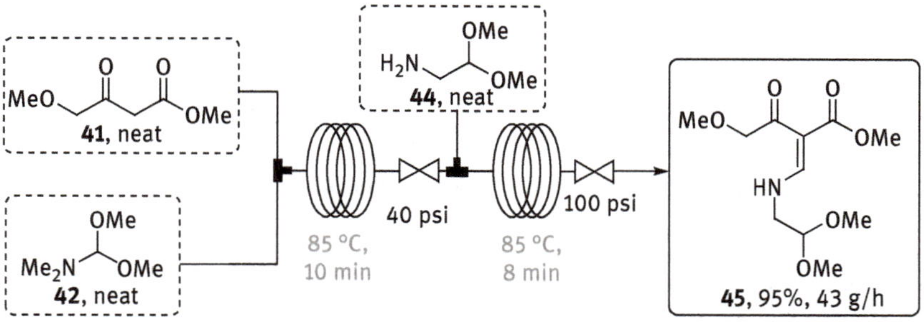

Fig. 4.22: Stage 1 of the flow process.

As the aminoacetaldehyde dimethylacetal reagent **44** is also liquid, this could be directly mixed with the stream of the first step (Fig. 4.22). Thus, a BPR (40 psi) was used as a one-way valve allowing to mix the crude mixture of the first step in a T-piece with a stream of **44** before entering a heated flow reactor coil (PFA, 10 mL, 85 °C). A residence time of 8 min was found sufficient to give the desired adduct **45** in high yield (95%) and with a throughput of 43 g h^{-1} after passing a further BPR (100 psi). Crucially, the high crystallinity of product **45** necessitated the elevated temperature (85 °C) to prevent reactor clogging.

Next, the conversion of adduct **45** into the dihydropyridone target (**47**) was studied as a separate step. As this was based on using two solid reagents, adduct **45** and dimethyl oxalate **46**, a solvent was required. Although initial tests indicated acetonitrile being a suitable option, this was subsequently altered in favor of methanol which provided greater solubility for both the substrates and NaOMe that was required as base. Using these adjustments rendered the target dihydropyridone in 91% yield under standard flow conditions (PFA reactor, 30 min, 85 °C).

Fully telescoped flow process: Having realized the optimization of the individual flow reactions, efforts toward the fully telescoped sequence were pursued (Fig. 4.23).

The driving force behind this was the wish to streamline the synthesis of the dihydropyridone target **47** together with obviating time-consuming isolation and purification operations. Minor alterations on the previous conditions were necessary including

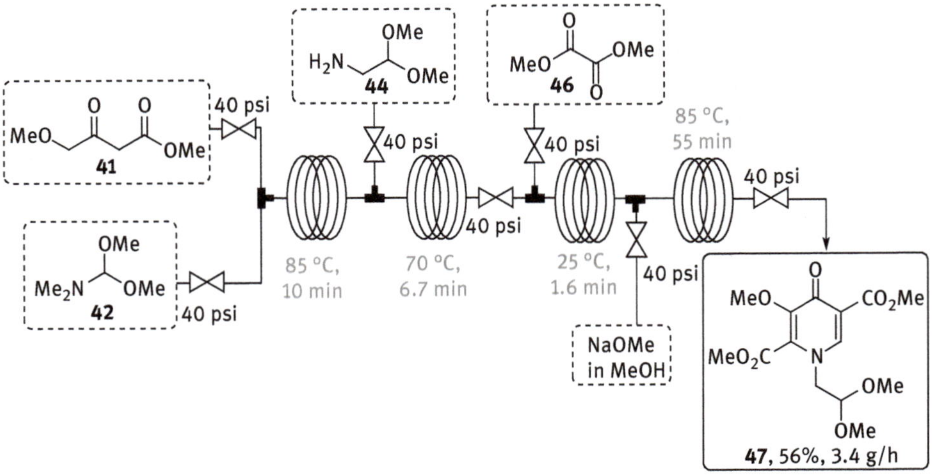

Fig. 4.23: Telescoped three-step flow synthesis of dihydropyridone **47**.

the use of several BPRs (40 psi) to act as check valves. As before streams of neat reagents (**41, 42,** and **44**) are mixed in T-pieces to initiate the sequence and render intermediate **45** after passing through two reactor coils. A methanolic solution of dimethyl oxalate (**46,** 3.0 equiv., 2.0 M) is then mixed via a further T-piece following a short reactor coil to ascertain complete mixing. Lastly a stream of NaOMe in MeOH (1.5 equiv., 4.5 M) is introduced prior to triggering cyclocondensation in a heated reactor coil (PFA, 85 °C) which required a longer residence time of 55 min. After passing a final BPR the target **47** was obtained after purification in 56% isolated yield, which is equivalent to a throughput of 3.4 g h^{-1} in a total residence time of 74 min.

4.14 Case study – in-line purification via liquid–liquid extraction in flow

Background: The continuous flow synthesis of target molecules is often combined with downstream unit operations to realize both the expedited generation and isolation of pure product. While several options for in-line purification are available (e.g., scavenger resins, distillation, and crystallization), liquid–liquid extraction is the most popular purification technique in flow mode with numerous reported applications [34]. As this case study will showcase, the integration of in-line extraction is very versatile and powerful yet requires a well-optimized flow process coupled with appropriate tools.

Flow process: A recent application by Kappe and coworkers [35] reports on a sustainable flow process that combines the synthesis of a versatile oxime building block (**50**, Fig. 4.24) with subsequent membrane separation technology to affect removal of spent reagents and by-products. Key to this is the use of a hydrophobic membrane (e.g., PTFE) with a small pore size that blocks polar molecules while allowing nonpolar species to pass and thus achieve separation of phases (aqueous–organic).

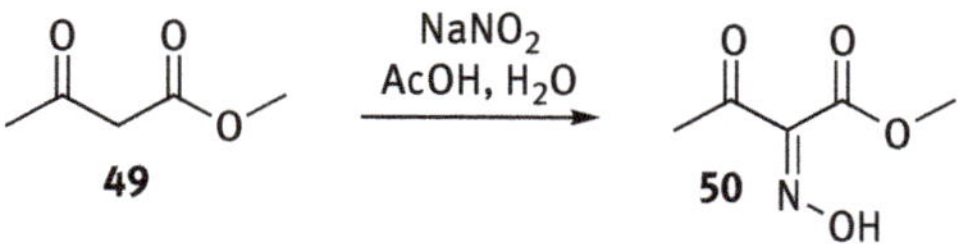

Fig. 4.24: General transformation toward oxime **50**.

Flow design: Specifically, a stream of neat methyl acetoacetate substrate is initially mixed in a T-piece with acetic acid (2.5 equiv., aq.) before combining this mixture with concentrated aqueous sodium nitrite solution (6 M, aq., 1.1 equiv.; Fig. 4.25).

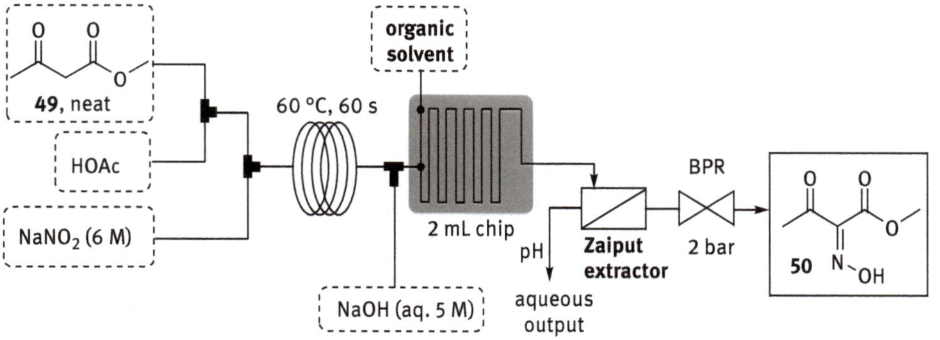

Fig. 4.25: Flow setup including liquid–liquid separation.

The resulting reaction mixture then enters a reactor coil at elevated temperature (60 °C, 1 mL volume, 1 min residence time) to ensure full conversion of nitrite into the reactive nitrosonium ion (NO^+) which reacts with the activated carbonyl substrate. Upon leaving this reactor the mixture was combined with a solution of NaOH (5 M, aq.) using a glass mixing chip (2 mL, ca. 70 s residence time, rt) to adjust the pH prior to entering a set of Zaiput membrane separators. Isopropyl acetate is then used as the organic solvent at equal flow rate (1.7 mL min^{-1}) in this process. A pH electrode controls the resulting aqueous phase to ensure the mixture is most effectively extracted, while the organic phase is obtained after passing a BPR (2 bar).

Optimization: The optimization of the flow process focused on increasing the extraction efficiency to minimize residual acetic acid in the product phase as this would necessitate a time-consuming evaporation process. It was established that at pH 6 ca. 7% of acetic acid was co-extracted, whereas at pH 8.2 less than 1% of acetic acid was present in the oxime solution. Conversely, the separation efficiency dropped from 96% to 72% as a consequence of the pH adjustment. To resolve this the authors employed a set of five consecutive Zaiput extractors (Fig. 4.26) and thus realized a robust in-line purification process that delivered the target product in 93% yield and 99% purity over a 5 h run.

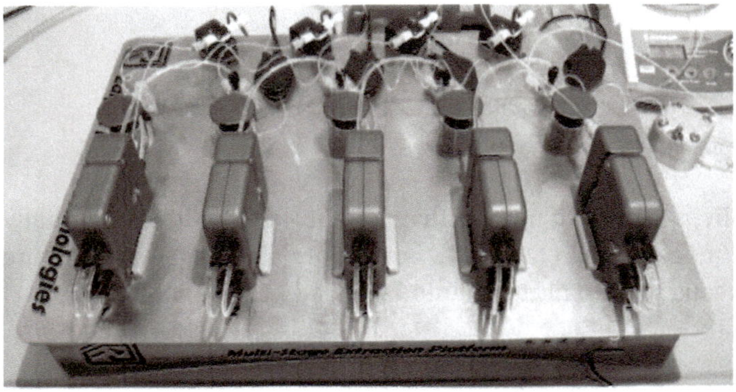

Fig. 4.26: Image of the multi-extraction stage using 5 Zaiput extractors.

Experimental procedure: Standard flow components (PFA polymer tubing with an inner diameter of 0.8 mm; 1/4–28 thread fittings, T-pieces, and PEEK or ETFE connectors) were used to assemble to platform. For the optimized continuous flow process, Syrris Asia syringe pumps were used to pump the feed solutions of methyl acetoacetate (0.249 mL min^{-1}), acetic acid (0.329 mL min^{-1}), sodium nitrite (0.422 mL min^{-1}) and the isopropyl acetate as shown in Fig. 4.25. After passing through a PFA coil reactor (60 °C, 1 min) the reaction mixture was combined in a T-piece with a quenching stream of NaOH (5 M aq., 0.7 mL min^{-1}, Vapourtec SF-10 peristaltic pump). Upon mixing with isopropyl acetate (1.7 mL min^{-1}) the material entered a glass static mixing chip (2 mL, Uniqsis) prior to entering the extraction stage. The pH of aqueous stream was continuously monitored, while the organic product solution was collected after passing the BPR (2 bar) and offline HPLC analysis.

i *(For further details on this issue please see Volume 1, Chapter 6, Title: Fundamentals of continuous manufacturing PAT: Sampling, analysis, and automation; Chapter 7, Title: Continuous gas-liquid and liquid–liquid separation.)*

4.15 Case study – kilogram synthesis of 6-nitrovanillin in flow

Background: The previous examples have not only highlighted the value of continuous flow processes when applied to a vast variety of transformations, but moreover demonstrated how adaptations in the experimental setup accommodate any necessary changes while exploiting the same underlying principles. Given the immense value of continuous flow processing, it is not surprising that analogous approaches are harnessed in the chemical industry where flow processing is pivotal for the manufacture of fine chemicals and pharmaceuticals [36, 37]; however, as the following case study will demonstrate, additional considerations are critical in realizing a safe continuous flow process on kilogram scale.

Flow process: Researchers from Bristol-Myers Squibb have recently disclosed a continuous flow nitration process to produce kilogram quantities of 6-nitrovanillin (**53**) [38], which is a valuable building block toward various bioactive targets (Fig. 4.27).

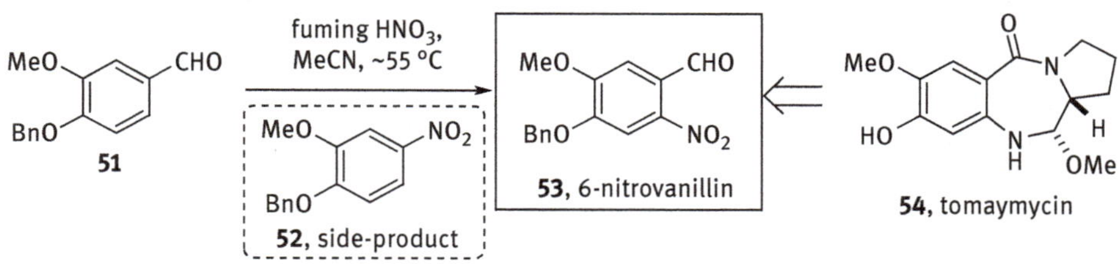

Fig. 4.27: Synthesis and use of 6-nitrovanillin.

As it is known that direct nitration of vanillin yields regioisomeric 5-nitrovanillin instead, protection of the free hydroxy group via a benzyl ether was necessary beforehand. Additionally, small-scale batch experiments revealed that acetonitrile is a possible solvent in combination with fuming nitric acid as the nitrating reagent to give high conversion of the substrate within acceptable reaction times at elevated temperatures (~55 °C). It was, however, inevitable to generate small amounts of side-product **52** that arose from competitive nitration followed by deformylation.

Cognizant of the risks of nitration reactions on larger scale (e.g., exothermicity, corrosive reagents) further tests were performed to evaluate both the energy profile and the compatibility of the reagent (fuming HNO_3) with the solvent. Through differential scanning calorimetry and other process analytical technologies, it was established that significant amounts of heat were released just above the intended reaction temperature in case of MeCN. This observation led to further solvent screening resulting in the identification of sulfolane as a replacement solvent for which significant heat release would only occur at temperatures above 150 °C. It was also established that the mixing process of substrate and fuming HNO_3 (both in sulfolane) was dose-dependent and 80% of the heat was released during this step, showing that this reaction would not be a viable long-term option in batch mode.

Flow design: The resulting flow process was realized via a plug flow reactor in which the substrate solution (**51**, 1 M, sulfolane, held at 35 °C) and a solution containing fuming nitric acid in sulfolane (33% w/w) were pumped via ceramic pumps and directed into a static mixer prior to entering a coiled flow reactor made of Hastelloy C-22 (i.d. 0.18 cm, length 9.1 m, 45 °C). This coiled reactor was equipped with a temperature sensor and a pressure gauge. The crude reaction mixture was directly collected into a vessel filled with cold water leading to the precipitation of the crude product.

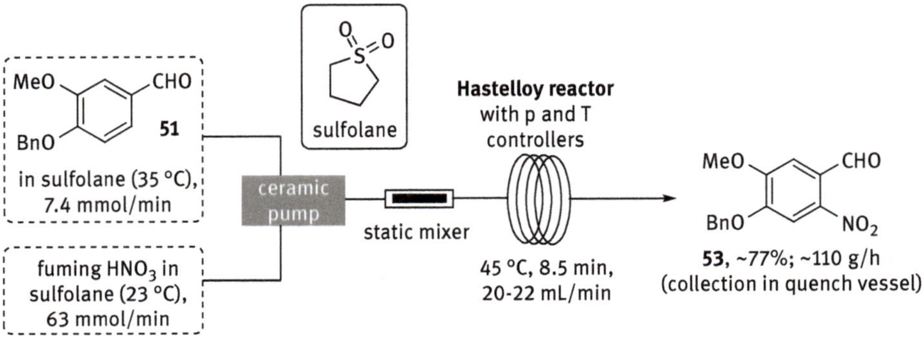

Fig. 4.28: Flow process for producing 6-nitrovanillin **53**.

Optimization: Upon optimization the flow reaction, it was found that an access of fuming HNO_3 (8.7 equiv.) was needed in combination with slightly elevated temperature (45 °C) to reach full conversion in a short residence time of 8.5 min. This flow process reached a throughput of ~110 g h^{-1} and was subsequently validated to produce 2 kg of product. Separation of the side-product was achieved in the subsequent batch debenzylation step using TFA in DCM that posed no safety risk and provide 6-nitrovanillin in pure form (62% over three steps from vanillin).

Experimental procedure: A solution of *O*-Bn vanillin **51** (1.0 kg, 4.1 mol, 1.0 equiv.) in sulfolane (4.0 L) was prepared in an all-glass reactor (Reactor-1), and maintained at 30–35 °C. In another similar reactor (Reactor-2), fuming nitric acid (2.3 kg, 36.0 mol, 8.7 equiv.) was charged slowly to sulfolane (3.5 L) maintained at 30–35 °C (during the mixing of fuming HNO_3 and sulfolane, a 5–10 °C exotherm was observed). These two solutions were pumped via an FMI-Ceramic pump (Ceram pump), and then through a plug flow reactor (C-22 Hastelloy coil) equipped with a temperature sensor (Hastelloy C) and a pressure gauge. The coil was immersed in a water bath maintained at 45–50 °C. The flow rate of each of the solutions was adjusted to 630 mL h^{-1} (10.5 mL min^{-1} each). The outlet stream from the plug flow reactor was drained into a glass quenching vessel containing chilled water (20 L) maintained at 5–10 °C. The solid that precipitated in the quenching vessel was filtered, and the filter cake was washed with water (5 L). The solid was unloaded and suspended in a fresh reactor with water (10 L) and stirred for 1 h at 30–35 °C (Note: this reslurry protocol was used to remove traces of sulfolane). The yellow solid mass was filtered, washed with water (5 L) and dried in vacuo at 50–55 °C for 16 h to provide 1.1 kg (71% assay corrected yield) of **53** as a pale-yellow solid. The material also contained ~20% of ipso product **52**. A total of 4.8 kg of crude **53** was synthesized using this protocol at a production rate of ~110 g h^{-1}.

i

(For further details on this issue please see Volume 1, Chapter 6, Title: Fundamentals of continuous manufacturing PAT: Sampling, analysis, and automation, Volume 2; Chapter 8, Title: Scale-up of flow chemistry system.)

?

Study questions

4.1 Why are fluorinated moieties important in modern drug discovery programs?
4.2 Highlight factors in this application that demonstrate advantages of flow chemistry over traditional batch processing.
4.3 Comment on the use of a relatively low partial pressure of oxygen gas of 9% in nitrogen, equivalent to 3 bar.
4.4 Name additional gases that are synthetically important in chemical syntheses.
4.5 Suggest other means of keeping precipitates from sedimentation in flow channels.
4.6 In which circumstances can it be desirable to generate solids in flow processes?
4.7 Comment on operational requirements for batch-based reactions above 150 °C.
4.8 Suggest important safety features required for high-temperature flow applications.
4.9 Why are precooling loops commonly utilized in low-temperature flow reactions?
4.10 Name reasons why metal plate reactors offer better heat transfer for mixing dependent reactions in comparison to standard T-mixers.
4.11 What are key advantages and disadvantages of immobilizing enzymes for flow applications?
4.12 What advantages offer packed-bed columns in view of mass transfer–limited reactions?
4.13 Explain how flow processing provides uniform irradiation within reactor coils or chips.
4.14 Comment on possible solutions to overcome issues with poorly light-absorbing substrates.
4.15 Comment on challenges associated with using batch-based electrosynthesis.
4.16 Explain why the use of electrons is attractive in view of sustainable chemical synthesis.
4.17 Comment on the advantages and risks of using neat reagents in flow.
4.18 What are key requirements for realizing effective multistep sequences in flow mode?
4.19 Propose a mechanism for the conversion of methyl acetoacetate into oxime product **50** under the above conditions (HOAc, $NaNO_2$).
4.20 Propose potential reasons for using isopropyl acetate over other solvents like EtOAc, THF, DCM, and toluene in the extraction process.
4.21 Propose a reaction mechanism that accounts for the formation of nitro-product **53** as well as ipso-product **52**.
4.22 What are the main safety concerns when using fuming nitric acid?

References

[1] Baumann, M, Baxendale, IR, The synthesis of active pharmaceutical ingredients (APIs) using continuous flow chemistry, Beilstein J Org Chem, 2015, 11, 1194–1219.

[2] Burcham, CL, Florence, AJ, Johnson, MD, Continuous manufacturing in pharmaceutical process development and manufacturing, Annu Rev Chem Biomol Eng, 2018, 9, 253–281.

[3] Hughes, DL, Applications of flow chemistry in drug development: Highlights of recent patent literature, Org Process Res Dev, 2018, 22, 13–20.

[4] Cranwell, P, Recent advances towards the inclusion of flow chemistry within the undergraduate practical class curriculum, SynOpen, 2020, 4, 96–98.

[5] Sans, C, Emerging trends in flow chemistry enabled by 3D printing: Robust reactors, biocatalysis and electrochemistry, Current Opin Green and Sust Chem, 2020, 25, 100367.

[6] Maier, MC, Valotta, A, Hiebler, K, Soritz, S, Gavric, K, Grabner, B, Gruber-Woelfler, H, 3D printed reactors for synthesis of active pharmaceutical ingredients in continuous flow, Org Process Res Dev, 2020, 24, 2197–2207.

[7] Rullière, P, Cyr, P, Charette, AB, Difluorocarbene addition to alkenes and alkynes in continuous flow, Org Lett, 2016, 18, 1988–1991.

[8] (a) Bychek, RM, Levterov, VV, Sadkova, IV, Tolmachev, AA, Mykhailiuk, PK, Synthesis of functionalized difluorocyclopropanes: Unique building blocks for drug discovery, Chem Eur J, 2018, 24, 12291–12297. (b) Wang F, Luo T, Hu J, Wang Y, Krishnan HS, Jog PV, Ganesh SK, Prakash GKS, Olah GA. Synthesis of gem-difluorinated cyclopropanes and cyclopropenes: Trifluoromethyltrimethylsilane as a difluorocarbene source, Angew Chem Int, Ed. 2011, 50, 7153–7157.

[9] Greene, JF, Hoover, JM, Mannel, DS, Root, TW, Stahl, SS, Continuous-flow aerobic oxidation of primary alcohols with a Copper(I)/TEMPO catalyst, Org Process Res Dev, 2013, 17, 1247–1251.

[10] Mallia, CJ, Baxendale, IR, The use of gases in flow synthesis, Org Process Res Dev, 2016, 20, 327–360.

[11] Brzozowski, M, O'Brien, M, Ley, SV, Polyzos, A, Flow chemistry: Intelligent processing of gas–liquid transformations using a tube-in-tube reactor, Acc Chem Res, 2015, 48, 349–362.

[12] Ye, X, Johnson, MD, Diao, T, Yates, MH, Stahl, SS, Development of safe and scalable continuous-flow methods for palladium-catalyzed aerobic oxidation reactions, Green Chem, 2010, 12, 1180–1186.

[13] Baumann, M, Moody, TS, Smyth, M, Wharry, S, Overcoming the hurdles and challenges associated with developing continuous industrial processes, Eur J Org Chem, 2020, 7398–7406.

[14] Filipponi, P, Gioiello, A, Baxendale, IR, Controlled Flow Precipitation as a Valuable Tool for Synthesis, Org Process Res Dev, 2016, 20, 371–375.

[15] Browne, DL, Deadman, B, Ashe, R, Baxendale, IR, Ley, SV, Continuous flow processing of slurries: Evaluation of an agitated cell reactor, Org Process Res Dev, 2011, 15, 693–697.

[16] Tsuong, J, Bogdan, AR, Kantor, S, Wang, Y, Charaschanya, M, Djuric, SW, Synthesis of fused pyrimidinone and quinolone derivatives in an automated high-temperature and high-pressure flow reactor, J Org Chem, 2017, 82, 1073–1084.

[17] Gould, RG, Jacobs, WA, The synthesis of certain substituted quinolines and 5,6-benzoquinolines, J Am Chem Soc, 1939, 61, 2890–2895.

[18] Razzaq, T, Kappe, CO, Continuous flow organic synthesis under high-temperature/pressure conditions, Chem Asian J, 2010, 5, 1274–1289.

[19] Gutmann, B, Cantillo, D, Kappe, CO, Continuous-flow technology—a tool for the safe manufacturing of active pharmaceutical ingredients, Angew Chem Int Ed, 2015, 54, 6688–6729.

[20] Yoshida, JI, Takahashi, Y, Nagaki, A, Flash chemistry: Flow chemistry that cannot be done in batch, Chem Commun, 2013, 49, 9896–9904.
[21] Von Keutz, T, Cantillo, D, Kappe, CO, Continuous flow synthesis of terminal epoxides from ketones using in situ generated bromomethyl lithium, Org Lett, 2019, 21, 10094–10098.
[22] Michnick, TJ, Matteson, DS, (Bromomethyl) lithium: Efficient in situ reactions, Synlett, 1991, 9, 631–632.
[23] Britton, J, Majumdar, S, Weiss, GA, Continuous flow biocatalysis, Chem Soc Rev, 2018, 47, 5891–5918.
[24] Contente, ML, Paradisi, F, Self-sustaining closed-loop multienzyme-mediated conversion of amines into alcohols in continuous reactions, Nature Catalysis, 2018, 1, 452–459.
[25] Sambiago, C, Noël, T, Flow photochemistry: Shine some light on those tubes!, Trends Chem, 2020, 2, 92–106.
[26] Di Filippo, M, Bracken, C, Baumann, M, Continuous flow photochemistry for the preparation of bioactive molecules, Molecules, 2020, 25, 356.
[27] Bracken, C, Baumann, M, Development of a continuous flow photoisomerization reaction converting isoxazoles into diverse oxazole products, J Org Chem, 2020, 85, 2607–2617.
[28] Wiebe, A, Gieshoff, T, Möhle, S, Rodrigo, E, Zirbes, M, Waldvogel, SR, Electrifying organic synthesis, Angew Chem Int Ed, 2018, 57, 5594–5619.
[29] Atobe, M, Tateno, H, Matsumura, Y, Applications of flow microreactors in electrosynthetic processes, Chem Rev, 2018, 118, 4541–4572.
[30] Noël, T, Cao, Y, Laudadio, G, The fundamentals behind the use of flow reactors in electrochemistry, Acc Chem Res, 2019, 52, 2858–2869.
[31] Green, RA, Pletcher, D, Leach, SG, Brown, RCD, N-heterocyclic carbene-mediated microfluidic oxidative electrosynthesis of amides from aldehydes, Org Lett, 2016, 18, 1198–1201.
[32] Britton, J, Raston, CL, Multi-step continuous-flow synthesis, Chem Soc Rev, 2017, 46, 1250–1271.
[33] Ziegler, RE, Desai, BK, Jee, J-A, Gupton, BF, Roper, TD, Jamison, TF, 7-step flow synthesis of the HIV integrase inhibitor dolutegravir, Angew Chem Int Ed, 2018, 57, 7181–7185.
[34] Weeranoppanant, N, Adamo, A, In-line purification: A key component to facilitate drug synthesis and process development in medicinal chemistry, ACS Med Chem Lett, 2020, 11, 9–15.
[35] Lebl, R, Murray, T, Adamo, A, Cantillo, D, Kappe, CO, Continuous flow synthesis of methyl oximino acetoacetate: Accessing greener purification methods with inline liquid–liquid extraction and membrane separation technology, ACS Sustainable Chem Eng, 2019, 7, 20088–20096.
[36] Baumann, M, Moody, TS, Smyth, M, Wharry, SA, Perspective on continuous flow chemistry in the pharmaceutical industry, Org Process Res Dev, 2020, 24, 1802–1813.
[37] Hughes, DL, Applications of flow chemistry in the pharmaceutical industry—highlights of the recent patent literature, Org Process Res Dev, 2020, 24, 1850–1860.
[38] Rakshit, S, Lakshminarasimhan, T, Guturi, S, Kanagavel, K, Kanusu, UR, Niyogi, AG, Sidar, S, Luzung, MR, Schmidt, MA, Zheng, B, Eastgate, MD, Vaidyanathan, R, Nitration using fuming HNO_3 in sulfolane: Synthesis of 6-nitrovanillin in flow mode, Org Process Res Dev, 2018, 22, 391–398.

Mara Guidi, Lucia Anghileri, Peter H. Seeberger and Kerry Gilmore

5 When and how to start flow chemistry

5.1 What is flow chemistry and what are its advantages

Flow chemistry has a number of intrinsic attributes – high surface-to-volume ratios [1], precision in delivery of reagents and reaction time, facility of pressurization without headspace – that can accelerate and simplify many chemical processes and even allow one to perform reactions that are inefficient or effectively impossible when using traditional methods [2]. The modularity of the technique presents additional opportunities, particularly in the design and use of flow chemical processes, as conditions are developed to induce specific chemical transformations. These can be used both in isolation and in a toolbox approach to effect transformations and perform syntheses based on transformation-specific – and not substrate-specific – flow reaction conditions.

These three key attributes make flow chemistry the ideal technique for a broad range of chemistries. There are two instances where surface-to-volume ratio plays a key role. The first is the large surface area between the solution and the applied conditions through which the solution is passed. This allows for heat to quickly and homogeneously be either applied or removed from the stream – advantageous for both exothermic and temperature-sensitive processes [1, 3, 4]. It also ensures complete irradiation of a solution without limitations imposed by the Lambert–Beer law, as well as allowing for electrochemical processes often without supporting electrolytes [5, 6]. A high surface-to-volume ratio is also beneficial for the interfacial areas of processes, both biphasic (gas–liquid, solid–liquid, liquid–liquid) and triphasic reactions, significantly accelerating and simplifying these chemistries (e.g., often, no phase-transfer catalysts in aqueous-organic Taylor flow systems).

The high degree of precision in the delivery and flow of reagents through a continuous flow system is advantageous, as it directly links reaction time with position in the reactor. For unstable/reactive products, the reaction stream can leave the applied conditions maximizing yield and/or selectivity and quenched inline via removal of conditions (e.g., light, charge, and heat) or addition of a suitable solution. The true power of flow's space–time relationship is exemplified by the generation and use of reactive/toxic intermediates, whose concentrations can be maximized at the specific point in the reactor, where the trapping reagent can be added. This has opened up entire areas of research – flash chemistry – that were impossible to perform otherwise [7].

Finally, pressurization without headspace can both accelerate chemistries and allow access to conditions impossible or unsafe in batch. Gas–liquid chemistries

https://doi.org/10.1515/9783110693676-005

benefit from this attribute, as the gas solubility can be safely increased through pressurization – applicable both to gaseous reagents and by-products formed. Reactive gases such as diazomethane can be easily generated and used in the flowing stream without risk of losing it to the headspace of the flask. Pressurization also allows for reactions to be performed well above the boiling point of solvents – opening access to reaction conditions inaccessible, otherwise [8].

The myriad of chemistries that can be performed as a direct result of these attributes has been described elsewhere [6] and are highlighted throughout this book. In this chapter, we will discuss the types of processes/transformation which *should* be performed in flow to best take advantage of these attributes. We will also point out where they should not – as not all chemistries either require the setup or are worth the effort, when the related batch process is easier.

5.2 Basic components in flow chemistry

A flow chemistry module is defined as a stable set of conditions inducing an overall effect on a flowing stream of reagent(s) [9]. Each module is comprised of pumps, mixers, reactors, and backpressure regulators (BPRs), along with reagent specific equipment such as mass flow controllers for gases or packed bed reactors for heterogeneous materials (Fig. 5.1). The details of these components are described in Volume 1, Chapter 3 (Technology overview/overview of the devices). The critical pieces – from a conceptual perspective – are the pumps, as they dictate the stoichiometry and overall flow rate (which is additive), and the reactors that provide the stable, controlled conditions through which the solution is passed.

As defined, a flow module is used to perform a single-step process. While the exiting stream can be collected for offline analysis/purification, it can also be an inlet to a subsequent module to perform two steps sequentially. When two or more modules are linked together, the multistep system is called a telescoped process. A wide range of target [10] – or functional core-specific [11] telescoped processes have been developed in both manual and automated syntheses. While telescoped syntheses are faster, greener, and safer, there are also several aspects/challenge points that need to be considered. First, all by-products/excess reagents/unreacted materials/catalysts from the previous steps are transferred to the next. While some impurities can be removed with inline work-up, it can generally be said that each subsequent reaction in a telescoped process should be more tolerant of impurities than those previous. Second, the flow rate will continue to increase with each additional pump/mass flow controller. This means that long residence times in late-stage modules require larger reactors than they would in a standalone module. Third, the larger the telescoped process, the longer it will take for the system to reach steady state, that is, more material will be wasted than in a collection of single-step modules.

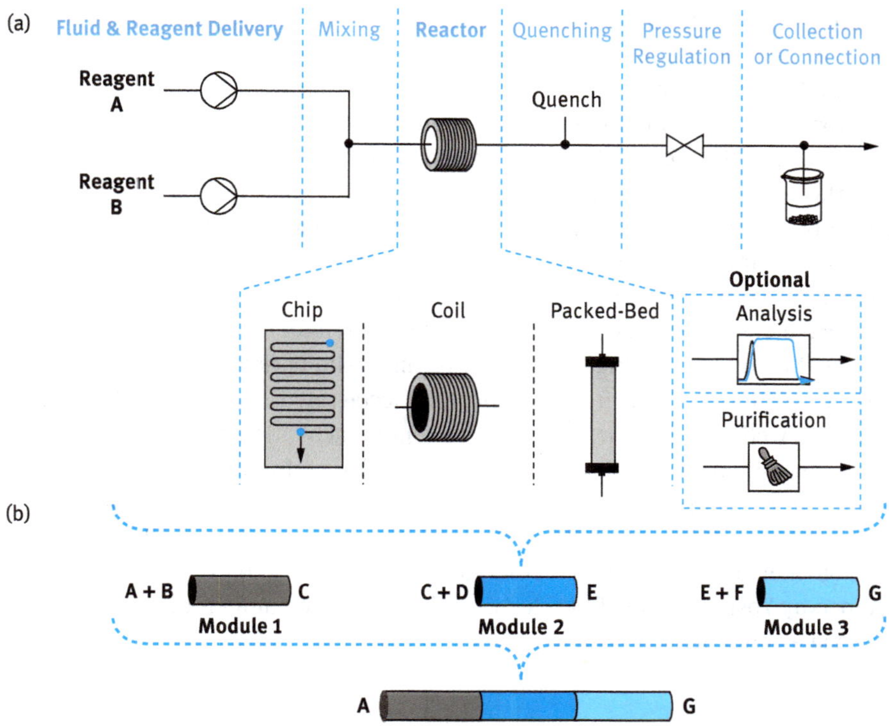

Fig. 5.1: **(a)** Fundamental zones of the standard two-feed flow setup, with zones key to standard process development in bold. **(b)** Combination of three flow modules in a linear fashion to perform a multistep process.

(For further details on this issue, please see Volume 1, Chapter 1, Title: Fundamentals of flow chemistry)

5.3 Why flow, and when you should not use it

While flow chemistry is a powerful technique that can enhance the performances of a broad spectrum of reactions, not all processes benefit enough from – or are suited for – the development of a flow system. The driving force behind turning to flow is to address the limitations/challenges of the related batch process [12, 13]. An efficient reaction exhibiting acceptable yield, safety, and control that does not involve toxic/unsafe reagents or intermediates, or one that is completed overnight is probably best performed in batch. Heterogeneous reagents/catalysts or reactions driven by precipitation are handled much easier in a flask than in tubing, due to clogging issues. It is important to identify the inherent issue or challenge in the batch process to ensure that it aligns with the advantages flow offers – as another technique may be better suited, or perhaps the chemistry needs to be redesigned. Flow requires a significant investment of both time and resources, and as such, a thorough analysis of the pros and cons of adopting flow techniques a priori is important.

Reactions well suited for flow are those that benefit from flow's core attributes: high surface-to-volume ratios, precision in delivery of reagents and reaction time, and facility of pressurization without headspace. As mentioned above, high surface-to-volume ratios give excellent temperature control – critical in exothermic [14] or temperature-sensitive [15] chemistries. Photochemistry [16] and electrochemistry [17] are two areas of research that have been both strengthened and more easily exploited as they rely on the high surface-to-volume ratio of flow reactors. Both these means of activation benefit from significantly accelerated reactions in the thinner channels as well as the ability to leave the zone of activation to prevent overreaction/decomposition of the product. This is true for performing functional group transformations, coupling reactions, and for the generation of reactive intermediates. Examples of the latter include the plethora of singlet oxygen chemistries in flow photochemical UV or LED reactors [18, 19] and the "cation flow" methodology developed by Yoshida, using flow chemistry for the on-demand production of carbocations via electrochemical oxidation [20] Fig. 5.2).

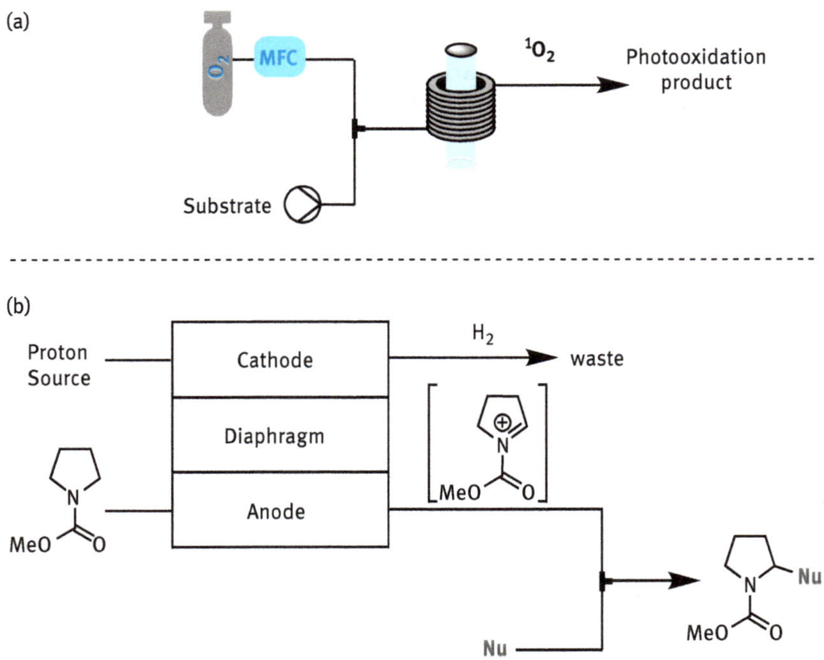

Fig. 5.2: (a) In situ production and use of singlet oxygen. (b) In situ production and use of carbocations.

Surface-to-volume ratios play a key role when two or more phases are involved. The resulting significant increase in interfacial area leads to improved mass and heat transfer, critical in reactions where the rate-limiting step is phase transfer. *Liquid-liquid* reactions are the simplest biphasic reactions to perform in flow. However, if no particular conditions are needed (extreme temperatures, irradiation, etc.) and scaling up is not an issue, liquid-liquid systems can be efficiently mixed in batch and do not benefit, particularly, from a flow apparatus. Efficient mixing between phases in flow requires either high flow or use of reactors packed with inert materials [21].

Gas–liquid reactions are often well suited for flow techniques due to the enhanced mass transfer and lack of headspace. Further, the higher pressures that can safely be achieved improve the solubility of the gas into the liquid phase where the transformation takes place. The stoichiometry of gas addition is controlled in flow using mass flow controllers, while avoiding the super-stoichiometric waste that is typical of batch [22].

Gaseous reagents can also be safely generated and used in flow, critical when hazardous gases such as phosgene and diazomethane [23] react. This capability opens the doors to applications that are otherwise carefully avoided.

Solid–liquid reactions benefit from continuous processing in two significant areas. They are generally performed by packing the solid material into a column (packed-bed) reactor. The liquid is then passed evenly through the heterogenous solid (e.g., catalyst), where it experiences a high effective molarity of the heterogeneous material (with turnover numbers increasing over the course of the volume passed through the column). The second advantage of packed-bed reactors is the combination of reaction and separation in a single unit, as the reaction mixture enters and exits the packed-bed unit as a liquid phase, and the heterogeneous material remains in the column. Moreover, catalysts often exhibit improved lifetime owing to their isolation from the outside environment [24].

However, nonimmobilized/packed solid reagents, or reagents/catalysts that are partially soluble, are not ideal as the solid material loosely passing through the thin tubing may cause clogging of the channels. The same risks and challenges are present for transformations that form precipitates. It is critical to choose reaction conditions (temperature, solvents, etc.) that prevent solid precipitation of complexes, products, by-products, or salts.

Gas–solid–liquid reactions benefit from the advantages of both individual classes discussed above, making heterogeneous catalysis transformations with gaseous reagents highly efficient in flow [25]. Hydrogenations using palladium on carbon catalyst are a well-studied transformation [26].

Processes best suited to flow:
- photochemical
- electrochemical
- gas/liquid reactions
- involving immobilized catalysts
- involving fast kinetics / unstable intermediates
- hazardous processes
- processes needing extreme conditions
- processes for which the scale-up is problematic

Microfluidic devices are well suited for fast, diffusion-limited processes, as they are extremely dependent on mixing. Poor mixing in these reactions can cause the accumulation of by-products or hazardous intermediates [27].

The area of *flash chemistry* deals with extremely fast transformations involving highly reactive and unstable species, with reaction times ranging from milliseconds to seconds (compared to the minutes/hours of traditional chemistry). Batch equipment cannot provide high-degree of control over such processes, and flow chemistry has opened up numerous transformations that are not feasible otherwise due to side-reactions and the decomposition of unstable intermediates [7, 28]. Evaluation of the kinetics of such reactions at the outset is important.

One example of the power of flash chemistry is the synthesis and use (or efficient avoidance) of benzyne (Fig. 5.3) [29]. A three-component coupling was achieved, where benzyne was formed and sequentially trapped with sequential reactions with a nucleophile and an electrophile. A careful choice of reactor and reaction conditions ensures that the initially formed aryl lithium species can be trapped with an electrophile, prior to the rapid elimination that forms benzyne. This precision enables synthetic chemists to truly control a reaction to prepare and use intermediates experimentally in much the same way they can be drawn on paper.

Exergonic reactions that require the slow addition of a reagent and/or cryogenic temperatures can be carried out in flow safely and, in some cases, even at room temperature due to the rapid heat dissipation of the thin tubing of a flow reactor. Slow transformations *can* benefit from process intensification when high temperatures and high pressures can be applied [30]. Finally, both fast and slow reactions can also benefit from the narrow temperature profile window of the system, and critical when product/side-product ratios are decided within a narrow window of transition state energy.

Overall, the three key attributes of flow increase the efficiency of – or allow access to – a broad range of different chemistries. However, flow represents neither a magic wand nor a required tool of chemical synthesis. It is important to understand the chemical process to be performed, and to use the best methodology for that process, (Fig. 5.4) whether that is flow or batch.

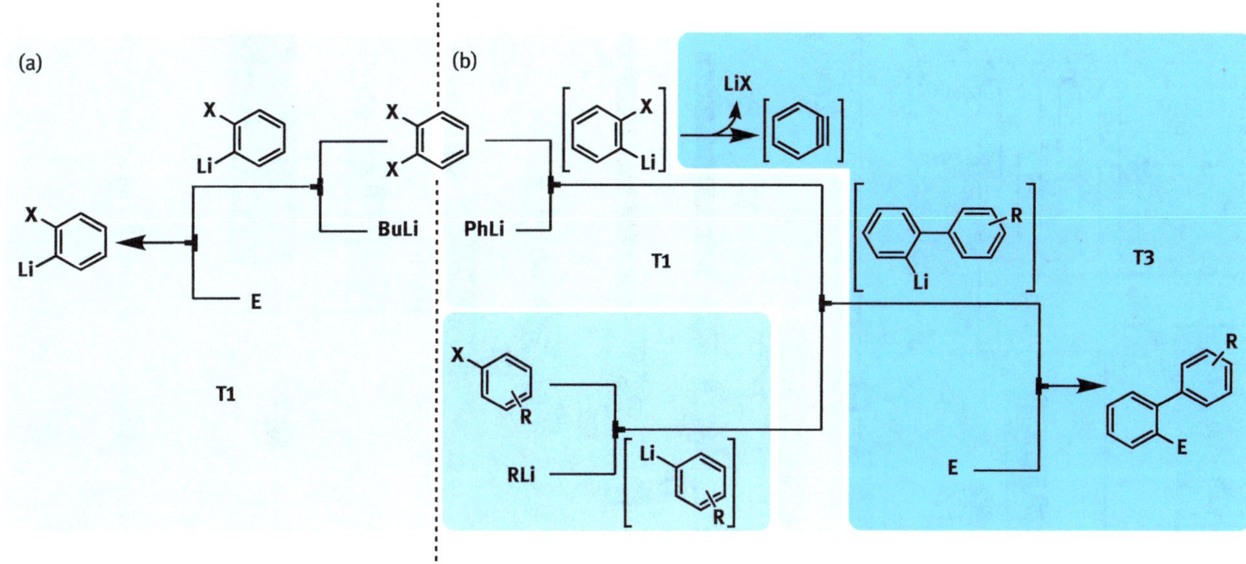

Fig. 5.3: Divergent scheme for (a) generation and direct utilization of carboanions and (b) generation and utilization of benzyne through rational design of a three-temperature zones system [29]. The three-temperature zones (T1, T2, T3) are shown with colored boxes.

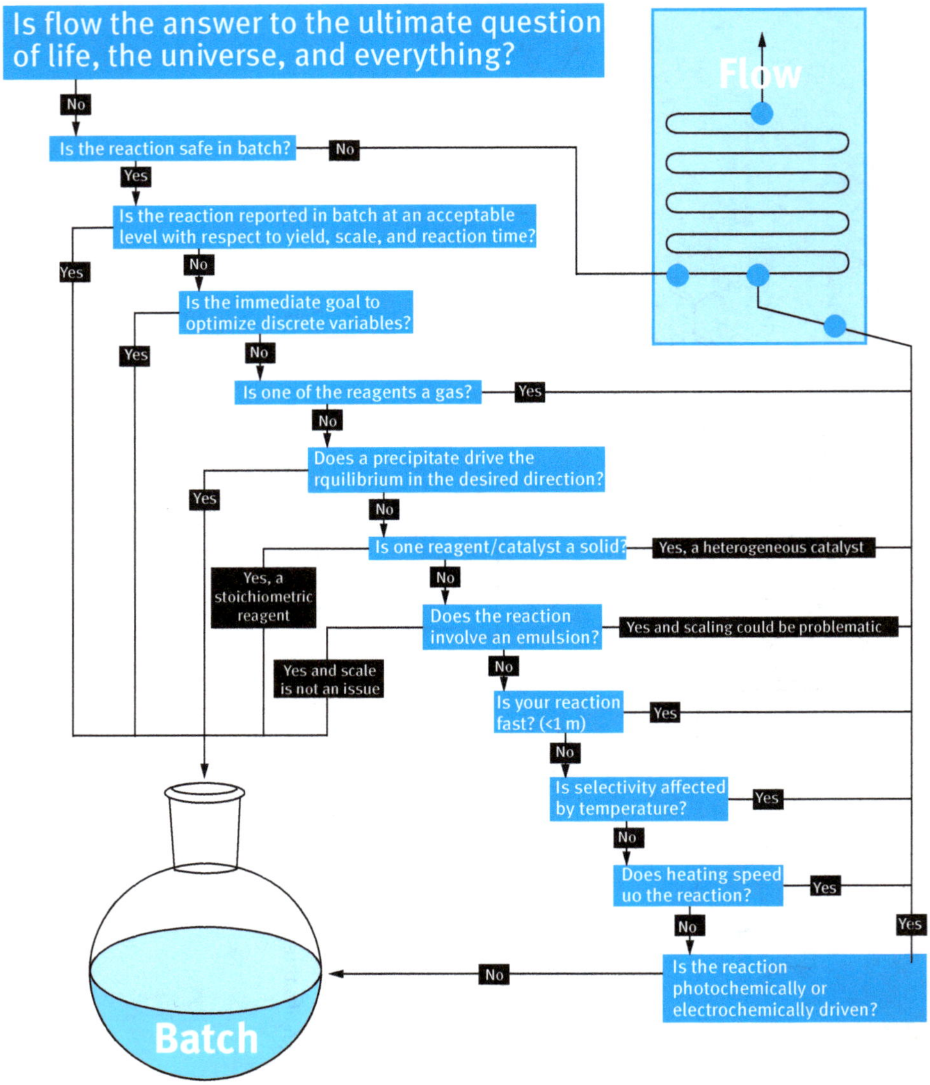

Fig. 5.4: Decision diagram for flow chemistry (from Plutschack et al., The Hitchhiker's guide to flow chemistry. Chem Rev. 2017).

5.4 How to start/develop a flow process

Once it has been established that it would be advantageous to perform a reaction in flow, we can begin to consider the implementation or how to do it. Several parameters must be kept in mind when moving from batch to flow [6]. While solvents must be chosen carefully to ensure the complete solubility of reagents, intermediates, by-products, and products, they can be selected irrespective of boiling point, as pressurization expands the process capabilities. Reaction time must be selected before the reaction begins, as opposed to starting the reaction and periodically checking reaction process via thin layer chromatography (TLC) until the starting material is consumed. Reaction time determination can be challenging for intensified processes (photo- or electrochemical, high temperature/pressure) as compared to the related batch procedure, as reaction times can be much shorter, and it is difficult to estimate how much so, a priori. Flow-specific errors, such as leaking, clogging, pinching of tubing, BPR clogging, and pump failure, can also occur in a process. While, ultimately, flow chemistry facilitates a range of chemical processes, it often requires more significant hands-on troubleshooting in the development stage as compared to batch processes.

In this section, we provide a guide on how a batch process can be translated into flow, starting from the evaluation of all variables that influence the chemical transformation. We will focus specifically on five variables: solvent/solubility, concentration/stoichiometry, temperature, pressure, and residence time. For each variable, we will highlight the differences between the considerations that have to be made when working in batch and in flow, before suggesting the best way to run their screening and optimization.

5.4.1 Solvent/solubility

The first issue to be considered when planning a flow reaction is reagent solubility. Solvent screening must be conducted in batch, since only homogeneous solutions of each reagent should be fed into the flow system. Slurries can be efficiently delivered and used in flow, via peristaltic pumps or serial micro-batch reactors (SMBR) [31] system. However, such approaches are challenging and require considerable flow chemistry experience. Batch solvent screening helps test many solvents and solvent mixtures in parallel and is the fastest way to determine the best candidate(s) for reaction media. The aim is to (a) identify the solvents for each reagent resulting in a homogeneous mixture; and (b) select a solvent that solubilizes all reagents and ideally has the physical properties desired for the transformation as the solvent of choice for the reaction(s). If no single solvent can be found, solvent mixtures have to be tested for the desired chemical transformation. After solvent(s) have been identified, the reaction is run in batch (if possible), to determine whether the products and by-products are soluble. Thereby, it will be ensured that no precipitation will occur in the flow system.

While identifying the best solvent, the ease of removal through workup, crystallization, or evaporation, and the environmental impact should be considered.

Practical example with multiple solvents: Imagine you have two reagents, A and B, for which the best solvents are identified as S1 and S2, respectively. The first step is to make two highly concentrated solutions: A in S1 and B in S2. Once starting solutions have been made, we can test their combination to investigate the compatibility of A and B (as well as products and by-products) in the S1/S2 mixture. All these operations are to be performed in batch, as we need to be sure that when we will stream such solutions into the narrow tubing of our flow system, no precipitation will occur. Any solubility issues that arise can ideally be addressed by lowering the concentration of the offending reagent (or the solution as a whole). A last resort is the addition of a third solvent to give the required properties to the overall solution.

Ex 1: During the solvent screening for the synthesis of 2,6-difluorobenzylazide, required for the synthesis of rufinamide [32], determining the correct solvent mixture turned out to be crucial. The solvent screening for the two starting materials was performed in batch and indicated acetonitrile as the best solvent for 2,6-difluorobenzylbromide and water for sodium azide. Initially, a solution of 2,6-difluorobenzylbromide (1.64 M in CH_3CN) and one of NaN_3 (1.97 M in water) were prepared and fed into the flow reactor to perform the temperature and residence time optimizations. Unfortunately, despite clean and full conversion to the desired 2,6-difluorobenzylazide, the outcome of this reaction was a two-phase system, impossible to be used in a telescoped process without inline separation. In an attempt to achieve a monophasic solution, it was established that the addition of methanol could avoid phase separation. Unfortunately, neither 2,6-difluorobenzylbromide or sodium azide are soluble in pure methanol; however, it was found to be an effective co-solvent for both reagents. As such, two new solutions were made, where a minimum amount of water and acetonitrile were used to solubilize the starting materials, and then methanol was added to reach the desired concentrations. Different solvent ratios, stoichiometry, concentrations, temperatures, and residence times were screened in flow. It was found that the monophasic, three-solvent solution exhibited faster reaction times than the acetonitrile/water mixture, resulting in complete conversion in five minutes at room temperature compared to 20 min at 100 °C necessary to get full conversion with the biphasic water/acetonitrile reaction medium.

5.4.2 Concentration/stoichiometry

We generally aim to work at the maximum concentration for the limiting reagent in order to maximize the throughput of our process. For other reagents, it is desirable to use the least amount in terms of equivalents, particularly for expensive and toxic chemicals, as this minimizes waste and potential isolation issues. Concentration and stoichiometry screening is performed inline, as the feed of each solution can be quickly and easily tuned by playing with flow rates or through the addition of an additional feed of solvent. It is important to keep in mind that flow rate adjustments affect not only the final concentration but also the residence time.

The best way to plan a stoichiometry screening in flow is to dedicate a pump to each reagent and a solvent line, thereby ensuring quick and straightforward modification of the isolated flow rates for optimization of both stoichiometries and the overall concentration. Each pump should be connected to stock solutions using the highest concentrations of all reagents. A starting stoichiometry that is dictated by the flow rates of the individual pumps and the concentrations of stock solutions is then selected and, once the reactor reaches steady state, material can be collected for analysis.

Practical example: Let us take our solutions of A in S1 and B in S2. When starting a screening to determine the optimal concentration, there are two factors that need to adapt to one another. The first is that our initial stock solutions are in the concentration and solvents found to avoid precipitation (or phase separation when not wanted) during the solvent screening. The second is the relative stoichiometry needed for the reaction that is the parameter under investigation. In the simplest case, A and B are at the same concentration in the stock solutions, and we can play with the relative flow rates to tweak stoichiometry, for example, from a 1:1 ratio to 1:1.4 (Fig. 5.5), by flowing the B line at higher and higher flow rates.

A good method is to start from previously reported values, if available, and try to maximize the ratio of the limiting reagent/excess reagent(s). It is important to keep in mind that the reactor volume is fixed, and as such all variations in the total flow rate affect residence time (Fig. 5.6a/b). However, variation in the relative flow rate of different feeds (in attempting to achieve a specific residence time) affects the final concentration (Fig. 5.6c/d), and as such, concentration of initial solutions may have to be adjusted.

5.4.3 Residence time

The residence time is the time from when a solution enters the reactor unit to when it leaves; it is the total time the solution is exposed to the conditions required for the transformation. The residence time is calculated by dividing the reactor volume (mL) by the flow rate (mL min^{-1}). For a given reactor volume, the overall flow rate can be increased or decreased to shorten or lengthen, respectively, the residence time. If certain flow rates are required, for example, fast flow rates for enhanced mixing, or when the flow rates are dictated by a previous step, the size of the reactor will have to be adjusted (Fig. 5.7).

One very important consequence of the direct correlation between reaction time and physical position in flow reactors is the possibility of identifying the physical location of a given reactive species generated in situ. When we write out a chemical reaction, reagents are drawn coming together, creating an intermediate, such as an organolithium species, that then reacts with an electrophile to give the desired product. In traditional batch chemistry, reagents, intermediates, and products are

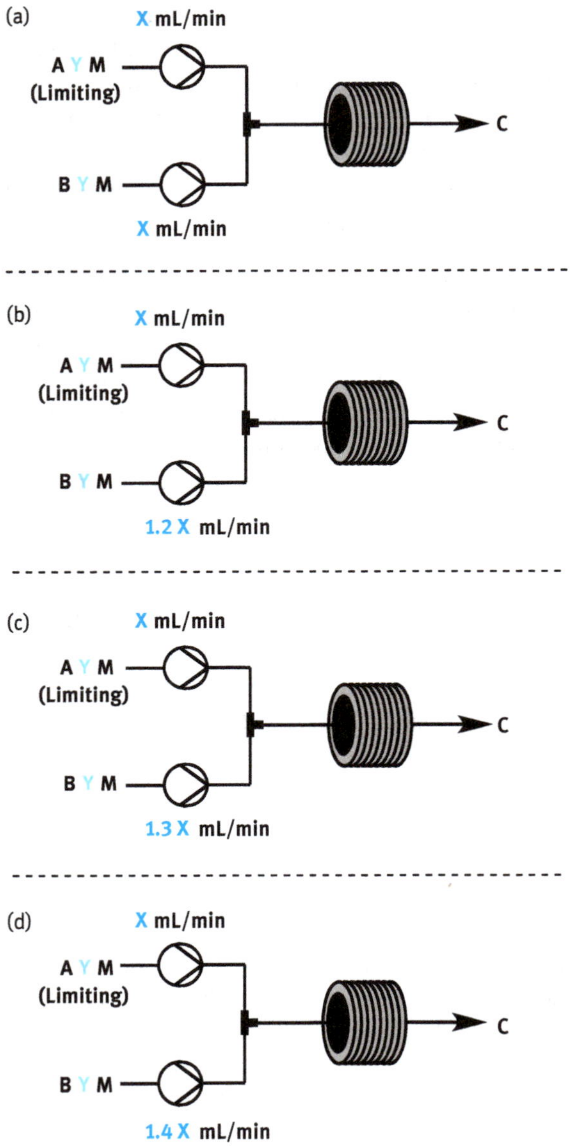

Fig. 5.5: Variations of relative flow rates to tweak stoichiometry between two reagent feeds.

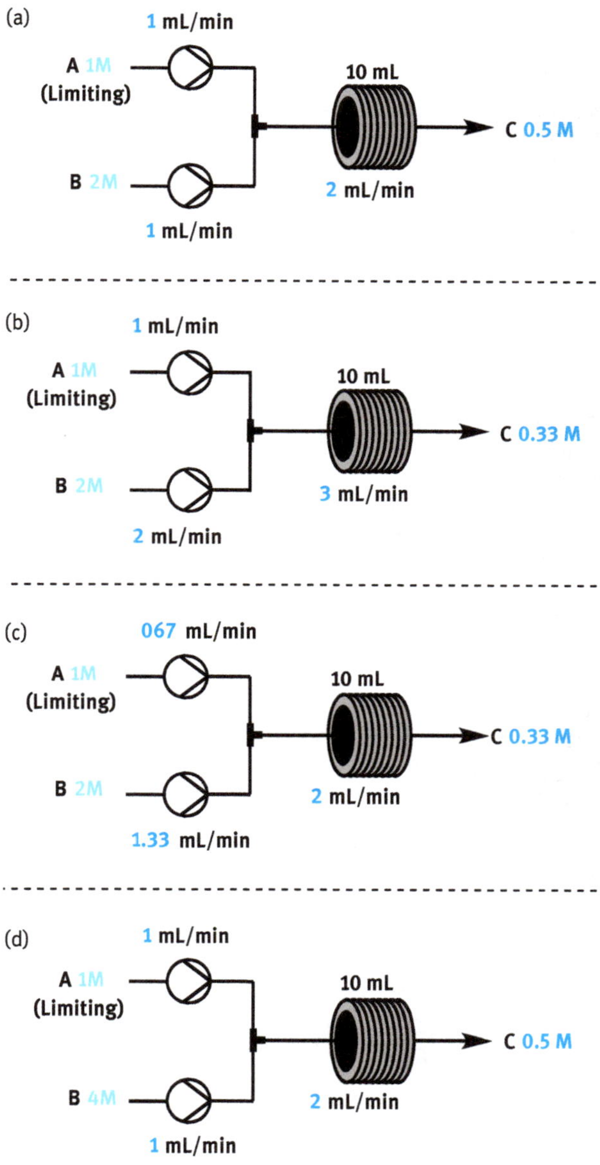

Fig. 5.6: Variations in the total flow rate (a, b) and variation in the relative flow rate and initial concentration of different feeds (c, d), with a fixed reactor volume.

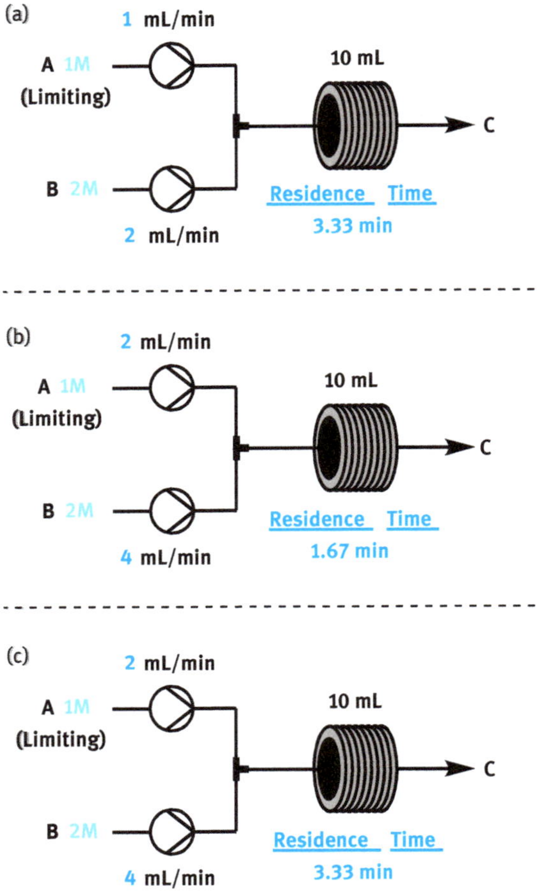

Fig. 5.7: Flow rate and reactor volume influence residence time.

present together at the same time in the reaction flask and can follow different competitive reaction pathways. In general, cryogenic conditions are required to harness reactivity, and there is a limit to the lowest achievable reaction time dependent on the manual velocity of the operator. However, flow systems can be designed to create a pathway identical to how the reaction is drawn (Fig. 5.8), thus controlling precisely the time of exposure of the reactive species to the reaction conditions, minimizing unwanted side reactions, and introducing the trapping species at the precise position where the concentration of the reactive species is highest.

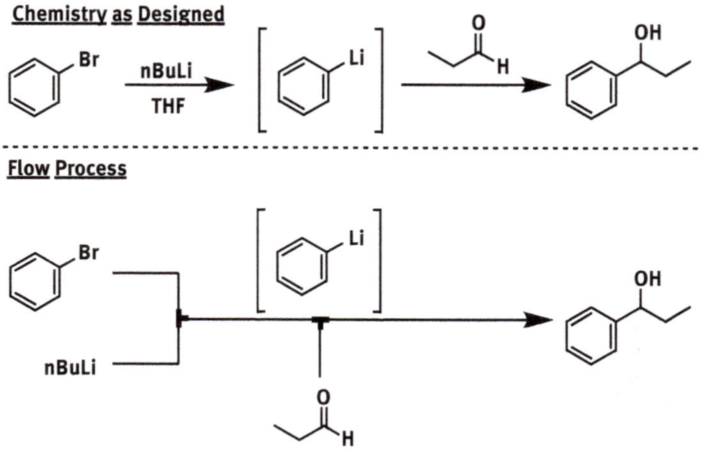

Fig. 5.8: (Top) A typical example of how chemists write out a reaction. (Bottom) Aryllithium generator, where an organolithium is generated cleanly and delivered directly to the electrophile, creating a flow system identical to the "ideal" written process.

Practical example: "Flash chemistry" of Yoshida best exemplifies the concept that perfectly exploits the space–time correlation of flow chemistry. Flash chemistry relies on very fast reactions (a second or less) involving highly reactive and unstable species that cannot be harnessed by traditional methods [7]. Residence time and temperature are key parameters to controlling the reactivity of carbanions. One example is the monofunctionalization of 1,2-dibromobenzene [33]. The flow setup for this reaction is composed of two microtube reactors for the lithiation and subsequent trapping and two T-shaped micromixers submerged in a cooling bath (−78 °C). This flow system is capable of very short residence times (0.8 s), and upon complete generation of *o*-bromophenyllithium, it can be efficiently trapped intermolecularly with an electrophile *before* the undesired intramolecular elimination occurs (Fig. 5.9).

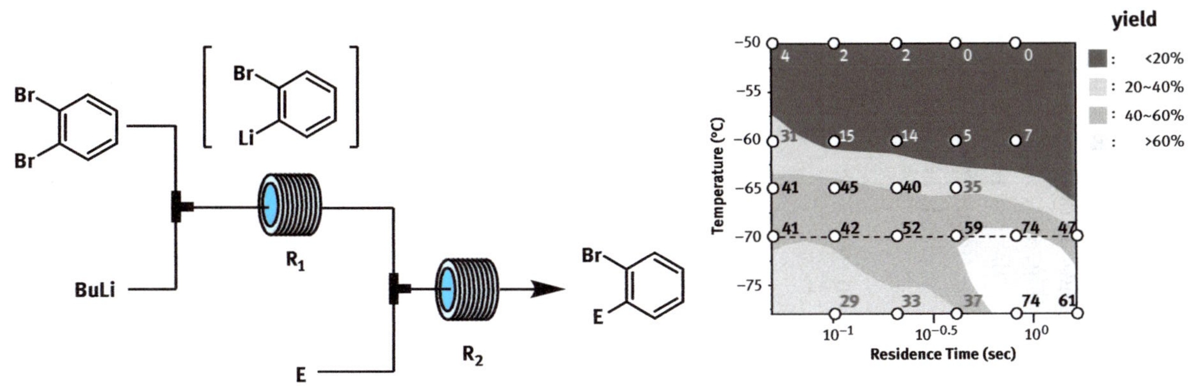

Fig. 5.9: (Left) Module for the monolithiation of 1,2-dibromobenzene [33]. (Right) Heat map of product yield with respect to variations in temperature and residence time for R1 of the described monolithation process.

5.4.4 Temperature

Two considerations are important when talking about temperature in flow systems. The first is related to the precise control that we have in flow on parameters such as heat transfer and heat dissipation, an intrinsic factor of the narrow tubing and/or channels, thereby the precise control over the temperature applied to the reaction in each zone of the flow system is achieved. This has important implications on selectivity and safe handling of exothermic reactions.

The second results from the ease at which flow systems can be pressurized, allowing one to work with solvents above their boiling point. For temperature-dependent reactions, it is often possible to accelerate the reaction time from several hours to minutes. Temperature screening is easier in flow than in batch, as the very good heat transfer in flow systems allows the application of rapid temperature ramps, with certainty that the same temperature is applied simultaneously to the entire flowing solution, with no gradient. The possibility to work at any temperature independently from the boiling point of the solvent makes such tests even easier, as there is no need to switch to a higher boiling solvent when increasing the temperature. All parameters can be kept identical, except for reaction temperature to generate accurate dependence curves of reaction conversion and by-product formation, with respect to this variable.

Practical example: Referring again to our system of A, S1, B, S2, if we wanted to screen temperatures that exceed the boiling point of the mixture of S1 and S2 in a batch process, we would have two choices. We could either switch from common glassware to a metal pressure reactor that can hold high pressure, or switch to solvents with higher boiling points that may not have the physical properties desired of the solvent (polarity, protic, etc.). However, in a flow process, we can apply a Back-Pressure Regulator (BPR) to easily pressurize the entire system. The limiting factor to the pressure that can be achieved is the strength of the tubing (based on the material, stainless steel can operate at high temperature/pressure) and the pumps (high-performance liquid chromatography pumps can work under high pressure, syringe pumps generally cannot).

5.4.5 Pressure

Applying and maintaining a certain pressure to a flow system is safe and easy, and it can prevent gas bubble formation when working with low boiling chemicals/solvents or with reactions that liberate gases (e.g., N_2 or CO_2). Working in a pressurized system increases gas solubility and is highly beneficial when working with gaseous reagents. The tools employed to generate and hold backpressure are called BPRs. The simplest BPR is a cartridge based on a spring mechanism that can be connected to the exit of the flow system, applying a pre-defined pressure to the entire system preceding it. Varying the pressure with this type of BPR requires disassembly of the system and exchanging the BPR cartridge, as each is set to a specific pressure value. There are, however, more flexible BPRs that allow for variation of the set pressure in a manual or automated manner during an ongoing process, without the need to stop the stream and dismantle the system [34].

Practical example: During the optimization of the flow synthesis of fluorinated α-amino acids [35], the optimal solvent mixture for the hydrolysis of fluorinated α-aminonitriles under acidic condition was acetic acid in 30% HCl_{aq} [1:4(v/v)]. This reaction required a temperature of 110 °C to go to completion, a temperature that is well above the boiling point of the required 30% HCl_{aq}. This was achievable by placing an 8-bar BPR at the end of the 22 mL flow reactor (Fig. 5.10).

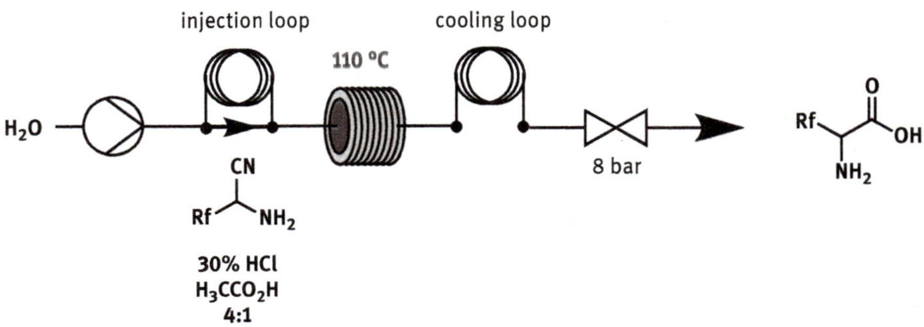

Fig. 5.10: Module for the hydrolysis of fluorinated α-aminonitriles, run above the boiling point of the 30% HCl_{aq} used.

5.4.6 Multistep processes

For multistep telescoped processes, several other factors need to be considered. In principle, each consecutive step can be optimized individually, but reaction conditions such as solvent, pH, and by-product tolerance often differ from one step to another [11]. However, there are often different factors that must be considered during the development of single modules and a series of connected modules creating a multistep system.

At the outset, the residence time necessary to accomplish each step needs to be considered. As previously mentioned, longer residence times require slower flow rates and larger reactor volumes. However, the flow rates in a multistep process are dependent on the rates of the added reagent solutions as well as the flow rate of the solution exiting the previous module. The flow rate is only going to increase for each step of process due to the continuous addition of reagents mixtures from the feed lines. The practical result of this constraint is that a multistep process requires a specific sequence of reactors with customized volumes and/or very careful management of the flow rate of every feed in order to achieve the required residence times (Fig. 5.11).

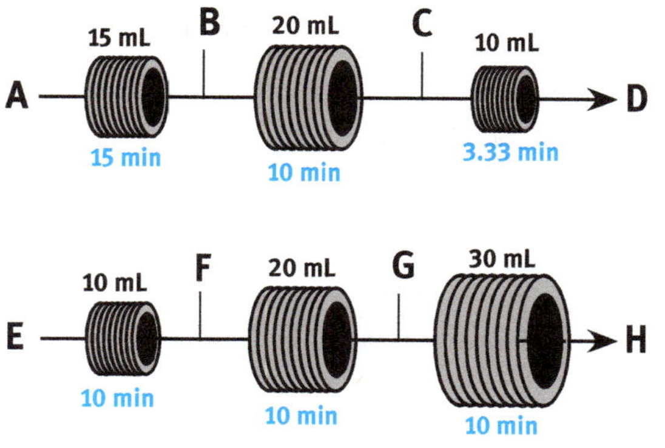

Fig. 5.11: Constraint between flow rate, reactor volume, and residence time. Different reactors with different volumes are needed in each step. For every new process, reactors must be replaced with others of the proper size.

An additional factor to be considered in multistep processes is the solvent compatibility *between* each step – both for solubility and reactivity issues. At the development stage, it is critical to identify, if possible, multiple solvents to perform each step. By-product formation should be minimized, although water-soluble by-products can be removed by in-line workup.

Practical example: Chemical assembly systems, where flow modules are combined in an interchangeable fashion, provide access to different routes and core structures [11]. The five modules composing the system were combined in three different orders to create continuous flow processes for the synthesis of β-amino acids, γ-amino acids, and γ-lactams (Fig. 5.12). Module 1 was a biphasic oxidation that worked well with a large variety of organic solvents. Module 2 was an olefination module that was also relatively solvent tolerant but required the addition of methanol. Module 3 (Michael addition) was not tolerant of methanol. Therefore, the methanol needed in module 2 had to be removed from the reaction mixture before entering module 3. Toluene was the most efficient solvent for the acidic in-line workup between the two modules that removed methanol from the solution, and thus toluene became the solvent chosen for the entire synthetic system.

(For basic and additional knowledge on these issues, please see Volume 1, Chapter 1, Title: Fundamentals of flow Chemistry, Chapter 2, Title: Principles of controlling reactions in flow chemistry and Chapter 4, Title: Practical aspects of performing continuous flow chemistry)

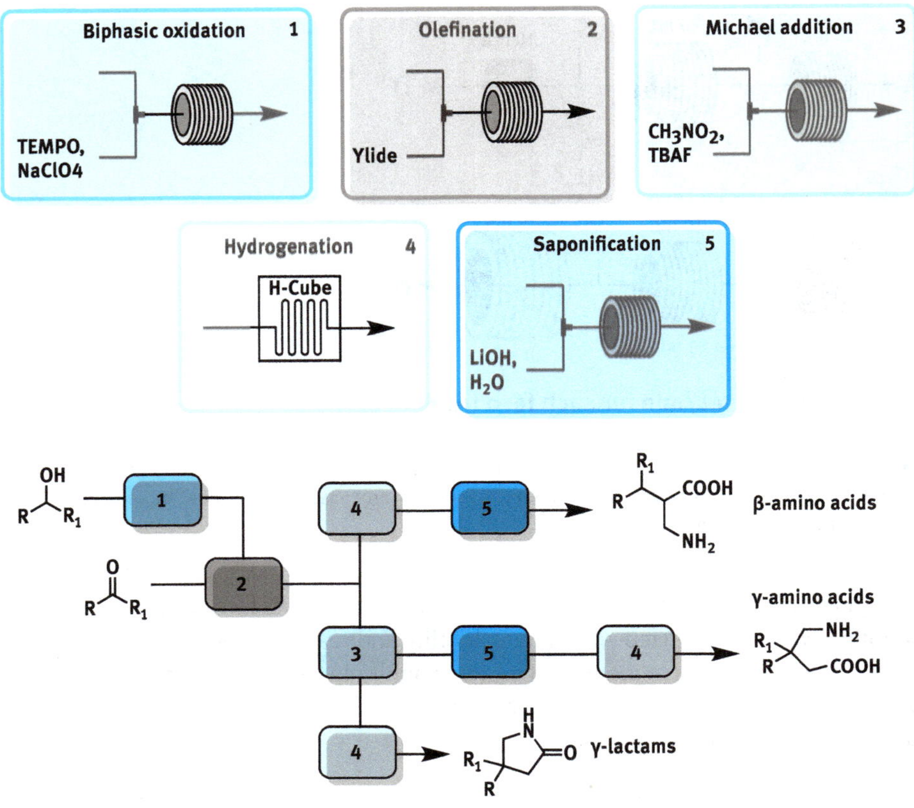

Fig. 5.12: CAS 5 – β-amino acids, γ-lactams, γ-amino acids: a divergent CAS synthesizes libraries of three different functional cores. Levels of control: Starting material choice, reagent choice, and order of modules.

5.5 Conclusion

Three main attributes of flow chemistry make it inherently advantageous: high surface-to-volume ratios, precision in reagent delivery and reaction time, and the ease of pressurization without headspace. Therefore, flow processes provide significantly more control over a procedure than comparable batch processes. Often, but not always, this level of control is required and either accelerates processes or even makes such processes possible. The development of a flow procedure, either adapted from a published batch procedure or de novo, necessitates a shift in thinking of not only which parameters need to be optimized, but also how they are screened. Here, we provided a walk-through of how a flow procedure can be developed by screening solvents, concentrations, stoichiometries, temperature, residence time, and pressure. While not applicable to all chemical scenarios, flow is a powerful tool for both synthetic and mechanistic chemistry. This chapter provides a starting point for those considering the use of flow chemistry for the first time.

Study questions

5.1

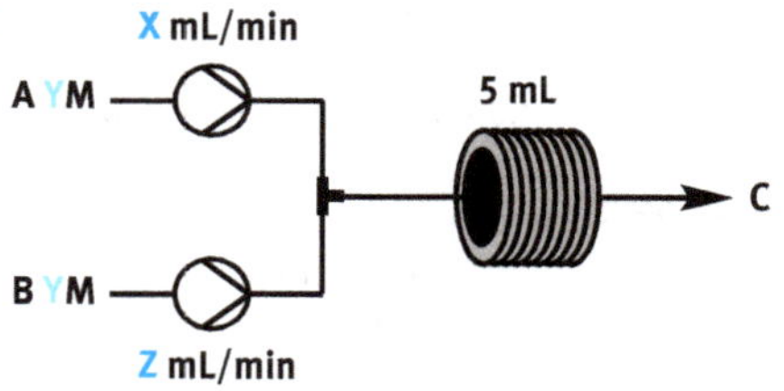

Solutions of A and B have the same concentration

Give values for the flow rates of the two feeds (X and Z) when the desired stoichiometry is A:B = 1:1.5 and the residence time is 40 min.

5.2 The pumps you are using to feed your reactor in problem 1 have a minimum flow rate value of 0.1 mL min^{-1}. What can you do to achieve the desired residence time?

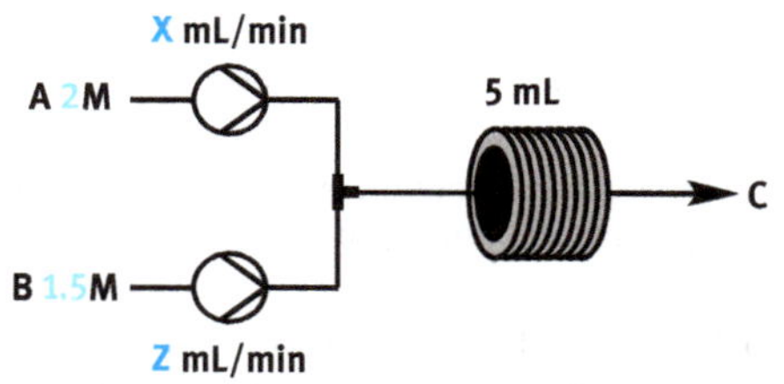

5.3 Give values for the flow rates of the two feeds (X and Z) when the desired stoichiometry is A: B = 1:2 and the residence time is 15 min.

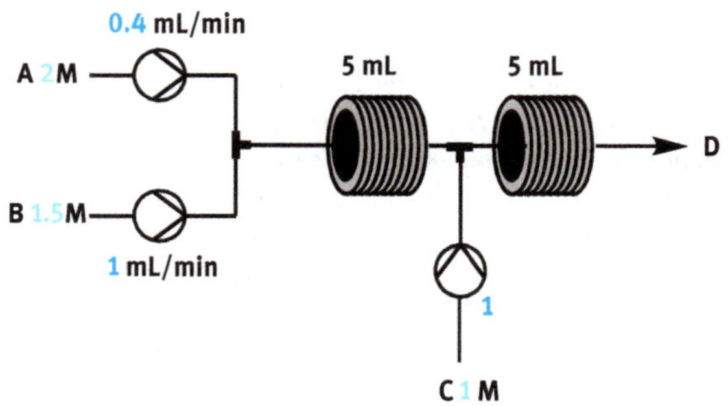

5.4 Consider that A is the limiting reagent of this reaction. If you collect the reaction mixture exiting the system for 10 min how many moles (or mmoles) of D will you collect?

References

[1] Wiles, C, Watts, P, Continuous flow reactors, a tool for the modern synthetic chemist, European J Org Chem, 2008, (10), 1655–1671. doi: 10.1002/ejoc.200701041.

[2] Gutmann, B, Cantillo, D, Kappe, CO, Continuous-flow technology – A tool for the safe manufacturing of active pharmaceutical ingredients, Angew Chemie – Int Ed, 2015, 54(23), 6688–6728. doi: 10.1002/anie.201409318.

[3] Watts, P, Wiles, C, Recent advances in synthetic micro reaction technology, Chem Commun, 2007, (5), 443–467. doi: 10.1039/b609428g.

[4] Razzaq, T, Kappe, CO, Continuous flow organic synthesis under high-temperature/pressure Conditions, Chem – An Asian J, 2010, 5(6), 1274–1289. doi: 10.1002/asia.201000010.

[5] Cambié, D, Noël, T, Solar photochemistry in flow, Top Curr Chem, 2018, 376(6), 1–27. doi: 10.1007/s41061-018-0223-2.

[6] Plutschack, MB, Pieber, B, Gilmore, K, Seeberger, PH, The Hitchhiker's guide to flow chemistry, Chem Rev, 2017, 117(18), 11796–11893. doi: 10.1021/acs.chemrev.7b00183.

[7] Yoshida, JI, Takahashi, Y, Nagaki, A, Flash chemistry: Flow chemistry that cannot be done in batch, Chem Commun, 2013, 49(85), 9896–9904. doi: 10.1039/c3cc44709j.

[8] Hone, CA, Kappe, CO, The use of molecular oxygen for liquid phase aerobic oxidations in continuous flow, Vol. 377, Springer International Publishing, 2019. doi: 10.1007/s41061-018-0226-z.

[9] Guidi, M, Seeberger, PH, Gilmore, K, How to approach flow chemistry, Chem Soc Rev Published online 2020. doi: 10.1039/c9cs00832b.

[10] Baumann, M, Baxendale, IR, The synthesis of active pharmaceutical ingredients (APIs) using continuous flow chemistry, Beilstein J Org Chem, 2015, 11, 1194–1219. doi: 10.3762/bjoc.11.134.

[11] Ghislieri, D, Gilmore, K, Seeberger, PH, Chemical assembly systems: Layered control for divergent, continuous, multistep syntheses of active pharmaceutical ingredients, Angew Chemie – Int Ed, 2015, 54(2), 678–682. doi: 10.1002/anie.201409765.

[12] Mason, BP, Price, KE, Steinbacher, JL, Bogdan, AR, McQuade, TD, Greener approaches to organic synthesis using microreactor technology, Chem Rev, 2007, 107(6), 2300–2318. doi: 10.1021/cr050944c.

[13] Vaccaro, L, Lanari, D, Marrocchi, A, Strappaveccia, G, Flow approaches towards sustainability, Green Chem, 2014, 16(8), 3680–3704. doi: 10.1039/c4gc00410h.

[14] Chentsova, A, Ushakov, DB, Seeberger, PH, Gilmore, K, Synthesis of α-nitro carbonyls via nitrations in flow, J Org Chem, 2016, 81(19), 9415–9421. doi: 10.1021/acs.joc.6b01634.

[15] Wang, Y, Huang, NY, Shen, JQ, Liao, PQ, Chen, XM, Zhang, JP, Hydroxide Ligands cooperate with catalytic centers in metal-organic frameworks for efficient photocatalytic CO 2 reduction, J Am Chem Soc, 2018, 140(1), 38–41. doi: 10.1021/jacs.7b10107.

[16] Cambié, D, Bottecchia, C, Straathof, NJW, Hessel, V, Noël, T, Applications of continuous-flow photochemistry in organic synthesis, material science, and water treatment, Chem Rev, 2016, 116(17), 10276–10341. doi: 10.1021/acs.chemrev.5b00707.

[17] Noël, T, Cao, Y, Laudadio, G, The fundamentals behind the use of flow reactors in electrochemistry, Acc Chem Res, 2019, 52(10), 2858–2869. doi: 10.1021/acs.accounts.9b00412.

[18] Pibiri, I, Buscemi, S, Palumbo Piccionello, A, Pace, A, Photochemically produced singlet oxygen: Applications and perspectives, ChemPhotoChem, 2018, 2(7), 535–547. doi: 10.1002/cptc.201800076.

[19] Ghogare, AA, Greer, A, Using singlet oxygen to synthesize natural products and drugs, Chem Rev, 2016, 116(17), 9994–10034. doi: 10.1021/acs.chemrev.5b00726.

[20] Suga, S, Okajima, M, Fujiwara, K, Yoshida, JI, Cation flow method: A new approach to conventional and combinatorial organic syntheses using electrochemical microflow systems, J Am Chem Soc, 2001, 123(32), 7941–7942. doi: 10.1021/ja015823i.
[21] Reichart, B, Kappe, CO, Glasnov, TN, Phase-transfer catalysis: Mixing effects in continuous-flow liquid/liquid O-and S-Alkylation processes, Synlett, 2013, 24(18), 2393–2396. doi: 10.1055/s-0033-1339839.
[22] Viviano, M, Glasnov, TN, Reichart, B, Tekautz, G, Kappe, CO, A scalable two-step continuous flow synthesis of nabumetone and related 4-aryl-2-butanones, Org Process Res Dev, 2011, 15(4), 858–870. doi: 10.1021/op2001047.
[23] Fuse, S, Tanabe, N, Takahashi, T, Continuous in situ generation and reaction of phosgene in a microflow system, Chem Commun, 2011, 47(47), 12661–12663. doi: 10.1039/c1cc15662d.
[24] Munirathinam, R, Huskens, J, Verboom, W, Supported catalysis in continuous-flow microreactors, Adv Synth Catal, 2015, 357(6), 1093–1123. doi: 10.1002/adsc.201401081.
[25] Irfan, M, Glasnov, TN, Kappe, CO, Heterogeneous catalytic hydrogenation reactions in continuous-flow reactors, ChemSusChem, 2011, 4(3), 300–316. doi: 10.1002/cssc.201000354.
[26] O'Brien, M, Taylor, N, Polyzos, A, Baxendale, IR, Ley, SV, Hydrogenation in flow: Homogeneous and heterogeneous catalysis using Teflon AF-2400 to effect gas-liquid contact at elevated pressure, Chem Sci, 2011, 2(7), 1250–1257. doi: 10.1039/c1sc00055a.
[27] Nagy, KD, Shen, B, Jamison, TF, Jensen, KF, Mixing and dispersion in small-scale flow systems, Org Process Res Dev, 2012, 16(5), 976–981. doi: 10.1021/op200349f.
[28] Yoshida, JI, Flash chemistry using electrochemical method and microsystems, Chem Commun, 2005, (36), 4509–4516. doi: 10.1039/b508341a.
[29] Nagaki, A, Ichinari, D, Yoshida, JI, Three-component coupling based on flash chemistry. Carbolithiation of benzyne with functionalized aryllithiums followed by reactions with electrophiles, J Am Chem Soc, 2014, 136(35), 12245–12248. doi: 10.1021/ja5071762.
[30] Glasnov, TN, Kappe, CO, The microwave-to-flow paradigm: Translating high-temperature batch microwave chemistry to scalable continuous-flow processes, Chem – A Eur J, 2011, 17(43), 11956–11968. doi: 10.1002/chem.201102065.
[31] Pieber, B, Shalom, M, Antonietti, M, Seeberger, PH, Gilmore, K, Continuous Heterogeneous Photocatalysis in Serial Micro-Batch Reactors, Angew Chemie – Int Ed, 2018, 57(31), 9976–9979. doi: 10.1002/anie.201712568.
[32] Chatterjee, S, Guidi, M, Seeberger, PH, Gilmore, K, Automated radial synthesis of organic molecules, Nature, 2020, 579(7799), 379–384. doi: 10.1038/s41586-020-2083-5.
[33] Usutani, H, Tomida, Y, Nagaki, A, Okamoto, H, Nokami, T, Yoshida, JI, Generation and reactions of o-bromophenyllithium without benzyne formation using a microreader, J Am Chem Soc, 2007, 129(11), 3046–3047. doi: 10.1021/ja068330s.
[34] Britton, J, Jamison, TF, The assembly and use of continuous flow systems for chemical synthesis, Nat Protoc, 2017, 12(11), 2423–2446. doi: 10.1038/nprot.2017.102.
[35] Vukelic, S, Ushakov, DB, Gilmore, K, Koksch, B, Seeberger, PH, Flow synthesis of fluorinated α-amino acids, European J Org Chem, 2015, 2015(14), 3036–3039. doi: 10.1002/ejoc.201500300.

Debasis Mallik, Wenyao Zhang, Mathieu Morin and
Michael G. Organ

6 Fundamentals of continuous manufacturing PAT: sampling, analysis, and automation

6.1 Introduction

The concept of continuous manufacturing is not new. Leiston Works in the county of Suffolk, UK, is believed to be the first organization to adopt assembly-line manufacturing of portable steam engines [1]; they called their assembly-line manifesting, "the long shop." As the size of the product shrunk from the bulk of the steam engine to molecules that are too small to manipulate with human motor skills, the complexity in the development of continuous manufacturing processes grew exponentially [2, 3]. Moving reagents reliably through an energized reactor space requires a host of configurable machinery that needs to remain functional throughout the lifecycle of a synthetic run, a routine task that can be better administered by a computer. However, automation in the manufacturing of fine chemicals in its present state requires supervision by humans to tackle challenges that are not routine. Continuous manufacturing, which is labeled by regulatory agencies like the USFDA (United States Food and Drug Administration) and EMA (European Medicines Agency) as an emerging technology that will lead pharmaceutical modernization, requires constant vigilance to ensure that all equipment required for fluid movement (e.g., pumps), fluid management (e.g., valves and chips, phase separators), and fluid energizing (e.g., heaters, coolers, sonicators, agitators, and light emitting sources) perform reliably [4]. Orchestrating such a large cluster of configurable tools to move streams of reactive matter (from viscous slurries to effervescent liquids) through an energized space in a controllable manner can be challenging. The machinery is destined to ultimately encounter mechanical glitch, leading to batch failures. Monitoring a process at the right frequency for identifying current failures as they occur and forecasting possible future glitches before they occur is critical to mitigating risks of spoiling an entire production batch.

Process analytical technology (PAT) is a field of process chemistry that helps process scientists invigilate a production campaign for a safe and sustainable operation. USFDA, EMA, and other regulatory agencies have made clear recommendations in the recent past for the inclusion of appropriate PATs in manufacturing methods to ensure that the dynamic platform of continuous manufacturing gets the much needed frequent monitoring, and produces quality product at all times [5].

A PAT tool comes with two core instrumentation technologies. The first includes a mechanism that extracts a representative portion of material(s), which is referred to as the IPC (in-process check) sample, from a reactor without causing any disturbance to the CPPs (critical process parameters) of the process in question. Extraction, which

https://doi.org/10.1515/9783110693676-006

is commonly referred to as sampling, must not alter the composition of matter of the material in the reactor. The second part of the PAT instrumentation includes an analyzer. An analyzer can only monitor attributes that are measurable by the analytical technique in question. For example, an analyzer that monitors the UV absorbance of a chemical entity cannot detect UV-inactive compounds. Consequently, more than one analyzer may be needed to inspect multiple quality attributes. Analytical techniques dictate how (e.g., in series or in parallel), where (e.g., in the middle or at the terminus of a reactor path), and in what combination (e.g., hyphenated technologies like liquid chromatography–gas chromatography (LC-GC) or IR-Raman), a group of analyzers can be used to monitor all CQAs (critical quality attributes) in a process.

6.2 Sampling

A sampling tool must be placed at the right location in the reactor to ensure that the PAT generates meaningful information on CQA of the process. Non-destructive PAT tools (e.g., IR) can be placed anywhere in the process line or on a bypass so that the material can be returned back to the main stream after analysis. Sometimes, destructive PAT tools (e.g., MS or LC, GC-coupled measurement tool) are essential for process monitoring. In those cases, the tool is placed at the terminus or on a bypass since the analysis involves a compositional change of the reaction mixture. In this chapter, PAT tools are classified into four categories based on their placement within the process line: in-line, on-line, at-line, and off-line (Fig. 6.1). Fundamental features of each class are presented in Tab. 6.1.

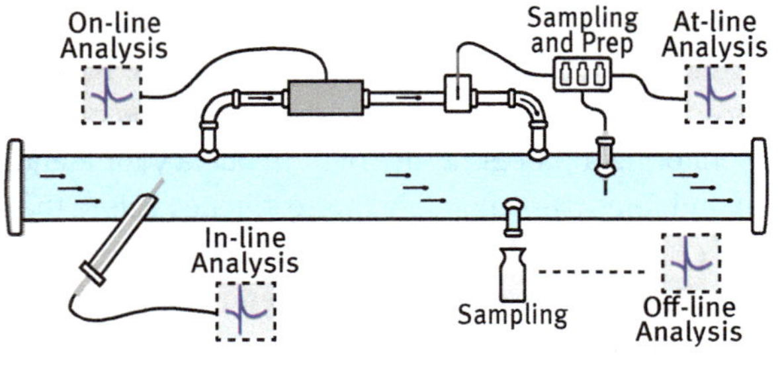

Fig. 6.1: Placement of in-line, on-line, at-line, and off-line PAT tools in a flow reactor.

Tab. 6.1: Classification of PAT tools.

In-line PAT	On-line PAT	At-line PAT	Off-line PAT
– Tool is located along the main stream of the process line – Method does not alter the composition of matter – Analyzed stream stays in the main stream – Amenable to automation	– Tool is located on a bypass line where a portion of the stream can be diverted – Method does not alter the composition of matter – Analyzed matter may reunite with the main stream after analysis. – Amenable to automation	– Tool is located at the terminus or on a bypass line, wherefrom the analyte does not return to the main stream – Method alters the composition of matter – Analyzed matter does not reunite with the main stream after analysis – Amenable to automation	– Tool is physically isolated from the production line; human intervention needed for sample transport and/or processing – Analyzed matter does not reunite with the main stream after analysis. – Requires human involvement

6.2.1 In-line and on-line sampling

In-line and on-line PATs do not require IPC samples to physically exit the reactor. A "window" made from materials transparent to the analytical method allows detection directly from a flow-through cell that forms a single lumen with the reactor [6–11]. This type of sampling occupies in a large section of the current in-line and on-line sampling practices since implementation is relatively straightforward. The cell is always in fluid communication with the process, thereby requiring no configurational modulation of the tool, during sampling. The material used for the window depends on the analytical method used, such as glass for NMR spectroscopy, or quartz for UV/vis spectroscopy. In some cases, a dedicated flow cell may not be necessary if the reactor material can accommodate the analytical technique (e.g., transparent PFA tube for Raman spectroscopy) [12].

6.2.2 At-line and off-line sampling

At-line and off-line modes of sampling, wherein flow paths of the analyzer need to be shielded from the harsh reaction environments, require a configurable device that allows a small portion of the reaction matter to enter the analytical flow paths only during sampling. The device can be a syringe needle that penetrates into a reactor through a pierceable membrane and aspirates a small portion of an IPC sample using a robotic means (Fig. 6.2) [13]. The robotics involved with this type of sampling action are relatively straightforward (typically, the movement of a robotic arm in the z-direction). The pressure tolerance of the pierceable membrane, however, sets a limit to how much pressure the flow-through cell, which forms a lumen with the main process stream, can withstand during the operation. The chemical compatibility of the membrane, especially if the process requires a high temperature, also raises concern. During iterative samplings, pierceable membranes are known to lose elasticity overtime, exposing the reaction matter to open atmosphere. Despite these concerns, the method does provide a simple, yet effective mechanism for probing a manufacturing process repetitively.

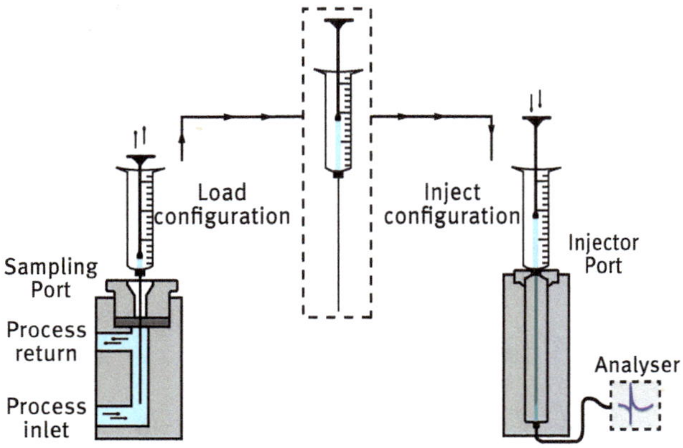

Fig. 6.2: Example of a liquid aspirating sampling device.

Alternatively, a liquid-holding cup that can be submerged in the reaction matter is opened during sampling (Fig. 6.3a) [14]. When the cup is filled with reaction matter, the entry to the cup closes, isolating the sample space from the reactor. The isolated IPC sample is transported to a downstream analyzer (Fig. 6.3b). If the sample is transported too fast, laminar flow in the transport lines could lead to random mixing between the reactor sample and the carrier solvent, in turn causing the analyte concentration to change randomly. For certain applications, IPC samples are quenched in the liquid holding cup itself to ensure that the progression of the reaction is stopped immediately after the matter exits the reactor. One concern with such a sample processing practice comes from the impact of the quenching reagents on the reactor sample, ranging from the formation of solid particulates to exotherms.

A configurable fluid-diverting device, commonly known as sampling valve, can also be used to isolate IPC samples from a continuous flow reactor. Rotary valves use two discrete positions between the stationary (stator) and the rotatable (rotor) components to maintain fluid connectivity within the reactor and analyzer during

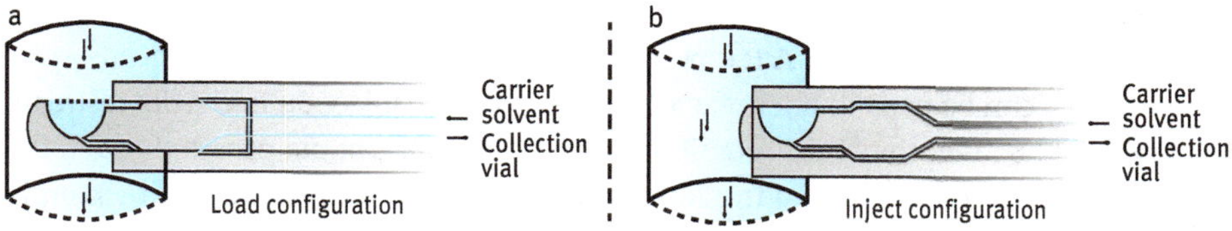

Fig. 6.3: Example of submerged sampling cup during (a) sample loading and (b) sample transport.

sampling. Typically, the valve toggles between two configurations. The first configuration, called the "load" configuration, allows the device to isolate a reactor sample in a sampling loop. The second configuration, called the "inject" configuration, dispenses the isolated sample from the loop to an analyzer. Figure 6.4 illustrates the load and inject configurations of a six-port sampling valve, which adopts the two configurations by rotating its rotor 60o clockwise and counter-clockwise against the stator. Process pressure and the back-pressure in the analytical flow path do not have to be the same. In fact, in most PAT cases, there is a pressure differential between the manufacturing and the analytical streams. When the sampling valve moves from the "load" to the "inject" configuration or vice versa, the fluid in the sampling loop experiences a change in pressure. If the pressure differential is small, the liquid in the sampling loop seamlessly transfers from the reactor stream to the analytical stream. Custom etchings of the configurable flow paths of the rotor are reported in literature to mitigate issues from possible pressure differentials [15, 16].

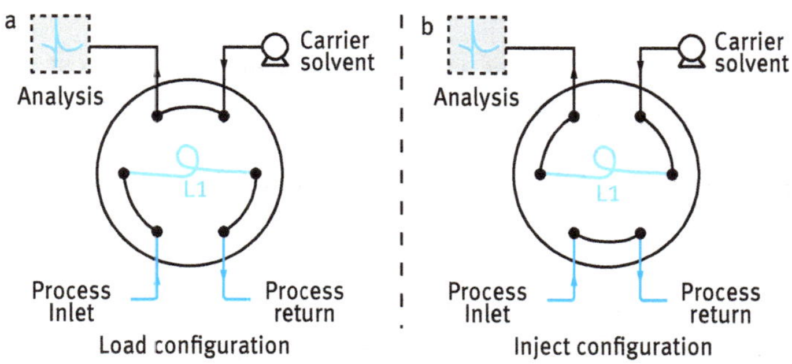

Fig. 6.4: Common six-port sampling valve. (a) Load configuration. (b) Inject configuration. L1: sample loop.

One disadvantage of using a rotary valve as a sampling tool comes from fouling of the configurable flow paths, which are, typically, extremely small. Unfortunately, increasing the physical dimension of these configurable flow paths causes the pressure tolerance of the device to drop. A rise in the process pressure may exceed the pressure threshold of the analytical device, causing the PAT tool to fail altogether. Many PAT tools involving GC, and some recent examples involving LC, have incorporated various head-on filtration mechanisms to remove solid particulates before the reactor samples are allowed to enter analytics. A ten-port sampling valve equipped with a head-on filtration device can be used to filter reactor samples prior to injection into a downstream analytics (Fig. 6.5). In the "load" configuration, the reactor sample passes through the filter before entering the sampling loop. In the "inject" configuration, the sample in the loop is injected into a downstream analytics. The filter membrane is back-flushed by a separate pump. When the reactor throughput involves excessive amount of solids, a multi-position valve sequence can be used to maintain a functional sampling valve [17, 18].

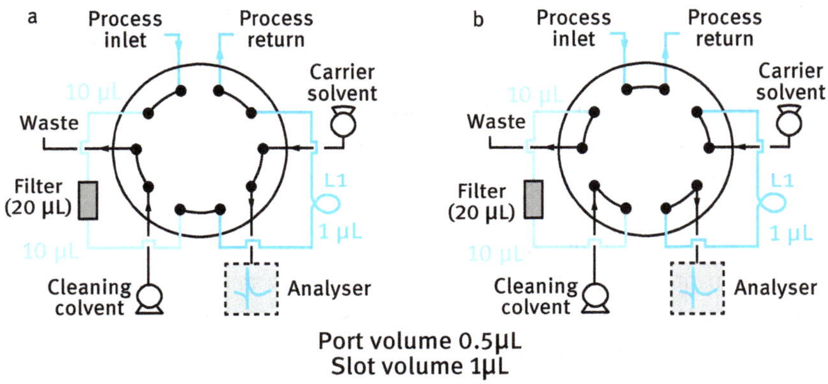

Fig. 6.5: Ten-port valve equipped for continuous sampling and operation. L1: sample loop.

Sample preparation is also an important step in integrated analysis. Almost all forms of analysis require dilution, at the very least, to obtain meaningful results from the downstream analyzer. A tube-in-tube sampling tool, wherein the IPC sample is pulled into an outer capillary tube using one pump, while a diluent is pushed through an inner capillary tube from a second pump, has been developed as a high sampling-rate PAT tool for pharmaceutical manufacturing (Fig. 6.6a) [19]. The push capillary is recessed within the pull capillary, preventing the diluent from contaminating the reaction vessel. The sample is drawn up into an eight-port valve containing two sampling loops (L1 and L2), for LC analysis. The sampling rate of this setup is reported to be as high as 0.1–6.0 µL min^{-1}.

In another example, a single cart PAT solution is reported for a continuous flow pharmaceutical process with a head-on junction between the main stream and the diluent stream (Fig. 6.6b). A small portion of the API stream is transferred from an overflowed tee to a configurable sampling zone that can be temporarily opened using an on-off valve. Subsequent actuation of a second on-off valve allows for the passage of this IPC sample into a second dilution chamber, wherefrom the diluted sample travels to an LC analyzer [20].

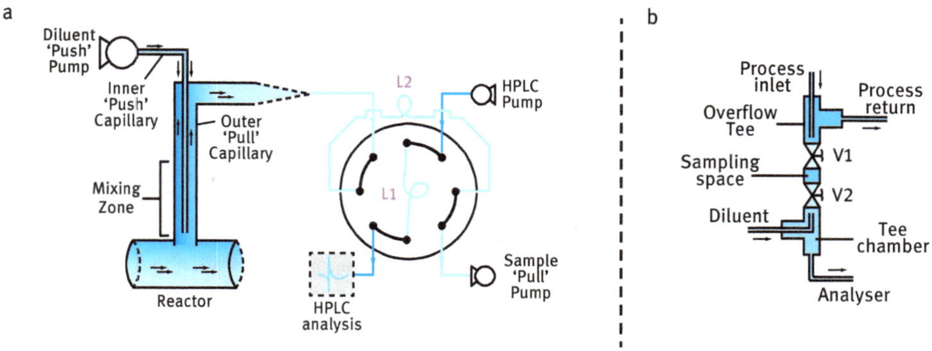

Fig. 6.6: (a) Push–pull sampling system using tube-in-tube design. L1–L2: sample loops. (b) A single-cart sampling system using an overflow-Tee. V1–V2: on–off valves.

6.3 Analysis

Choosing the right analytical tool is critical for monitoring a specific attribute of a process. The dynamic format of continuous production makes analytical methods with short data acquisition time (i.e., fast scan rate) essential in processes that are sensitive to transient failures. High resolutions methods may be necessary to detect small quantities of a catalyst or reactive intermediate present in the reaction mixture, or to validate the overall health of the process or the quality of the reagent streams. A summary of analytical methods and their suitability is presented in Tab. 6.2.

There are many examples of laboratory-scale in-line, on-line, at-line, and off-line tools that use IR, Raman, UV–vis, NMR, and MS as PATs [6, 7, 9, 21–29]. Attenuated total reflectance variants of IR and Raman techniques can allow for reliable data acquisition from a large variety of reaction matrices, including ones that are optically opaque. These optical methods use flow-through cells that do not require configurational adjustment (e.g., switching between load and inject positions). The flow cells can be selected from of a variety of designs that are fabricated specifically for a spectroscopic technique in question and cover a wide range of temperatures, pressures, and chemical compatibility.

An example of in-line measurement that uses fast scan rates involves the real-time concentration measurement of a Grignard reagent (Fig. 6.7) [30]. In this particular example, an in-line IR tool was used to measure the concentration of the Grignard reagent formed in the upstream reactor. The short acquisition time and rapid quantification of aryl magnesium **2** allowed for the accurate dosing of the carbonyl electrophile **3** downstream from the in-line IR detector to achieve maximum conversion towards **4**. This example illustrates the ability of IR spectroscopy to detect a reactive intermediate

Tab. 6.2: Some analytical methods with suitability assessment.

Analytical technique	Critical factors that determine suitability as an integrated PAT tool
Infrared spectroscopy	– Nondestructive; in-line/on-line measurements possible – Short acquisition time; suitable for detecting analytes in trace quantities – Does not provide structural information; analyte identification can be carried out using spectrum libraries – Can be used in conjunction with an integrated separation tool (e.g., at-line LC-IR); inclusion of a separation method extends analysis time – Mostly transparent to reaction matrices as long as the reaction matrices do not carry an IR-active functional group(s) same as the analytes; minimum sample preparation is needed for in-line/on-line measurements; LC injections may require additional sample preparation to safeguard the separation tool – Low interference from heterogeneous matrices
Raman spectroscopy	– Generally nondestructive – some high-powered excitation lasers are reported to degrade compounds; in-line/on-line measurements possible – Short acquisition time; weak Raman signals require long acquisition time – Can be used in conjunction with an integrated separation tool (e.g., at-line LC-Raman); inclusion of a separation method extends analysis time – Does not provide structural information; analyte identification can be carried out using spectrum libraries – Low interference from heterogeneous matrices – Mostly transparent to reaction matrices
UV/vis spectroscopy	– Nondestructive; in-line/on-line measurements possible – Short acquisition time; suitable for detecting analytes in trace quantities – Does not provide structural information; a specific set of reagent(s) and product(s) may allow use of this PAT without a separation tool if chromophores in question do not overlap – Popular PAT with an integrated separation tool (e.g., at-line LC-UV or LC-DAD); inclusion of a separation method extends analysis time – Integrated LC can be used for orthogonal separation in multidimensional, multiattribute analysis from a single injection – Generally suitable for clear analytes – Mostly transparent to reaction matrices as long as the reaction matrices do not carry chromophore(s) same as the analytes; minimum sample preparation is needed for in-line/on-line measurements. LC injections may require additional sample preparation to safeguard the separation tool
NMR spectroscopy	– Nondestructive; in-line/on-line measurements possible – Acquisition time is relatively longer than IR, Raman, or UV technique. Modern high-field NMR instruments can acquire sufficient data for relatively slow reactor throughput (in µL min^{-1} range)

Tab. 6.2 (continued)

Analytical technique	Critical factors that determine suitability as an integrated PAT tool
	– Can be used with an integrated separation tool (e.g., at-line LC-NMR) or without (stand-alone flow NMR); inclusion of a separation method further extends analysis time – Provides structural information at a molecular level; the ability to obtain structural characterization of analytes makes this PAT useful in applications where the structural identification of by-products is critical – Moderate interference from heterogeneous matrices; "shimming," an analytical system-preparation step required before data acquisition, is less troublesome when a matrix does not consist of particulates – High interference from signals originating from the reaction matrix. Typically, deuterated solvents are used to minimize solvent-signal intensities. Nondeuterated solvents can be used with a solvent-signal suppression pulse sequence if no analyte signal appears at the same region of the solvent signal in the Fourier-transformed spectrum; LC-NMR requires eluting solvents to produce signals not overlapping with analytes for routine solvent suppression techniques – A number of NMR-active nuclei can be detected with a multi-nuclear NMR probe; Certain naturally abundant nuclei (e.g., ^{11}B, ^{19}F, ^{13}C, ^{29}Si) allow this PAT to monitor specific analytes bearing these atoms in their chemical structure without interference from reaction matrices [11] – Various multinuclear NMR pulse techniques are available for structural elucidation of complex organic molecules (e.g., HSQC and HMBC) but are accompanied by a long data acquisition time
Mass spectrometry	– Analytes are dissociated into ions; suitable for at-line or off-line measurements – Short acquisition time when coupled as a stand-alone at-line tool; suitable for detecting analytes in trace quantities – Can provide structural information at a molecular ion level; the structural characterization of by-product(s) possible – Can be used with an integrated separation tool (e.g., LC-MS or GC-MS) or without (stand-alone MS); the inclusion of a separation method extends analysis time; various modes of mass spectrometric analysis (ESI, APCI, EI, or CI) allows chemists to couple MS techniques with an appropriate separation tool and also to match the method of ionization in tune with CQA of the process (e.g., GC-CI or LC-APCI is generally used to detect the entire molecular ions, whereas GC-EI or LC-ESI generally uses library to perform/assist structural characterization) – Integrated LC can be used for orthogonal separation in multidimensional multiattribute analysis from a single injection; parent ion/daughter ion scans from a triple quadrupole mass spectrometer can also help with advanced separation of ions and thorough structural identification

in-line from a complex reaction matrix. There are many other examples of similar IR, Raman, and UV PATs in the literature [27–29, 31].

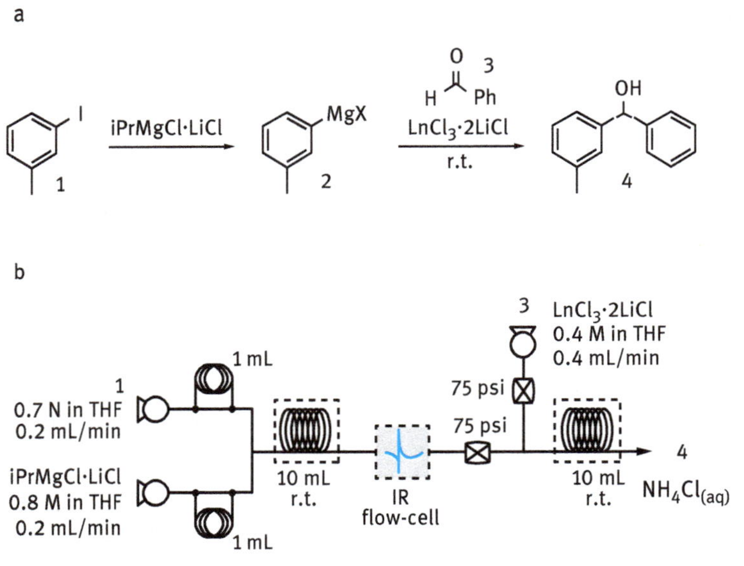

Fig. 6.7: (a) Grignard formation and 1,2 addition. (b) Flow diagram of reactor used in in-line IR monitoring of Grignard reagent, followed by 1,2 addition.

In Fig. 6.8, illustrates a continuous flow method for the nitration of salicylic acid (**5**) that was optimized using NMR spectroscopy [10]. This allowed reaction conditions to be modulated to favor the formation of 3-nitrosalicyclic acid (**7**). The ability of NMR spectroscopy to identify molecular structures helped chemists adjust process conditions toward increasing the consumption of **5** and minimizing the formation of **6** and **8**.

A miniaturized mass spectrometer was shown to provide valuable mechanistic insight in the optimization of process temperature and flow rate for the Diels–Alder reaction between anthranilic acid (**9**) and furan (**10**, Fig. 6.9) [32]. The system made use of a six-port sampling valve with a 5 μL sampling loop to connect the at-line mass spectrometer to the system. Like NMR spectroscopy, mass spectrometry also provides structural information on the by-product(s) during process development.

LC and GC , which must be performed at-line or off-line, are two popular separation techniques that can be coupled to an analysis tool as an integrated at-line PAT. Although chromatographic data provides a wealth of information on the chemical state of a process, long chromatographic runtime can reduce the sampling rate. A Mizoroki–Heck reaction between 4-chlorobenzotrifluoride (**12**) and 2,3-dihydrofuran

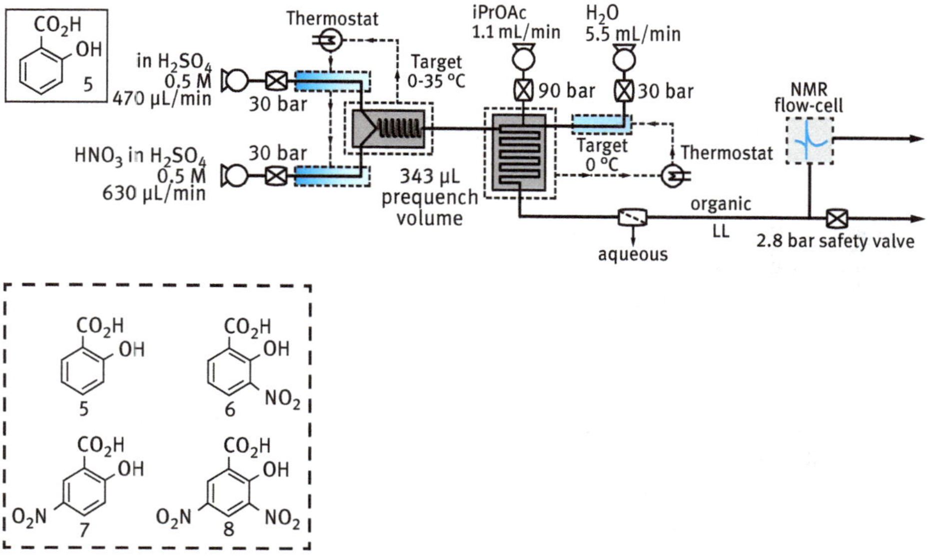

Fig. 6.8: Flow diagram of reactor used in nitration of salicylic acid, monitored by in-line NMR spectroscopy.

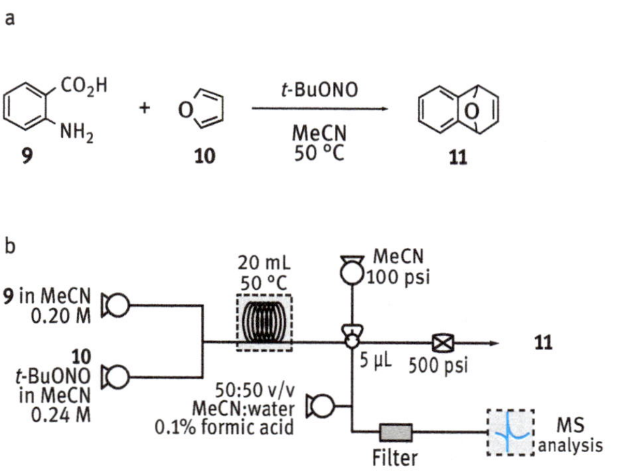

Fig. 6.9: (a) Dehydro-Diels–Alder (DDA) reaction. (b) Flow diagram of reactor used in DDA reaction monitored with at-line mass spectrometry PAT tool.

(**13**) was reported to use at-line LC as a PAT tool for mesoscale scale preparation of compound **14** (Fig. 6.10) [33].

At-line GC was similarly used for the kinetic measurement of a C–H activation reaction using a slug flow technique, where search algorithms were used to self-optimize process parameters from the GC PAT (Fig. 6.11) [34]. In another example, an

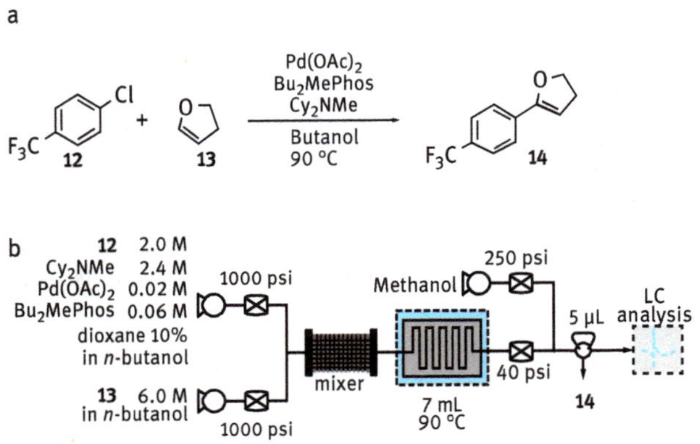

Fig. 6.10: (a) Mizoroki–Heck reaction under flow conditions. (b) Flow diagram of reactor used in at-line LC monitoring of Mizoroki–Heck reaction.

integrated GC PAT is reported in the optimization of benzimidazole from 1,2-diaminobenzene [35]. The second example also includes a detailed description of how a standard liquid handling robot can be customized to obtain reaction specimens from a pressurized reactor without altering CPPs of the chemical transformation in question.

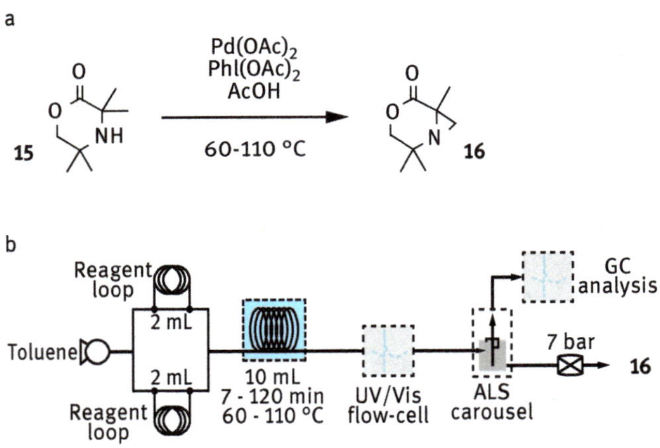

Fig. 6.11: (a) C–H amination reaction in flow conditions. (b) Flow diagram of reactor used in at-line GC monitoring of C–H activation. ALS: automatic liquid sampler.

Quite often, multiple reaction steps are telescoped into a single continuous sequence. The ability to obtain efficient conversions to products in each reaction step depends on the reaction in the previous step. Multiple PATs with appropriate analytical techniques can be put in place to ensure each reaction step forward maximum quantity of the product in the best possible purity to the next step, where the product of the previous step becomes the reagent. An example of an Aldol condensation reaction is shown in Fig. 6.12. A discussion on suitable PATs for this process can be found in the end-of-the-chapter study question section.

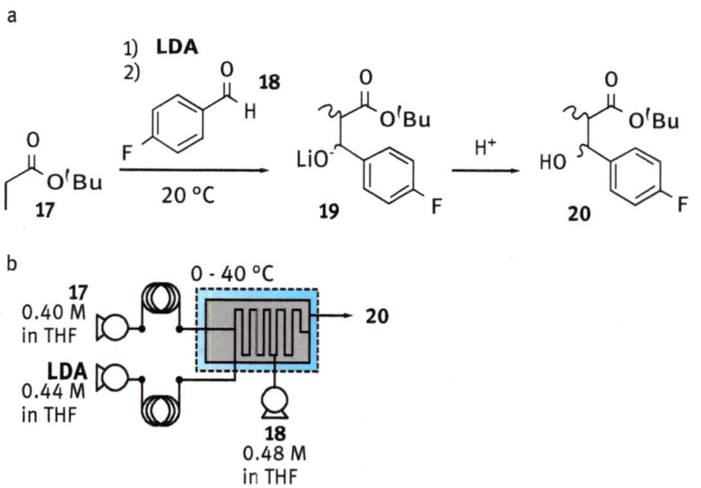

Fig. 6.12: (a) Aldol reaction under flowing conditions. (b) Flow diagram of reactor used for Aldol reaction with multiple PAT tools. LDA: lithium di-isopropylamide.

An example of high-throughput experimentation (HTE) in flow conditions uses two LCMS analyzers integrated to a slug-flow reactor in series (Fig. 6.13). The setup was shown to be effective for the analysis of about 1,500 reactions a day. Slug flow allows for HTE to be carried out reliably at the nanomole scale.

The traditional method of optimizing process parameters, one attribute at a time using one dimensional analysis, can be time-consuming at times. With the advent of multidimensional analysis, especially in the field of chromatography, the assessment of multiple process attributes from a single analytical run is possible. An example of such a two-dimensional LC analysis can be accomplished using a design from Fig. 6.14a [37]. In this 2-dimensional LC design, a pump moves an IPC specimen through the achiral column to reveal chemical purity of the IPC sample first. A second pump then moves the eluent bearing the chromatographed chiral entity from the first chromatogram through a chiral LC column to reveal enantiomeric excess of the chiral API.

The on-analysis stability (or a lack thereof) of analytes under harsh chromatographic conditions can be problematic for certain analytes. While automation and HTE techniques made positive impacts in expediting QC throughputs,

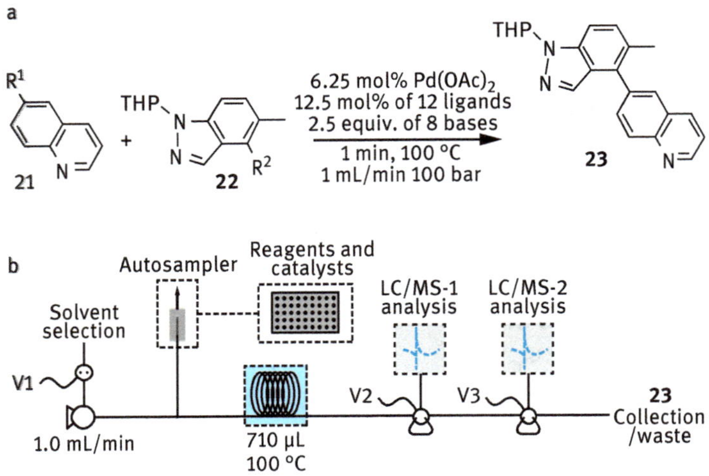

Fig. 6.13: (a) Screening conditions of C–C coupling under flow conditions. (b) Flow diagram of a slug-flow reactor used in reaction screening and synthesis with integrated dual LC-MS at-line PAT tools. V1: solvent selector valve; V2–V3: sampling valves for LC-MS.

there is a need for close monitoring of chromatographic artifacts during analysis. A multidimensional LC technology that recirculates analytes through two LC columns for the simultaneous measurement and validation of multiple process attributes was recently reported [38]. In this recirculatory multidimensional LC design (Fig. 6.14b), analytes of interest are isolated within heart-cut loops after separation and are then recirculated through LC columns to investigate possible artefacts from the analysis. The multidimensional LC method is also shown to provide a large set of analytical results during process developments, using very small quantities of reagents, in a short period of time.

The key strength of continuous manufacturing stems from its ability to transform reaction time into reactor length, providing chemists a means to precisely modulate reactivity by controlling reactor residence time through space. PATs that are capable of measuring product concentrations at various lengths of a continuous flow reactor can rapidly generate important kinetic information for a continuous process. Various PAT tools are reported for such multi-location analysis at constant (steady state) and variable (nonsteady state) flow rates for kinetic studies. A flow reactor, which intersects a 24-port sampling valve four times at various reactor lengths, has been reported recently for reaction kinetics under steady-state conditions (Fig. 6.15). This system allows the simultaneous isolation of four samples from an aza-Michael addition reaction from one valve actuation (Fig. 6.15d). The incorporation of a rotary valve as the PAT tool promises that similar designs may be applied in future manufacturing processes that require high process pressure.

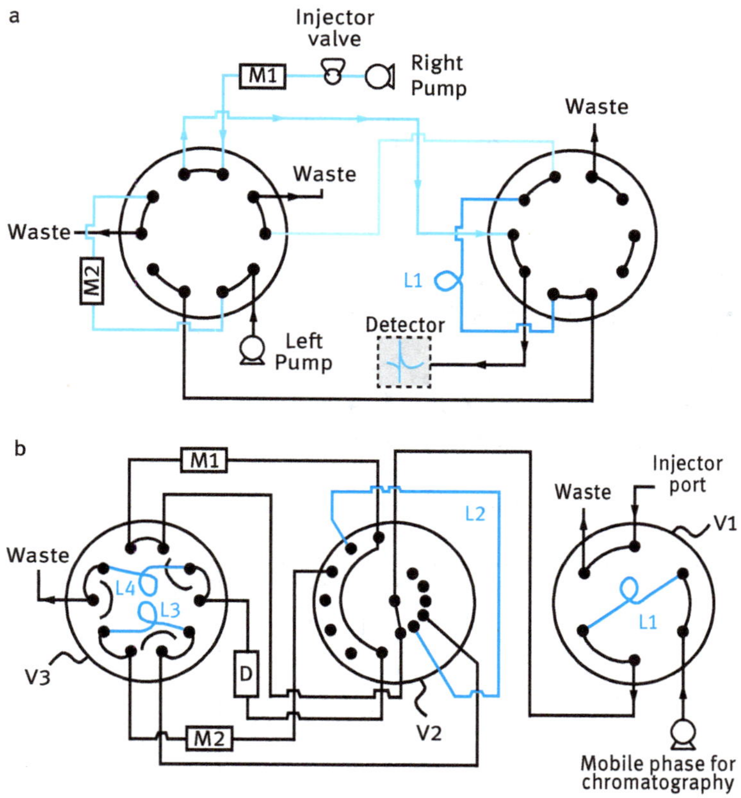

Fig. 6.14: (a) Multidimensional liquid chromatography system. M1: achiral LC column; M2: chiral LC column. (b) Multidimensional LC recapture valve configuration. V1: LC injector valve; V2: column-selector valve; V3: heart-cut valve; L1: injection loop; L2: priming loop; L3 and L4: heart-cut loops; M1: achiral LC column; M2: chiral LC column. D: detector.

An alternative approach to the measurement of reaction kinetics under steady-state conditions involves conducting the flow experiment under a variable flow rate with a constant gradient, as follows:

$$\tau_{\mathrm{ins}} = \tau_{\mathrm{init}} + \alpha \cdot t \tag{6.1}$$

where τ_{ins} and τ_{init} are reactor residence times at time t and at time 0, t is the clock time, and α is the flow-rate gradient

When a single-point analyzer is located at a distance with swept volume of V_{d}, the time of measurement (t_{m}) can be equated to the reactor residence time τ, as follows:

$$\tau = (1 - e^{-\alpha})e^{-\frac{V_{\mathrm{d}}}{V_r}\alpha}\left(t_{\mathrm{m}} + \frac{\tau_{\mathrm{init}}}{\alpha}\right) \tag{6.2}$$

where V_{r} is the volume of the flow reactor.

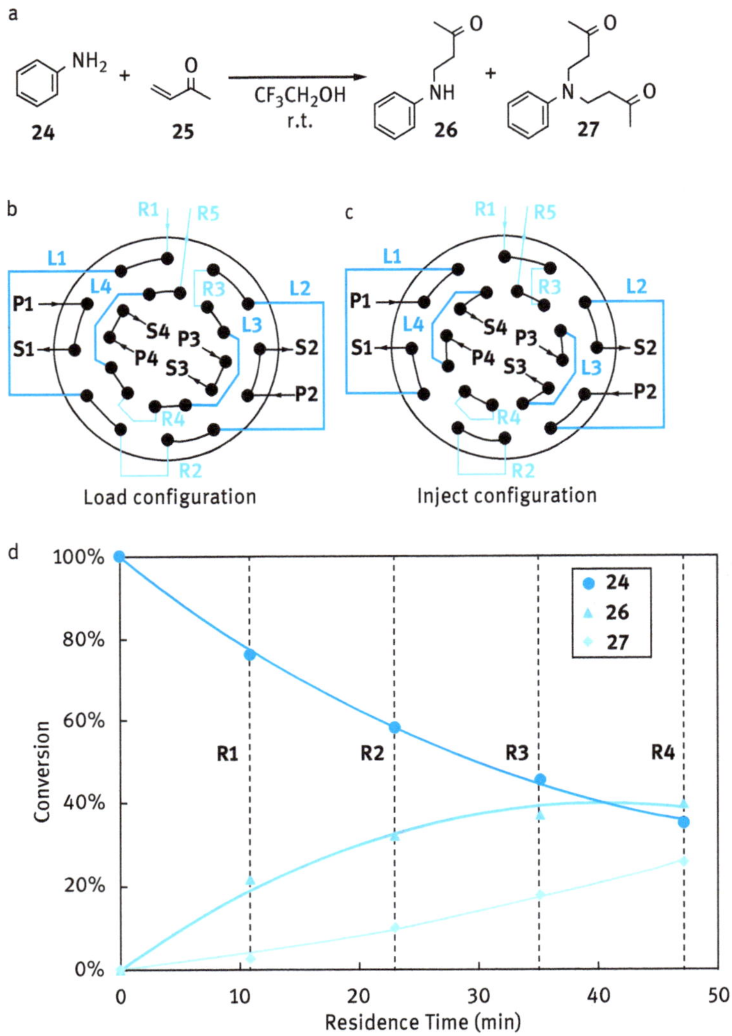

Fig. 6.15: Aza-Michael addition with 24-port multipoint sampling valve PAT device. (a) Reaction scheme. (b) Load configuration. (c) Inject configuration. L1–L4: sampling loops; R1–R5: reactor segments; P1–P4: sampling pumps. S1–S4: samples. (d) Reaction progression from single valve actuation (average of 5).

When reactor residence time τ is plotted against product concentrations, which are obtained from the downstream analyzer at different reaction temperatures, the activation energy of a chemical process can be estimated. A Raman spectroscopy-based movable PAT tool is reported in the monitoring of the Michael addition reaction between piperidine (**28**) and ethyl acrylate (**29**) at eight discrete locations of a microfluidic flow reactor (Fig. 6.16) using this kinetic strategy. Measurements at constant flow rates under steady state and at gradient flow rates under nonsteady state were compared. Two hundred data points were obtained in less than one hour, employing reactor residence times between 0.3 and 0.49 s using the variable flow rate method described earlier. The incorporation of such a rapid kinetic tool could help pave the way towards rapidly obtaining large analytical datasets for the self-optimization of process parameters using machine learning.

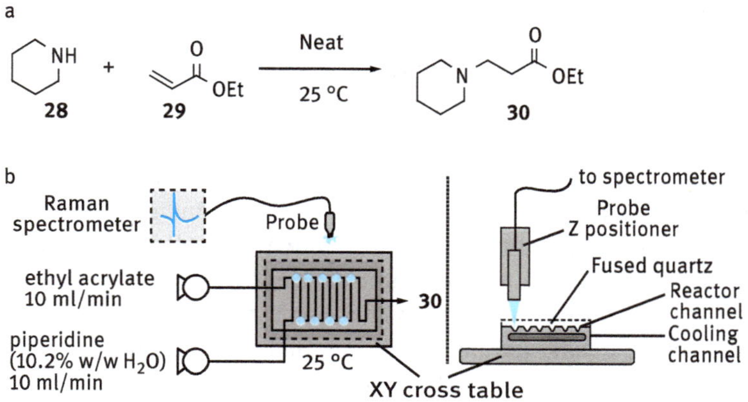

Fig. 6.16: (a) Aza-Michael addition under flow conditions. (b) Flow diagram of reactor used in single- and multipoint monitoring of aza-Michael addition with Raman spectroscopy.

6.4 Computer-aided continuous manufacturing

PAT tools are routinely used as an integral part of the production equipment for quality control and quality assurance in manufacturing [39]. Human resources are usually on alert to tackle unforeseen circumstances due to failure in the machinery or inconsistency in the quality of raw feeds. The integration of PAT is straightforward in batch manufacturing because the developer of the process garners sufficient knowledge from typical laboratory-scale validation experiments before choosing appropriate PAT parameters in a production campaign. Once raw feed is committed to a batch production, the manufacturer cannot afford a large number of process deviations, as the deviations could result in batch failure. The dynamic format of continuous manufacturing makes the probability of spotting a process failure directly proportional to the sampling frequency. Therefore, regulatory agencies recommend samplings as frequently as possible in continuous manufacturing in order to catch all out-of-specification (OoS) occurrences. The sooner one catches a production glitch, the better chance that the operator has to fix the problem, divert OoS lot(s) away from the main stream, and save a campaign. Making such complex impromptu decisions is tough for computers and human alike because straightforward if–else logic often leads to conflicting thoughts that do not lead to a resolution. In batch, humans apply their knowledge from past experiences and make thoughtful decisions with ample time at hand. In continuous manufacturing, the flow does not stop. The decision/correction has to be made as soon as possible, and machine-derived decisions can be no less accurate than human operators. Fortunately, computers can be trained through experiential learning to make such smart and prompt decisions. Historical knowledge of an ongoing process can be used to feed computers with mathematical formulae that connect input process parameters and output process attributes to help computers make intelligent, and hopefully, productive decisions. Artificial intelligence (AI) is an area of science that deals with training of machines for making intelligent decisions based on the machine's experiential learning. AI-enabled PAT is expected to usher in a renaissance of machine-operated autonomous manufacturing of commodities in the near future by equipping PAT tools with mathematical strategies, commonly called algorithms. Algorithms can be tasked with various objectives, ranging from finding an optimal solution to maintaining a process within a boundary, around that optimal solution. It is important to recognize that algorithms are mathematical tools suitable for undertaking a specific task using a set strategy that is built prior to the execution of the task. If the strategy happens to have the ability to find a solution for a particular problem, the algorithm will be able to rapidly act on the task. If the operative strategy in the algorithm does not posses the ability to solve the problem in question, the algorithm could iterate aimlessly without converging to a meaningful decision.

A metaheuristic search algorithm looks for best-possible answer(s) by quantifying the distance between the current position of a candidate solution and the position of a desirable destination, and modifies the candidate solution appropriately to reach the destination iteratively. Typically, a search algorithm deals with not one, but a group of candidate solutions collectively called the population. All candidate solutions individually express certain characteristics and collectively display a group behavior. Individual candidates are modified (or dropped) by the algorithm from iteration to iteration to converge the population toward the desired objective. Illustrations of how algorithms target a problem and arrive at an answer is presented in following sections, which explain the basic principles of algorithmic search through visualized examples so one can appreciate the complex forms of similar search operations in real-life experiments.

6.4.1 Search algorithms

Imagine a manufacturing method where reagents A and B give product C; the quality of which depends on the quality of raw feeds, A and B. This can be represented as follows:

$$\begin{aligned} &A + B \rightarrow C \\ &z = f(ax^m,\ by^n) \end{aligned} \tag{6.3}$$

where z, x, and y are chemical purities of product C, reagent A, and reagent B, respectively, f is an unknown function to be discovered by an algorithm through metaheuristic, and search, a, b, m, and n are real positive numbers.

Algorithms can be used to find minimum purity requirements for raw feed streams, A and B, that give the product C at a minimum purity of 90% ($z > 90\%$). A successful search would help cut the cost of the raw feed without compromising the quality of product C. It is important to recognize that the algorithm needs feedback from the process in order to discover function f of equation (6.3). The algorithm accomplishes the task through systematic manipulations of x and y based on experimentally observed responses for z. Mathematical equations governing the search space of real-life problems are often not smooth or predictable, thus forcing algorithms to perform many iterations and examine input-output relationships in all regions of the searchable space. The differential in inputs between iterations defines the velocity of the search, which dictates how quickly a search can be completed. On the other hand, the vector, a mathematical function that determines the direction in search, adds diversity, and ensures thoroughness in the search. There are many population-based search algorithms that are capable of conducting a diversified search for process optimization.

Particle swarm optimization (PSO) is one of the popular population-based search algorithms that is inspired by the migratory habits of birds. Quite often, birds are seen to form elaborate patterns in the sky as a flock of follower birds take the lead from a leader bird and moves to a destination for better habitat. The vector and the velocity of the leader bird dictate how likely and how soon the flock will arrive at the destination. The leader bird, which is presumably at a coordinate closest to the destination, represents the global best solution at a given point in time. The remaining birds will adjust their individual direction and speed to follow the leader bird. The flock will travel together in a general direction of the destination while maintaining diversity in individual flight paths to conduct a thorough search for the habitat. Flight path diversification is important because sometimes solutions (the habitat) can be hidden inside an uneven search space that the leader bird may accidentally miss. It is also important to recognize that the search algorithm uses probability. Every time, the algorithm may not return the exact set of candidate solutions. But the algorithm is expected to ultimately arrive at the same solution, with enough iterations. The four-iteration PSO search has been presented in Tab. 6.3 to demonstrate this search principle. Mathematically, the mode of PSO operation can be summarized as

$$\begin{aligned} V_{\text{new}} = c1 * V_{\text{Current}} &+ c2 * (\text{personalBest} - \text{currentPosition}) \\ &+ c3 * (\text{globalBest} - \text{currentPosition}) \end{aligned} \tag{6.4}$$

where $c1$, $c2$, and $c3$ are positive numbers, $V_{current}$ and V_{new} are velocities of a candidate solution before and after an iteration cycle, personalBest and currentPosition are coordinates of the best position and the current position of a candidate solution, and globalBest is the best position of the leading solution.

Tab. 6.3: Search history of first four iteration for a PSO run with the lead candidate highlighted.

Solutions	Iteration 1			Iteration 2			Iteration 3			Iteration 4		
	x%	*y*%	*z*%	*x*%	*y*%	*z*%	*x*%	*y*%	*z*%	*x*%	*y*%	*z*%
Candidate I	25	5	3	55	35	21	75	65	49	82	95	79
Candidate II	15	30	6	45	60	28	75	88	67	86	99	86
Candidate III	25	20	5	55	50	28	75	80	60	83	98	82
Candidate IV	70	70	49	71	71	51	72	72	51	73	79	58
Candidate V	25	20	5	55	50	28	85	75	64	95	94	90

In the first iteration, the algorithm identifies the fourth candidate of Tab. 6.3 as the leading solution because the product purity (z) is the highest among all candidates ($z = 49$). The algorithm will then modify all x and y entries (candidates I, II, III, and V) in iteration 2 to follow suit from the leading candidate of iteration 1 ($x = 70$, $y = 70$). The rate of change (i.e., the velocity) in those modifications will be determined by the differential between the input parameters of the leading candidate ($x = 70$, $y = 70$) and those of the following candidates (I, II, III, and V, iteration 1). It is important to note that the $c1$ coefficient, which in this illustrative example is an evolving coefficient that starts at 0.9, linearly loses its strength all the way to a value of 0.2 by the end of the fourth iteration. The remaining coefficients ($c2$ and $c3$) are kept constant at 2. This allows candidate solutions to rely more on the global best and their own best positions (local best), and less on their absolute current positions as the search matures. Figure 6.17a–d demonstrates how candidate solutions improve the value of z (i.e., the product purity), iteration by iteration, toward finding the optimal reagent purities of reagents A and B. Candidate IV (Tab. 6.3, iteration 2), candidate II (Tab. 6.3, iteration 3), and candidate V (Tab. 6.3, iteration 4) lead the race in iterations 2, 3, and 4, respectively. The following candidates would change the direction of search, with each iteration, in accordance with the leader. It is important to recognize that all candidate solutions are expected to ultimately converge towards candidate V of iteration 4 ($z = 90\%$), if the algorithm is allowed to iterate further. Algorithms can also be equipped with relevant boundary conditions in order to find a solution that is executable from real experiments. For example, if an algorithm is tasked to find an optimal flow rate for a continuous manufacturing process, a minimum and maximum flow rate can be set as boundary conditions for the search. It is also worth noting that in a reaction optimization experiment, where 5 candidate solutions are tested for up to 3 iterations, a total

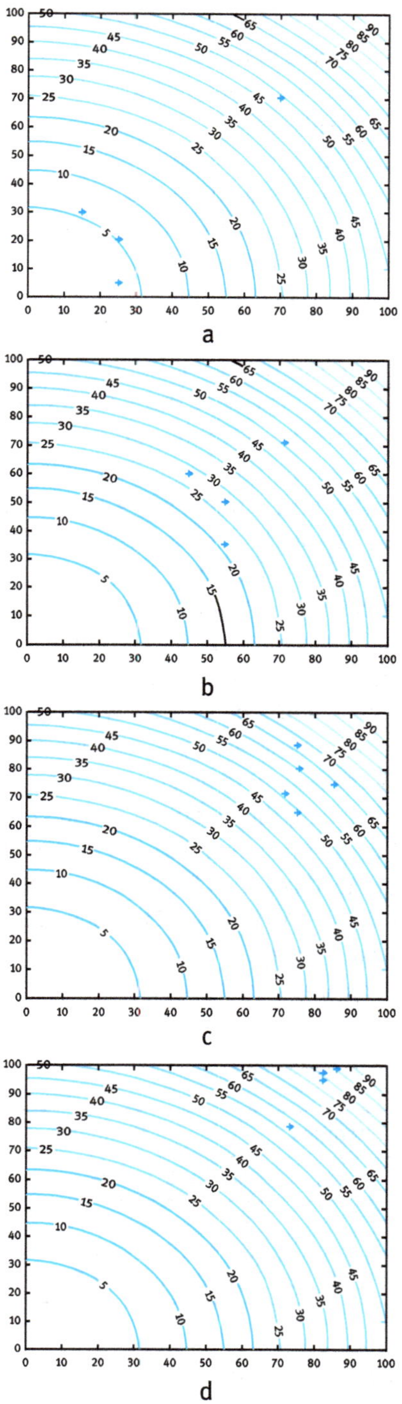

Fig. 6.17: Graphical representation of four iterations of a search by the PSO algorithm. Contours in each figure represent a constant value of product purity (z). Percent purities of A and B are shown in the X and Y directions. Dots in each figure represent the location of the candidate solutions in each iteration as per Tab. 6.3.

of 15 experiments will have to be run for the optimization routine. The more candidate solutions and iterations are assessed, the better the chances of locating a global maximum. However, this comes at a cost of more optimization time, higher cost of chemicals, and strain on the computing network running the algorithm.

The propagation mechanism of algorithmic searches can be further visualized graphically by another population-based search algorithm that uses geometric movement (e.g., reflection of candidate solution(s) against a virtual plane or line in the design of experiment (DoE) space). This algorithm is known as the Nelder–Mead algorithm. Here, the search begins with a minimum of 3 randomly chosen candidate solutions. Once all 3 results (the outputs) for all 3 candidate solutions are obtained from laboratory experiments, the best and the worst input parameters are connected by a virtual line in the DoE space. The first three candidate solutions of Tab. 6.3 will be used to visually demonstrate the evolution of the Nelder–Mead search. Inputs and outputs, shown up to two decimal places, of these experiments are shown in Tab. 6.4. In the first iteration, the third candidate ($x = 25$, $y = 20$) is reflected across the virtual line between the best (II) and the worst (I) candidates to come up with a new candidate for the second iteration ($x = 15$, $y = 15$). The second-generation population (iteration 2, Tab. 6.4) now includes candidates I and II of iteration 1 and a new candidate III solution. Candidate III of iteration 2 gives purity of 2.2, which continues to be the worst candidate from iteration 1 to iteration 2. In this instance, the algorithm employs another search propagation tool, which involves the "contraction" in the reflection method described earlier. In this example, the reflection has been cut by a factor of 2 (set prior to the start of the search) to obtain a new candidate III ($x = 17.5$, $y = 16.25$) in iteration 3. Upon contraction, the algorithm would continue its search with a reflection across the line of the best (candidate II) and the worst (candidate III) solutions to obtain a new candidate (candidate I of Tab. 6.4) for the fourth iteration. An improvement, albeit small, in the purity of product C is reported for candidate I ($z = 8.8$) after the fourth iteration. A similar strategy for the expansion of the reflection method also exists in the Nelder–Mead protocol as situations warrant.

Tab. 6.4: Candidate solutions for first four iterations of Nelder–Mead optimization.

Solutions	Iteration 1			Iteration 2			Iteration 3			Iteration 4		
	x%	*y*%	*z*%	*x*%	*y*%	*z*%	*x*%	*y*%	*z*%	*x*%	*y*%	*z*%
Candidate I	25	5	3.2	25	5	3.2	25	5	3.2	7.5	41.25	8.8
Candidate II	15	30	5.6	15	30	5.6	15	30	5.6	15	30	5.6
Candidate III	25	20	5.1	15	15	2.2	17.5	16.25	2.9	17.5	16.25	2.9

A complex version of the previously described simplex Nelder–Mead method was reported in the optimization of several CPPs (reaction temperature, reagent concentration, and reagent residence time inside a flow reactor) in the hydration of 3-cyanopyridine (Fig. 6.18) [40]. In this algorithm, reflection, expansion, and contraction protocols were used to move candidate solutions towards the best production condition for making amide **32** from nitrile **31**. This particular search strategy started its search with four candidate solutions in a population. Using this Nelder–Mead strategy, the nitrile hydration reaction was found to produce the best response ratio between the MS signal from the product and that from the starting nitrile at 91 °C for a residence time of 12.88 min at an inlet concentration 0.202 M of nitrile. Optimization was carried out over 12 experiments over 17 h. An at-line GC-MS analyzer was used to obtain feedback from the nitrile hydrolysis process.

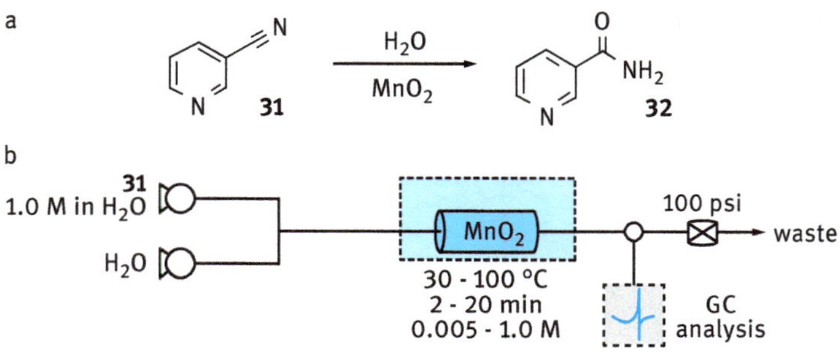

Fig. 6.18: (a) Nitrile hydration reaction under flow conditions. (b) Flow diagram of reactor used in the optimization of a nitrile hydration by Nelder–Mead algorithm.

The Stable Noisy Optimization by Branch and Fit (SNOBFIT) is another population-based search algorithm that divides the search space into smaller sub-spaces and conducts sub-space searches in parallel. The algorithm identifies trends in the search pattern and chooses a sub-search that follows a trend. The sub-space that is not selected during an iteration is replaced by another sub-space using a branch prediction method. The SNOBFIT algorithm was shown to work well in the optimization of four process parameters (reaction temperature, reagent flow rate, reagent stoichiometry, and reagent residence time inside a flow reactor) in the synthesis of the acrylamide **36** from aniline **33** (Fig. 6.19). A total of 40 experiments were needed to find the optimal reaction condition, which gave 92% yield of the production at 117.8 °C using 16 equivalent. of Et_3N and 1.7 equivalent. of **34** with a reactor residence time of 12.2 min.

In Fig. 6.20, 2-chloronitrobenzene (**38**) was reacted with nine different amines to give the corresponding aniline derivative [41]. The reaction parameters were optimized by SNOBFIT within 12 h using 34 experiments to give product **35** at 94% yield and at 320 mg h^{-1} throughput in the presence of 10 equivalent of tributyl amine at 100 °C with a reactor residence time of 1.82 min.

Every algorithm follows its own mathematical strategy to converge into an optimal solution. However, the mathematical formulation of the optimizable attribute

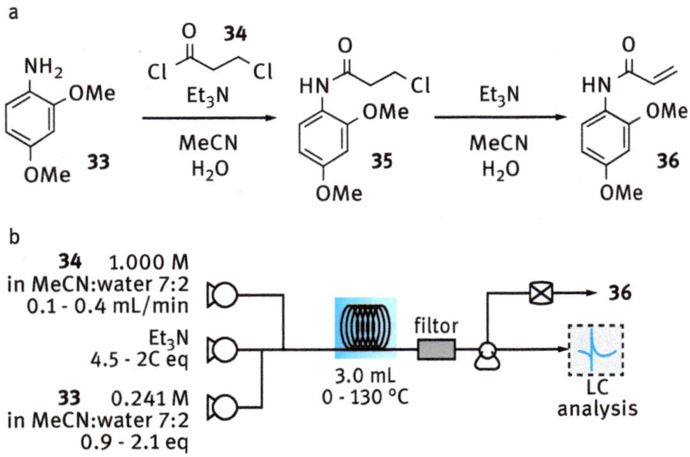

Fig. 6.19: (a) Acrylamide synthesis under flow conditions. (b) Flow diagram of modular reactor used for Nelder–Mead optimization.

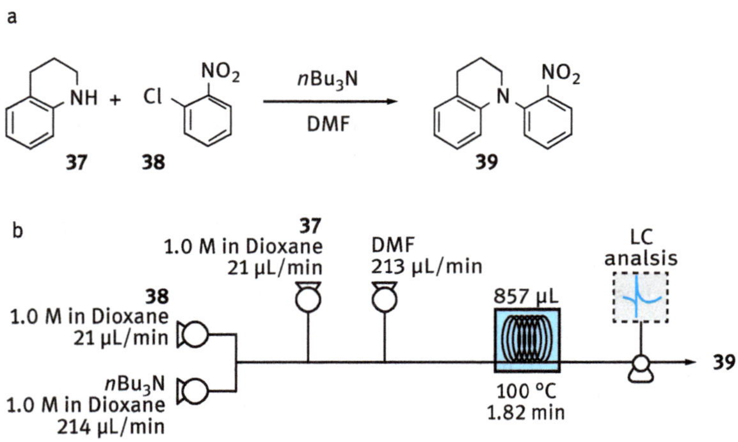

Fig. 6.20: S_NAr reaction with at-line HPLC optimized using a SNOBFIT algorithm.

sometimes dictates how rapidly the convergence would occur. An example of such a modulation of the optimizable attributes is shown in the bromination of a secondary benzylic alcohol using carbon tetrabromide and triphenylphosphine (i.e., the Appel reaction) [40]. A set of five process parameters were optimized for three key quality attributes, which were throughput, conversion to the production, and consumption of starting materials. Figure 6.21 contains two equations, both of which include all three quality attributes for the Appel reaction as sums of individual attributes with different weight factors. Equation (6.5) combines IR responses from the starting alcohol and forming phosphine oxide with consumptions of the brominating reagent and phosphine in a single mathematical function. Equation (6.6) separates IR responses from the consumption of the brominating agent and starting phosphine. The optimization, using equation (6.6), was shown to converge much more effectively than when using Equation (6.5).

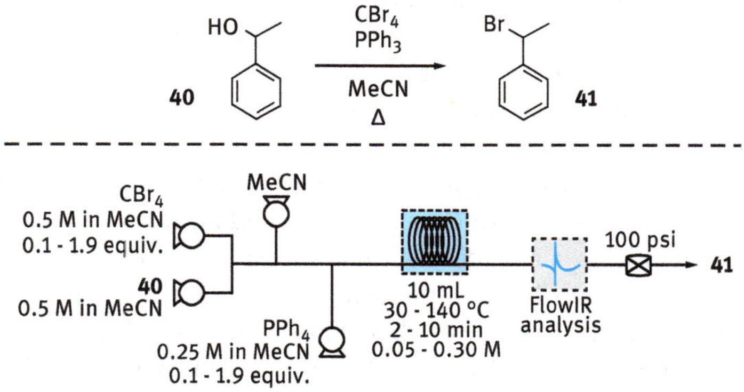

(Eq. 6.5)

$$F1() = 0.25\left(\frac{1}{t} + z\right) + \frac{16p}{s + 0.25(x+y)}$$

Throughput, Conversion, Consumption, Conversion

(Eq. 6.6)

$$F2(\ldots) = 0.15\left(\frac{1}{t} + z\right) + \frac{p}{s} + 0.05\frac{1}{x+y}$$

Throughput, Conversion, Consumption

Where
t = residence time
z = overall concentration
p = PPh_3O IR response
s = alcohol IR response
x = equiv. of CBr_4
y = equiv. of PPh_3
F1 = the first evaluation function

Where
t = residence time
z = overall concentration
p = PPh_3O IR response
s = alcohol IR response
x = equiv. of CBr_4
y = equiv. of PPh_3
F2 = the second evaluation function

Fig. 6.21: Appel reaction monitored by in-line IR analysis and optimized using Nelder–Mead algorithm.

It is also important to understand that equations designed for the optimization of process attributes do not always represent the kinetics (or rate law) of the reaction in question. Certain process attributes are combined in the equation with relative importance (i.e., weight factors) of each process attributes (e.g., conversion, throughout, or purity), so the algorithm can measure the progress of the optimization in a quantifiable manner.

All three process attributes (conversion, consumption, and throughput) of the example are connected to a singular chemical transformation. Analytical results obtained from the measurement of the reaction stream and the knowledge of input stoichiometry lead to reaction conditions that gives the best of all three attributes. The described example is an excellent demonstration of how one could achieve such a multiple-attribute assessment from a single PAT measurement. There are processes where attributes are not related. In fact, they can be counter-productive at times. For example, in a synthesis method for a chiral API, the chemical purity and the optical purity, two unrelated CQAs may not get optimized using a single optimization exercise. The reaction condition that yields the best product conversion may not produce the best optical purity. Similarly, in the production of radiotracers, lengthy reaction conditions that give good conversion to the desired product may not be suitable for short-lived radioisotopes. In such examples, it is difficult for machines or humans to take decisive actions on the process parameters without understanding a global relationship of such seemingly orthogonal attributes under various synthetic conditions.

Pareto front is a graphical representation that identifies what is possible in such multiple objective optimization (MOO) exercises. When a set of attributes is plotted under various reaction conditions, a boundary of what is feasible versus what is not feasible begins to take shape. Knowledge of the boundary allows the search algorithm to systematically alter more than one attribute in an MOO of the Pareto front. An example of such a plot is shown in Fig. 6.22a. Two process attributes, f_1 and f_2, are plotted in the *x*- and *y*-axis of the plot. Data points (blue and red circles) originate from real measurements of f_1 and f_2. One of the two attributes (f_1 or f_2) has a local maximum in data points shown by red circles. The shaded red area (feasible region) represents combinations of f_1 and f_2 that are attainable by laboratory experiments. The area outside this zone (utopian region) represents combinations of f_1 and f_2 that are unachievable within the current experimental conditions. Experiments from the shaded area can yield high values for one of two attributes (f_1 or f_2), but not both. For example, there is no experimental condition that yields high values (e.g., 0.8) for both f_1 and f_2. Red circle A comes with f_1 of 0.8, which is a local maximum, and an f_2 of 0.08. Similarly, red circle B has a local maximum in f_2 (0.8), but a low f_1 (0.09). Conducting Pareto optimization to maximize both f_1 and f_2 requires selecting trial conditions from the right region (i.e. top-right section of the red shaded region of Fig. 6.22a) and further optimizing the selected conditions. An example of MOO using Pareto optimization is reported for the alkylation of morpholine (**39**, Fig. 6.23) [41]. In this example, space–time yield (STY) and environmental factor (E-factor) are optimized for the S_NAr reaction between 2,4-difluoronitrobenzene and morpholine. The algorithm was able to find the Pareto front after a total of 68 experiments, 26 of which yielded data points that formed the Pareto front (Fig. 6.22b). The reaction was found to be optimal at 140 °C for a reactor residence time of 30 s with a 4.7:1 stoichiometric ratio of the reagents. The optimized condition gave 13,120 kg m^{-3} h^{-1} of STY and an E-factor of 1.57.

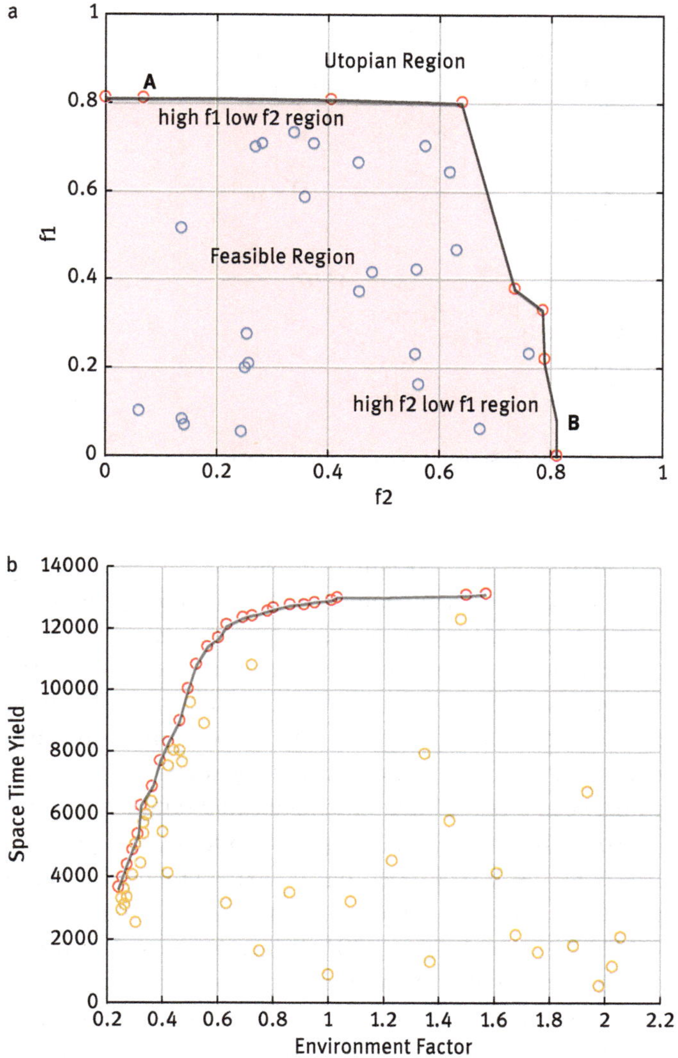

Fig. 6.22: Graphical representation of Pareto front from a multiobjective optimization. (a) An illustrative example of a Pareto front. (b) Pareto front found by algorithm in an alkylation reaction of morpholine (Fig. 6.23).

6.4.2 Neural network

An algorithm may not be equipped with all the mathematical functions necessary to find the solution to a problem. Evolving a mathematical tool in accordance with the problem can be daunting. The algorithm reacts differently to a response that is based on a context. For example, if the increase in the reaction temperature leads to

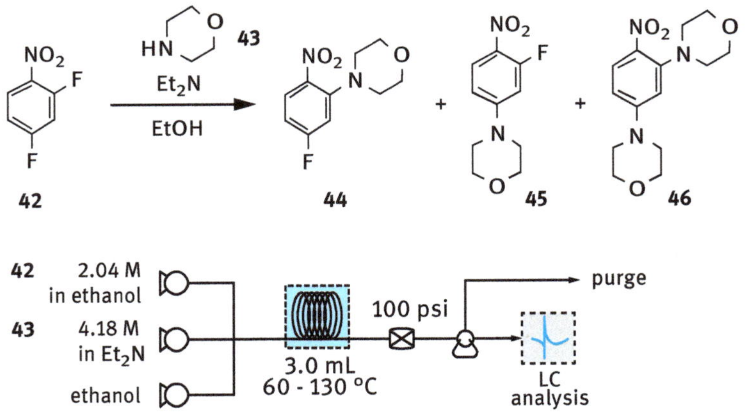

Fig. 6.23: Multiobjective optimization of an S_NAr reaction using Pareto front.

a better conversion of the starting reagents to the product, the algorithm may find that increasing the reaction temperature is a positive influence on the reaction. On the other hand, if the heightened reaction temperature reduces selectivity, the same algorithm may interpret the outcome as a negative experience. Humans learn from the interaction with the environment and constantly evolve decisions based on the situation. Training a computer to learn from the behavior of a problem and interpret based on the context is possible, but challenging. The principle of reinforced learning can be applied to the training of a network for undertaking such decisions. The network can be interlaced between a set of input parameters and output attributes in the form of a neural network (NN), as shown in Fig. 6.24. Training a neural network for making machine-guided decisions based on the experiential learning of the computer infrastructure itself is commonly termed as machine learning. Lines connecting inputs (e.g., CPPs) and outputs (e.g., CQAs) intersect at nodes, which are mathematical operators that are capable of interrogating outputs and modifying inputs accordingly. The higher the number of nodes and layers, the better is the machine's ability to make human-like decisions. Training can be supervised or unsupervised. Supervised learning is more suited when some degree of knowledge is available of a manufacturing process. Unsupervised learning is more complex as it allows the machine to define the nature of attributes from the process behavior. The central idea of reinforcement learning is to let the algorithm learn from the environment. There are two key steps in setting up a reinforcement learning method. First is an action set where the action indicates how the algorithm will conduct training of the machine. The second is a rewarding system and each action will correspond to its own reward. The machine is given freedom to build a relationship between the action and the score from the action. A comparative study between Nelder–Mead simplex method, SNOBFIT, and a reinforced learning method, which is termed as the deep reinforcement optimizer (DRO), show that the reinforced learning strategy has the ability to outperform traditional search methods in the area of chemical process optimization [42].

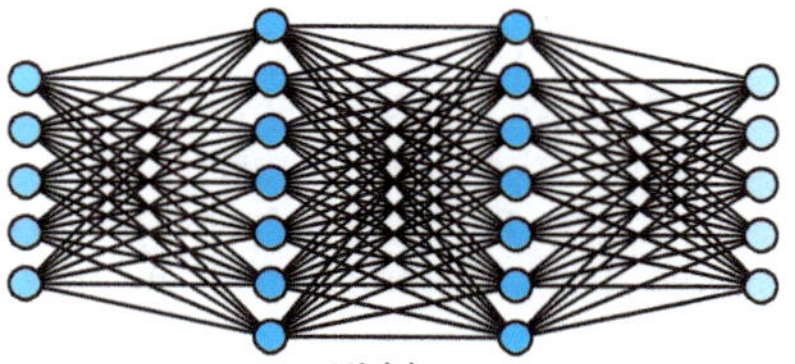

Fig. 6.24: Connections between nodes of different layers in a neural network.

6.5 Conclusions

The goal of this chapter was to make readers aware of the PAT strategies and techniques available for continuous manufacturing. The sampling and analysis sections are specifically designed to educate readers on the science behind PAT and its implementation in manufacturing processes. Introduction to search algorithms and neural networks is expected to help synthetic chemists familiarize themselves with future direction of PATs that involve automation, robotics, and computer-aided manufacturing.

Further readings

- Cerda V, Ferrer L, Avivar J, Cerda A. Flow Analysis. First edition. Elsevier Science/2014.
- Trojanowicz M. Flow chemistry vs. flow analysis. Talanta 2016;146, 621–40.
- Manka D, Editor Automated stream analysis for process control. First edition. Academic Press/1982
- Price GA, Mallik D, Organ MG. Process analytical tools for flow analysis: A perspective. J. Flow. Chem. 2017;7(3–4): 82–6.

Study questions

6.1 Injecting IPC samples from a low-pressure reactor flow path into a high-pressure analyzer flow path (e.g., UPLC-coupled detectors) is possible due to the low compressibility of liquids in question (i.e., IPC samples). Under ambient pressure, when samples inside the loop are exposed to a pressurized UPLC system from the inject configuration of the injector valve, the seal in the sampling valve experiences stress due to the sharp spike in pressure on the stator, causing damage to the stator and the valve seal. Suggest a rotor design that mitigates such an issue.

6.2 An environmental research team is using a 10-port valve (Fig. 6.5) for sampling from a slurried stream of volatile organic substances that are contaminated with fine dust particulates from the atmosphere using a GC-TCD analyzer. The time to acquire data from a single GC injection is 3.5 min (run-time). Calculate the maximum number of data points the team can acquire per day using a sample size of 1 μL from a liquid stream that flows at 3 μL min^{-1}. Assume that the amount of cleaning solvent required to remove particulates from the filter is less than 50 µL and the cleaning pump is capable of pumping at a flow rate of 500 µL min^{-1}. What would be the maximum number of data points the team could have collected, had the sample not been contaminated with particulates?

6.3 A research team is performing an reaction between ketone **17** and aldehyde **18** using LDA as a base (Fig. 6.12) [36]. This synthetic method can be subdivided into three key discrete events: (1) deprotonation of the ketone to form a reactive enolate, (2) addition of the enolate to aldehyde **18** to give intermediate **19**, and (3) protonation to give the final target product (**20**). **A.** Explain how these three key steps can be monitored using appropriate PAT. **B.** Suggest a downstream reactor design for the continuous collection of product **20** with no OoS lots.

6.4 Consider the following product yields from the optimization of reaction temperature and the flow rate of a chemical transformation in flow. **A.** What temperature and flow rate would be chosen in the next experiment by the Nelder–Mead algorithm using the reflection method. **B.** Explain how the algorithm will propagate its search if the yield from the fourth experiment falls between 2% and 80%.

Tab. 6.5: First three solutions from the Nelder–Mead algorithm.

Experiments	Flow rate (µL min^{-1})	Temperature (°C)	Yield (%)
I	50	50	2
II	100	100	80
III	20	5	52

6.5 Imagine an example of supervised learning using a simplified neural network of a 2 × 2 matrix as "*Z*" from a reaction between A and B to make C and D. Concentrations of reagents A and B are represented by set *x*. Concentrations of products C and D are represented by set *y*. The mathematical representation of the simplified neural network is shown below:

$$Z * x = y \tag{6.7}$$

The experimental data shows the paired values (tabulated as the actual concentrations times 10 in molar concentration (M)) from *x* and *y* sets:

$$(x, y) \in \left\{ \left(\begin{bmatrix} 3 \\ 5 \end{bmatrix}, \begin{bmatrix} 1 \\ 1 \end{bmatrix} \right), \left(\begin{bmatrix} 6 \\ 10 \end{bmatrix}, \begin{bmatrix} 2 \\ 2 \end{bmatrix} \right), \left(\begin{bmatrix} 7 \\ 11 \end{bmatrix}, \begin{bmatrix} 1 \\ 3 \end{bmatrix} \right), \left(\begin{bmatrix} 7 \\ 12 \end{bmatrix}, \begin{bmatrix} 3 \\ 2 \end{bmatrix} \right), \left(\begin{bmatrix} 14 \\ 23 \end{bmatrix}, \begin{bmatrix} 4 \\ 5 \end{bmatrix} \right) \right\}$$

a. Find missing elements in *Z* where

$$Z = \begin{bmatrix} z_1 & 2 \\ 2 & z_2 \end{bmatrix} \tag{6.8}$$

Where z_1 and z_2 are unknown elements of *Z*.

b. Predict yields of C and D when concentrations of A and B are [10, 18] T.

References

[1] The Long Shop Museum [Internet]. [cited 2020 Nov 24];Available from: https://www.longshopmuseum.co.uk/

[2] Wood, C. Scientists make digital breakthrough in chemistry that could revolutionize the drug industry [Internet]. CNBC. 2020 [cited 2020 Nov 24];Available from: https://www.cnbc.com/2020/10/24/how-a-digital-breakthrough-could-revolutionize-drug-industry.html

[3] Magazine KS Nature. Machine stitches complex molecules at touch of a button [Internet]. Scientific American. [cited 2020 Nov 24];Available from: https://www.scientificamerican.com/article/machine-stitches-complex-molecules-at-touch-of-a-button/

[4] Center for Drug Evaluation and Research. Modernizing the Way Drugs Are Made: A Transition to Continuous Manufacturing [Internet]. FDA. 2019 [cited 2020 Aug 4];Available from: https://www.fda.gov/drugs/news-events-human-drugs/modernizing-way-drugs-are-made-transition-continuous-manufacturing

[5] Chatterjee, S, FDA perspective on continuous manufacturing, Raw Material, 22.

[6] Hamlin, TA, Leadbeater, NE, Raman spectroscopy as a tool for monitoring mesoscale continuous-flow organic synthesis: Equipment interface and assessment in four medicinally-relevant reactions, Beilstein J Org Chem, 2013, 9, 1843–1852.

[7] Lebl, R, Cantillo, D, Kappe, CO, Continuous generation, in-line quantification and utilization of nitrosyl chloride in photonitrosation reactions, React Chem Eng, 2019, 4(4), 738–746.

[8] Glotz, G, Kappe, CO, Design and construction of an open source-based photometer and its applications in flow chemistry, React Chem Eng, 2018, 3(4), 478–486.

[9] Carter, CF, Lange, H, Ley, SV, Baxendale, IR, Wittkamp, B, Goode, JG, Gaunt, NL, ReactIR flow cell: A new analytical tool for continuous flow chemical processing, Org Process Res Dev, 2010, 14(2), 393–404.

[10] Sagmeister, P, Poms, J, Williams, JD, Kappe, CO, Multivariate analysis of inline benchtop NMR data enables rapid optimization of a complex nitration in flow, React Chem Eng, 2020, 5(4), 677–684.

[11] Schotten, C, Howard, JL, Jenkins, RL, Codina, A, Browne, DL, A continuous flow-batch hybrid reactor for commodity chemical synthesis enabled by inline NMR and temperature monitoring, Tetrahedron, 2018, 74(38), 5503–5509.

[12] Braun, F, Schwolow, S, Seltenreich, J, Kockmann, N, Röder, T, Gretz, N, Rädle, M, Highly Sensitive raman spectroscopy with low laser power for fast in-line reaction and multiphase flow monitoring, Anal Chem, 2016, 88(19), 9368–9374.

[13] Amara, Z, Poliakoff, M, Duque, R, Geier, D, Franciò, G, Gordon, CM, Meadows, RE, Woodward, R, Leitner, W, Enabling the scale-up of a key asymmetric hydrogenation step in the synthesis of an api using continuous flow solid-supported catalysis, Org Process Res Dev, 2016, 20(7), 1321–1327.

[14] Brown, P, Schober, M Continuous Flow Process Development: Sampling Reactions Over Extended Periods [Internet]. 2019;Available from: https://www.mt.com/dam/0autochem/RapidFlowProcessDevelopment_AN_en_20190212_Original_58061.pdf

[15] Hochgraeber, H, Ruegenberg, G Autosampler for high-performance liquid chromatography [Internet]. 2012 [cited 2020 Nov 24];Available from: https://patents.google.com/patent/US8196456/en

[16] Tilley, M, Li, G, Savel, P, Mallik, D, Organ, MG, Intelligent continuous collection device for high-pressure flow synthesis: Design and implementation, Org Process Res Dev, 2016, 20(2), 517–524.

[17] Kwak, JS, Zhang, W, Tsoy, D, Hunter, HN, Mallik, D, Organ, MG, A multiconfiguration valve for uninterrupted sampling from heterogeneous slurries: an application to flow chemistry, Org Process Res Dev, 2017, 21(7), 1051–1058.

[18] Waters, T, Analytical Sampling from Industrial Processes [Internet], In: Meyers, RA, editor, Encyclopedia of Analytical Chemistry, Chichester, UK, John Wiley & Sons, Ltd, 2017, cited 2020 Nov 24, 1–41. Available from:: http://doi.wiley.com/10.1002/9780470027318.a9618.

[19] Chisolm, CN, Evans, CR, Jennings, C, Black, WA, Antosz, FJ, Qiang, Y, Diaz, AR, Kennedy, RT, Development and characterization of "push–pull" sampling device with fast reaction quenching coupled to high-performance liquid chromatography for pharmaceutical process analytical technologies, J Chromatogr A, 2010, 1217(48), 7471–7477.

[20] Lambertus, GR, Webster, LP, Braden, TM, Campbell, BM, Groh, JM, Maloney, TD, Milenbaugh, P, Spencer, RD, Sun, W-M, Johnson, MD, Development of universal, automated sample acquisition, preparation, and delivery devices and methods for pharmaceutical applications, Org Process Res Dev, 2019, 23(2), 189–210.

[21] Sans, V, Porwol, L, Dragone, V, Cronin, L, A self optimizing synthetic organic reactor system using real-time in-line NMR spectroscopy, Chem Sci, 2015, 6(2), 1258–1264.

[22] Gomez, MV, De La Hoz, A, NMR reaction monitoring in flow synthesis, Beilstein J Org Chem, 2017, 13, 285–300.

[23] Nordon, A, Diez-Lazaro, A, Wong, CWL, Mcgill, CA, Littlejohn, D, Weerasinghe, M, Mamman, DA, Hitchman, ML, Wilkie, J, Consideration of some sampling problems in the on-line analysis of batch processes by low-field NMR spectrometry, Analyst, 2008, 133(3), 339.

[24] Foley, DA, Bez, E, Codina, A, Colson, KL, Fey, M, Krull, R, Piroli, D, Zell, MT, Marquez, BL, NMR flow tube for online NMR reaction monitoring, Anal Chem, 2014, 86(24), 12008–12013.

[25] Giraudeau, P, Felpin, F-X, Flow reactors integrated with in-line monitoring using benchtop NMR spectroscopy, React Chem Eng, 2018, 3(4), 399–413.

[26] Cortés-Borda, D, Wimmer, E, Gouilleux, B, Barré, E, Oger, N, Goulamaly, L, Peault, L, Charrier, B, Truchet, C, Giraudeau, P, Rodriguez-Zubiri, M, Grognec, EL, Felpin, F-X, An autonomous self-optimizing flow reactor for the synthesis of natural product carpanone, J Org Chem, 2018, 83(23), 14286–14299.

[27] Musio, B, Gala, E, Ley, SV, Real-time spectroscopic analysis enabling quantitative and safe consumption of fluoroform during nucleophilic trifluoromethylation in flow, ACS Sustainable Chem Eng, 2018, 6(1), 1489–1495.

[28] Chaplain, G, Haswell, SJ, Fletcher, PDI, Kelly, SM, Mansfield, A, Development and evaluation of a raman flow cell for monitoring continuous flow reactions, Aust J Chem, 2013, 66(2), 208.

[29] Lichtenegger, GJ, Tursic, V, Kitzler, H, Obermaier, K, Khinast, JG, Gruber-Wölfler, H, The plug & play reactor: A highly flexible device for heterogeneous reactions in continuous flow, Chemie Ingenieur Technik, 2016, 88(10), 1518–1523.

[30] Brodmann, T, Koos, P, Metzger, A, Knochel, P, Ley, SV, Continuous preparation of arylmagnesium reagents in flow with inline IR monitoring, Org Process Res Dev, 2012, 16(5), 1102–1113.

[31] Krishnadasan, S, Brown, RJC, deMello, AJ, deMello, JC, Intelligent routes to the controlled synthesis of nanoparticles, Lab Chip, 2007, 7(11), 1434.

[32] Browne, DL, Wright, S, Deadman, BJ, Dunnage, S, Baxendale, IR, Turner, RM, Ley, SV, Continuous flow reaction monitoring using an on-line miniature mass spectrometer, RCM, 2012, 26(17), 1999–2010.

[33] McMullen, JP, Stone, MT, Buchwald, SL, Jensen, KF, An integrated microreactor system for self-optimization of a heck reaction: from micro- to mesoscale flow systems, Angew Chem Int Ed, 2010, 49(39), 7076–7080.

[34] Echtermeyer, A, Amar, Y, Zakrzewski, J, Lapkin, A, Self-optimisation and model-based design of experiments for developing a C–H activation flow process, Beilstein J Org Chem, 2017, 13, 150–163.

[35] Sauks, JM, Mallik, D, Lawryshyn, Y, Bender, T, Organ, M, Continuous-Flow Microwave, A, Reactor for conducting high-temperature and high-pressure chemical reactions, Org Process Res Dev, 2014, 18(11), 1310–1314.

[36] Sagmeister, P, Williams, JD, Hone, CA, Kappe, CO, Laboratory of the future: a modular flow platform with multiple integrated PAT tools for multistep reactions, React Chem Eng, 2019, 4(9), 1571–1578.

[37] Liu, Q, Jiang, X, Zheng, H, Su, W, Chen, X, Yang, H, On-line two-dimensional LC: A rapid and efficient method for the determination of enantiomeric excess in reaction mixtures, J Sep Sci, 2013, 36(19), 3158–3164.

[38] Kwak, JS, Bizzari, N, Sharif, S, Zhang, WP, Mallik, D, Organ, MG, The synthesis of warfarin using a reconfigurable-reactor platform integrated to a multiple-variable optimization tool, Chem Eur J, 2020, 26, 15505–15508.

[39] Bordawekar, S, Chanda, A, Daly, AM, Garrett, AW, Higgins, JP, LaPack, MA, Maloney, TD, Morgado, J, Mukherjee, S, Orr, JD, Reid III, GL, Yang, B-S, Ward, II, Industry, HW, Perspectives on process analytical technology: tools and applications in API manufacturing, Org Process Res Dev, 2015, 19(9), 1174–1185.

[40] Fitzpatrick, DE, Battilocchio, C, Ley, SV, A novel internet-based Reaction monitoring, control and autonomous self-optimization platform for chemical synthesis, Org Process Res Dev, 2016, 20(2), 386–394.

[41] Bédard, A-C, Adamo, A, Aroh, KC, Aroh, KC, Russell, MG, Bedermann, AA, Torosian, J, Yue, B, Jensen, KF, Jamison, TF, Reconfigurable system for automated optimization of diverse chemical reactions, Science, 2018, 361(6408), 1220–1225.

[42] Zhou, Z, Li, X, Zare, RN, Optimizing chemical reactions with deep reinforcement learning, ACS Cent Sci, 2017, 3(12), 1337–1344.

Nopphon Weeranoppanant, Lorenzo Milani and Andrea Adamo

7 Continuous gas–liquid and liquid–liquid separation

7.1 Introduction

As the field of flow chemistry continues to grow, the abundance of reactions carried out in flow can benefit from direct integration with downstream processes. In-line liquid–liquid or gas–liquid separation integrated with the chemical synthesis offers a broader set of opportunities for telescoping reactions that use different solvents and that require workup. Continuous separations allow the overall process to be more efficient and smaller. The smaller internal volume of in-line separators provides greater safety, particularly with toxic, explosive, or highly volatile substances. In this chapter, we will provide an overview of technologies for the in-line separation together with examples of their applications.

7.2 Continuous separators as an alternative for conventional separator

In the scope of this chapter, we will cover only technologies that separate immiscible phases (i.e., gas–liquid and liquid–liquid), where different phases are present in a significant amount. If a content of one phase (residual) is much lower than that of the other phase, other technologies, such as adsorption or distillation, can be used, and they are not discussed in this chapter.

A conventional separator almost completely relies on the gravity effect. The separator can be in different shapes, either horizontal or vertical. It is typically large in size to allow for a sufficient residence time for complete phase separation. An average residence time in the conventional separators can vary from 5 to 90 min [1]. For example, the residence time for hydrocarbon/water separation is 3–5 min. The residence time for ethylene glycol/hydrocarbon separation is 20–60 min. Droplet settling theory is an important foundation for designing and sizing the conventional phase separators. During the separation, droplets will move upward or downward into a layer of another phase. There are three major forces acting on a droplet: buoyancy, drag, and gravity [2]. The force balance can be performed to estimate a terminal velocity of a droplet. The terminal velocity can then be used to size a separator for a desired residence time. It is important to note that the drag force is related to a drag coefficient, which depends on particle's shape and Reynold's number of the

https://doi.org/10.1515/9783110693676-007

fluid. Therefore, the sizing of the equipment must be designed according to the properties of the system (e.g., droplet's diameter, viscosity, and density).

Continuous separators, on the other hand, can be operated by gravity or surface effects. Due to the continuous nature, the residence time is short (a few seconds to 5 min). As a result, the continuous separators are relatively small. They can be simply installed in a fume hood and integrated a flow synthesis. It is important to note that the continuous gravity-based separator has a similar design to a conventional separation vessel. In contrast, as discussed later in this chapter, the design of the surface-force-based separators requires information about an interfacial force between different phases. The comparison between conventional and continuous separation is summarized in Tab. 7.1.

Tab. 7.1: Comparison between conventional and continuous separation (gas–liquid, liquid–liquid).

	Conventional separation	Continuous separation
Main separation mechanism	Gravity effect (settling)	Gravity or surface effect
Equipment size	Medium to large (m^2-scale space required)	Small (cm^2-scale space required)
Residence time	5–90 min	A few seconds to 5 min
Operational mode	Batch/semi-batch	Continuous
Suitable samples	Crude samples (medium to high solid content)	Relatively clean samples (minimum solid content)
Examples of applications	Natural gas processing, industrial biotechnology, rare earth extraction	Flow chemistry, continuous manufacturing of fine chemicals and pharmaceuticals, analytical chemistry

7.3 Liquid–liquid separation technologies

Liquid–liquid separation is an important unit operation in flow chemistry as all flow syntheses use liquid as a main medium. Typically, liquid–liquid separation is required after a biphasic reaction, or to carry out an in-line extraction. For the liquid–liquid extraction, one of the liquid phases is an extraction solvent. The extraction involves mixing and phase separation. A mixing step allows for mass transfer of the chemicals that are soluble in both liquids; finally, a separation step will split the phases providing a raffinate stream and an extract stream. Liquid–liquid extraction is typically used to purify a liquid stream after reaction. Generally, most multistep flow syntheses require some in-line purification unit between sequential steps to remove excess reagents or unwanted by-products. Continuous extraction in flow has been shown with efficient mixing [3–5]. In the following sections, we will discuss different approaches to liquid–liquid separation and provide examples of use.

7.3.1 Gravity-based separation

Gravity-based separators leverage gravity as the driving force for separation. This type of devices requires that the two immiscible liquids must have different densities. The separation unit must be large enough to ensure that gravity is a dominant force over surface effects. This type of separation is similar to an industrial decanter, although the size may be much smaller for the flow setting.

In flow chemistry applications, this separation unit not only has a smaller size but also needs automation to enable continuous operation. Typically, the two liquids are continuously added into a separation vessel, which has a sufficiently large volume for the two phases to settle under gravity. Then, the liquid–liquid interface position is maintained to be within the vessel at all times with the addition of some type of sensors together with related automation hardware to control the outflow from the vessel. Several examples of sensors to detect the interface have been presented. For instance, the liquid–liquid interface could be detected leveraging differences of refractive indexes of the different phases [6] (Fig. 7.1). A fully automated liquid–liquid extraction system was enabled by gradually drawing liquids from an extraction vial. The liquids flowed through a differential refractometer equipped with two photocells. This signal provided by is tool was zero when the liquids inside the two cells are the same, and became nonzero when the cells housed different liquid phases, thus identifying the location of the liquid–liquid interface.

Another sensing method to identify the interface is the use of a camera. A commercially available webcam could be applied to directly detect the interface [7]. A green plastic ball, which had a density between that of light and heavy phases, was added to highlight the position of the interface. Software written in Python programming language was developed to calculate the height of the interface and to control the flow rate of the light-phase outlet stream (Fig. 7.2).

Another technique relies on different dielectric constants of different liquids. A separator, whose shape was similar to a Dean-Stark apparatus, was designed with a capacitive sensor [8]. The interface was detected using the sensor attached to the exterior surface of the settling vessel. A LabVIEW algorithm was written to control the flow rate of the heavy-phase outlet so as to maintain the position of the interface. A similar concept was presented using an impedance probe [9], and the system was found to be robust to the presence of solids. (Fig. 7.3 and Fig. 7.4).

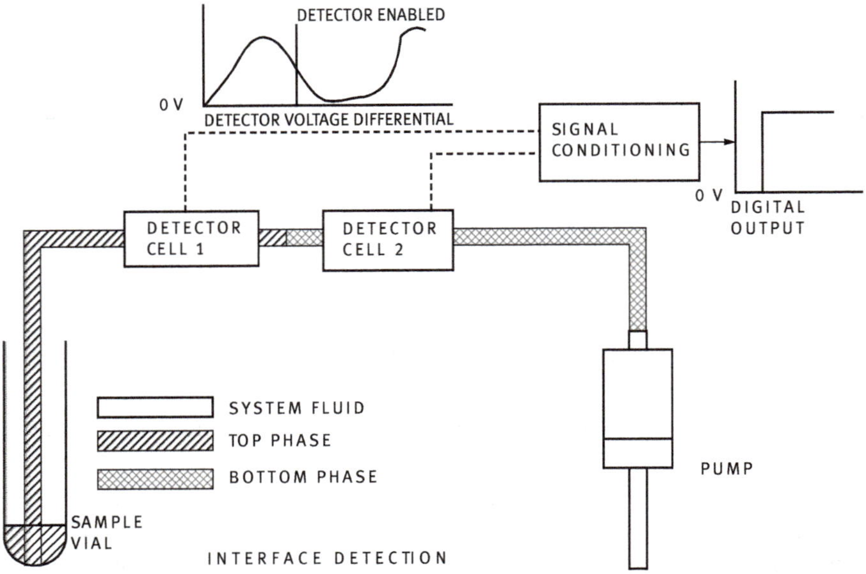

Fig. 7.1: Differential refractometers for the detection of the liquid–liquid interface [6].

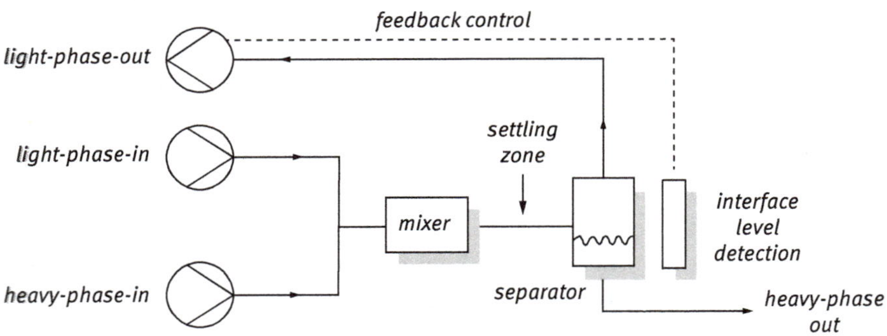

Fig. 7.2: Setup for the camera-based liquid–liquid separator [7].

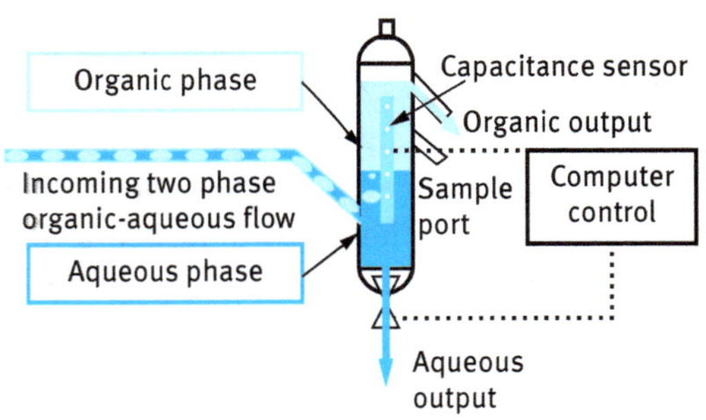

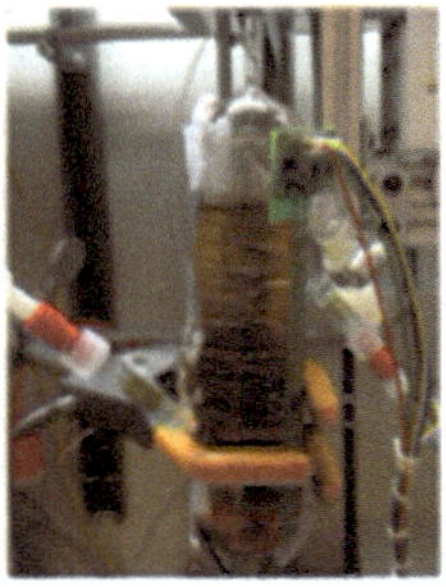

Fig. 7.3: A gravity-based separator using a capacitance sensor [8].

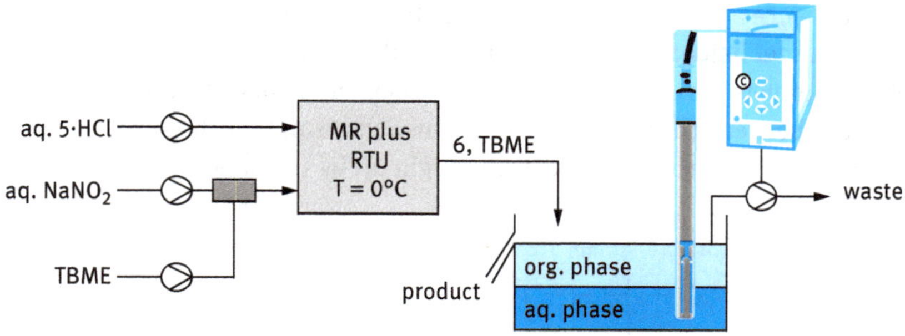

Fig. 7.4: A gravity-based separator using an impedance probe [9].

7.3.2 Surface force–based separation

Surface force–based separations are closely associated with the concept of device miniaturization. The microstructured units offer significantly high surface-to-volume ratios as well as short diffusion and thermal conduction length scales [10]. Although the focus of this chapter is on the fluidic phase separation, there are many other important microstructured unit to accomplish operations such as extraction [11–13], mixing [14], absorption [15], and distillation [16]. A detailed review of these devices can be found [10].

When the system is small, the surface force becomes an important player. The relative importance of gravity to surface effects can be represented by the dimensionless Bond number [17]:

$$\mathrm{Bo} = \frac{\rho a L^2}{\gamma}$$

where Bo is the Bond number, ρ is a density of a medium, a is an acceleration such as gravity, L is a characteristic length scale, and γ is surface tension. As the equipment size becomes smaller, the value of L becomes smaller, and so is the Bond number. Therefore, as most micro- and milli-fluidic systems have a small size, the Bond number is small ($\mathrm{Bo} \ll 1$) and the surface effects dominates onto gravity effect.

7.3.3 Surface force–based separation: droplet coalescer

Surface forces are leveraged to achieve either droplet coalescence or phase separation. In 2004, a simple microchannel device was proposed for droplet coalescence [18] (Fig. 7.5). The channel was sandwiched between two flat plates made of glass and PTFE. By studying the droplet size distribution, the authors proposed a mechanism of the droplet coalescence as shown in Fig. 7.5. First, the dispersed droplet must be larger than the microchannel height. As the continuous and dispersed phases flow through the channel, the droplets are deformed. As the dispersed droplets wet on the PTFE surface, the velocities between the continuous (aqueous) phase and the disperse (organic) phase differ. Finally, the climbing droplets are caught up by surrounding droplets.

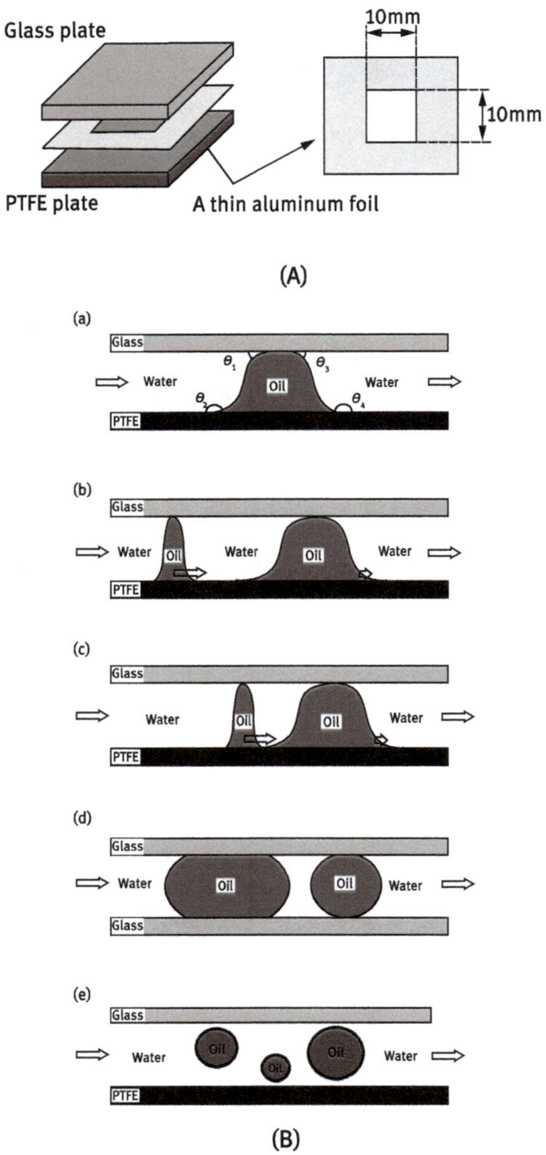

Fig. 7.5: (a) Droplet coalescence to facilitate the liquid–liquid separation and (b) droplet coalescence mechanism [18].

Subsequently, a similar plate-type coalescer was constructed [19], as shown in Fig. 7.6. Two plates were made of different materials, including PTFE and stainless steel. They used a larger channel height as well as a longer channel than those reported previously [18]. In this case, they found that complete liquid–liquid separation could be obtained in all cases where droplets were larger than the channel height. Stable operation at the throughput of 180 L h^{-1} was achieved by using the channel sizes of 100–200 μm.

7.3.4 Surface force–based separation: flow splitter

Other examples of surface force-driven separation involve the use of a simple junction connecting to channels of different materials or channels with surface modifications [20–22]. A Y-junction flow splitter was designed [23] (Fig. 7.7). To demonstrate aqueous–organic separation, the aqueous and organic phases had preferential affinities for steel and PTFE, respectively. They found that when the flow ratio was one (equal flow rates between the two phases), the splitting performance was almost independent of the total flow rate. The capillary size also did not have a significant effect on the splitting performance. The most significant factor was the flow ratio because the wetting force of the low-flow phase could be dominated by the inertial force of the high-flow phase, resulting in the undesired entrainment of different phases (e.g., organic phase entrained into steel outlet).

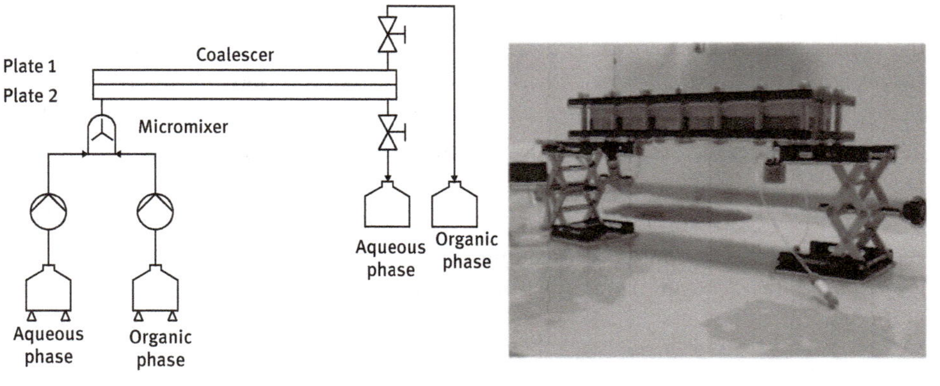

Fig. 7.6: Plate-type coalescer for a continuous liquid–liquid separation [19].

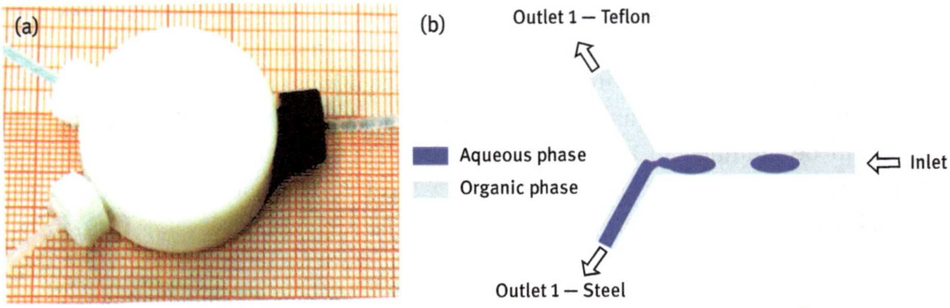

Fig. 7.7: (a) A Y-junction flow splitter and (b) the splitter consist of two channels that are made of different materials for preferential wetting [23].

7.3.5 Surface force–based separation: membrane separator

Another type of liquid–liquid separator that takes advantage of the surface forces is the membrane-based separator [24–27]. This concept was first introduced in a microfluidic device. A microfabricated chip was constructed with two glass plates sandwiching a microporous membrane [28]. The aqueous and organic streams were flowed on different glass plates, and the membrane was to maintain the interface between the two phases. This type of microfluidic device could be used as an extraction unit and a sample pretreatment before a chemical analysis (e.g., gas chromatography and HPLC (high-performance liquid chromatography)). A membrane can also be used to directly separate two phases from a single flow stream. One of the early commercial examples is Syrris flow liquid–liquid extraction (FLLEX) module (www.syrris.com). A similar device was designed to separate aqueous–organic flow [24]. In this device (Fig. 7.8), the separator contains a hydrophobic PTFE membrane, which allows for the organic phase to permeate through the membrane while the aqueous is retained.

In this study, the authors developed also a systematic theory for the operation of the device. Let the pressures on the aqueous (retentate) and organic (permeate) sides be P_1 and P_2, respectively. Then, the pressure drop through the membrane (ΔP_{mem}) is

$$P_1 = \Delta P_{\text{mem}} + P_2$$

There are two conditions to be met for successful separation. First, the value of ΔP_{mem} should be less than the capillary pressure; otherwise, the aqueous phase cannot be retained by the membrane:

$$\Delta P_c > \Delta P_{\text{mem}}$$

where ΔP is the capillary pressure. The capillary pressure can be estimated by the Young–Laplace equation:

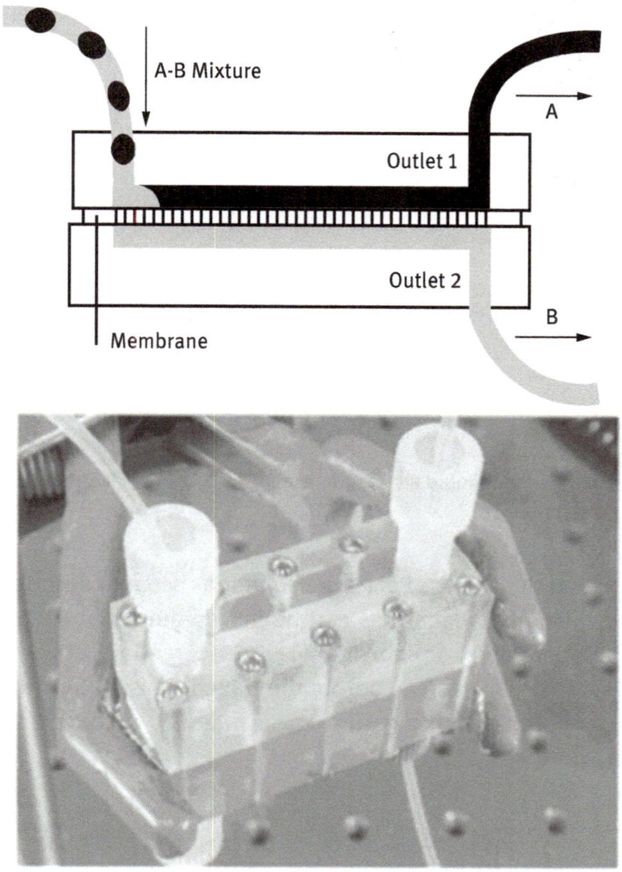

Fig. 7.8: A separator employing a hydrophobic PTFE membrane for continuous aqueous–organic separation [24].

$$\Delta P_c = \frac{2\gamma}{R_{mem}} \cos\theta$$

where y is the interfacial tension, R_{mem} is the pore size, and θ is the wetting angle.

Second, the entire flow of the organic (permeate) phase needs to be able to travel through the membrane; hence, its pressure drop through the membrane needs to be less than the pressure difference across the membrane (ΔP_{mem}). As a result:

$$\Delta P_{per} < \Delta P_{mem} < \Delta P_c$$

The pressure drop of the organic can be estimated by using Hagen–Poiseuille equation [26].

The theory highlights the fact that successful separation in the porous membrane-based approach depends on the ability to properly control pressure. More specifically if the value of ΔP_{mem} falls below ΔP_{per}, then the partial flow of the permeate phase will be retained (i.e., not permeated). The problem is called "retention," as shown in Fig. 7.9. If the value of ΔP_{mem} becomes higher than ΔP_{c}, then some of the retentate phase will permeate through the membrane. The problem is called "breakthrough," as shown in Fig. 7.9.

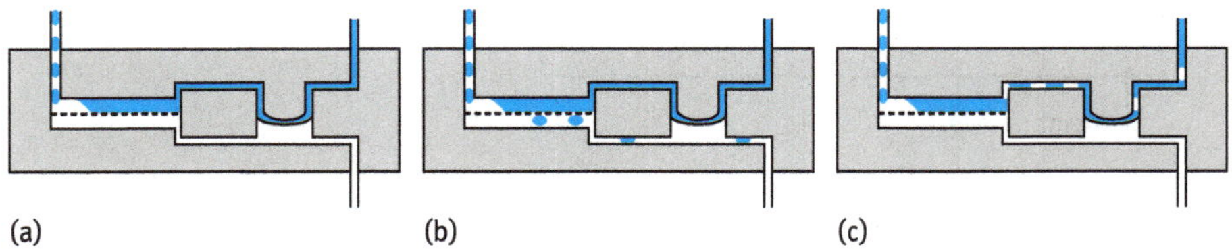

Fig. 7.9: Three possible situations of the liquid–liquid separation in a membrane separator: (a) good, (b) breakthrough, and (c) retention [26].

Although the Young–Laplace and the Hagen–Poiseuille equations seem to describe the operating range quite well, the two equations can be inaccurate when ΔP_{mem} comes close to the lower and upper limits. As shown in Fig. 7.10, retention occurs at the pressure higher than ΔP_{per}, which is estimated from the Hagen–Poiseuille equation. The breakthrough also happens at the pressure lower than ΔP_{c} predicted by the Young–Laplace equation.

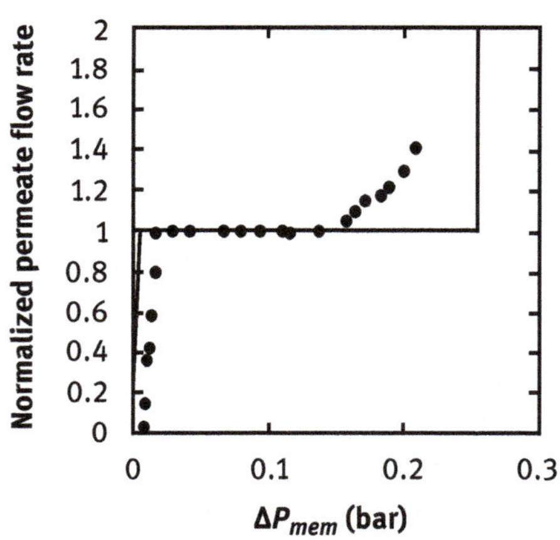

Fig. 7.10: Normalized permeate flow rate represents the ratio of the actual permeate flow rate to the desired permeated flow rate. The graph shows the comparison between theoretical limits (curve) and experimental measurements (circles) [26].

The new models were proposed by modifying ΔP_{per} and ΔP_{c} with a tortuosity factor (C_{tor}) and the pore size distribution [29]. These modifications were able to describe the experimental results accurately.

Model	Retention (lower limit of ΔP_{mem})	Breakthrough (upper limit of ΔP_{mem})
Original	$\Delta P_{per} = \frac{8\mu_{org}Q_{org}}{n\pi R_{mem}^4} L_{mem}$	If $\Delta P_{mem} > \Delta P_C$, then the breakthrough happens
Yang et al. [16]	$\Delta P_{per} = \frac{8C_{tor}\mu_{org}Q_{org}}{n\pi R_{mem}^4} L_{mem}$	$Q_{aq,2} = \frac{\pi}{8C_{tor}\mu_{aq}L_{mem}} \Delta P_{mem} \int_{\frac{2\gamma}{\Delta P_{mem}}}^{\infty} n(R)R^4 dR$

where C_{tor} is a tortuosity factor of the membrane pore. $Q_{aq,2}$ is the flow rate of aqueous phase that permeate through the membrane. $Q_{aq,2}$ is a result of the breakthrough situation, and its value varies due to a pore size distribution. From these theoretical formulations, the operating regimes are well defined. However, the use of the membrane separator had relied much on the manual controls of the pressures [25]. Recently, automated control of the pressures was developed by monitoring the light transmitted through fluid outlet streams and adjusting a needle valve on one outlet accordingly until the complete separation was obtained [30, 31] (Fig. 7.11).

To eliminate need for electrical parts and sensors and simplify the use of the membrane separator, the self-tuning pressure element was developed [26]. The element was made of an elastic diaphragm, as shown in Fig. 7.12.

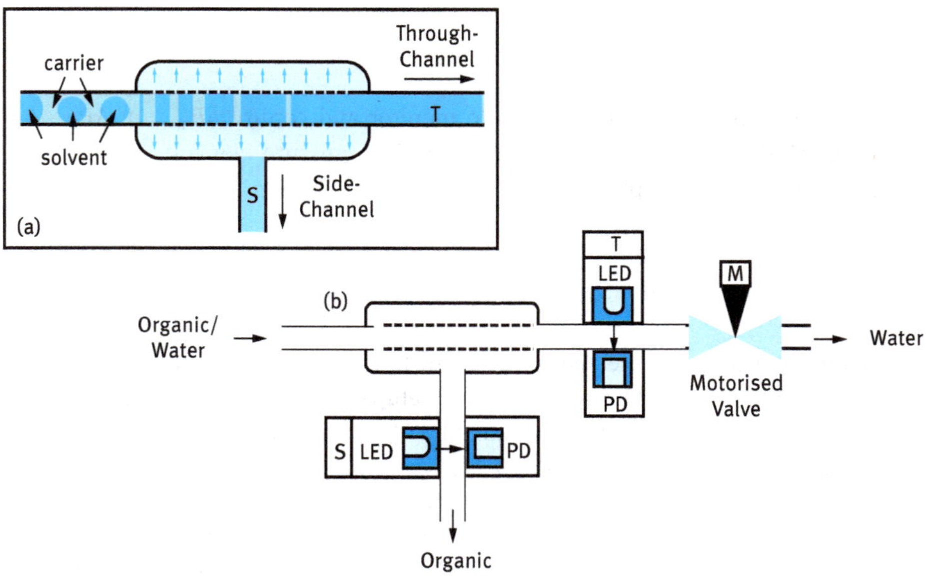

Fig. 7.11: A capillary-based separator with automated adjustment of the needle valve at the outlet in response to the signals from photodetectors [30].

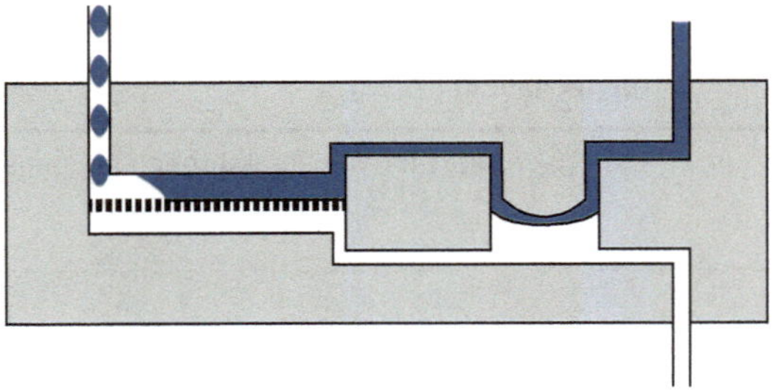

Fig. 7.12: A membrane separator integrated with a self-tuning pressure control element [26].

When ΔP_1 was too low, the diaphragm sealed against the retentate flow path such that the liquid on the retentate could not flow. Dynamically, the retentate liquid had to build up a pressure until ΔP_1 was larger than ΔP_2 ($\Delta P_1 > \Delta P_2$), so the retentate liquid could flow out to the outlet. The difference between ΔP_1 and ΔP_2 ($\Delta P_1 - \Delta P_2$) depended on the tension of the diaphragm (P_{dia}):

$$\Delta P_{\text{dia}} = \Delta P_1 - \Delta P_2$$

Therefore,

$$\Delta P_{\text{dia}} = \Delta P_{\text{mem}}$$

Their careful design (e.g., the diaphragm material and the element's structure) offered ΔP_{dia} to be within the operating range:

$$\Delta P_{\text{per}} < \Delta P_{\text{dia}} < \Delta P_{\text{C}}$$

With this element, one does not need to monitor and control ΔP_{mem} or $\Delta P_1 - \Delta P_2$ anymore. The diaphragm served as a self-tuning pressure element to ensure the complete liquid–liquid separation.

7.3.6 Multistage countercurrent extraction

The self-tuning pressure element was useful not only as an in-line separator but also as a modular element for more complex setups such as a multistage separation. The membrane separators can be arranged as to provide a multistage countercurrent extraction [32], as shown in Fig. 7.14. The challenge of the multistage separation is that a flow rate between stages is not known, and, thereby, it is difficult to set flow rates of the interstage pumps. However, this challenge can be addressed by the self-tuning pressure element. To verify this concept, the membrane separator was tested with a downstream pump on the retention side [32], as shown in Fig. 7.13(a). Q_{set} was the set flow rate of the downstream pump. $Q^{\text{ret}}_{\text{out},i}$ and $Q^{\text{per}}_{\text{out},i}$ were the actual flowrates of retentate and permeate phases upon the complete separation at stage *i*. The membrane separator showed complete separation when Q_{set} was equal or higher than $Q_{out,i}^{ret}$, corresponding to Fig. 7.13(b) and (c). The valve could dynamically open and close the retentate channel, and the self-tuning pressure element worked as desired. On the other hand, when $Q_{set} < Q_{out,i}^{ret}$, the diaphragm in the self-tuning element was pushed down. The valve was unable to dynamically close the retentate channel, and the breakthrough occurred due to the failure of the pressure controller, as depicted in sub-Fig. (d) in Fig. 7.13. Therefore, the rule of thumb is to set the rates of the pumps between stages to be higher than the estimated rate of the retentate phase in each stage. With this method, the multistage extraction was found to provide an equilibrium stage for each physical stage.

Owing to their robustness and flexibility, the membrane separators [26] have now been commercialized by Zaiput Flow Technologies (Zaiput) (Fig. 7.14). Zaiput

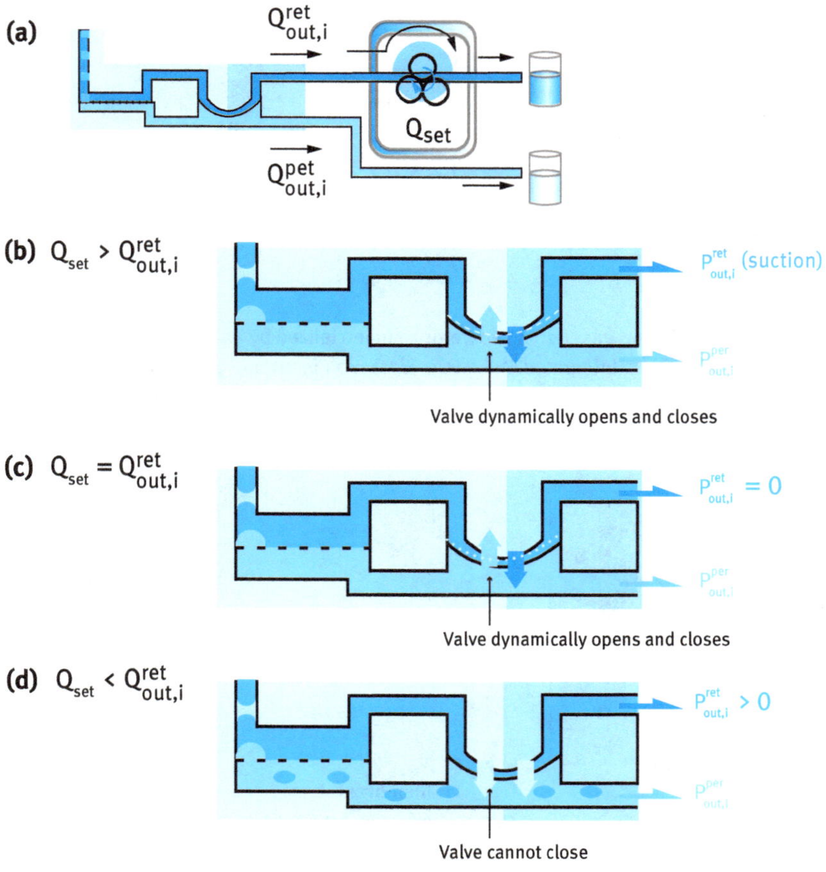

Fig. 7.13: Three possible scenarios of interstage pumps in the multistage separation setup [32].

(www.zaiput.com) has continued the development effort of separation technology. Its main contributions from a technological perspective are:

(1) Demonstration of scalability of the membrane-based separator (a wide range of flow rates from 50 μL min^{-1} to several L min^{-1})
(2) Redesign of the countercurrent extraction platform to simplify its operation and increase its robustness. The new platform has a diaphragm pump on each countercurrent line that is controlled by a proprietary flow sensor and related control hardware and script. With this design approach the platform provides scalable and robust operation (Fig. 7.14).
(3) Demonstration of the ability to separate emulsions. A membrane-based approach to separation is effective for the separation of emulsions because the removal of the continuous phase stimulates further coalescence of the dispersed phase (Fig. 7.15).

Fig. 7.14: In line liquid–liquid and gas–liquid separator commercialized by Zaiput Flow Technologies. Flow chemistry product line (left), Integrated Multistage Extraction Platform– Laboratory scale (middle), Integrated Multistage Extraction Platform – Pilot plant scale (right) (www.zaiput.com).

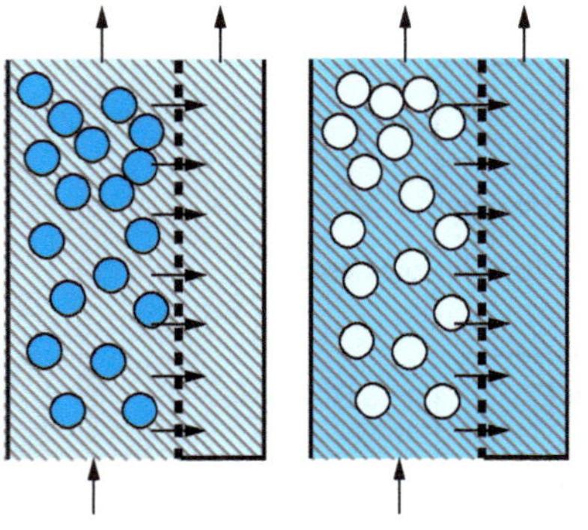

Fig. 7.15: Membrane-based separation provided by Zaiput is effective for emulsions separation. The selection of a membrane that is wet by the continuous phase ensures that its removal and hence forces the coalescence of the dispersed phase.

7.4 Applications of continuous liquid–liquid separation

Continuous liquid–liquid extraction has found its use in many applications. In this section, a number of examples will be discussed and categorized based on the fields of work.

7.4.1 Flow chemistry for organic synthesis

Organic synthesis has benefited greatly from flow chemistry. In flow, reagents are likely to be mixed more efficiently and safer to handle than those in batch. Also, an intermediate or a product can be directly purified with in-line separation devices. Liquid–liquid extraction is one of the most common purification methods for flow chemistry. The diazotization of amino acid in flow was demonstrated along with continuous aqueous workups [33]. The flow synthesis produced a high-value hydroxyacid, which was subsequently purified by extraction with ethyl acetate. However, due to its low partition coefficient (ca. 0.74), multiple times of extraction were needed. Therefore, as shown in Fig. 7.16, the flow synthesis was integrated with three stages of continuous liquid–liquid extraction and separation.

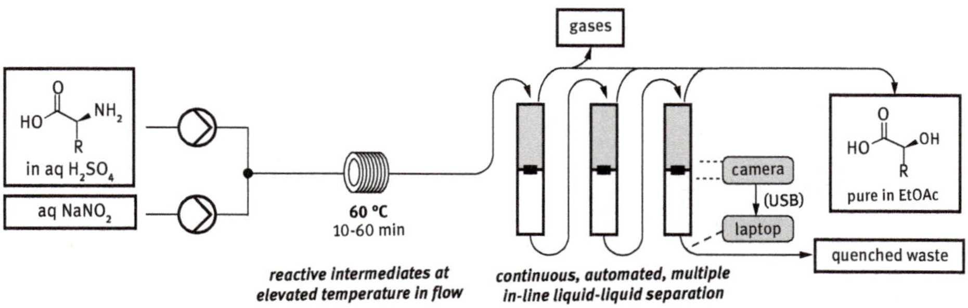

Fig. 7.16: A camera-based separator for the diazotization of amino acid in flow [33].

This type of camera-based separation was also applied in some other organic syntheses [34]. For example, It was used to continuously extract the unreacted *tert*-butyl carbazate (starting reagent) from the product stream during the flow synthesis of carbazate hydrazones [35]. The same group also used the device for the iodination [36] and the benzamide formation [37]. The flow setup was developed for synthesizing acyl azide from hydrazine [9]. Acyl azide is potentially explosive, so it is unsafe to scale up the synthesis in batch. Thus, flow chemistry, which typically needs minimal hold-up volumes, emerges as a promising alternative. In their setup, the flow synthesis and the isolation of the acyl azide into an organic solvent were integrated (Fig. 7.17). Then,

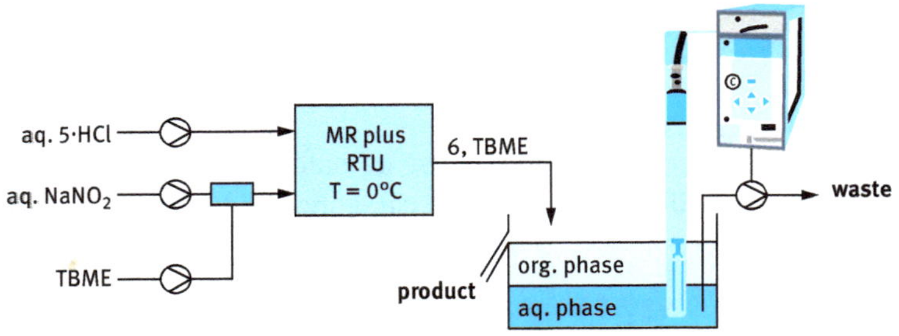

Fig. 7.17: The flow synthesis of acyl azide, which is directly isolated by flow extraction with TBME and separation using a continuous separator with an impedance probe [9].

the continuous aqueous–organic separation was implemented. The separation was automated with the impedance controller as discussed previously.

A membrane separator serves as a plug-and-play device in many flow syntheses. One of the early examples is the synthesis of an allylic ether in microcapillary flow disk reactor [38], which was followed by an in-line aqueous extraction (Fig. 7.18). The extraction was to remove salt and excess base from the product. A Syrris FLLEX membrane-based module was used to enable continuous separation. A similar flow scheme was demonstrated for the generation of diazomethane [39].

(a)

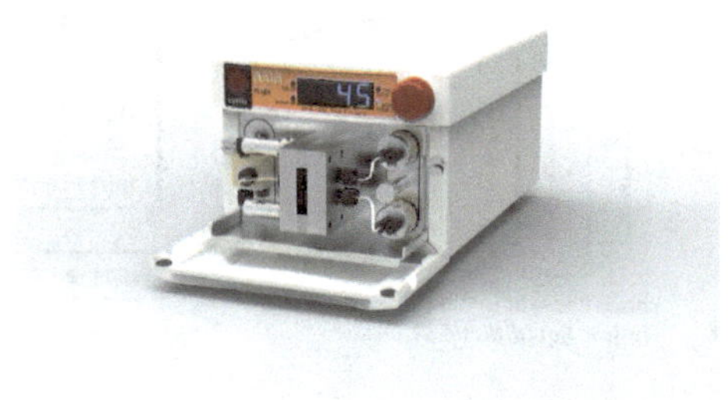

(b)

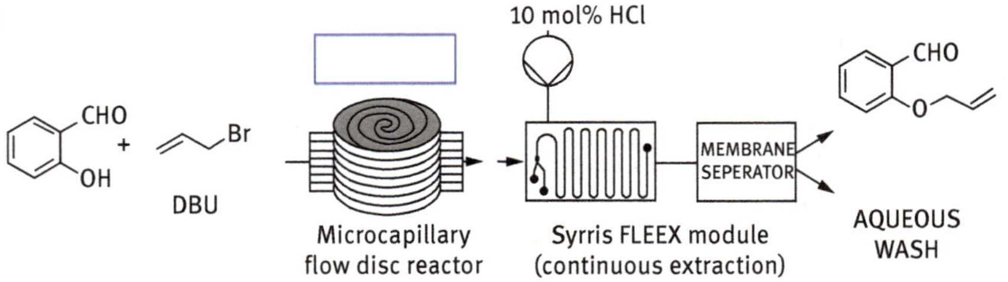

Fig. 7.18: (a) Syrris FLLEX module (www.syrris.com) and (b) a flow synthesis of an allylic ether with in-line membrane separator [38].

Quenching is an important step that benefits from flow chemistry. The in-line quenching was demonstrated during the preparation of oxazolines [40]. This flow protocol improved safety greatly compared to the batch synthesis. The reactive stream in dichloromethane with residual HF was quenched with the aqueous solution of sodium bicarbonate, forming a biphasic stream. The two phases were then separated by using the Zaiput continuous membrane–based liquid–liquid separator (Fig. 7.19). In addition, a membrane separator could be used for the in-line solvent switch [41].

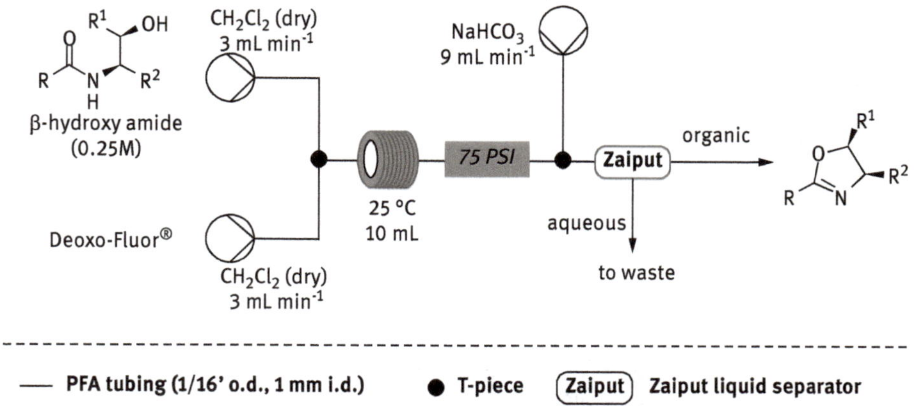

Fig. 7.19: An in-line quenching via a membrane separator during the preparation of oxazolines [40].

The liquid–liquid separator is a convenient tool to collect products after the synthesis. The use of continuous liquid–liquid separation was reported for the generation of chlorine [42]. In their work, NaOCl and HCl were mixed in the first junction to produce chlorine. Then, an organic solvent such as $CHCl_3$, CH_2Cl_2, hexane, and chlorobenzene was added to dissolve chlorine gas. A subsequent step was the aqueous–organic separation using a membrane separator (Fig. 7.20).

An automated self-optimizing system was built [43] with the in-line separation to provide a direct separation of a biphasic stream after the Claisen–Schmidt condensation reaction (Fig. 7.21). Following the separator, the organic phase was analyzed with an online HPLC. The online data was then used in the machine learning algorithm to propose sampling points at each iteration. With their system, they were able to optimize with three objectives (purity, space–time yield, and reaction mass efficiency) within only 65 h, compared to multiple weeks of the conventional batch-wise optimization. From this example, the in-line workup was proven to be an essential tool to enable the automated optimization for chemical processes.

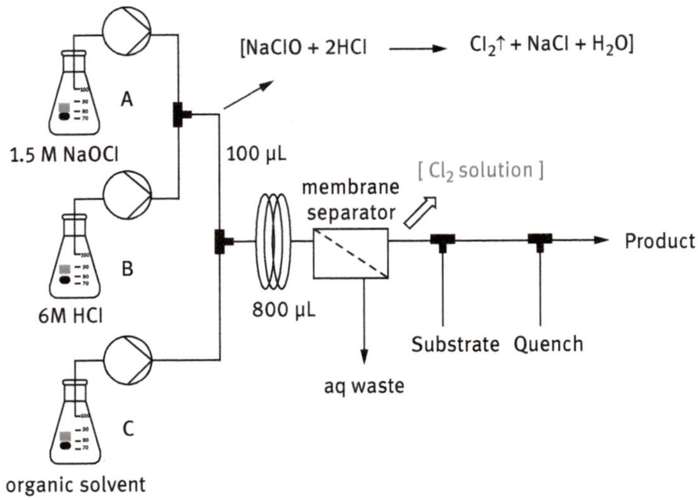

Fig. 7.20: A scheme for continuous generation of chlorine [42].

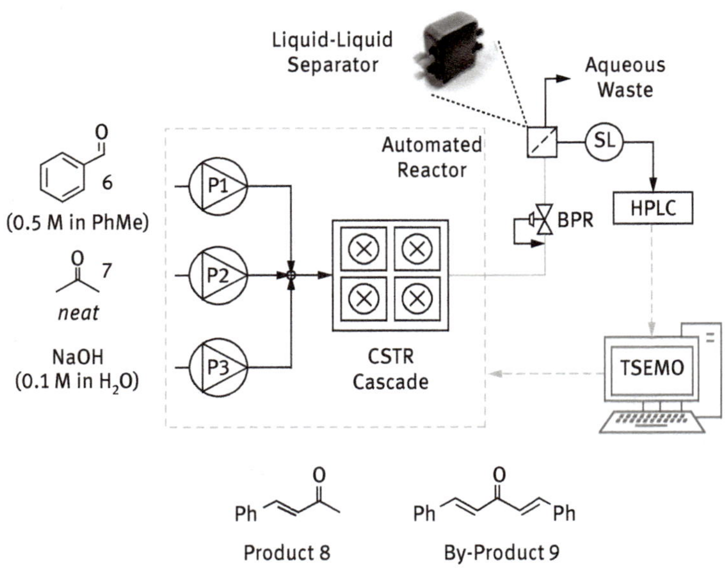

Fig. 7.21: An automated self-optimizing system enabled by an online HPLC and an in-line continuous liquid–liquid separator [43].

7.4.2 Continuous manufacturing of pharmaceuticals

Flow chemistry has provided an interesting paradigm for continuous manufacturing of pharmaceuticals [44]. Most pharmaceutical syntheses are multistep [45], so they often require workup steps in between sequential reactive steps [46]. The following examples demonstrate the use of continuous liquid–liquid separation in continuous synthesis of pharmaceuticals. The three-minute synthesis of ibuprofen in flow was developed [47]. The synthesis consisted of three steps, starting with the Friedel–Crafts reaction. In the first reaction, the aluminum reagent was used, but it was not compatible with the second step. Therefore, quenching with HCl was required, and the in-line liquid–liquid separation (Fig. 7.22) was to separate Al and remaining HCl from the desired intermediate.

As opposed to the ibuprofen example, the flow synthesis of atropine did not require in-line purification of intermediates between reactions [48]. However, the final stream contained a significant amount of by-products, which were structurally similar to atropine. Since by-products and atropine had different pK_a, acid- and alkali-based extraction was possible. To take advantage of the flow setup, the in-line separations were cascaded such that each stage of extraction was at a particular pH, and could remove certain by-products (Fig. 7.23).

Another remarkable example was the on-demand production of pharmaceuticals at MIT [8, 49, 50]. Particularly, the flow synthesis of fluoxetine consisted of three reactions (Fig. 7.24), which were not compatible with one another. In-line workups were to remove excess reagents as well as to separate water from an organic phase. The membrane-based separators performed effectively even under a pressurized condition (1.7 MPa), allowing for a fully continuous pharmaceutical synthesis.

The liquid–liquid separation was used to prepare benzyl chloride, which could be used directly in subsequent reactions to synthesize the final antihistaminic agents [51]. The in-line separator enabled a simple integration of four-step synthesis with a high yield (Fig. 7.25).

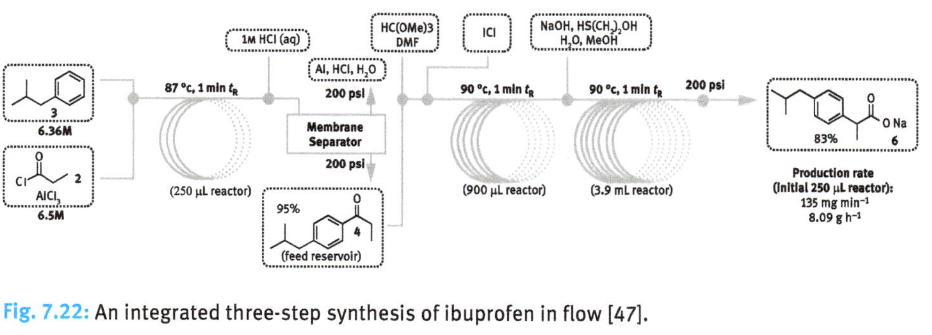

Fig. 7.22: An integrated three-step synthesis of ibuprofen in flow [47].

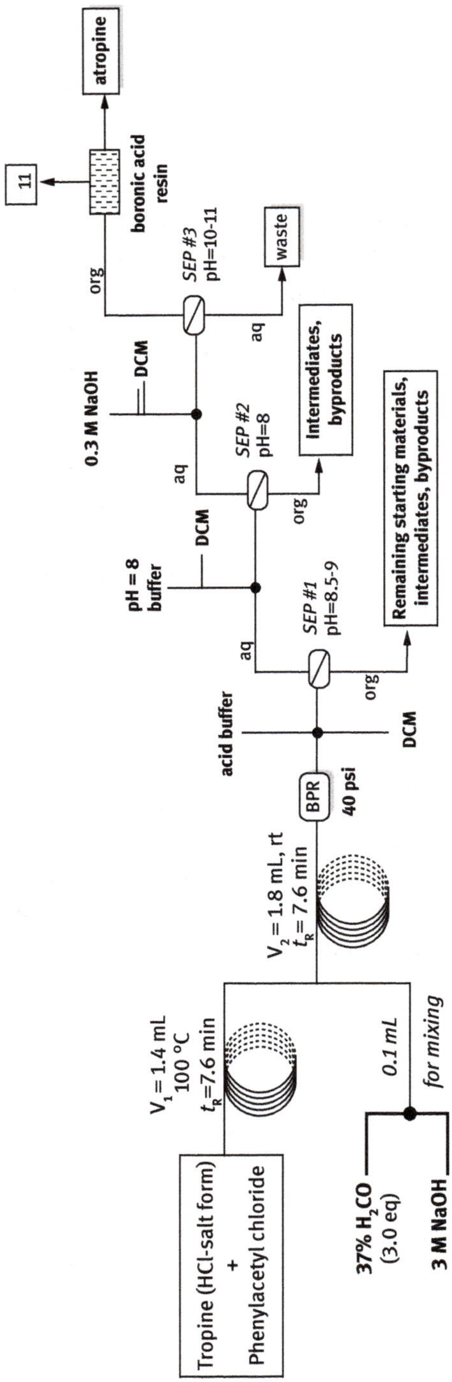

Fig. 7.23: A flow synthesis of atropine with cascaded continuous liquid–liquid extractions [48].

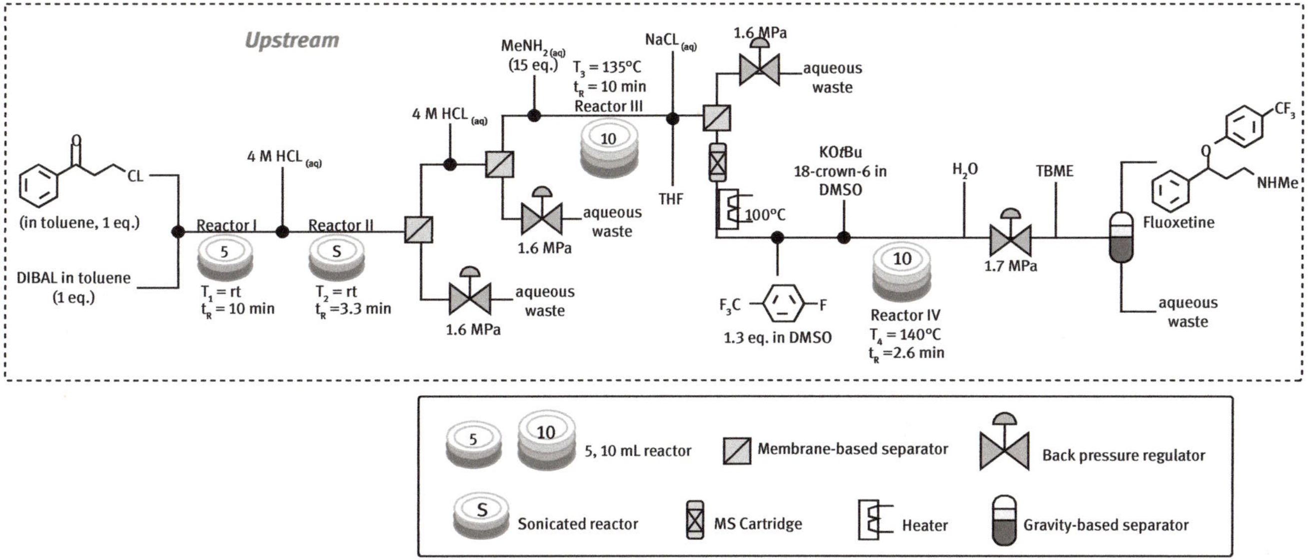

Fig. 7.24: A continuous-flow multistep synthesis of fluoxetine [8].

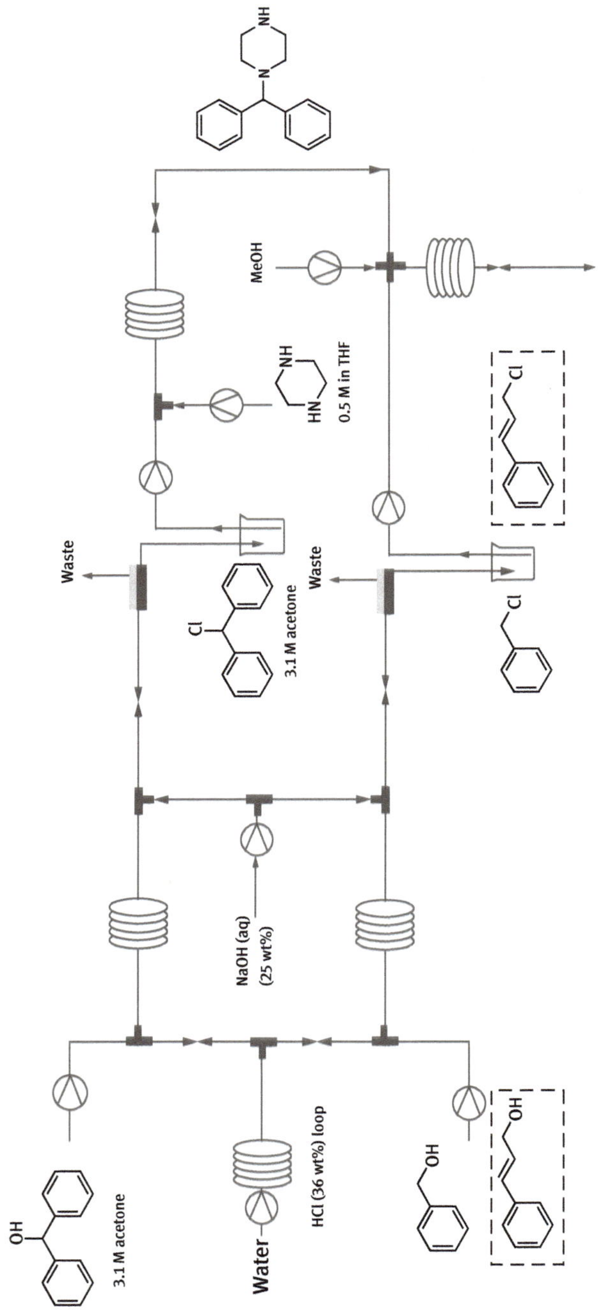

Fig. 7.25: The flow synthesis of antihistaminic agents enabled by continuous liquid–liquid separators [51].

7.4.3 Continuous preparation of nanomaterials

Continuous liquid–liquid separation also found applications in the field of nanomaterial preparation. Nanocrystals, such as quantum dots, can be prepared by using excess precursors. Purification is typically required to remove unreacted precursors before a subsequent step such as shell growth. Although flow syntheses of metal and semiconductor nanoparticles have already been demonstrated in many examples, a continuous purification was rarely reported. A multistage extraction setup was constructed to purify CdSe quantum dots and Au nanoparticles [52] (Fig. 7.26). The membrane-based liquid–liquid separation was implemented for each stage to separate the two phases from the extraction. Through process optimization (e.g., a number of stages, solvent-to-feed ratio), more than 90% of excess precursors could be removed. The setup is fully continuous, so it can directly be integrated to the flow synthesis.

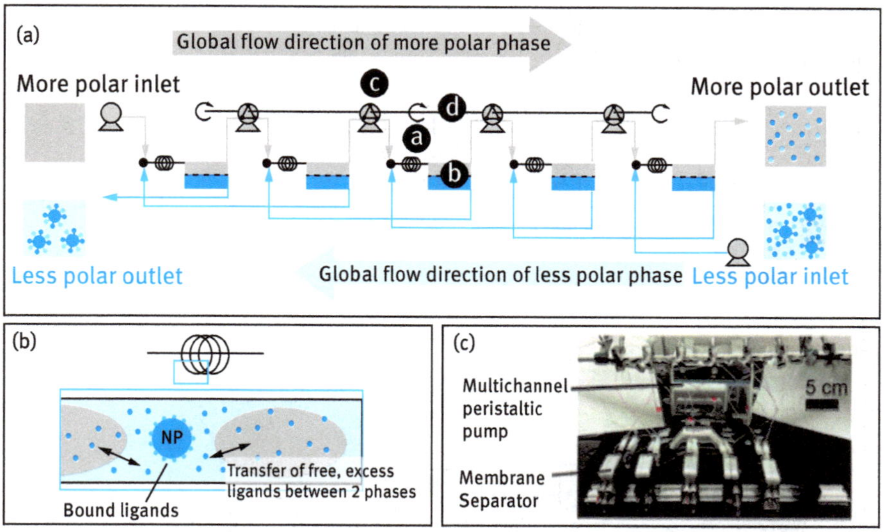

Fig. 7.26: A multistage extraction for purification colloidal nanoparticles [52].

Following the similar approach, the continuous liquid–liquid extraction was implemented to remove any excess capping ligands (sodium dodecyl sulfate, in this case) from the stream of flow-synthesized silver nanoparticles [53] (Fig. 7.27). From their work, the continuous extraction, which was integrated to the flow synthesis, yielded more than 70% removal of the excess capping ligands.

Similarly, the liquid–liquid separation was used to wash tetraoctylammonium bromide (TOAB) from TOAB-stabilized Au nanoparticles [54] (Fig. 7.28). To allow for complete washing, three stages of extraction with an aqueous solution of Na_2SO_4 were implemented. In addition, in this work, the liquid–liquid separation had another role

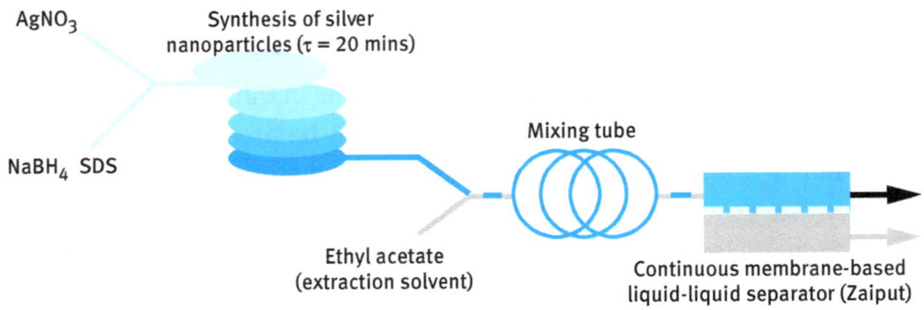

Fig. 7.27: Flow synthesis of silver nanoparticles integrated with continuous purification [53].

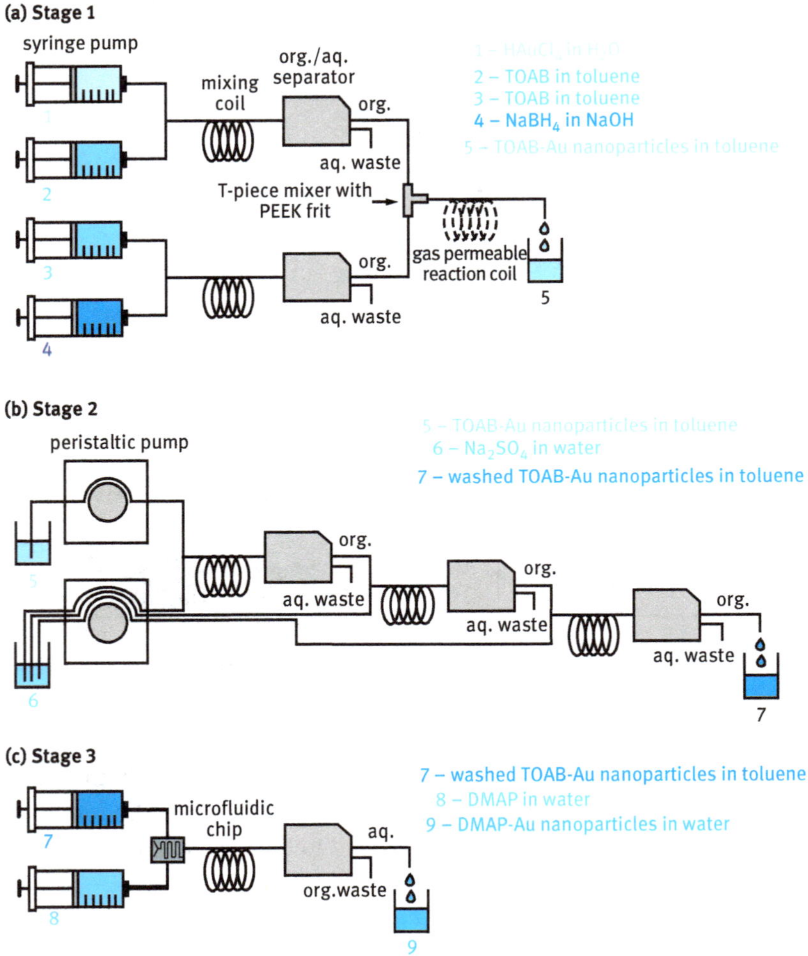

Fig. 7.28: Membrane separators as washing of TOAB from Au nanoparticles and transfer of Au nanoparticles from organic to aqueous phases [54].

during the transfer of Au-nanoparticles from the organic phase to an aqueous phase. During this step, 4-(dimethylamino)pyridine (DMAP) was added to facilitate the stabilization of Au nanoparticles in the aqueous phase.

Apart from the previous examples, continuous liquid–liquid separation was also applied to many other continuous-flow applications including continuous separation of catalysts [55, 56], continuous solvent extraction of rare earth elements [57] and nuclear materials [58], continuous radioisotope purification [59–61].

7.4.4 Scalability and commercial use

Scalability of a process from bench to production is a critical aspect in the context of commercial applications. Flow chemistry, as a technological approach, offers solutions that are either faster to scale up or, depending on equipment selection, that are inherently scalable meaning that the engineering of scaled up versions of the equipment has already taken into account scaling up challenges, thus providing an opportunity for seamless growth of production with very limited effort.

In this context, flow chemistry separation equipment selection also needs to be mindful, already at the laboratory scale, of options available as production needs grow. Currently scalable extraction/separation is achieved either with the devices commercialized by Zaiput or with a batch-based approaches, typically executed by leveraging existing batch equipment/infrastructure. On occasions, adaptation of other existing methods such as centrifuges or extraction columns may be encountered to either address rather specific needs or because resources are available at hand.

7.5 Gas–liquid separation

While in batch-based operations removal of a gas phase is seldom an issue as gravity drives it quickly to the top of a reactor, in flow there are several examples of cases where in-line removal may bring advantages. Technology used for gas–liquid separation hinges on the same type of approaches used for gas–liquid separation:

- gravity-driven separation is effective given the large density difference between the two phases;
- surface forces–based separation. This separation is typically achieved with the liquid being the wetting phase that permeates through a porous membrane/structure as presented earlier in this chapter. If the liquid is very polar (or very nonpolar) separation also can be achieved by having the gas phase permeate through a porous structure that is not-wetted by the liquid. While this latter method can be equally successful, it requires the liquid and the surface to have very different surface properties as to decrease chances of improper wetting that would impair separation.

7.5.1 Applications of continuous gas–liquid separation

Typical applications of continuous gas–liquid separation include in-line purification of a product stream, removal/recycle of an unreacted gas reagent, stripping of a volatile component, and elimination of gas for better analytical analysis.

7.5.2 Removal of N_2 from various liquids before analysis via UV–vis spectrometer

A study on the hydrodynamics of gas–liquid flow was conducted in micropacked beds using a tracer and nitrogen gas with various liquids such as water, propanol, cyclohexanone, and methanol [62] (Fig. 7.29). The gas and liquid feeds entered the packed bed at one end and exited together at the other. In order to detect the tracer in the liquid stream without gas interference by a high-resolution UV–vis–NIR spectrometer, nitrogen gas was separated out using a surface-force-based separator with a hydrophobic membrane that is permeable to the gas phase but not the liquid.

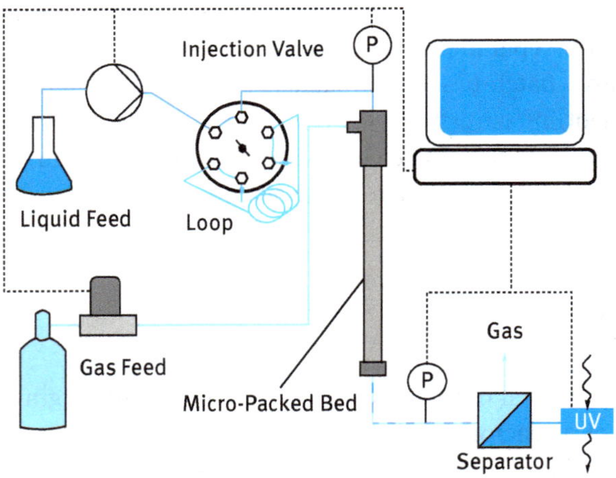

Fig. 7.29: Schematic overview of experimental setup with in-line separator used to eliminate gas from liquid before analytical instrument [62].

7.5.3 Removal of H_2 from product during hydrogenation of α-tetralone

The hydrogenation of α-tetralone was carried out by recycling a ruthenium diphosphine/diamine catalyst using continuous nanofiltration [63]. Their flow system mainly consisted of a packed bed followed by a nanofiltration module and then a membrane-based separator as illustrated in Fig. 7.30. After reaction, the gas–liquid mixture proceeds to the nanofiltration unit where separation occurs by gravity, and hydrogen is vented using 2.76 MPa of backpressure. In this work, the gravity-based nanofiltration along was not sufficient because significant hydrogen outgassing in the liquid stream was observed. Therefore, in order to fully eliminate hydrogen in the liquid stream, a gas–liquid Zaiput separator was also implemented to generate a permeate stream only filled with liquid. This allowed to remove hydrogen from the product, as well as to quantify its concentrations by analytical tools.

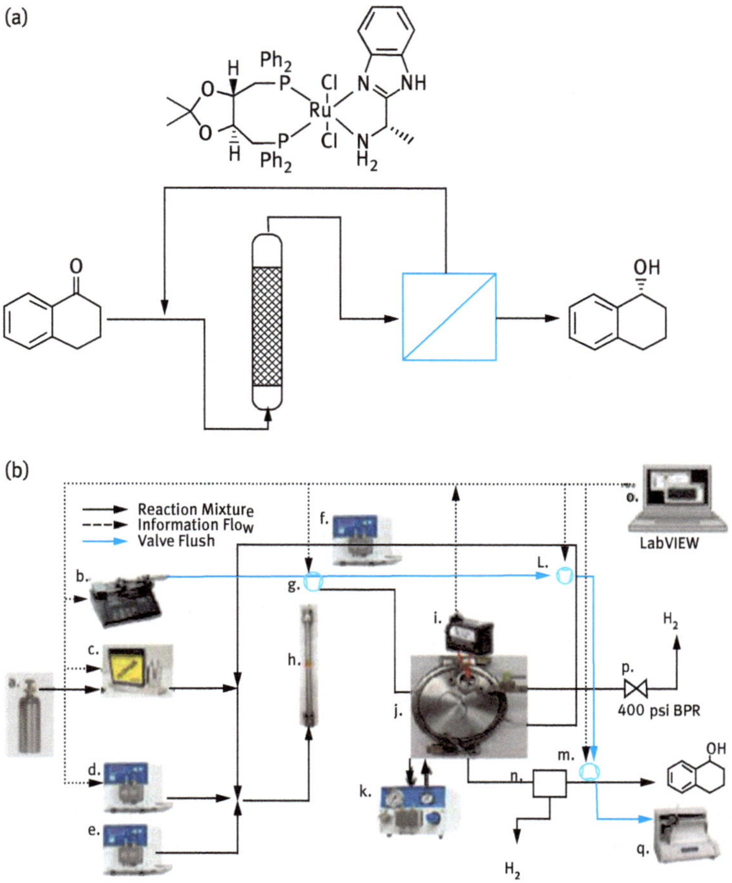

Fig. 7.30: (a) An overall schemetic and (b) experimental setup with the coupling of a gravity-based nanofiltration unit followed by a surface-tension-based separator to fully elimiinate hydrogen gas from permeate liquid stream [63].

7.5.4 Removal of CO_2 and catalyst recycle in photo-oxidation of α-terpinene and citronellol

The performance of fluorous biphasic catalyst for continuous photo-oxidation of organic compounds was investigated [64]. Experiments were conducted safely using supercritical carbon dioxide as the choice of suitable non-flammable and non-toxic solvent for efficient scale-up of such photo-oxidations. In this study, the organic substrate was mixed with carbon dioxide and oxygen as well as a photocatalyst (F8/HFE) recycle loop before entering the reactor as illustrated in Fig. 7.31. After the reaction, the products and gases were separated using a gravity-based separation approach by implementing a glassware flask with an ultrasonic bath to accelerate phase separation which allowed recycling the catalyst while venting carbon dioxide and unreacted oxygen gas.

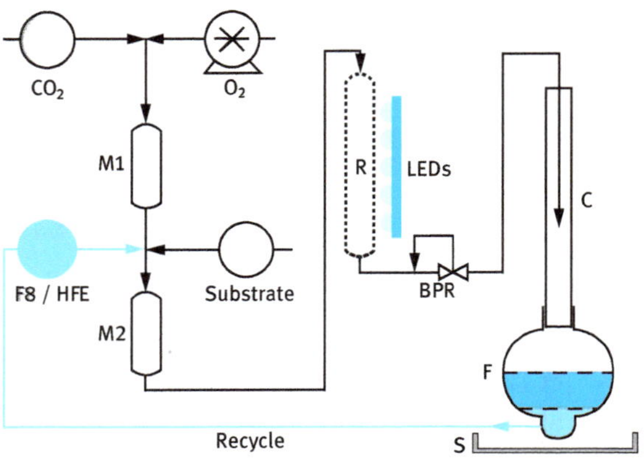

Fig. 7.31: Process schematic of the experimental setup with the gravity-based glassware flask used to accelerate phase separation and venting of gases [64].

7.5.5 Synthesis of diazomethane

Recently researchers at Snapdragon Chemistry [65] demonstrated a safe and scalable synthesis of diazomethane, a very powerful reagent that is little used due to the hazards it presents. The authors reduced the chemical inventory of diazomethane to <53 mg by using PFR reactors and membrane-based gas liquid separators (Fig. 7.32). The purification of diazomethane was completed by using the commercially available membrane separators produced by Zaiput Flow Technologies to extract diazomethane from the generation reaction mixture using nitrogen to dilute the explosive gas. This allows access to diazomethane on multiple scales ranging from laboratory to larger scales by using the appropriate separator. This system demonstrated a synthetically useful scale of 0.44 mol h^{-1}. After diazomethane generation, both methylation and palladium-catalyzed cyclopropanation reactions are successfully carried out, thereby validating that the purity of the reagent is high enough to be compatible with palladium catalysis.

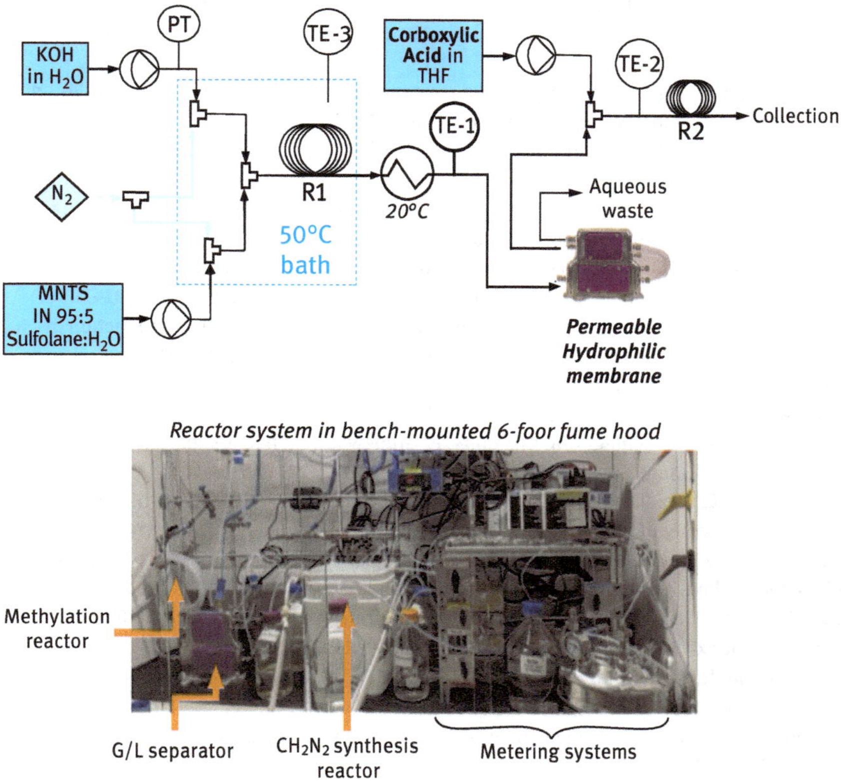

Fig. 7.32: Synthesis of diazomethane [65].

Study questions

7.1 What are the applications of continuous gas–liquid separation in flow chemistry?

7.2 What are the applications of continuous liquid–liquid separation in flow chemistry?

7.3 What are different types of sensors that can be implemented for gravity-based separators?

7.4 For the continuous separation of organic-aqueous stream, what are the upper and lower limits of the operating range in the membrane-based separator? Suppose that the membrane is hydrophobic, and an organic phase preferentially wets the membrane.

7.5 What are the challenges of using back-pressure regulators to control a pressure difference across a membrane during a continuous separation?

References

[1] Bahadori, A, Natural Gas Processing: Technology and Engineering Design, Gulf Professional Publishing, 2014.
[2] De Nevers, N, Silcox, GD, Fluid Mechanics for Chemical Engineers, New York, McGraw-Hill, 1991.
[3] Assmann, N, Ładosz, A, Rudolf Von, RP, Continuous micro liquid-liquid extraction, Chem Eng Technol, 2013, 36, 921–936.
[4] Aota, A, Mawatari, K, Takahashi, S, et al., Phase separation of gas–liquid and liquid–liquid microflows in microchips, Microchimica Acta, 2009, 164, 249.
[5] Wang, K, Luo, G, Microflow extraction: A review of recent development, Chem Eng Sci, 2017, 169, 18–33.
[6] Maslana, E, Schmitt, R, Pan, J, A fully automated liquid–liquid extraction system utilizing interface detection, J Autom Methods Manag Chem, 2000, 22, 187–194.
[7] O'Brien, M, Koos, P, Browne, DL, Ley, SV, A prototype continuous-flow liquid–liquid extraction system using open-source technology, Org Biomol Chem, 2012, 10, 7031–7036.
[8] Adamo, A, Beingessner, RL, Behnam, M, et al., On-demand continuous-flow production of pharmaceuticals in a compact, reconfigurable system, Science, 2016, 352, 61–67.
[9] Sprecher, H, Payán, MNP, Weber, M, Yilmaz, G, Wille, G, Acyl azide synthesis and curtius rearrangements in microstructured flow chemistry systems, J Flow Chem, 2012, 2, 20–23.
[10] Kenig, EY, Su, Y, Lautenschleger, A, Chasanis, P, Grünewald, M, Micro-separation of fluid systems: A state-of-the-art review, Sep Purif Technol, 2013, 120, 245–264.
[11] TeGrotenhuis, WE, Cameron, RJ, Butcher, MG, Martin, PM, Wegeng, RS, Microchannel devices for efficient contacting of liquids in solvent extraction, Sep Sci Technol, 1999, 34, 951–974.
[12] Berthier, J, Tran, V-M, Mittler, F, Sarrut, N, The physics of a coflow micro-extractor: Interface stability and optimal extraction length, Sens Actuators A Phys, 2009, 149, 56–64.
[13] Aota, A, Nonaka, M, Hibara, A, Kitamori, T, Countercurrent laminar microflow for highly efficient solvent extraction, Angewandte Chemie, 2007, 119, 896–898.
[14] Hessel, V, Löwe, H, Schönfeld, F, Micromixers – a review on passive and active mixing principles, Chem Eng Sci, 2005, 60, 2479–2501.
[15] Constantinou, A, Gavriilidis, A, CO2 absorption in a microstructured mesh reactor, Ind Eng Chem Res, 2010, 49, 1041–1049.
[16] Yang, R-J, Liu, -C-C, Wang, Y-N, Hou, -H-H, Fu, L-M, A comprehensive review of micro-distillation methods, Chem Eng J, 2017, 313, 1509–1520.
[17] Weeranoppanant, N, Integrated Continuous-Flow Chemistry Enabled by Multistage Separations, Massachusetts Institute of Technology, 2016.
[18] Okubo, Y, Toma, M, Ueda, H, Maki, T, Mae, K, Microchannel devices for the coalescence of dispersed droplets produced for use in rapid extraction processes, Chem Eng J, 2004, 101, 39–48.
[19] Kolehmainen, E, Turunen, I, Micro-scale liquid–liquid separation in a plate-type coalescer, Chem Eng Process, 2007, 46, 834–839.
[20] Jahromi, PF, Karimi-Sabet, J, Amini, Y, Fadaei, H, Pressure-driven liquid-liquid separation in Y-shaped microfluidic junctions, Chem Eng J, 2017, 328, 1075–1086.
[21] Kamrani, S, Mohammadi, A, Controlling the microscale separation of immiscible liquids using geometry: A computational fluid dynamics study, Chemical Engineering Science, 2020, 115625.
[22] Liu, D, Wang, K, Wang, Y, Wang, Y, Luo, G, A simple online phase separator for the microfluidic mass transfer studies, Chem Eng J, 2017, 325, 342–349.
[23] Kashid, M, Harshe, Y, Agar, DW, Liquid–liquid slug flow in a capillary: An alternative to suspended drop or film contactors, Ind Eng Chem Res, 2007, 46, 8420–8430.

[24] Kralj, JG, Sahoo, HR, Jensen, KF, Integrated continuous microfluidic liquid–liquid extraction, Lab Chip, 2007, 7, 256–263.
[25] Cervera-Padrell, AE, Morthensen, ST, Lewandowski, DJ, Skovby, T, Kiil, S, Gernaey, KV, Continuous hydrolysis and liquid–liquid phase separation of an active pharmaceutical ingredient intermediate using a miniscale hydrophobic membrane separator, Org Process Res Dev, 2012, 16, 888–900.
[26] Adamo, A, Heider, PL, Weeranoppanant, N, Jensen, KF, Membrane-based, liquid–liquid separator with integrated pressure control, Ind Eng Chem Res, 2013, 52, 10802–10808.
[27] Imbrogno, J, Rogers, L, Thomas, DA, Jensen, KF, Continuous purification of active pharmaceutical ingredients utilizing polymer membrane surface wettability, Chem Commun, 2018, 54, 70–73.
[28] Cai, Z-X, Fang, Q, Chen, H-W, Fang, Z-L, A microfluidic chip based liquid–liquid extraction system with microporous membrane, Anal Chim Acta, 2006, 556, 151–156.
[29] Yang, L, Weeranoppanant, N, Jensen, KF, Characterization and modeling of the operating curves of membrane microseparators, Ind Eng Chem Res, 2017, 56, 12184–12191.
[30] Bannock, JH, Lui, TYM, Turner, ST, deMello, JC, Automated separation of immiscible liquids using an optically monitored porous capillary, React Chem Eng, 2018, 3, 467–477.
[31] Harvie, AJ, Herrington, JO, deMello, JC, An improved liquid–liquid separator based on an optically monitored porous capillary, React Chem Eng, 2019, 4, 1579–1588.
[32] Weeranoppanant, N, Adamo, A, Saparbaiuly, G, et al., Design of multistage counter-current liquid–liquid extraction for small-scale applications, Ind Eng Chem Res, 2017, 56, 4095–4103.
[33] Hu, DX, O'Brien, M, Ley, SV, Continuous multiple liquid–liquid separation: Diazotization of amino acids in flow, Org Lett, 2012, 14, 4246–4249.
[34] Ley, SV, Ingham, RJ, O'Brien, M, Browne, DL, Camera-enabled techniques for organic synthesis, Beilstein J Org Chem, 2013, 9, 1051–1072.
[35] O'Brien, M, Cooper, DA, Mhembere, P, The continuous-flow synthesis of carbazate hydrazones using a simplified computer-vision controlled liquid–liquid extraction system, Tetrahedron Lett, 2016, 57, 5188–5191.
[36] O'Brien, M, Cooper, DA, Dolan, J, Continuous flow iodination using an automated computer-vision controlled liquid-liquid extraction system, Tetrahedron Lett, 2017, 58, 829–834.
[37] O'Brien, M, Hall, A, Schrauwen, J, Made J, VD, An open-source approach to automation in organic synthesis: The flow chemical formation of benzamides using an inline liquid-liquid extraction system and a homemade 3-axis autosampling/product-collection device, Tetrahedron, 2018, 74, 3152–3157.
[38] Hornung, CH, Mackley, MR, Baxendale, IR, Ley, SV, A microcapillary flow disc reactor for organic synthesis, Org Process Res Dev, 2007, 11, 399–405.
[39] Martin, LJ, Marzinzik, AL, Ley, SV, Baxendale, IR, Safe and reliable synthesis of diazoketones and quinoxalines in a continuous flow reactor, Org Lett, 2011, 13, 320–323.
[40] Glöckner, S, Tran, DN, Ingham, RJ, et al., The rapid synthesis of oxazolines and their heterogeneous oxidation to oxazoles under flow conditions, Org Biomol Chem, 2015, 13, 207–214.
[41] Hamlin, TA, Lazarus, GM, Kelly, CB, Leadbeater, NE, A continuous-flow approach to 3, 3, 3-trifluoromethylpropenes: Bringing together grignard addition, peterson elimination, inline extraction, and solvent switching, Org Process Res Dev, 2014, 18, 1253–1258.
[42] Strauss, FJ, Cantillo, D, Guerra, J, Kappe, CO, A laboratory-scale continuous flow chlorine generator for organic synthesis, React Chem Eng, 2016, 1, 472–476.
[43] Clayton, AD, Schweidtmann, AM, Clemens, G, et al., Automated self-optimisation of multi-step reaction and separation processes using machine learning, Chem Eng J, 2020, 384, 123340.
[44] Bogdan, AR, Dombrowski, AW, Emerging trends in flow chemistry and applications to the pharmaceutical industry, J Med Chem, 2019, 62, 6422–6468.

[45] Baumann, M, Baxendale, IR, The synthesis of active pharmaceutical ingredients (APIs) using continuous flow chemistry, Beilstein J Org Chem, 2015, 11, 1194–1219.

[46] Weeranoppanant, N, Adamo, A, In-line purification: A key component to facilitate drug synthesis and process development in medicinal chemistry, ACS Med Chem Lett, 2019, 11, 9–15.

[47] Snead, DR, Jamison, TF, A Three-minute synthesis and purification of ibuprofen: Pushing the limits of continuous-flow processing, Angew Chem Int Ed, 2015, 54, 983–987.

[48] Dai, C, Snead, DR, Zhang, P, Jamison, TF, Continuous-flow synthesis and purification of atropine with sequential in-line separations of structurally similar impurities, J Flow Chem, 2015, 5, 133–138.

[49] Monbaliu, J-CM, Stelzer, T, Revalor, E, Weeranoppanant, N, Jensen, KF, Myerson, AS, Compact and integrated approach for advanced end-to-end production, purification, and aqueous formulation of lidocaine hydrochloride, Org Process Res Dev, 2016, 20, 1347–1353.

[50] Zhang, P, Weeranoppanant, N, Thomas, DA, et al., Advanced continuous flow platform for on-demand pharmaceutical manufacturing, Chem Eur J, 2018, 24, 2776–2784.

[51] Borukhova, S, Noël, T, Hessel, V, Continuous-flow multistep synthesis of cinnarizine, cyclizine, and a buclizine derivative from bulk alcohols, ChemSusChem, 2016, 9, 67–74.

[52] Shen, Y, Weeranoppanant, N, Xie, L, et al., Multistage extraction platform for highly efficient and fully continuous purification of nanoparticles, Nanoscale, 2017, 9, 7703–7707.

[53] Khositanon, C, Adpakpang, K, Bureekaew, S, Weeranoppanant, N, Continuous-flow purification of silver nanoparticles and its integration with flow synthesis, J Flow Chem, 2020, 10, 353–362.

[54] Chow, E, Raguse, B, Della Gaspera, E, et al., Flow-controlled synthesis of gold nanoparticles in a biphasic system with inline liquid–liquid separation, React Chem Eng, 2020, 5, 356–366.

[55] Peer, M, Weeranoppanant, N, Adamo, A, Zhang, Y, Jensen, KF, Biphasic catalytic hydrogen peroxide oxidation of alcohols in flow: Scale-up and extraction, Org Process Res Dev, 2016, 20, 1677–1685.

[56] Gürsel, IV, Noël, T, Wang, Q, Hessel, V, Separation/recycling methods for homogeneous transition metal catalysts in continuous flow, Green Chem, 2015, 17, 2012–2026.

[57] Leblebici, ME, Kuhn, S, Stefanidis, GD, Van Gerven, T, Milli-channel mixer and phase separator for solvent extraction of rare earth elements, Chem Eng J, 2016, 293, 273–280.

[58] Darekar, M, Chaurasiya, RK, Singh, K, Mukhopadhyay, S, Shenoy, K, In-line phase separator for microfluidic solvent extraction of uranium, J Radioanal Nucl Chem, 2020, 1–11.

[59] Pedersen, KS, Imbrogno, J, Fonslet, J, Lusardi, M, Jensen, KF, Zhuravlev, F, Liquid–liquid extraction in flow of the radioisotope titanium-45 for positron emission tomography applications, React Chem Eng, 2018, 3, 898–904.

[60] Pedersen, KS, Nielsen, KM, Fonslet, J, Jensen, M, Zhuravlev, F, Separation of radiogallium from zinc using membrane-based liquid-liquid extraction in flow: Experimental and COSMO-RS studies, Solvent Extr Ion Exch, 2019, 37, 376–391.

[61] Martini, P, Adamo, A, Syna, N, et al., Perspectives on the use of liquid extraction for radioisotope purification, Molecules, 2019, 24, 334.

[62] Zhang, J, Teixeira, AR, Kögl, LT, Yang, L, Jensen, KF, Hydrodynamics of gas–liquid flow in micropacked beds: Pressure drop, liquid holdup, and two-phase model, AIChE J, 2017, 63, 4694–4704.

[63] O'Neal, EJ, Lee, CH, Brathwaite, J, Jensen, KF, Continuous nanofiltration and recycle of an asymmetric ketone hydrogenation catalyst, ACS Catal, 2015, 5, 2615–2622.

[64] Hall, JF, Han, X, Poliakoff, M, Bourne, RA, George, MW, Maximising the efficiency of continuous photo-oxidation with singlet oxygen in supercritical CO 2 by use of fluorous biphasic catalysis, Chem Commun, 2012, 48, 3073–3075.

[65] Sheeran, JW, Campbell, K, Breen, CP, et al., Scalable on-demand production of purified diazomethane suitable for sensitive catalytic reactions, Org Process Res Dev, 2020.

Yuesu Chen, Martin Cattoen and Jean-Christophe M. Monbaliu

8 Mitigation of chemical hazards under continuous flow conditions

8.1 Introduction: safety considerations in chemical processes

Chemical processes are hazardous by their nature. From laboratories to chemical plants, people carrying out chemical reactions are exposed to the dangers of fire, explosion, and toxic release that are particularly related to the use of hazardous materials and extreme process conditions [1]. Accidents associated with chemicals and chemical processes cause the loss of lives and properties every year, especially in highly industrialized countries [2–4]. Recently, a tank explosion at Midland Resource Recovery (Philippi, West Virginia, USA) killed two people and severely injured one; the release of methanethiol at DuPont la Porte Facility (la Porte, Texas, USA) killed four workers and injured a fifth; the explosion and fire at Pryor Trust (Pittsburg County, Oklahoma, USA) fatally injured five workers [5]. The World Health Organization estimated 65,000 people died of technological disasters between 2009 and 2018 [6].

8.1.1 Hazards of chemicals

Hazardous compounds could enter human body by inhalation, dermal contact, or ingestion, causing acute or chronic damages to our health. The concentrations of toxic substances in the air at the workplace should be maintained below their threshold limits [7]. Therefore, safe operation procedures must always be followed taking into account the properties of the substances in order to minimize exposure and prevent fire or explosion [8].

The possession, use, storage, and transportation of chemicals are regulated by laws. In the European Union, producers and importers of chemicals in quantities $>1\ t\ a^{-1}$ are obliged to register their products with safety data according to the EU Regulation 1907/2006 (REACH-regulation) before marketing (REACH Article 6). The chemicals are classified and labeled according to the Globally Harmonized System (GHS) (Fig. 8.1), which is also implemented in EU Regulation 1272/2008 (CLP regulation). Based on REACH and CLP regulations, the management of hazardous chemicals is further detailed by national laws and regulations of individual countries [7, 9].

https://doi.org/10.1515/9783110693676-008

Globally **H**armonized **S**ystem of Classification and Labeling of Chemicals (**GHS**)

Physical hazards	explosive, flammable gas/liquid/solid, aerosol, oxidizing gas/liquid/solid, gas under pressure, self-reactive, pyrophoric, self-heating, in contact with water emit flammable gases, organic peroxides, corrosive to metals
Health hazards	acute toxic, skin-corrosive/irratant, eye damage/irritation, respiratory/skin sensitisation, germ cell mutagenic, carcinogenie, reproductive toxic, specifie target organ toxic, aspiration hazard
Environmental hazards	hazardous to the aquatic environment / to the ozone layer

National rules implementied GHS (selected examples)	EU: Regulation (EC) No 1272/2008 (CLP Regulation) CLP = Classification, Labelling, and Packaging US: Occupational Safety & Health Administration standards CN: GB/T 22234-2008

Fig. 8.1: The classification of the chemical hazards: the Globally Harmonized System (GHS, by the United Nations).

8.1.2 Hazards of chemical processes

Safety considerations must be embedded from the very beginning of the process design. Apart from exposure to chemicals in regular operations, chemical reactions and extreme conditions have posed additional threats to the safety of the personnel. For example, exothermic reactions may cause temperature overshooting; overheated solvent and gas released from reactions could build up pressure very rapidly; autoclaves under high temperatures and pressures have the risk of explosion. Hence, multiple layers of protections are added around the chemical processes to reduce chemical risk (with, ideally, its complete suppression by removing the chemical hazard and the exposure to chemicals) and the frequency of chemical incidents, as well as to minimize the damage through safer design features, process control technologies, and administrative control (Fig. 8.2) [10].

Despite the huge cost of establishment and maintenance of protective measures, the restrictions imposed by increasingly strict laws and regulations (e. g., REACH annex XVII) are forcing chemists and chemical engineers to redesign the established routes that rely on some scheduled compounds, striving for inherently safer processes for the synthesis of many fine chemicals, especially pharmaceuticals [11].

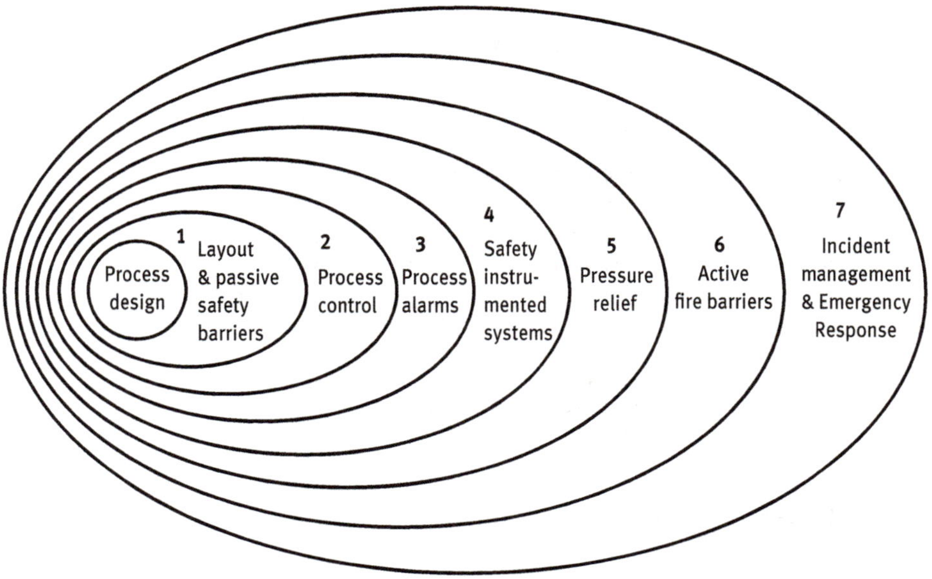

Fig. 8.2: Layers of protection for the process design [10].

8.1.3 The inherently safer process design

The *inherently safer design* of chemical processes aims to eliminate or reduce the risk at source (Fig. 8.3) by controlling both the hazard and the likelihood of exposure [12, 13]. The *substitution* of the hazardous step with benign surrogates seems to be ideal (Fig. 8.3a), but this is not always feasible, since many harsh reaction conditions or the use of reactive intermediates provide the most efficient and straightforward transformation. In these cases, people could try to *minimize* the inventory of the noxious material by using a smaller reactor (Fig. 8.3b) to reduce hazard hold-up, so that the quantity of material could, therefore, no longer support a severe fire, explosion, or serious toxic release.

Considering three typical ideal reactors (Fig. 8.4) having the same productivity (space-time yield), the plug flow tubular reactor (PFR) is smaller in hold-up volume than the batch reactor (BR) (taking the time for intermediate loading and unloading into account) and the continuous stirred tank reactor (CSTR) [14]. This could be one of the safety incentives to swap from batch and tank reactors to tubular flow.

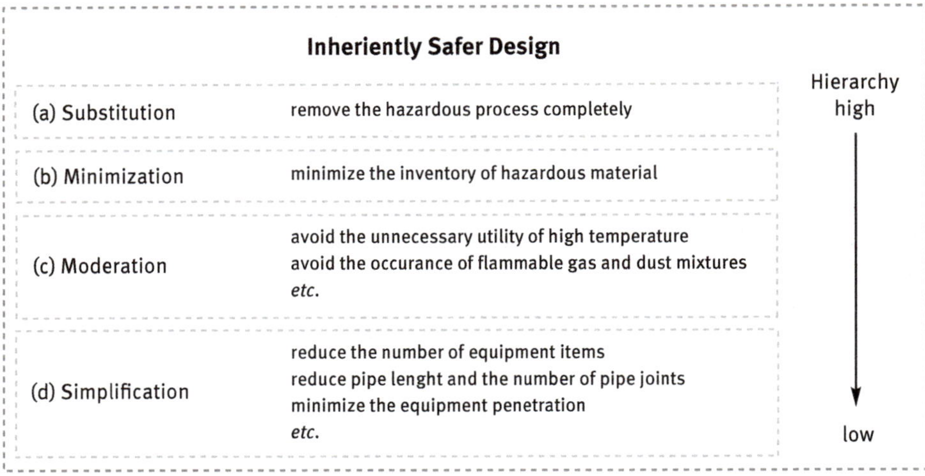

Fig. 8.3: The hierarchy of inherently safer design [13].

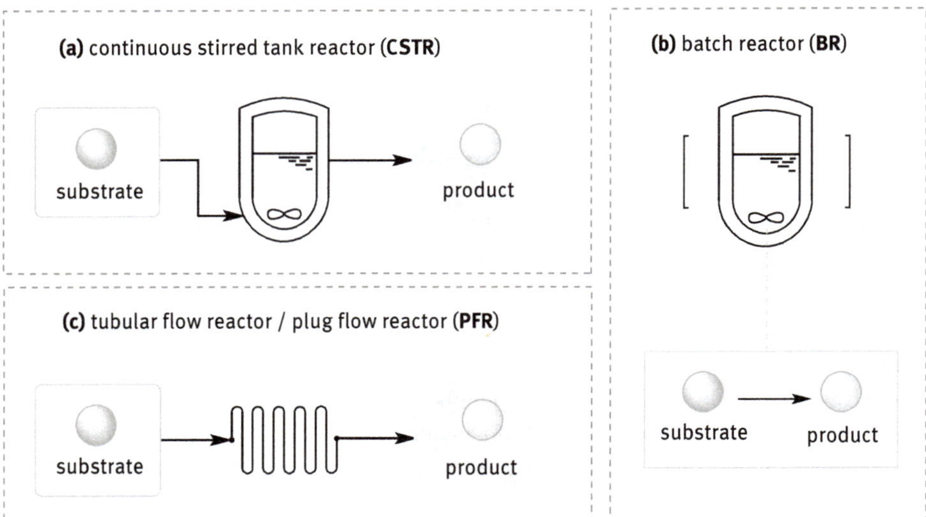

Fig. 8.4: Three basic types of ideal reactors. (In this chapter, the pumping systems are denoted by round-cornered rectangle frames at the inlets of the flow systems).

8.2 Continuous flow reactors – a step toward inherent safety

The micro- and mesofluidic reactors used for chemical synthesis can generally be considered as tubular flow reactors (Fig. 8.4c), which have at least one dimension within (sub-) millimeter range [15, 16]. With features such as intensified transport processes, structural robustness, low hold-up volume, and continuous flow operation, micro- and mesofluidic reactors have become powerful tools in managing hazardous chemicals and harsh reaction conditions [17–19]. In this section, basic principles underlying the inherent safety of flow reactors are highlighted.

8.2.1 Intensified heat and mass transport processes

As the majority of industrial reactions are exothermic ($\Delta_r H_m < 0$) [20], temperature overshooting occurs if heat generation is faster than heat removal (often referred to as *thermal runaway*). When the cooling system fails, the accumulated heat could further accelerate the reaction to generate more heat, and such a vicious circle results, ultimately, in a thermal run-away accident [21, 22].

The rate of heat generation (dQ/dt) from a chemical reaction is proportional to its scale and measured by the volume of the solution (V) (equation (8.1)). Combining it with the equation of heat removal through the reactor wall (equation (8.2)), the required specific area (A/V) can be calculated (equation (8.3)); under the same cooling and reaction conditions, reactors of larger scale require a higher specific area because of the increased wall thickness (Δx_{wall}) for mechanical strength [23].
Heat generation:

$$\frac{dQ}{dt} = \Delta_r H_m \cdot \frac{d\xi}{dt} = \frac{V \cdot \Delta_r H_m}{\nu} \cdot \frac{dc}{dt} \tag{8.1}$$

Heat removal:

$$\frac{dQ}{dt} = -\frac{A \cdot \Delta T}{\frac{\Delta x_{wall}}{k_{wall}}} \tag{8.2}$$

Specific area:

$$\frac{A}{V} = \frac{\frac{\Delta x_{wall}}{k_{wall}}}{\Delta T} \cdot \frac{(-\Delta_r H_m)}{\nu} \cdot \frac{dc}{dt} \tag{8.3}$$

Symbol – physical quantity [unit]: Q – heat [J], t – time [s], $\Delta_r H_m$ – molar reaction enthalpy [J mol^{-1}], ξ – extent of reaction [mol], V – volume of the solution [m^3], ν – stoichiometric coefficient [1], c – concentration of a reactant or product [mol m^{-3}], A – area for heat transfer [m^2], ΔT – temperature difference [K], Δx – distance of heat conduction [m], and k – thermal conductivity [W (m$^{-1}\cdot$K^{-1})]. (The equations are simplified for a qualitative discussion, assuming the reaction mixture is being thoroughly agitated.)

However, the specific area of conventional reactors decreases with increasing size. For example, when the volume of a round bottomed flask increases from 10 to 2,000 mL, the specific area for heat transfer decreases by 83% (Fig. 8.5a). Therefore, large flasks and tank reactors (Fig. 8.4a and b) are not ideal for scale-up due to thermal safety reason. The rate of heat generation is limited by applying lower temperature and controlled addition of one of the reactants, inexorably posing problems in productivity and the control of stoichiometry.

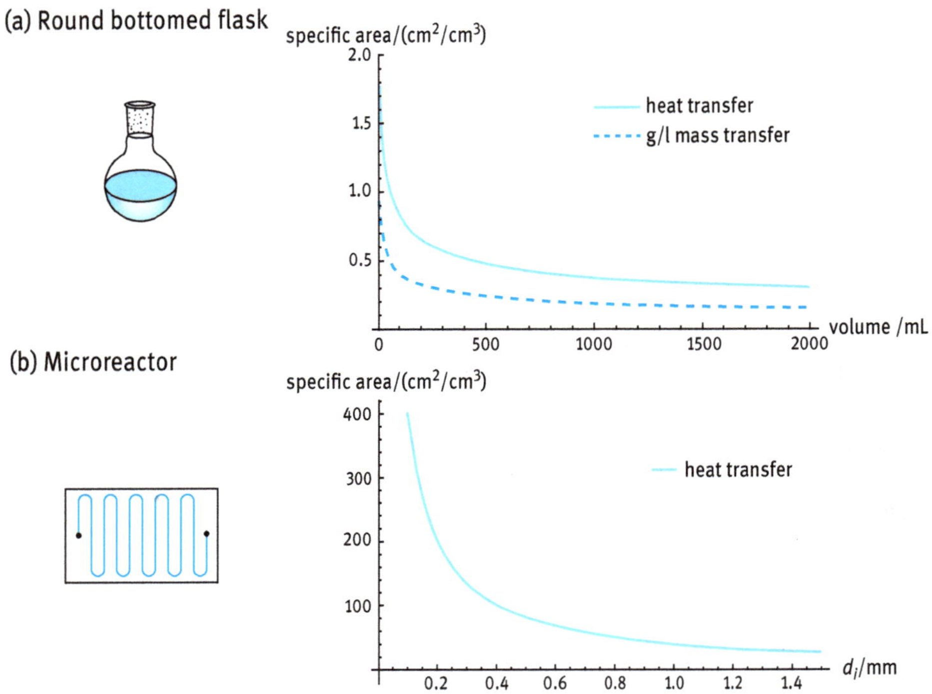

Fig. 8.5: Specific area for heat transfer in flasks and microreactors (solid curves).

In micro- and mesofluidic reactors, the reaction mixture is provided with a tremendous specific area for heat transfer, which is greater than that of flasks and tanks (Fig. 8.5b) by orders of magnitude. With fixed inner diameter (d_i) of the flow channel, the internal volume can be modified by varying the channel length without changing the specific wall area. This important feature of micro- and mesofluidic reactors makes the scale-up in continuous flow safer and easier than in batch.

Similar to exothermic reactions, the scale-up of gas/liquid reaction in batch is often problematic because of the decreasing specific interface (Fig. 8.5a, dashed curve). In flow channels, the gas/liquid mass transfer process could be significantly accelerated by virtue of the enormous interface created by diverse gas/liquid flow patterns [24]. Using mass flow controllers, gases can be dosed in stoichiometric quantity. The finely dispersed gas bubbles shrink and dissolve during the reaction, leaving no headspace behind, whereas the headspace in batch reactors always exists and has to be filled with excessive gaseous reactants, which is a potential source of explosion when a flammable mixture is formed.

For further details on this issue, please see Volume 1, Chapter 1, Title: Fundamentals of flow chemistry i

---

8.2.2 Structural robustness

High pressure containers are prone to explosion, so they are manufactured with thick walls made of robust materials (characterized by a large yield stress σ). Even so, explosions of gas bottles and pressurized reactors are still emerging as common lab and industry accidents. For a long cylinder-shaped container (like tubes), the maximal pressure is inversely proportional to its inner diameter (d_i) [25]:

$$p_{\max} = \frac{2\,\sigma t}{d_i} \tag{8.4}$$

Symbol – physical quantity [unit]: $p_{\max}$ – maximal pressure [Pa], σ – yield stress [Pa], t – wall thickness [m], and d_i – inner diameter [m].

Microreactors could attain a high structural robustness with a thin wall (low t), by virtue of their tiny channel diameter (typically $d_i < 1$ mm) (Fig. 8.6). Stainless steel capillaries have been used for reactions under 900 bar at 450 °C [26], reliable enough for many high pressure synthetic reactions [27, 28], especially gas/liquid reactions [29, 30].

8.2.3 Enclosed small volume under continuous flow

Supported by the intensified transport process, the small reaction volume enclosed by robust structure has made the microreactor an ideal platform for safe handling of hazardous compounds and harsh conditions. The inherent processes safety is ensured by the low hazard hold-up and its correspondingly minimized inventory (cf. Figure 8.3). Chemists could therefore make use of process conditions that are far from conventional practice (so-called *novel process windows* (NPW)) on laboratory

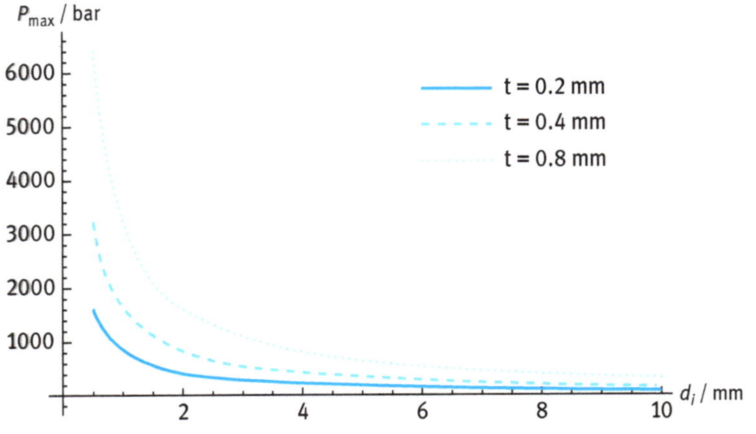

Fig. 8.6: Maximum pressure for stainless steel tubing with different inner diameters (d_i) and wall thicknesses (t). (Yield stress $\sigma = 200$ MPa was assumed for the calculation here. The yield stress of stainless steel is normally close to or greater than this value).

and production scales [28, 31]. The six intensification pathways of NPW (Fig. 8.7) are supported with safety guaranteed by the flow reactors, especially the reactions under high temperature/pressure/concentration and within explosive regime (so-called *forbidden chemistry* [32]), which are considered to be unsafe or unattainable in batch reactors.

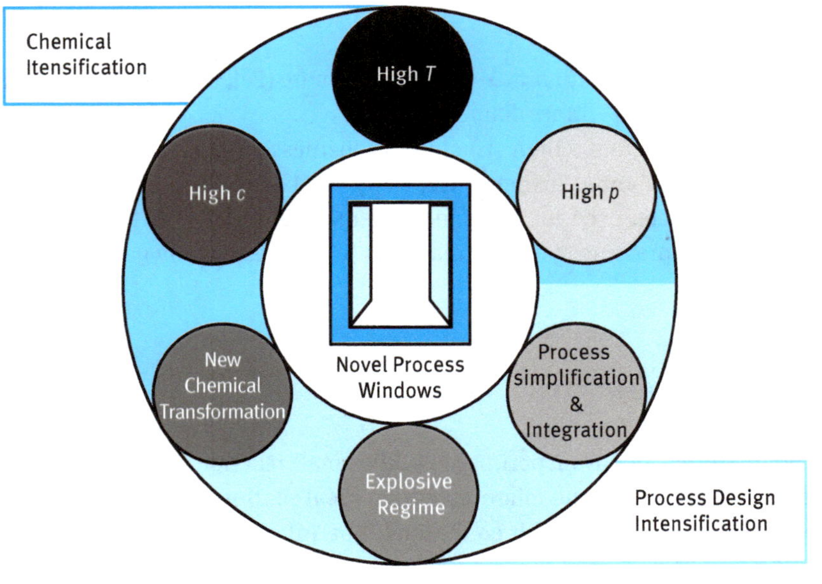

Fig. 8.7: Novel process windows – process intensification by applying extreme reaction conditions in microreactors [28].

Apart from the seamless scalability with smaller inventory footprint, the continuous flow operation of micro- and mesofluidic reactors enables the integration of several synthetic steps in one flow (also known as concatenation of telescoping), without the isolation and purification of intermediates [33, 34]. This feature is of significance in performing reactions involving highly reactive, explosive, toxic, and short-lived reactants: they can be generated in an upstream reactor (so-called *chemical generator*) from less hazardous precursors and subsequently consumed in the synthetic reaction downstream (Fig. 8.8b) [35]. The hazard present in the reactor system is minimized and requires no container for their storage, necessitating no obligation of registration or approval for the reagent (REACH Article 9) [7]. Besides, the quenching of hazardous materials remaining in the reaction mixture could be concatenated in one flow to give a more benign mixture for the workup and product isolation.

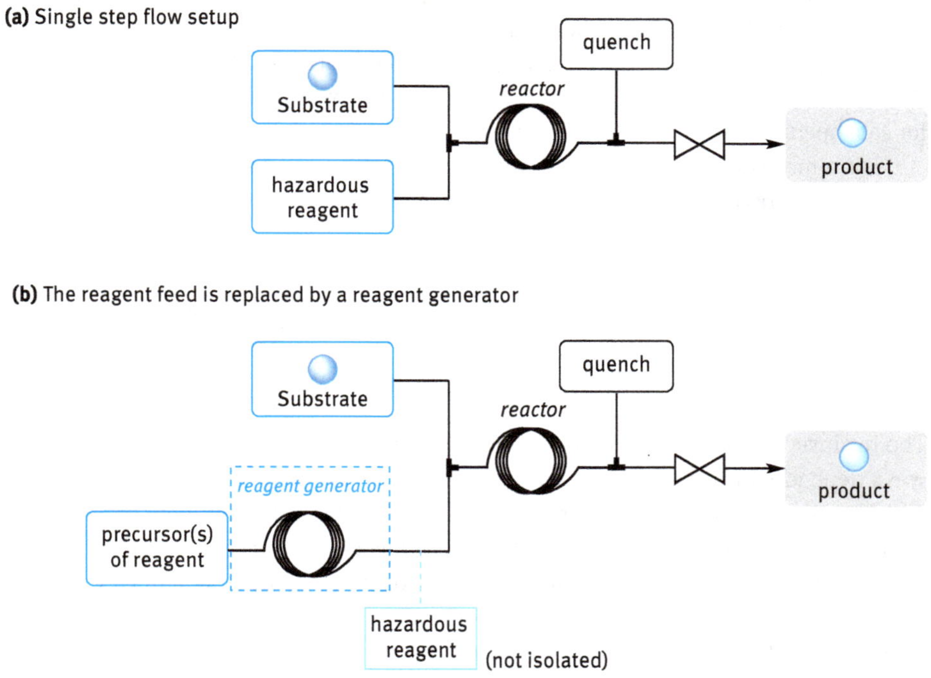

Fig. 8.8: The concept of chemical generator: the in-line generation and consumption of hazardous chemicals [35].

The combination of flow reactor with sensors and process analytical technologies provides further possibilities in the control of process safety, which could serve as the inner layers of protection (layers 2–5 in Fig. 8.2). In the cases of anomaly or emergency, the protective program could be triggered to stop heating, release pressure, and cut off the material supply.

i *For further details on this issue, please see Volume 1, Chapter 1, Title: Fundamentals of flow chemistry*

8.2.4 Limitations of flow reactors in overall safety

Micro- and mesofluidic reactors contribute to making the reaction processes inherently safer by their own features, but they cannot change the hazardous nature of chemicals and process conditions. The layout of flow equipment and related technology relative to the storage facilities should be taken into account during the design of the chemical plant.

An inherently safer reactor cannot replace the proper training of workers, since many reaction process-related accidents are caused by operator errors, management failures, and lack of organized operating procedures [21]. Workers in laboratories and factories should be imparted with knowledge on the properties of hazardous materials (REACH Article 7). In storage of chemicals and operation of flow setups, the guidelines for safe operation and emergency measure must always be followed [20].

For a more detailed and quantitative discussion on the safety of flow processes, the following review is recommended to the reader [36].

8.3 Mitigating chemical risks and hazards in continuous flow

The intrinsic merits of micro- and mesofluidic reactors in safety manifest in different ways depending on the specific hazards. Flow processes reduce the risks associated with reactive compounds and harsh conditions, thereby unlocking new process windows for more efficient chemical synthesis. In this section, recent examples of flow reactions (published since 2014) are presented to showcase their beneficial influence on safety and illustrate the concepts presented in the previous sections. Since one synthetic reaction can have different types of hazards, the examples in this section are classified by their most significant safety concerns during their development. Certain hazardous reagents are highlighted with GHS pictograms and hazard statements according to the CLP regulation.

8.3.1 Highly exothermic reactions

Nitration is a useful and well-known transformation reputed for its high exothermicity and aptitude towards thermal runaway. The reagent nitric acid (**1**) is a highly oxidative strong acid, often used as concentrated solution (in water or conc. H_2SO_4) or in neat form (fuming HNO_3); its highly corrosive nature and strong oxidation power could either rapidly affect the integrity of auxiliaries and reactors or invoke fire and explosion upon contact with organic substrates (Fig. 8.9a). A variety of organic molecules has been safely nitrated in microreactors, improving safety, selectivity, and scalability [37].

The nitration of (trifluoromethoxy)benzene with fuming HNO_3/H_2SO_4 has been optimized in a microreactor to improve its selectivity to *p*-nitro isomer (**2**); the reaction was scaled up to 1 kg h^{-1}, using a packed tubular flow reactor (Fig. 8.9b) [38]. The double nitration of *o*-toluic acid with the same nitrating agent in a stainless steel microreactor (SS316L, $\sigma = 205$ MPa) was able to deliver up to 1.6 kg h^{-1} of product (**3**) in 96% yield and 99.5% purity under isothermal and adiabatic conditions (Fig. 8.9c) [39]. When more nitro groups are introduced, the product could potentially become explosive, along with an increased sensitivity to heat and shock [40]. If the polynitrated compound is not the final product, subsequent reactions can be concatenated at downstream to circumvent the storage of such explosive intermediates. In the synthesis of aromatic precursor (**6**) of a metal chelating agent 1,3,5-triamino-1,3,5-trideoxy-*cis*-inositol (TACI) (Fig. 8.9d), a highly explosive trinitro compound (**5**) was generated from the nitration of phloroglucinol (**4**) in the aqueous phase. The solution of **5** was continuously delivered into a packed-bed hydrogenator (ThalesNano® H-Cube® Fig. 8.17b) without isolating this dangerous material [41]. The same nitration/reduction strategy has been applied in the synthesis of an 1,4-benzoxazinone intermediate of a herbicidal compound [42].

For nitration and other exothermic reactions, a large amount of heat is produced during the mixing of reactants; the effectiveness of mixing under temperature control is critical to enable a successful scale-up. Osimertinib (**7**) is a third-generation epidermal growth factor receptor tyrosine kinase inhibitor (EGFR-TKI) developed by AstraZeneca. Researchers from Research Center Pharmaceutical Engineering (RCPE GmbH) and Lonza AG developed a flow nitration protocol for the production of the key building block, 4-fluoro-2-methoxy-5-nitroaniline (**10**), in the synthesis of Osimertinib (Fig. 8.10) [43]. The aniline substrate (**8**) was protected by *N*-acetyl to prevent oxidation. The exothermic mixing of *N*-protected substrate (**9**) and nitric acid took place on a Lonza® FlowPlate (reactor II, manufactured by Ehrfeld), a modularized microreactor with mixing structure and temperature control. Good yield and purity of **10** were achieved in the lab scale (25 mmol h^{-1}), and similar performance was maintained when the reaction was scaled-up for the pilot plant production (2 mol h^{-1}), by using a larger mesofluidic device (Lonza® FlowPlate A5).

The **oxidation** of organic molecules can also be highly exothermic in many cases. The mixture of oxidating agent and flammable materials (solvent, substrate, etc.) tend to undergo an exothermic redox reaction, which could cause fire and explosion; therefore, their mixed storage is prohibited by regulations (e.g., TRGS510 in

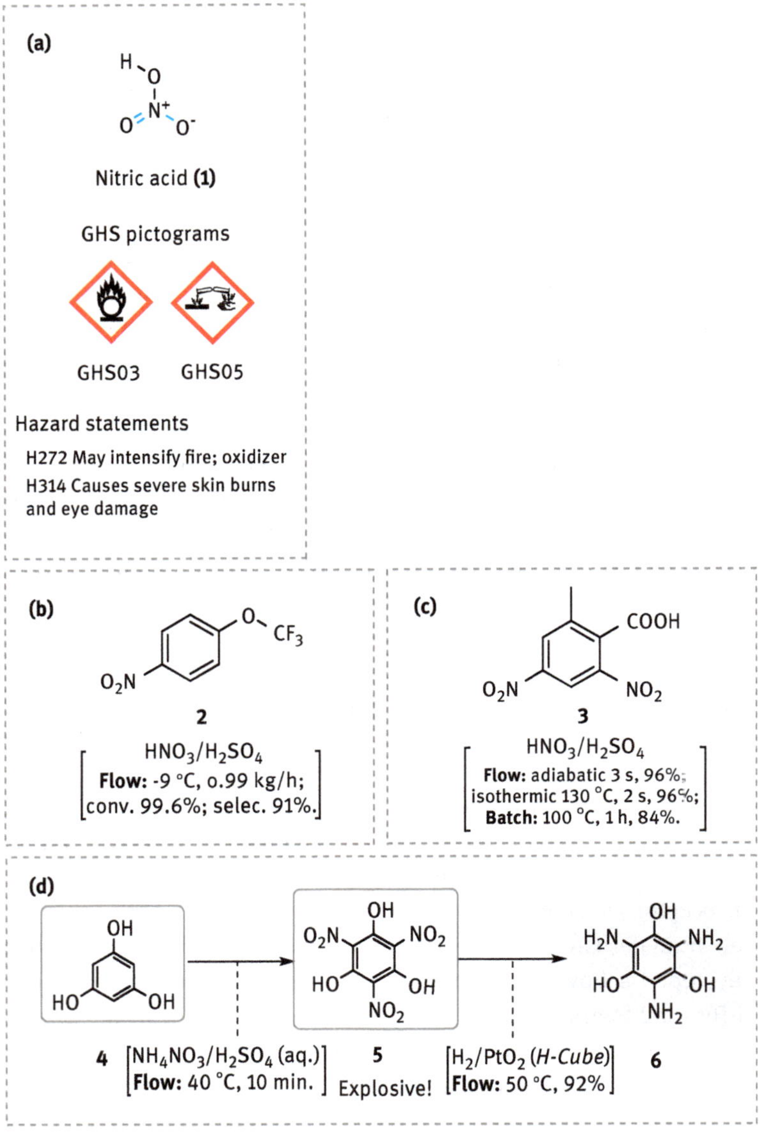

Fig. 8.9: Nitric acid (a) and mono- (b), di- (c), and trinitration (d) in microreactors.

Germany). Flow reactors have also been frequently applied for oxidation reactions [44], especially when molecular oxygen is the oxidant [45, 46].

In a continuous flow synthesis of ketamine (**11**) (Fig. 8.11), a hydroxylation step using molecular oxygen (reactor I) is involved [47]. The scale-up attempts of this strongly exothermic oxidation in batch could not provide any reproducible outcome,

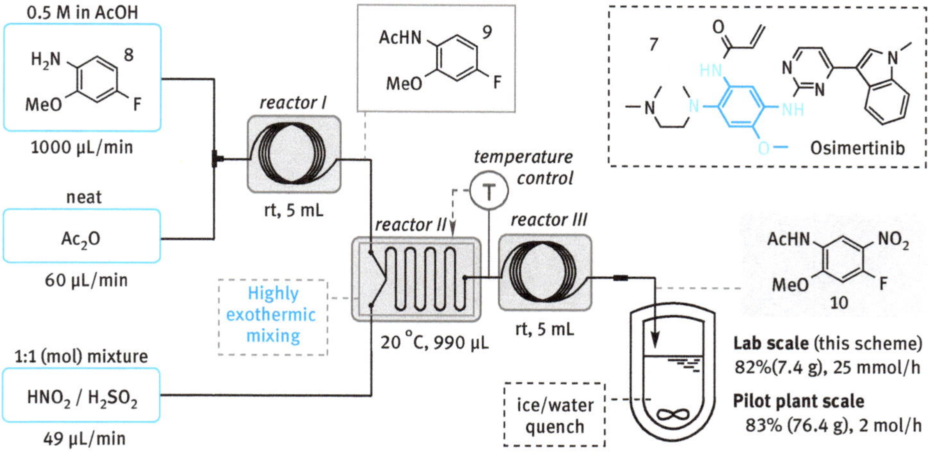

Fig. 8.10: Continuous flow nitration for the production of a precursor of Osimertinib [43].

whereas the flow reaction using a commercial flow reactor with mixing structure (Corning® Advanced-Flow) achieved improved yield within shortened reaction time [48], with a productivity of 1.3 kg day^{-1}. Two additional downstream transformations (imination and thermal backbone rearrangement) were telescoped with the preliminary hydroxylation step, hence providing a fully integrated continuous flow process toward API ketamine.

Many organic rearrangement reactions are exothermic. Caprolactam, the monomer of Nylon-6, is produced from the Beckmann rearrangement of cyclohexanone oxime, which is highly exothermic and proceeds very rapidly. Researcher from Tsinghua University evaluated the influence of reactor type and acid catalyst (fuming H_2SO_4 and trifluoroacetic acid) quantitatively and found that the process using microreactor is inherently safer than CSTR [49].

Mechanistically similar to Beckmann rearrangement, the Schmidt reaction also involves a substituent migration to nitrogen atom. The reagent, hydrazoic acid (HN_3), is the simplest azide whose structure was characterized in 2011 [50]. This highly explosive, toxic, and volatile liquid (b.p. = 37 °C) can be generated in situ by treating azide salts with an acid. The electrophilic amination of arenes (Fig. 8.12a) and Schmidt reaction of aromatic carboxylic acids (Fig. 8.12b) have been carried out in flow [51]. These exothermic reactions release stoichiometric amount of gas, which could result in violent boiling in a batch or tank reactor. Using the PTFE coil as reactor and mixing element, both reactions completed within 5 min in superheated solvent, giving a dilute aqueous solution for the product isolation.

Researchers from AstraZeneca investigated the Staudinger ketene–imine cycloaddition for the preparation of a spirocyclic β-lactam, a key intermediate in the synthesis of a melanin-concentrating hormone (MCH) receptor antagonist [52]. This reaction involves the in situ generation of a ketene (from acid chloride and base)

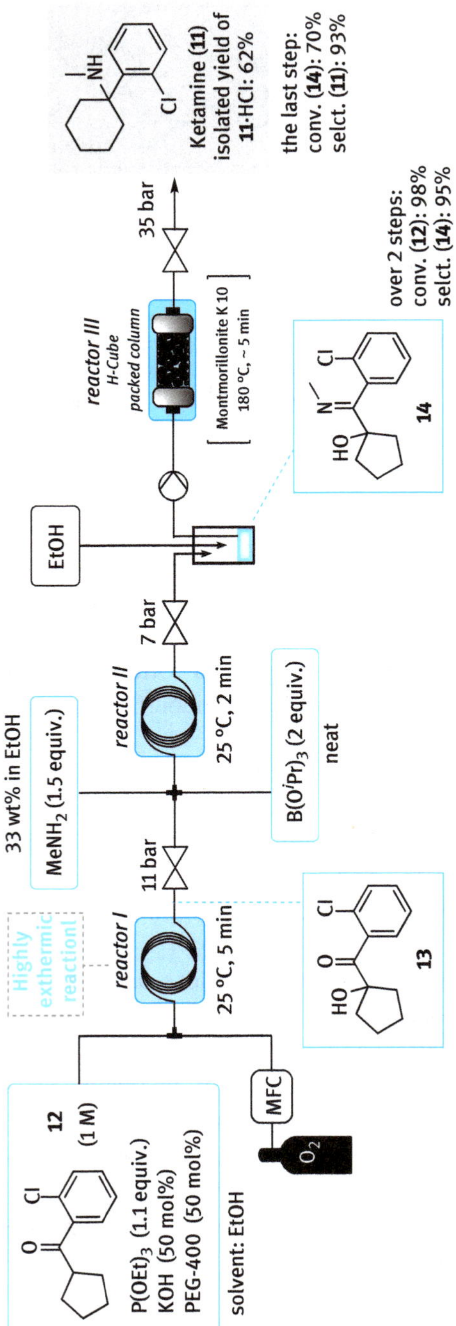

Fig. 8.11: Flow synthesis of ketamine involving a strongly exothermic oxidation step (reactor I) [47].

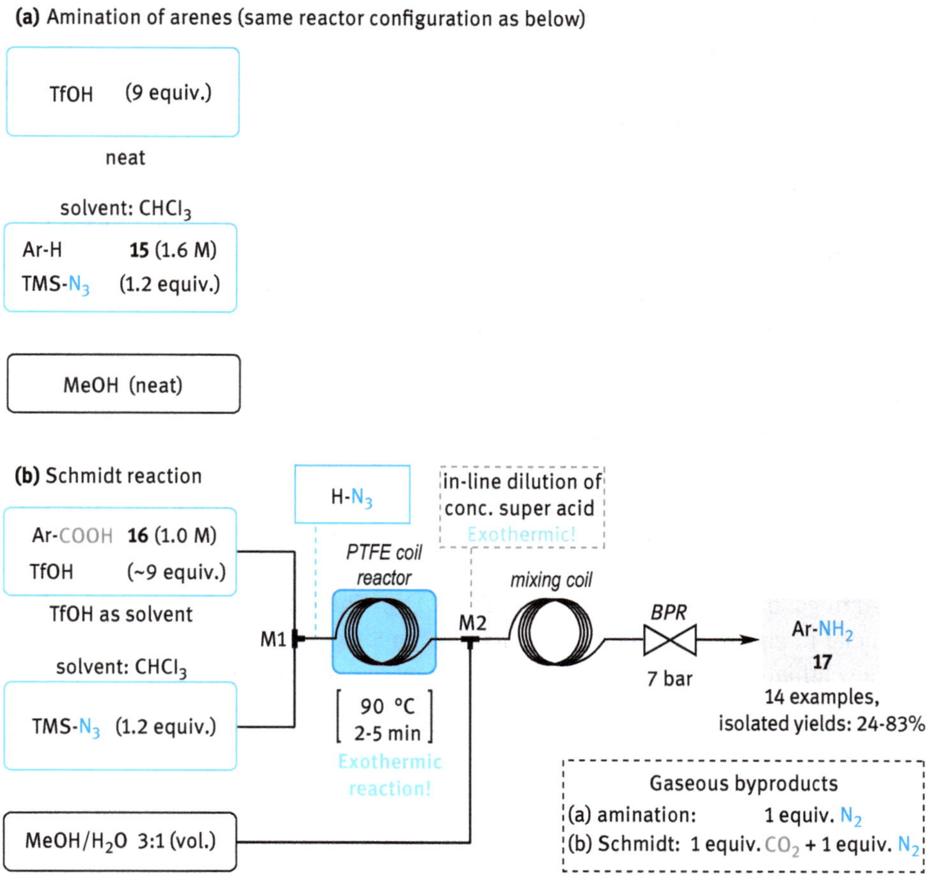

Fig. 8.12: The electrophilic amination of arenes (a) and Schmidt reaction of carboxylic acids (b) in super acidic media [51].

and its [2 + 2] addition with the imine. Due to the high exotherm, this step was performed in flow for a better control without cryogenic condition, although the yield is lower due to the use of a more soluble base.

8.3.2 Pyrophoric reagents

Organometallics remain a cornerstone of modern synthetic chemistry, but many synthetically useful organometallic reagents (e.g., organolithium, organomagnesium, etc.) are unstable and react so vigorously with air and moisture that they can cause fire and explosion; therefore, air- and moisture-free cryogenic conditions must be used to avoid undesired side-reactions and mitigate fire hazards. In conventional batch reactors, some useful reactivities are masked by the instability of the reagents. The exotherm arising from the reactivity of organometallics requires precise control of temperature and mixing performance.

The accurate control of short reaction time (≤1 s) ensured by the effective mixing performance has enabled microreactors to generate and consume highly unstable intermediates for synthetic purposes (*flash chemistry*, a concept introduced and largely documented by the late Professor Yoshida [53]). Cryogenic reactions in batch can therefore be performed at or close to room temperature, accessing process windows which are unfeasible in batch due to functional group incompatibilities (i.e., new chemical transformation in Fig. 8.7). Besides, the absence of head space in microreactors liberates the chemists from the vacuum gas manifold (i.e., the Schlenk line) for the maintenance of the headspace atmosphere. The abovementioned advantages of microreactors in safely handling reactive organometallics led to the widespread adoption of flow technology in synthetic organometallic chemistry [54].

In one of the most impressive applications of microreactor devices to **organolithium** chemistry, a microchip reactor was used to outpace an anionic Fries rearrangement of *o*-lithiated arylcarbamates (**19**) (Fig. 8.13a) [55]. Within a submillisecond residence time (0.33 ms) for the lithiation step, unrearranged products (**22**) were obtained in good to high yields (compare Fig. 8.13b). The high flow rate (4.5 mL min^{-1}) is crucial to ensure fast mixing, producing **22** in 5.3 g h^{-1}. For carbamate substrate, the flow lithiation could be performed at ambient temperature (25 °C).

Trapping the in-line generated lithium reagents by other electrophiles could access other molecular scaffolds, such as epoxides, cyanides, and esters [56], as well as carbamoylanions [57] and azides [58]. Interestingly, the approach was also amenable for the synthesis of heterotelechelic polymer [59]. Recently, the reactivity of dimetalated arenes has been accessed by the consecutive lithium halogen exchanges in flow [60]. Similar approaches are also available for the synthesis of boronic acids [61], ferrocenyl azides [62], and α-substituted 1,3,4-oxadiazoles [63].

Verubecestat is a potential API for the treatment of Alzheimer's disease. Researchers from Merck & Co. (USA) have developed the kilogram-scale manufacturing of a precursor (**23**) of Verubecestat in flow (Fig. 8.14) [65]. The Mannich-type addition of an alkyllithium species (**27**) had been suffering from a competitive deprotonation of the substrate (**24**) in batch [65]. Conducting the reaction in flow was essential to outpace this side-reaction by improved mixing performance, giving **23** in high assayed yield (87–91%), while allowing the reaction to be performed under higher temperatures (≥ −20 °C).

The flow method is not limited to lithium as shown by a number of published examples on other pyrophoric organometallic reagents. The lithium–halogen exchange/transmetalation for the generation of **organomagnesium** and **organozinc** reagents (**29**) has been studied in batch and flow (Fig. 8.15) [66]. The reaction proceeded at 0 °C in flow (batch −78 °C) with higher yields. Similar procedures based on halogen–magnesium exchange with commercial Grignard reagents have also been described [67, 68]. Alternatively, the use of packed magnesium or zinc column

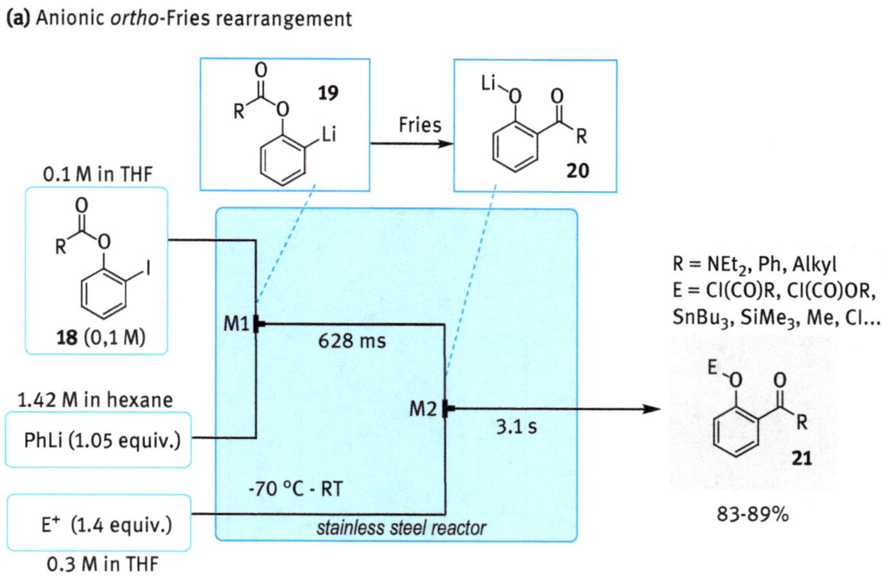

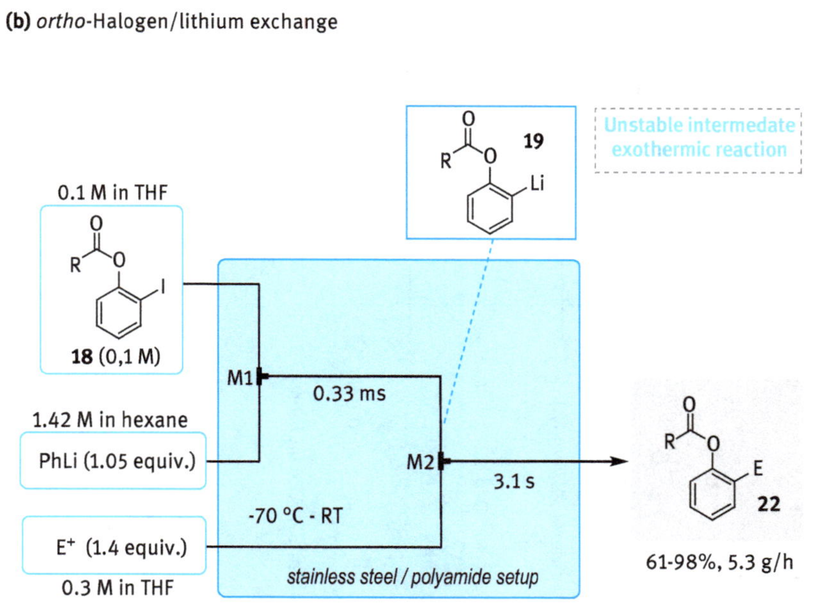

Fig. 8.13: Lithium–halogen exchange in sub-second (a) and sub-millisecond (b) timescale (E^+ = electrophile) [47].

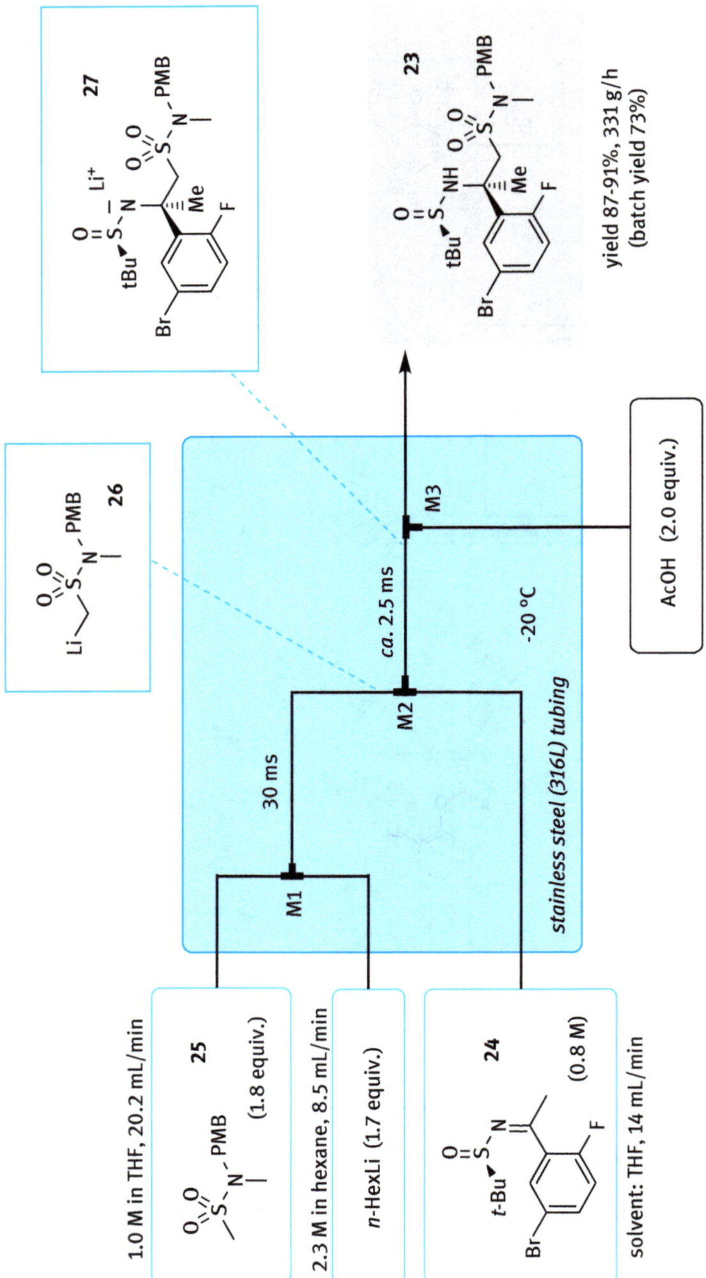

Fig. 8.14: The continuous flow synthesis of Verubecestat precursor [65].

in flow allows the generation of the corresponding organometallic reagent, without pyrophoric precursors [69–71]. For the **organosodium** reagents generated from the sodiation of arenes and pyridines with $^{i}Pr_2NNa$ (NaDA), a broader substrate scope than batch was accessed under flow conditions [72]. Finally, it is worth noting that microreactor devices have been used routinely for pyrophoric reductants such as diisobutylaluminum hydride (DIBAL) [73, 74] and boranes [75].

For further details on this issue, please see Volume 1, Chapter 2, Title: Principles of controlling reactions in flow chemistry. i

8.3.3 Reactions under high pressure and temperature

The structural robustness of microreactors provides a convenient access to elevated pressures and temperatures, comparable to or more than in batch autoclaves. Generally, high temperature can significantly accelerate the reactions; and high pressure increases the solubility of gases in the liquid phase. In microreactors, the process intensification by high T/p conditions can be realized in a more controlled, safer, and scalable fashion, giving improved selectivity within shorter reaction time [31]. In addition, high-boiling solvents can be replaced by superheated low-boiling solvents close to or within supercritical regime to alleviate the difficulties in workup and product isolation.

Chemists at AbbVie studied the properties of the sequential thermal ring opening/Diels–Alder of substituted benzocyclobutenes (**31** and **35**) in flow (Fig. 8.16) [76]. The reactants were heated at 300 °C, 120 bar for 30 s to complete the reaction; the generated dienes (**34**) were captured in situ either by dienophiles (**32**) or intramolecularly (**35**). The isolation of the products was considerably easier than in batch, where high-boiling solvent was used. Similar flow systems have also been implemented for the synthesis of heterocycles via nucleophilic aromatic substitution [77] and Gould–Jacobs-type cyclization [78] at T/p up to 450 °C, 150 bar; the latter reaction was normally carried out on flash vacuum pyrolysis equipments [79].

Hydrogenation under high pressure is notorious for explosion because of the flammability of hydrogen gas (Fig. 8.17a), the presence of metal catalyst as ignition source, and the exotherm of this transformation itself. Microreactors provide a safe platform for hydrogenations due to their robust structure and intensified gas/liquid mass transfer [80]. The efficient mixing of the gas, solution, and catalyst in flow can be ensured through various strategies such as static mixing in segmented flow regimes, gas-permeable membranes (e.g., “tube-in-tube” setup), and packed-bed reactors. The commercial flow reactor ThalesNano® H-Cube for hydrogenation under

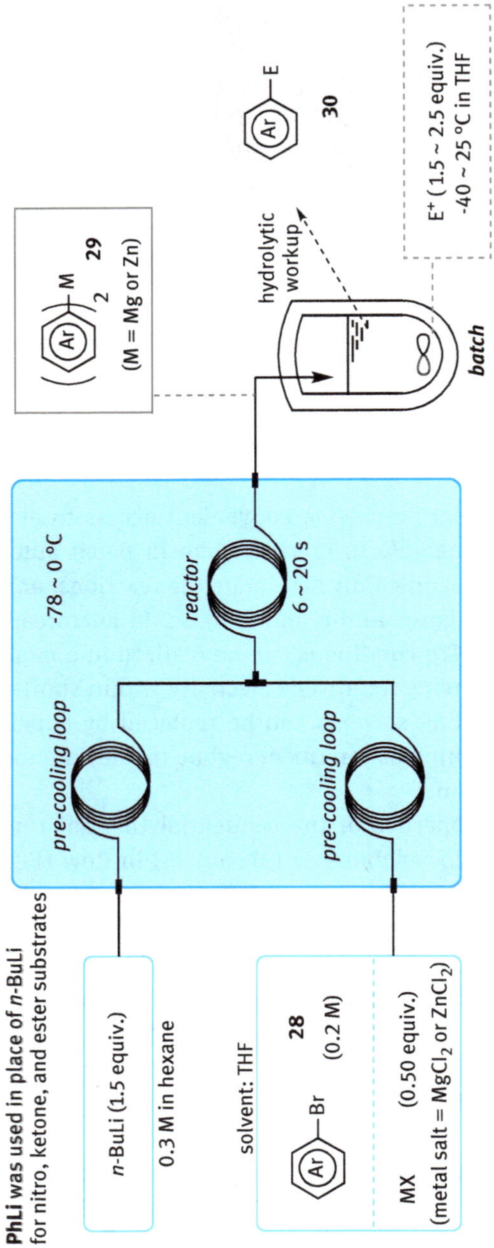

Fig. 8.15: The lithium–halogen exchange/transmetalation strategy for the generation of organomagnesium and organozinc reagents [66].

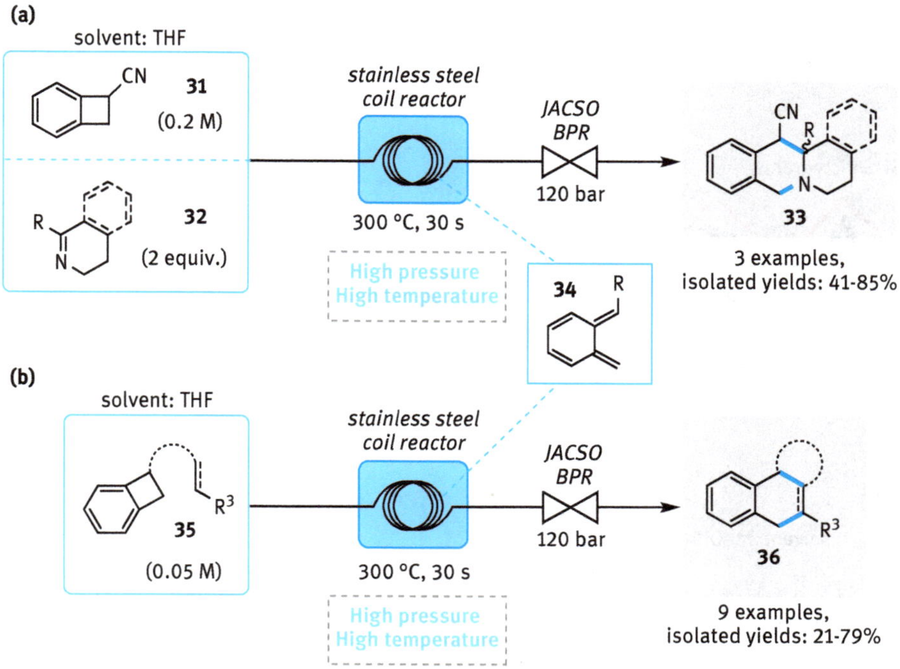

Fig. 8.16: Thermal ring opening and Diels–Alder cycloadditions under high *T*/*p* conditions.

high *T*/*p* conditions integrates replaceable cartridges packed with metal catalysts and a hydrogen generator based on water electrolysis (Fig. 8.17b); therefore, a stand-by hydrogen cylinder is no longer necessary.

The intensification of the hydrogenation and semihydrogenation of a propargylic alcohol (**39**) has been investigated using a stainless coil coated with catalyst on its inner surface. It was found that increasing the temperatures from 70–90 °C to ca. 150 °C could accelerate the reaction by 2.5- to 8-fold (Fig. 8.18) [82]. This rate improvement was only achievable in flow, whereas significant substrate decomposition occurred at higher temperature in batch [82, 83]. In another project, using the catalyst-coated microreactors, the same group of researchers showed that potential trade-off of the high throughput with other parameters such as that the catalyst deactivation under harsh conditions can reduce the yield [84] should be considered.

Other flammable gasses have also been used in flow under high pressure. **Synthesis gas** (CO/H_2) has been used in flow for an asymmetric hydroformylation at 70 °C under pressure up to 55 bar [85]. The selectivity was improved under higher pressures. Using a 1:3 mixture of CO and H_2, aryl bromides could be converted to aryl aldehydes using palladium catalyst (120 °C, 12 bar, ca. 45 min) [86]. Similar conditions are applicable to phenol-derived aryl fluorosulfonates (120 °C,

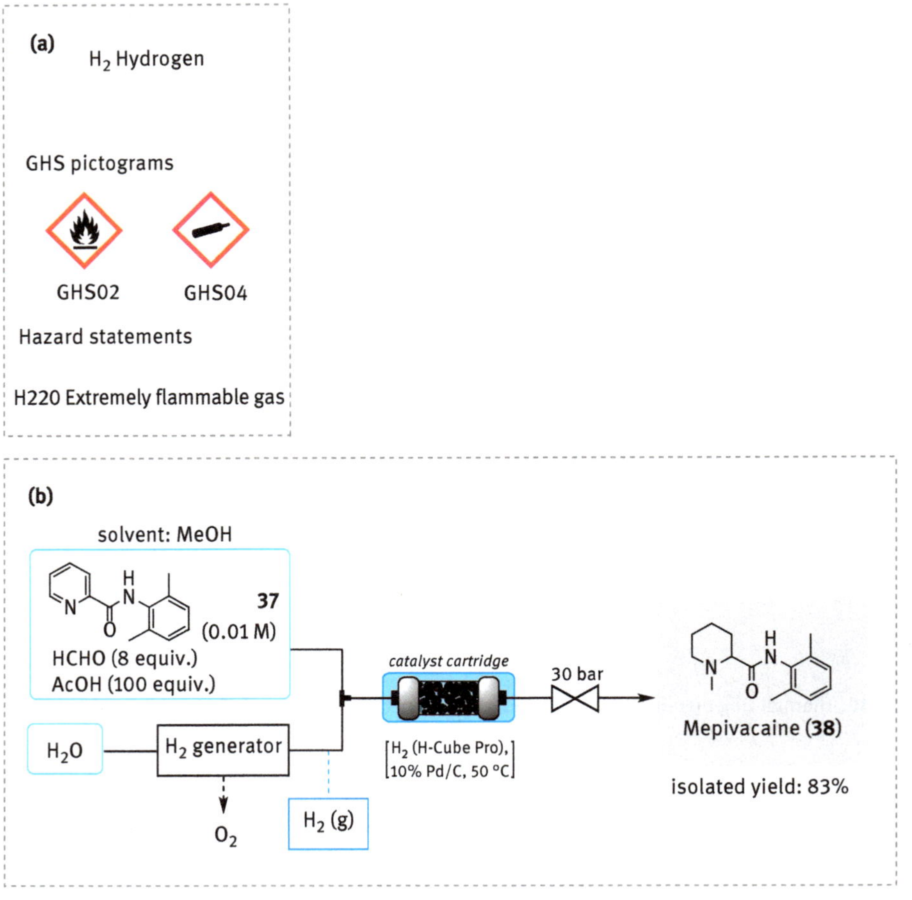

Fig. 8.17: (a) The hazardous properties of H_2; (b) ThalesNano® H-Cube: packed-bed reactor for hydrogenation under high pressure (example [81]).

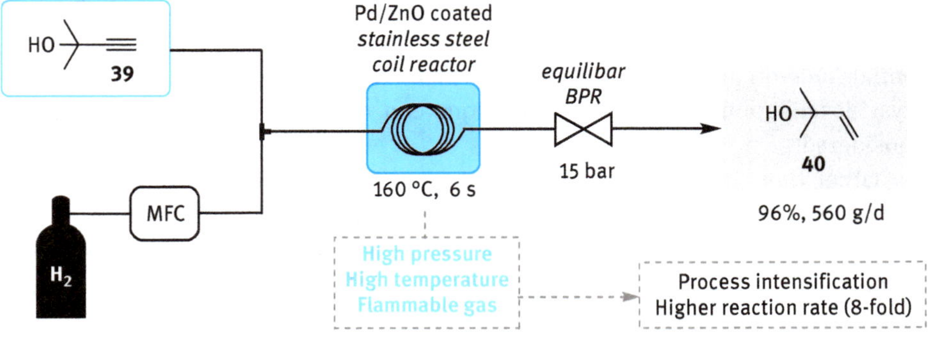

Fig. 8.18: The continuous flow hydrogenation using a catalyst-coated coil reactor [82].

20 bar) [87]. Propyne ($CH_3C{\equiv}CH$) gas has been employed as a Sonogashira cross-coupling partner at 160 °C (2 bar, 10 min) for the synthesis of the building block of a potential drug molecule [88].

For the reduction of carbonyl group to methylene group, the Huang modification of Wolff–Kishner reduction uses **hydrazine** (N_2H_4) and a strong base to realize this transformation in a high boiling solvent (e.g., triethylene glycol) at elevated temperature [89]. Since stainless steel capillary could catalyze the decomposition of hydrazine, and the strong base could corrode glass, a flow reactor made of silicon carbide (SiC) plates by Chemtrix® was employed to perform the reduction in superheated methanol, under high T/p condition (200 °C, 50 bar) (Fig. 8.19b) [90].

8.3.4 Explosive reagents and conditions

Explosive compounds and mixture can present significant risks to facilities and personnel. Hence, the use of highly explosive reagents (e.g., azides and peroxides) is sometimes precluded, despite their synthetic relevance, especially on large scale, to the point that these chemicals are sometimes referred to as "forbidden chemistry." Although microreactors are not immune to explosion hazards under all circumstances, the low hold-up of hazard makes them particularly advantageous in handling explosive reagents, since the explosive power is estimated to be proportional to 1/3 power of the mass of the explosive material [36]. In many cases, performing the reaction in flow is the only possible practical means to deal with high-energy reagents and intermediates.

The aromatic diazonium salt is a versatile intermediate of particular importance in the preparation of dyes and other aromatic compounds. Due to the explosiveness of dry diazonium salts, they are usually prepared in solution by diazotization of anilines at low temperatures (≤5 °C), and then consumed immediately in the subsequent transformation. A special case is the Balz–Schiemann reaction, since it requires the isolation and thermolysis of dried diazonium tetrafluoroborates. Researchers from Zhejiang University of Technology disclosed a Balz–Schiemann synthesis of *m*-difluorobenzene (**45**) from *m*-phenylenediamine (**43**) (Fig. 8.20a) [91]. The diazotization in flow (reactor II) afforded the diazonium salt (**44**) in 94% yield, after desiccation in vacuum. The thermolysis of dry diazonium **44** in batch afforded the difluoride (**45**) in 85% yield, based on **43**, whereas the fully batch procedure gave only 56% yield over two steps. Endeavors have been made to preclude the isolation of diazonium salt by concatenating the diazotization and thermolysis steps (Fig. 8.20b) [92]. The amines were diazotized in anhydrous media (^{t}BuONO, CF_3COOH, $LiBF_4$ in ^{n}BuOAc) at room temperature; the resulting diazotization mixture was heated in a feed-batch mode to afford the fluoride. The flow setup has to be sonicated to prevent clogging by diazonium tetrafluoroborate precipitate.

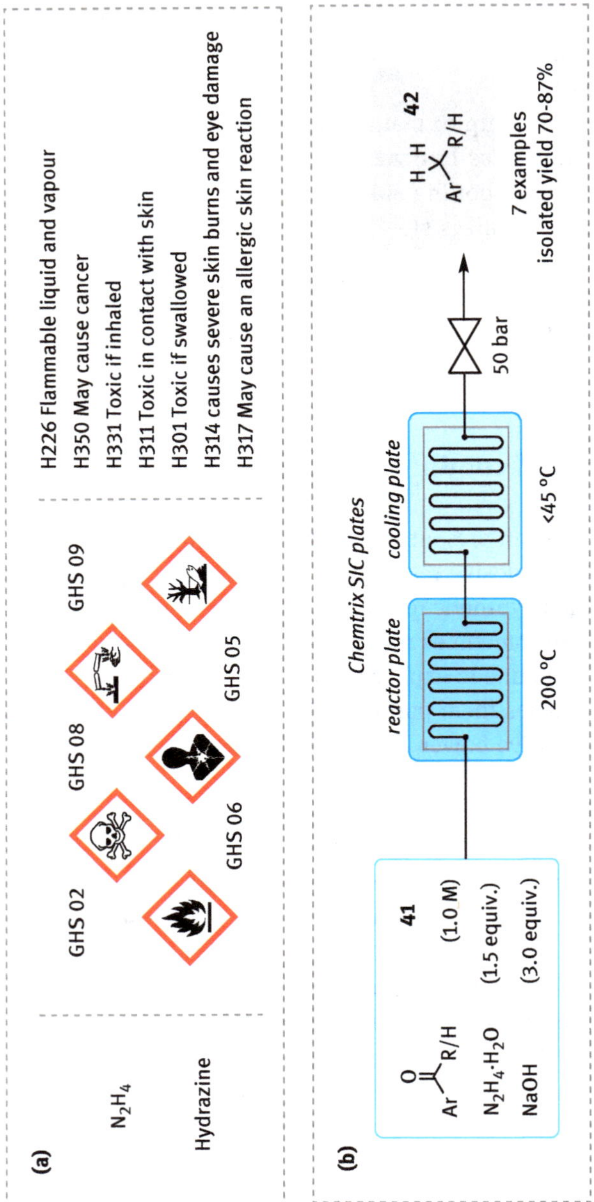

Fig. 8.19: (a) The hazardous properties of hydrazine and (b) the carbonyl reduction using hydrazine under high *T*/*p* condition on SiC plates [90].

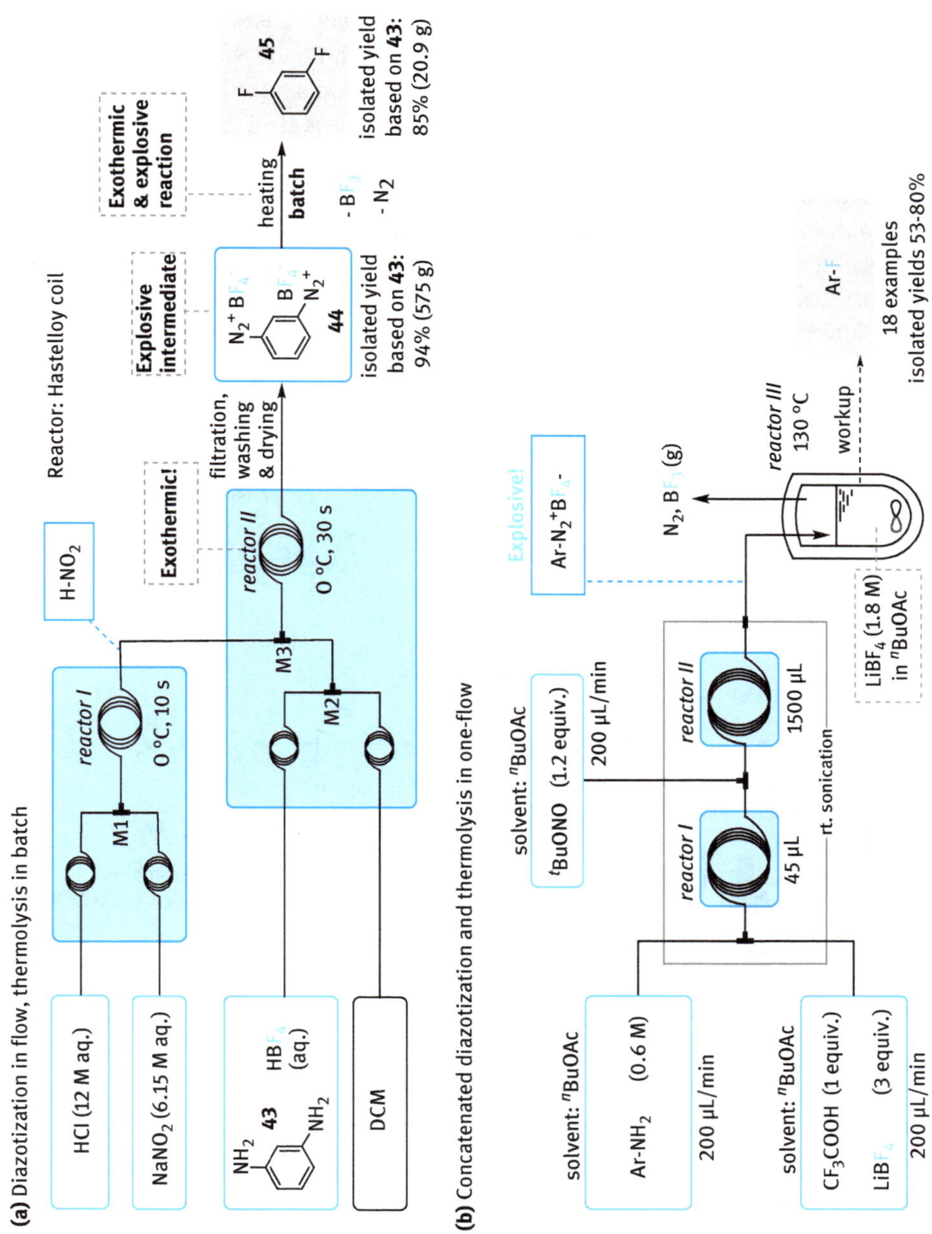

Fig. 8.20: Balz–Schiemann reaction in flow: with (a) [91] and without (b) [92] the isolation of dry diazonium salt.

The formation of **explosive mixture** of flammable gases and oxygen or air should be avoided in the design of batch processes (cf. Figure 8.3c), since such mixture could easily explode upon contact with an ignition source [46] (e.g., the metal catalyst [93]). Using the structural robustness of stainless steel capillary ($d_i = 1$ mm), the palladium-catalyzed oxidative carbonylation of *o*-aminophenols and aminoalcohols could be performed using CO and O_2 as a gaseous feedstock (Fig. 8.21) [94]. The reaction proceeded smoothly with an explosive gas mixture (50 vol.% CO in O_2) in the heated stainless steel coil, giving useful five- and six-membered carbonylated heterocycles (**47**) with retained chirality. This highly efficient protocol provides a green alternative to phosgenation, while reducing the risks and generating lesser waste compared to its batch counterpart, where the gas mixture had to be diluted to a composition outside of the explosive regime [95].

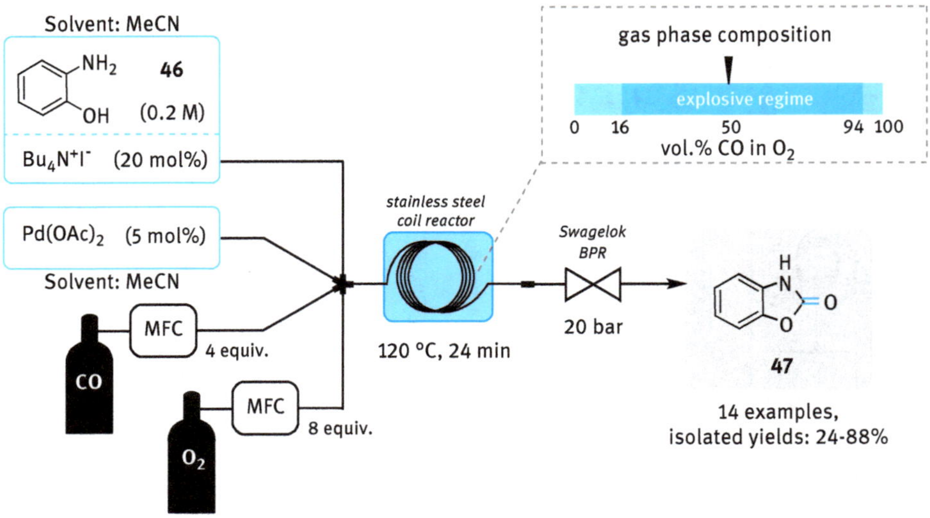

Fig. 8.21: Oxidative carbonylation in flow using an explosive gas mixture of CO and O_2 [94].

Diazomethane (CH_2N_2) is a versatile and reliable C_1-building block with well-documented applications in synthetic chemistry. However, its toxicity, volatility, and instability necessitate its fresh generation and distillation prior to use, with a number of specialized kits that are commercially available. Accompanied by the development of microreactor and membrane technology, a number of flow protocols have been developed to generate diazomethane on site and consume it before accumulation [96]. Membrane separators have been used in a telescoped flow system, starting from a benign material, *N*-methylurea (**48**) (Fig. 8.22) [97]. The methylurea was *N*-nitrated with sodium nitrite, followed by an in-line liquid/liquid extraction and separation to give *N*-methyl-*N*-nitrosourea (**49**). This carcinogenic compound was subsequently treated with potassium hydroxide to generate diazomethane, which was finally consumed in the *O*-methylation of carboxylic acids (**50**), to afford their methyl ester (**51**) in >95% yields. An alternative "tube-in-flask" approach starting from Diazald (*N*-methyl-*N*-nitroso-*p*-toluene sulfonamide) was reported [98]. After mixing this precursor with potassium hydroxide solution in flow, the coil was placed in a chamber containing the substrate solution, and the membrane allowed diffusion of the gaseous reagent into the substrate solution.

Ozone (O_3) is an appealing reagent for the cleavage of double and triple bonds, due to its selective reactivity and its convenient and affordable generation from oxygen. Nonetheless, the primary ozonide intermediate presents notable explosion hazard, and the hydrolytic quenching, thereof, is also exothermic. Many research groups have developed flow approaches to avoid the accumulation of explosive intermediates by in-line quenching. The reaction could be carried out at a higher temperature of −10 °C in microreactors (−78 °C in batch) [99]. In a published flow synthesis of Ivacaftor (**52**) (Fig. 8.23), an API for the treatment of cystic fibrosis, the formation of the ozonide (**54**) was found to be much faster than the subsequent quenching, making it challenging to telescope this protocol [100]. Therefore, a CSTR was inserted in the flow stream for this slower step, ensuring an overall throughput of 7.2 g day^{-1}. Similar operation modes such as flow (feed-batch) and flow (recycling flow) have been applied in the cleavage of quinoline [101]. The ozonization in flow can also be employed for the preparation of ozonide drug candidate [102].

Peroxides (ROOR) present a more general class of oxidants, with a long history of use in various industries from polymers to organic synthesis, while likewise posing explosive hazards under thermal conditions or in the presence of minute amounts of metals. Flow reactors are thus useful both in the production and reaction of peroxides. For instance, the combination of oxygen and hydrogen over a catalytic packed-bed column has been described for the generation of hydrogen peroxide, including at relevant industrial concentrations of up to 10 wt.%, and remains an active research area [103, 104]. Hydrogen peroxide may also serve an intermediate in preparing reactive peracids used for more specific transformations. The performic acid (HCOOOH, detonation at 80 °C) generated in situ was used for the hydroxylation of oripavine (**57**) in the synthesis of analgesic drug oxymorphone (**59**), which is also a precursor of naloxone (**60**) (Fig. 8.24) [105]. Despite the low detonation limit of HCOOOH (80 °C), the authors were able to intensify the process to complete the 14-hydroxylation within 7 min at 100 °C, 4 bar (reactor I).

8.3.5 Extremely toxic reagents

Halogenations using **elemental halogens** pose various challenges including the exotherm and the difficulty in selectivity control, as well as the corrosive and toxic properties of the reagents (Fig. 8.25). Much work thus concerns halogenations in

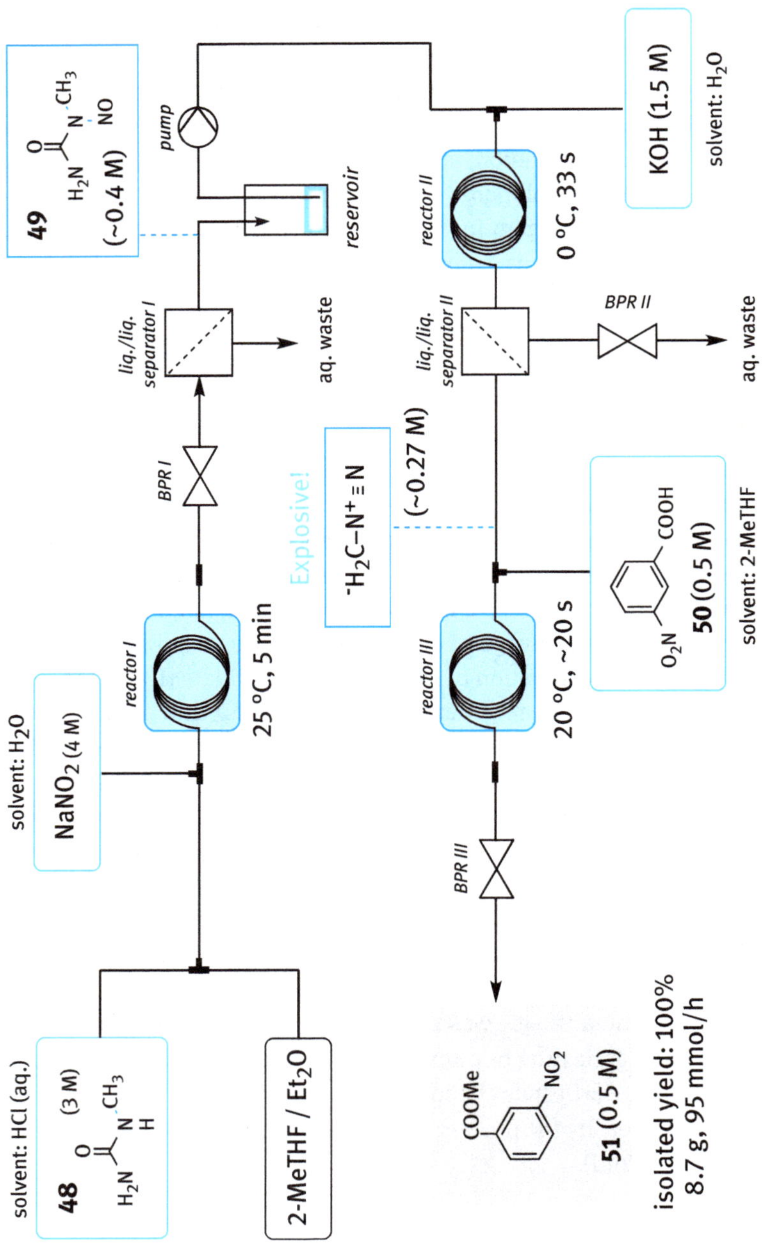

Fig. 8.22: In-line generation and consumption of diazomethane in flow [97].

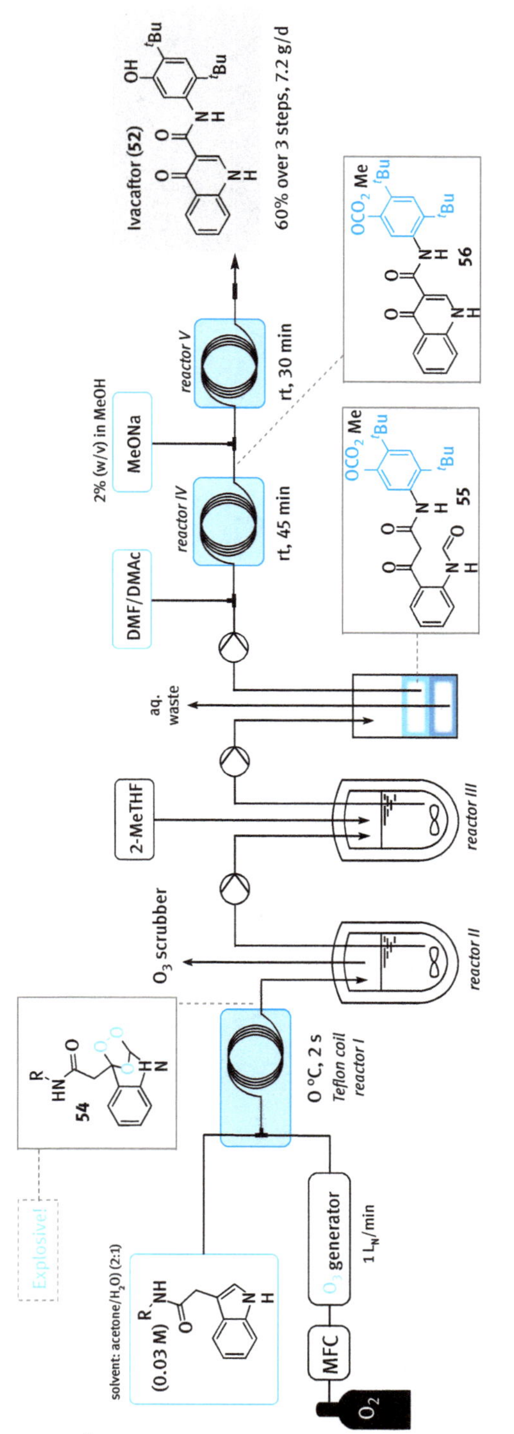

Fig. 8.23: The ozonolysis of an indole substrate (reactor I) in the one-flow synthesis of Ivacaftor [100].

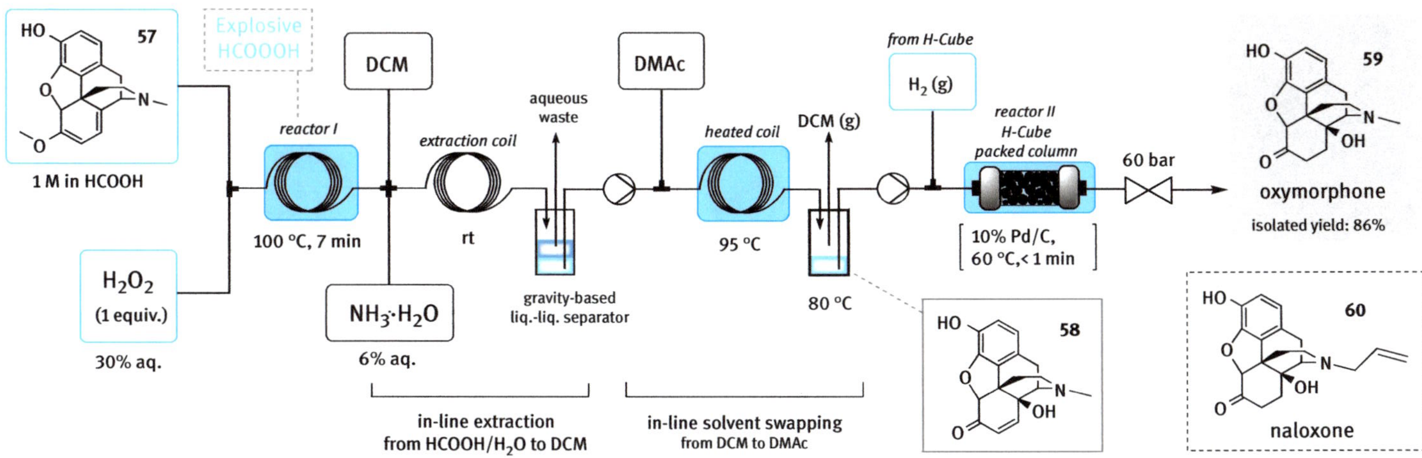

Fig. 8.24: Continuous flow synthesis of oxymorphone: the 14-hydroxylation of oripavine using performic acid (reactor I) generated in situ and the hydrogenation using ThalesNano® H-Cube (reactor II) [105].

microreactor devices, often based on the in situ generation of the halogen from more benign reagents [106].

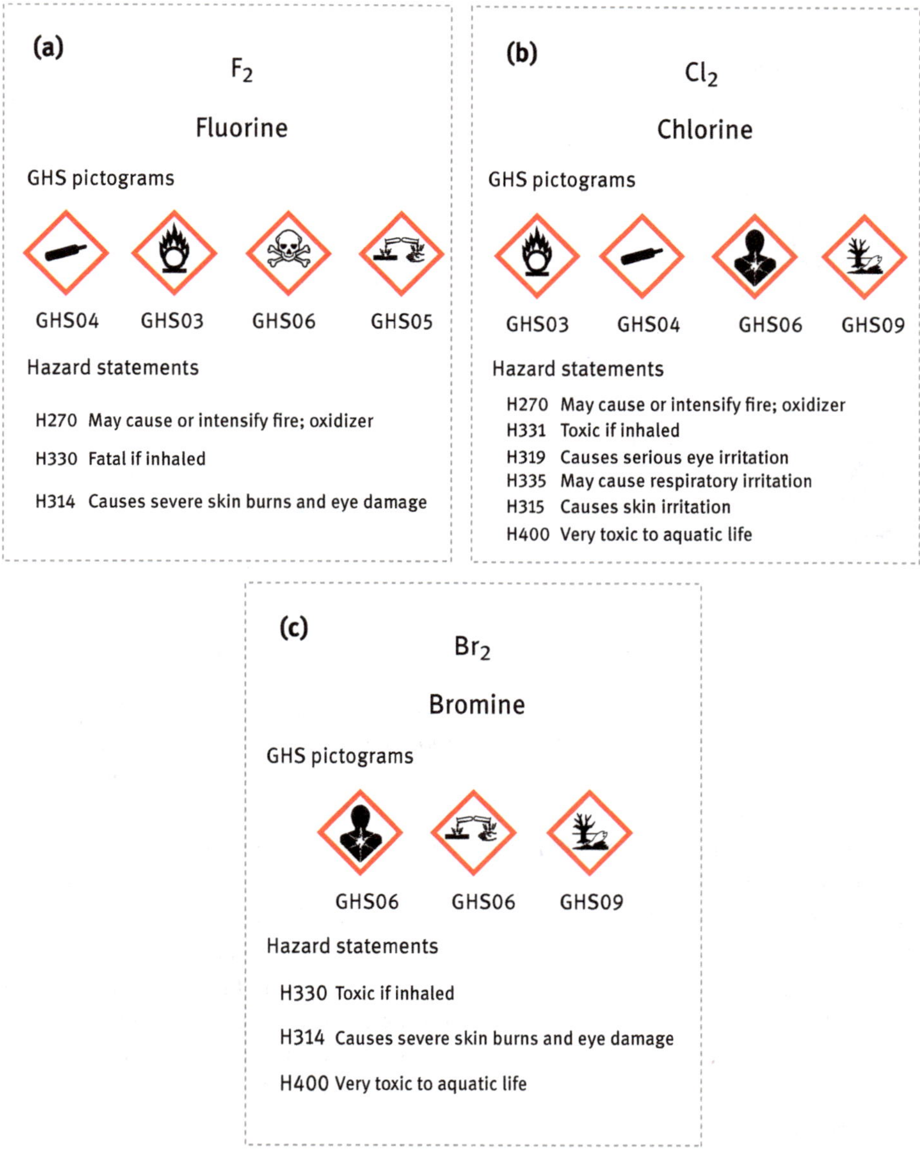

Fig. 8.25: Hazardous properties of halogens.

Fluorine is a highly reactive and atom-economic fluorinating agent. A very low concentration (25 ppm) of fluorine in the air could cause irritation of the eyes and nose [107]. A continuous flow procedure was developed to prepare the World Health

Organization (WHO) essential medicine, flucytosine (**62**), by the fluorination of cytosine (**61**) with up to 58 g h^{-1} produced (Fig. 8.26) [108]. An improvement of the yield from 38% to 83% was observed by virtue of the selectivity improvement in flow. Furthermore, the persistent demand for fluorinated molecules in the pharmaceutical industry is expected to benefit from the safer handling of fluorine gas in flow. The deployment of other elemental halogens in flow microreactors is also an active research area, for instance, with bromine [109].

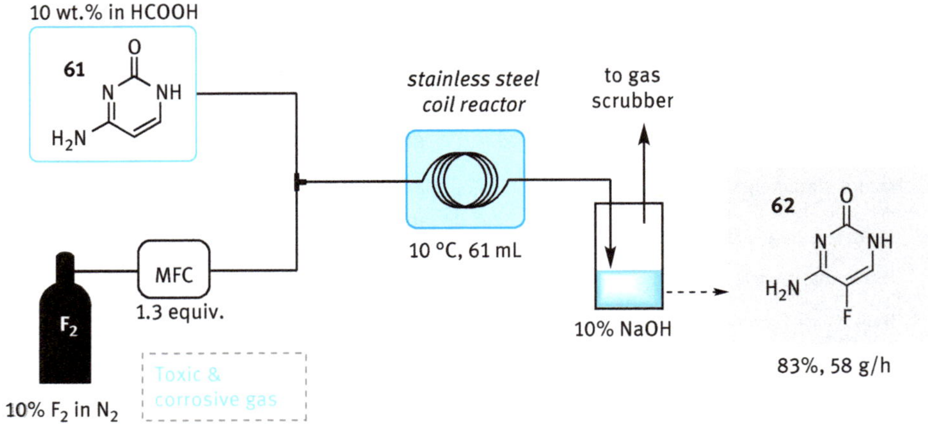

Fig. 8.26: Synthesis of flucytosine: fluorination using elemental fluorine in flow [108].

Safer procedures have been described based on the on-site generation of halogens in flow [106]. Researchers from RCPE GmbH reported a benzylic bromination on pilot scale (4 kg h^{-1} produced) [110]. The molecular bromine for the photochemical bromination was generated from hydrobromic acid (HBr) and sodium bromate ($NaBrO_3$). The reaction was terminated by the in-line quench with sodium thiosulfate ($Na_2S_2O_3$). Both bromination and quenching are exothermic, and the combination of microreactor and thermostat enabled their safe execution under high throughput. When the bromination was scaled-up to a throughput of multikilogram per hour, a detailed process risk assessment was carried out based on possible observations, including the possible risks/failures, how the failure would be observed during the run, its consequences, and the required actions (such as safe shutdown) [110]. Similarly, chlorine generators utilizing the redox reaction between hydrochloric acid and bleach solution (NaClO) were also developed for the quantitative generation of chlorine gas [111]. Continuous in-line extraction and membrane separation were employed for the reactions in organic phase. Apart from halogens; other small molecule reagents like **sulfur dioxide** (from Na_2SO_3 + H_2SO_4) [112] can be prepared in an upstream generator to access their chemistry in a safer way.

Carbon monoxide (CO) is an odorless, toxic gas widely utilized in the bulk and fine chemical industries for carbonylation reactions. Its versatility has prompted

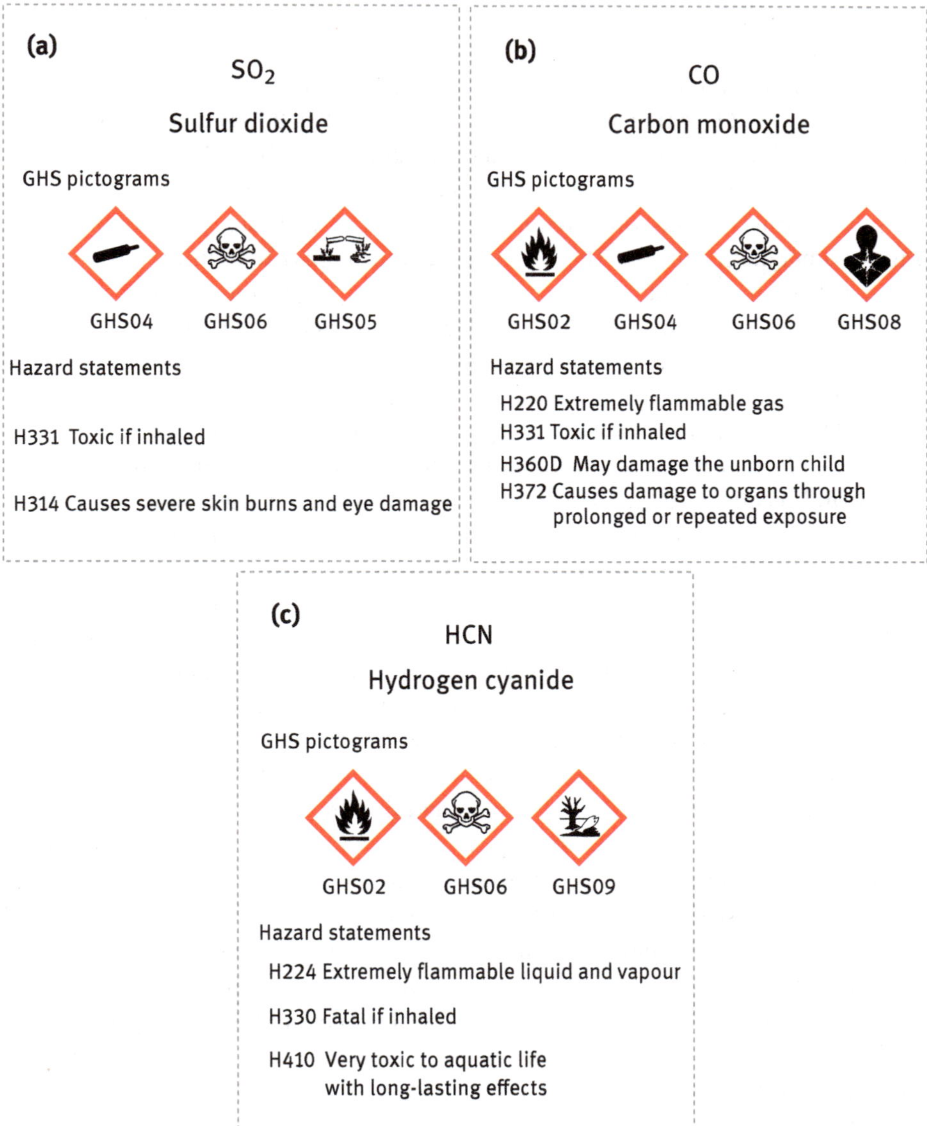

Fig. 8.27: The hazardous properties of SO_2, CO, and HCN.

ongoing research on its use in industrial-scale microreactor devices [85, 113]. The tube-in-tube reactors based on gas-permeable Teflon AF-2400 membranes offer convenient solutions, although the risk of membrane failure remains a concern [114]. Carbonylations with CO, either supplied by gas cylinders (Fig. 8.28a) or generated in-line from oxalyl chloride (**63**) (Fig. 8.28b), have been reported [115]. The treatment with sodium hydroxide provided a steady flow of the gas. Interestingly, the use of immiscible solvents (toluene and water) required a mixing unit made of a glass column fitted with two magnetic stirrers. The dosing of gas through a tube-in-tube reactor enabled a clean, high-yielding alkoxycarbonylation of vinyl iodides (**64**).

Hydrogen cyanide (HCN) also displays high toxicity levels (it is lethal above 100 ppm) combined with a low boiling point (26 °C). The use of tube-in-flask and tube-in-tube setups provides a safe and convenient method for the lab use of HCN (Fig. 8.29) [116]. The treatment of an aqueous sodium cyanide solution with sulfuric acid gave a feed of hydrogen cyanide, which diffused through the Teflon AF-2400 tubing and reacted with imine electrophiles (**66**). The membrane tube setup limits exposure of the workers to HCN allowing them to access the chemistry of anhydrous HCN in a convenient manner. More recently, a continuous flow cyanation was implemented in the course of the multigram scale synthesis of Remdesivir [117]. Under the optimized conditions, almost 500 kg of the intermediate were prepared (2 kg h^{-1}). While TMSCN was used as the cyanide source, acidic conditions entailed the presence of hydrogen cyanide in the reaction mixture, making the protocol useful in minimizing the accumulation of HCN.

Chemical warfare agents (CWA) have filled the world with dread since their emergence during the First World War, prompting the enactment of the *Chemical Weapons Convention* to prohibit their possession, manufacture, and use. One of the challenges posed by CWA is the safe destruction of existing stockpiles. Considering the embedded toxicity of these compounds, in order to facilitate their on-site rapid neutralization, flow technology is uniquely positioned to afford transportable, benign, and easily operable devices. A French collaboration led by Legros studied the oxidation of bis(2-chloroethyl) sulfide (**68**, mustard gas) to the corresponding sulfoxide in a flow system. Working on model compounds (one of which is **69**), the team showed that treating with urea/hydrogen peroxide as oxidant in the presence of methanesulfonic acid (CH_3SO_3H) led to full conversion to the neutralized product at a rate of ca. 780 g day^{-1} [118]. More recently, milder conditions were reported, based on singlet oxygen, which was generated by visible light irradiation using methylene blue (MB) as a photosensitizer (Fig. 8.30) [119].

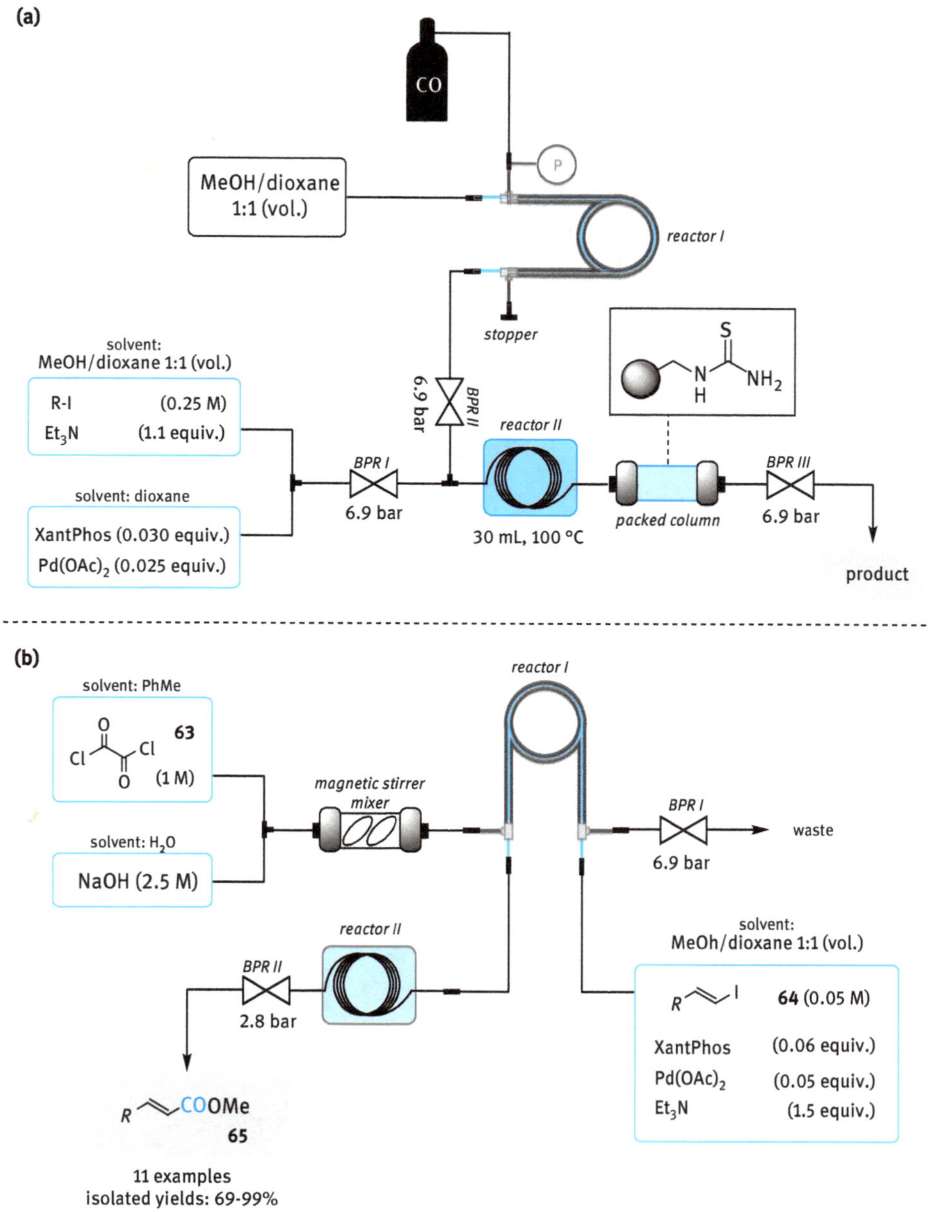

Fig. 8.28: Tube-in-tube reactors for the dosage of gaseous CO (a) and the in-line generation of CO (b) [115].

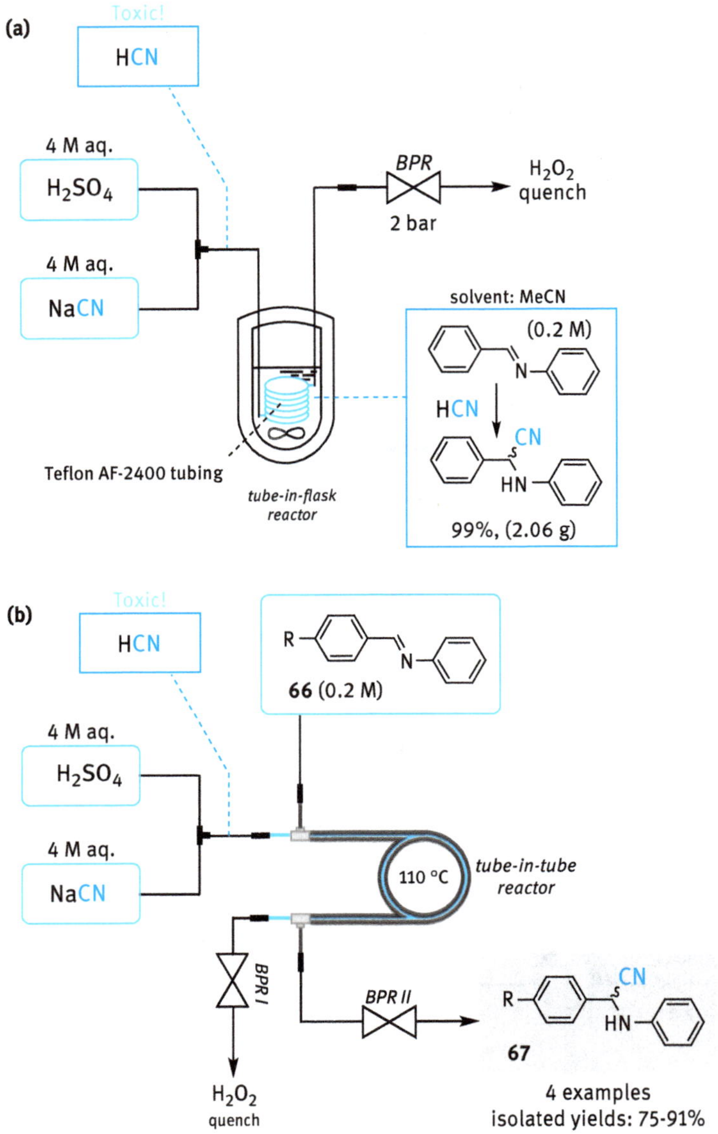

Fig. 8.29: Continuous generation of anhydrous HCN in tube-in-flask (a) and tube-in-tube (b) reactors [116].

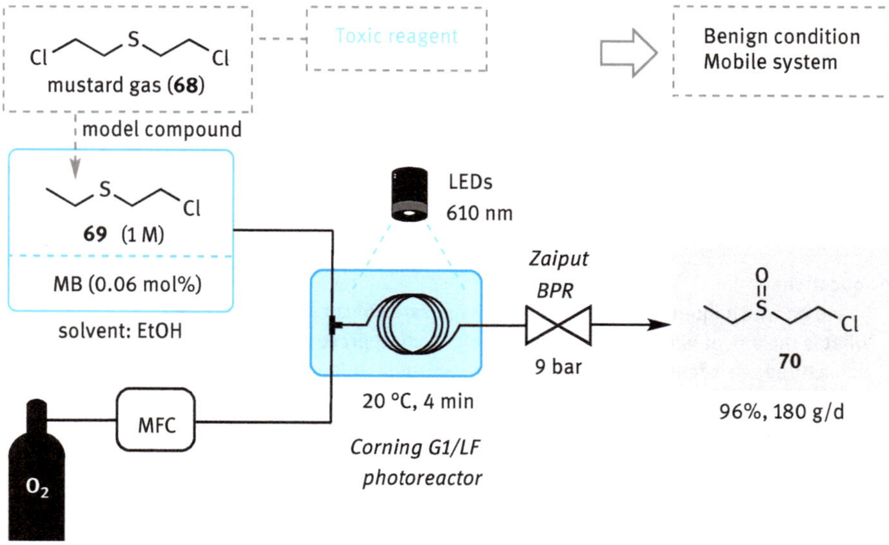

Fig. 8.30: Photochemical neutralization of chemical warfare agents in flow [119].

8.4 Conclusions

Micro- and mesofluidic reactors make a reaction process inherently safe by its miniaturized structure: only minimal amounts of hazardous species are present in a robust structure. Reactions under extreme conditions and dangerous reactants can be safely handled by flow reactors and applied for laboratory preparation and industrial synthesis. The continuous flow operation allows on-site, on-demand generation of hazardous reagents, enabling better control of the process safety.

Further readings

For a quick introduction of process safety, Chapter 28 of the following book is recommended:

- Smith, R. (2016) *Chemical Process Design and Integration*, John Wiley & Sons, Ltd, Chichester, UK.

For an introduction to the CLP and REACH regulations, the following literature recommended:

- Bender, H.F. (2018) Sicherer Umgang mit Gefahrstoffen: unter Berücksichtigung von REACH und GHS, Wiley-VCH, Weinheim. (textbook)
- Boss, M.J., Boss, B., Boss, C., and Day, D.W. (eds.) (2017) Handbook of Chemical Regulations: Benchmarking, Implementation, and Engineering Concepts, CRC Press,Taylor & Francis Group, Boca Raton. (handbook)

Review and monographs on the safety of chemical processes:

- Kockmann, N., Thenée, P., Fleischer-trebes, C., Laudadio, G., and Noël, T. (2017) Safety assessment in development and operation of modular continuous-flow processes. *React. Chem. Eng.*, **2**, 258–280.

- Stoessel, F. (2020) *Thermal Safety of Chemical Processes: Risk Assessment and Process Design*, Wiley-VCH, Weinheim.
- Center for Chemical Process Safety (2019) *Guidelines for Inherently Safer Chemical Processes: A Life Cycle Approach*, Wiley-AIChE, Hoboken, NJ.

Monograph on the application of microreactors in extreme conditions:
- Hessel, V., Kralisch, D., and Kockmann, N. (2015) *Novel Process Windows: Innovative Gates to Intensified and Sustainable Chemical Processes*, Wiley-VCH, Weinheim.

i

Study questions

8.1 Based on which international system are the hazards of chemicals classified?

8.2 What is the aim of *inherently safer design* of chemical processes?

8.3 Name three basic features of microreactors regarding process safety.

8.4 What is *novel process windows*? Name four of the intensification pathways which are of direct safety concern.

8.5 Explain the advantage of microreactors in mitigating hazards associated with exothermic reactions.

8.6 What is a reagent generator and how is it useful from a safety perspective?

8.7 Suggest reactions for the in-line generation of hydrazoic acid, diazomethane, and aryl lithium species.

8.8 Suggest at least three additional reagents whose hazards can be mitigated using microreactor technology.

8.9 What mitigating measures should be taken to minimize risks in microreactors setups?

8.10 What are the risks associated with particulates formation in microreactors? Suggest solutions to deal with this potential issue.

8.11 Name two aspects of chemical hazards that are NOT mitigated by microreactor technology.

References

[1] Smith, R, Chemical Process Design and Integration, Chichester, UK, John Wiley & Sons, Ltd, 2016.

[2] Lee, S, Chang, SR, Suh, Y, Developing concentration index of industrial and occupational accidents: The case of European countries, Saf Health Work, 2020, 11(3), 266–274.

[3] Wang, B, Li, D, Wu, C, Characteristics of hazardous chemical accidents during hot season in China from 1989 to 2019 : A statistical investigation, Saf Sci, 2020, 129(2020), 104788.

[4] Jung, S, Woo, J, Kang, C, Analysis of severe industrial accidents caused by hazardous chemicals in South Korea from January 2008 to June 2018, Saf Sci, 2020, 124(December 2019), 104580.

[5] U.S. Chemical Safety and Hazard Investigation Board (2020) https://www.csb.gov/.

[6] World Health Organization (2018) https://www.who.int/health-topics/chemical-incidents#tab=tab_1.

[7] Bender, HF, Sicherer Umgang mit Gefahrstoffen: unter Berücksichtigung von REACH und GHS, Weinheim, Wiley-VCH, 2018.

[8] Carlson, PA, Mumford, CJ, Hazardous Chemicals Handbook, Oxford, Butterworth-Heinemann, 2002.

[9] Boss, MJ, Boss, B, Boss, C, Day, DW, eds., Handbook of Chemical Regulations: Benchmarking, Implementation, and Engineering Concepts, Boca Raton, CRC Press,Taylor & Francis Group, 2017.
[10] Center for Chemical Process Safety (CCPS), Layer of Protection Analysis: Simplified Process Risk Assessment, New York, Wiley-AIChE, 2001.
[11] Gerardy, R, Monbaliu, J-CM, Multistep continuous-flow processes for the preparation of heterocyclic active pharmaceutical ingredients, Top Heterocycl Chem, 2018, 56, 1–102.
[12] Kletz, TA, Amyotte, P, Process Plants: A Handbook for Inherently Safer Design, CRC Press, Taylor & Francis group, 2010.
[13] Center for Chemical Process Safety, Guidelines for Inherently Safer Chemical Processes: A Life Cycle Approach, Hoboken, NJ, Wiley-AIChE, 2019.
[14] Davis, ME, Davis, RJ, Fundamentals of Chemical Reaction Engineering, Boston, McGraw-Hill, 2003.
[15] Reschetilowski, W, (ed.) Microreactors in Preparative Chemistry: Practical Aspects in Bioprocessing, Nanotechnology, Catalysis and more, Weinheim, Wiley-VCH, 2013.
[16] Wirth, T, ed., Microreactors in Organic Chemistry and Catalysis, Weinheim, 2013.
[17] Gutmann, B, Cantillo, D, Kappe, CO, Continuous-flow technology – A tool for the safe manufacturing of active pharmaceutical ingredients, Angew Chemie – Int Ed, 2015, 54(23), 6688–6728.
[18] Plutschack, MB, Pieber, B, Gilmore, K, Seeberger, PH, The Hitchhiker's guide to flow chemistry, Chem Rev, 2017, 117(18), 11796–11893.
[19] Movsisyan, M, Delbeke, EIP, Berton, JKET, Battilocchio, C, Ley, SV, Stevens, CV, Taming hazardous chemistry by continuous flow technology, Chem Soc Rev, 2016, 45(18), 4892–4928.
[20] Nolan, PF, Barton, JA, Some lessons from thermal-runaway incidents, J Hazard Mater, 1987, 14(2), 233–239.
[21] Saada, R, Patel, D, Saha, B, Causes and consequences of thermal runaway incidents – Will they ever be avoided?, Process Saf Environ Prot, 2015, 97, 109–115.
[22] Stoessel, F, Thermal Safety of Chemical Processes: Risk Assessment and Process Design, Weinheim, Wiley-VCH, 2020.
[23] Fogler, HS, Elements of Chemical Reaction Engineering, Bosten, Prentice Hall, Pearson Education, 2020.
[24] Sobieszuk, P, Aubin, J, Pohorecki, R, Hydrodynamics and mass transfer in gas-liquid flows in microreactors, Chem Eng Technol, 2012, 35(8), 1346–1358.
[25] Coker, AK, Ludwig's Applied Process Design for Chemical and Petrochemical Plants, Gulf Professional Publishing, 2007.
[26] Tidona, B, Urakawa, A, Rudolf von Rohr, P, High pressure plant for heterogeneous catalytic CO2 hydrogenation reactions in a continuous flow microreactor, Chem Eng Process Process Intensif, 2013, 65, 53–57.
[27] Razzaq, T, Kappe, CO, Continuous flow organic synthesis under high-temperature/pressure : Conditions, Chem – An Asian J, 2010, 5(6), 1274–1289.
[28] Hessel, V, Kralisch, D, Kockmann, N, Noël, T, Wang, Q, Novel process windows for enabling, accelerating, and uplifting flow chemistry, ChemSusChem, 2013, 6(5), 746–789.
[29] Liu, Y, Chen, G, Yue, J, Manipulation of gas-liquid-liquid systems in continuous flow microreactors for efficient reaction processes, J Flow Chem, 2020, 10(1), 103–121.
[30] Mallia, CJ, Baxendale, IR, The use of gases in flow synthesis, Org Process Res Dev, 2016, 20(2), 327–360.
[31] Hessel, V, Kralisch, D, Kockmann, N, Novel Process Windows: Innovative Gates to Intensified and Sustainable Chemical Processes, Weinheim, Wiley-VCH, 2015.

[32] Gutmann, B, Kappe, CO, Forbidden chemistries go flow in API synthesis, Chim Oggi/Chem Today, 2015, 33(June), 18–25.
[33] Britton, J, Raston, CL, Multi-step continuous-flow synthesis, Chem Soc Rev, 2017, 46(5), 1250–1271.
[34] Webb, D, Jamison, TF, Continuous flow multi-step organic synthesis, Chem Sci, 2010, 1(6), 675–680.
[35] Dallinger, D, Gutmann, B, Kappe, CO, The concept of chemical generators: on-site on-demand production of hazardous reagents in continuous flow, Acc Chem Res, 2020, 53(7), 1330–1341.
[36] Kockmann, N, Thenée, P, Fleischer-Trebes, C, Laudadio, G, Noël, T, Safety assessment in development and operation of modular continuous-flow processes, React Chem Eng, 2017, 2(3), 258–280.
[37] Kulkarni, AA, Continuous flow nitration in miniaturized devices, Beilstein J Org Chem, 2014, 10, 405–424.
[38] Wen, Z, Jiao, F, Yang, M, Zhao, S, Zhou, F, Chen, G, Process development and scale-up of the continuous flow nitration of trifluoromethoxybenzene, Org Process Res Dev, 2017, 21(11), 1843–1850.
[39] Yu, Z, Xu, Q, Liu, L, Wu, Z, Huang, J, Lin, J, Su, W, Dinitration of o-toluic acid in continuous-flow: process optimization and kinetic study, J Flow Chem, 2020, 10(2), 429–436.
[40] Gustin, J, Runaway reaction hazards in processing organic nitro compounds, Org Process Res Dev, 1998, 2(1), 27–33.
[41] Cantillo, D, Damm, M, Dallinger, D, Bauser, M, Berger, M, Kappe, CO, Sequential nitration/hydrogenation protocol for the synthesis of triaminophloroglucinol: Safe generation and use of an explosive intermediate under continuous-flow conditions, Org Process Res Dev, 2014, 18(11), 1360–1366.
[42] Cantillo, D, Wolf, B, Goetz, R, Kappe, CO, Continuous flow synthesis of a key 1,4-benzoxazinone intermediate via a nitration/hydrogenation/cyclization sequence, Org Process Res Dev, 2017, 21(1), 125–132.
[43] Köckinger, M, Wyler, B, Aellig, C, Roberge, DM, Hone, CA, Kappe, CO, Optimization and Scale-Up of the Continuous Flow Acetylation and Nitration of 4-Fluoro-2-methoxyaniline to Prepare a Key Building Block of Osimertinib, Org Process Res Dev, 2020, 24(10), 2217–2227.
[44] Gemoets, HPL, Su, Y, Shang, M, Hessel, V, Luque, R, Noël, T, Liquid phase oxidation chemistry in continuous-flow microreactors, Chem Soc Rev, 2016, 45(1), 83–117.
[45] Hone, CA, Kappe, CO (2019) The Use of Molecular Oxygen for Liquid Phase Aerobic Oxidations in Continuous Flow.
[46] Hone, CA, Roberge, DM, Kappe, CO, The use of molecular oxygen in pharmaceutical manufacturing: Is flow the way to go?, ChemSusChem, 2017, 10(1), 32–41.
[47] Kassin, V-EH, Gérardy, R, Toupy, T, Collin, D, Salvadeo, E, Toussaint, F, Van Hecke, K, Monbaliu, J-CM, Expedient preparation of active pharmaceutical ingredient ketamine under sustainable continuous flow conditions, Green Chem, 2019, 21(11), 2952–2966.
[48] Kassin, V-EH, Toupy, T, Petit, G, Bianchi, P, Salvadeo, E, Monbaliu, J-CM, Metal-free hydroxylation of tertiary ketones under intensified and scalable continuous flow conditions, J Flow Chem, 2020, 10(1), 167–179.
[49] Zhang, JS, Wang, K, Zhang, CY, Luo, GS, Safety evaluating of Beckmann rearrangement of cyclohexanone oxime in microreactors using inherently safer design concept inherent risk of design index likelihood index of inherently safer de,Chem Eng Process Process Intensif, 2016, 110(2016), 44–51.
[50] Evers, J, Göbel, M, Krumm, B, Martin, F, Medvedyev, S, Oehlinger, G, Steemann, FX, Troyan, I, Klapötke, TM, Eremets, MI, Molecular structure of hydrazoic acid with hydrogen-bonded tetramers in nearly planar layers, J Am Chem Soc, 2011, 133(31), 12100–12105.

[51] Chen, Y, Gutmann, B, Kappe, CO, Continuous-flow electrophilic amination of arenes and schmidt reaction of carboxylic acids utilizing the superacidic trimethylsilyl azide/triflic acid reagent system, J Org Chem, 2016, 81(19), 9372–9380.
[52] Karlsson, S, Bergman, R, Löfberg, C, Moore, PR, Pontén, F, Tholander, J, Sörensen, H, Development of a large-scale route to an MCH 1 receptor antagonist: investigation of a staudinger ketene–imine cycloaddition in batch and flow mode, Org Process Res Dev, 2015, 19(12), 2067–2074.
[53] Yoshida, J-I, Flash Chemistry: Fast Organic Synthesis in Microsystems, Chichester, John Wiley & Sons, Ltd, 2008.
[54] Colella, M, Nagaki, A, Luisi, R, Flow technology for the genesis and use of (Highly) reactive organometallic reagents, Chem – A Eur J, 2020, 26(1), 19–32.
[55] Kim, H, Min, K-I, Inoue, K, Im, DJ, Kim, D-P, Yoshida, J-I, Submillisecond organic synthesis: Outpacing Fries rearrangement through microfluidic rapid mixing, Science (80-), 2016, 352(6286), 691–694.
[56] Nagaki, A, Yamashita, H, Hirose, K, Tsuchihashi, Y, Yoshida, J, Alkyllithium compounds bearing electrophilic functional groups: A flash chemistry approach, Angew Chemie Int Ed, 2019, 58(12), 4027–4030.
[57] Nagaki, A, Takahashi, Y, Yoshida, J, Generation and reaction of carbamoyl anions in flow: applications in the three-component synthesis of functionalized α-ketoamides, Angew Chemie Int Ed, 2016, 55(17), 5327–5331.
[58] Ichinari, D, Ashikari, Y, Mandai, K, Aizawa, Y, Yoshida, J, Nagaki, A, A novel approach to functionalization of aryl azides through the generation and reaction of organolithium species bearing masked azides in flow microreactors, Angew Chemie Int Ed, 2020, 59(4), 1567–1571.
[59] Nagaki, A, Yamashita, H, Tsuchihashi, Y, Hirose, K, Takumi, M, Yoshida, J, Generation and reaction of functional alkyllithiums by using microreactors and their application to heterotelechelic polymer synthesis, Chem – A Eur J, 2019, 25(60), 13719–13727.
[60] Ashikari, Y, Kawaguchi, T, Mandai, K, Aizawa, Y, Nagaki, A, A synthetic approach to dimetalated arenes using flow microreactors and the switchable application to chemoselective cross-coupling reactions, J Am Chem Soc, 2020, 142(40), 17039–17047.
[61] Usutani, H, Cork, DG, Effective utilization of flow chemistry: Use of unstable intermediates, inhibition of side reactions, and scale-up for boronic acid synthesis, Org Process Res Dev, 2018, 22(6), 741–746.
[62] Kleoff, M, Schwan, J, Boeser, L, Hartmayer, B, Christmann, M, Sarkar, B, Heretsch, P, Scalable synthesis of functionalized ferrocenyl azides and amines enabled by flow chemistry, Org Lett, 2020, 22(3), 902–907.
[63] Wong, JYF, Tobin, JM, Vilela, F, Barker, G, Batch versus flow lithiation–substitution of 1,3,4-oxadiazoles: Exploitation of unstable intermediates using flow chemistry, Chem – A Eur J, 2019, 25(53), 12439–12445.
[64] Thaisrivongs, DA, Naber, JR, Rogus, NJ, Spencer, G, Development of an organometallic flow chemistry reaction at pilot-plant scale for the manufacture of verubecestat, Org Process Res Dev, 2018, 22(3), 403–408.
[65] Thaisrivongs, DA, Naber, JR, McMullen, JP, Using flow to outpace fast proton transfer in an organometallic reaction for the manufacture of verubecestat (MK-8931), Org Process Res Dev, 2016, 20(11), 1997–2004.
[66] Ketels, M, Ganiek, MA, Weidmann, N, Knochel, P, Synthesis of polyfunctional diorganomagnesium and diorganozinc reagents through in situ trapping halogen-lithium exchange of highly functionalized (Hetero)aryl halides in continuous flow, Angew Chemie Int Ed, 2017, 56(41), 12770–12773.

[67] von Keutz, T, Cantillo, D, Kappe, CO, Organomagnesium based flash chemistry: continuous flow generation and utilization of halomethylmagnesium intermediates, Org Lett, 2020, 22(19), 7537–7541.
[68] von Keutz, T, Williams, JD, Kappe, CO, Continuous Flow C -glycosylation via metal–halogen exchange: process understanding and improvements toward efficient manufacturing of remdesivir, Org Process Res Dev, 2020, 24(10), 2362–2368.
[69] Huck, L, De La Hoz, A, Díaz-Ortiz, A, Alcázar, J, Grignard reagents on a tab: Direct magnesium insertion under flow conditions, Org Lett, 2017, 19(14), 3747–3750.
[70] Lau, SH, Bourne, SL, Martin, B, Schenkel, B, Penn, G, Ley, SV, Synthesis of a precursor to sacubitril using enabling technologies, Org Lett, 2015, 17(21), 5436–5439.
[71] Huck, L, Berton, M, de la Hoz, A, Díaz-Ortiz, A, Alcázar, J, Reformatsky and Blaise reactions in flow as a tool for drug discovery. One pot diversity oriented synthesis of valuable intermediates and heterocycles, Green Chem, 2017, 19(6), 1420–1424.
[72] Weidmann, N, Ketels, M, Knochel, P, Sodiation of Arenes and Heteroarenes in Continuous Flow, Angew Chemie Int Ed, 2018, 57(33), 10748–10751.
[73] Uhlig, N, Martins, A, Gao, D, Selective DIBAL-H Monoreduction of a Diester Using Continuous Flow Chemistry: From Benchtop to Kilo Lab, Org Process Res Dev, 2020, 24(10), 2326–2335.
[74] Fukuyama, T, Chiba, H, Kuroda, H, Takigawa, T, Kayano, A, Tagami, K, Application of Continuous Flow for DIBAL-H Reduction and n -BuLi Mediated Coupling Reaction in the Synthesis of Eribulin Mesylate, Org Process Res Dev, 2016, 20(2), 503–509.
[75] Ötvös, SB, Kappe, CO, Continuous-flow amide and ester reductions using neat borane dimethylsulfide complex, ChemSusChem, 2020, 13(7), 1800–1807.
[76] Tsoung, J, Wang, Y, Djuric, SW, Expedient Diels–Alder cycloadditions with ortho-quinodimethanes in a high temperature/pressure flow reactor, React Chem Eng, 2017, 2(4), 458–461.
[77] Charaschanya, M, Bogdan, AR, Wang, Y, Djuric, SW, Nucleophilic aromatic substitution of heterocycles using a high-temperature and high-pressure flow reactor, Tetrahedron Lett, 2016, 57(9), 1035–1039.
[78] Tsoung, J, Bogdan, AR, Kantor, S, Wang, Y, Charaschanya, M, Djuric, SW, Synthesis of fused pyrimidinone and quinolone derivatives in an automated high-temperature and high-pressure flow reactor, J Org Chem, 2017, 82(2), 1073–1084.
[79] Lengyel, LC, Sipos, G, Sipöcz, T, Vágó, T, Dormán, G, Gerencsér, J, Makara, G, Darvas, F, Synthesis of condensed heterocycles by the gould-jacobs reaction in a novel three-mode pyrolysis reactor, Org Process Res Dev, 2015, 19(3), 399–409.
[80] Yu, T, Jiao, J, Song, P, Nie, W, Yi, C, Zhang, Q, Li, P, Recent progress in continuous-flow hydrogenation, ChemSusChem, 2020, 13(11), 2876–2893.
[81] Suveges, NS, de Souza, ROMA, Gutmann, B, Kappe, CO, Synthesis of Mepivacaine and Its Analogues by a Continuous-Flow Tandem Hydrogenation/Reductive Amination Strategy, European J Org Chem, (2017), 2017(44), 6511–6517.
[82] Cherkasov, N, Bai, Y, Rebrov, E, Process intensification of alkynol semihydrogenation in a tube reactor coated with a Pd/ZnO Catalyst, Catalysts, 2017, 7(12), 358.
[83] Bai, Y, Cherkasov, N, Huband, S, Walker, D, Walton, R, Rebrov, E, Highly selective continuous flow hydrogenation of cinnamaldehyde to cinnamyl alcohol in a Pt/SiO2 coated tube reactor, Catalysts, 2018, 8(2), 58.
[84] Exposito, AJ, Bai, Y, Tchabanenko, K, Rebrov, EV, Cherkasov, N, Process intensification of continuous-flow imine hydrogenation in catalyst-coated tube reactors, Ind Eng Chem Res, 2019, 58(11), 4433–4442.
[85] Abrams, ML, Buser, JY, Calvin, JR, Johnson, MD, Jones, BR, Lambertus, G, Landis, CR, Martinelli, JR, May, SA, McFarland, AD, Stout, JR, Continuous liquid vapor reactions part 2:

Asymmetric hydroformylation with rhodium-bisdiazaphos catalysts in a vertical pipes-in-series reactor, Org Process Res Dev, 2016, 20(5), 901–910.
[86] Hone, CA, Lopatka, P, Munday, R, O'Kearney-McMullan, A, Kappe, CO, Continuous-flow Synthesis of Aryl Aldehydes by Pd-catalyzed Formylation of Aryl Bromides Using Carbon Monoxide and Hydrogen, ChemSusChem, 2019, 12(1), 326–337.
[87] Köckinger, M, Hanselmann, P, Hu, G, Hone, CA, Kappe, CO, Continuous flow synthesis of aryl aldehydes by Pd-catalyzed formylation of phenol-derived aryl fluorosulfonates using syngas, RSC Adv, 2020, 10(38), 22449–22453.
[88] Kashani, SK, Jessiman, JE, Newman, SG, Exploring homogeneous conditions for mild Buchwald-Hartwig amination in batch and flow, Org Process Res Dev, 2020, 24(10), 1948–1954.
[89] Huang, M, A simple modification of the Wolff-Kishner reduction, J Am Chem Soc, 1946, 68(12), 2487–2488.
[90] Znidar, D, O'Kearney-Mcmullan, A, Munday, R, Wiles, C, Poechlauer, P, Schmoelzer, C, Dallinger, D, Kappe, CO, Scalable Wolff-Kishner reductions in extreme process windows using a silicon carbide flow reactor, Org Process Res Dev, 2019, 23(11), 2445–2455.
[91] Liu, J, Wang, D, Xu, Q, Yu, H, Zhou, J, Yu, Z, Su, W, Continuous-flow double diazotization for the synthesis of m-difluorobenzene via Balz-Schiemann reaction, J Flow Chem, 2020, 10, 589–596.
[92] Park, NH, Senter, TJ, Buchwald, SL, Rapid Synthesis of Aryl Fluorides in Continuous Flow through the Balz-Schiemann Reaction, Angew Chemie, 2016, 128(39), 12086–12090.
[93] Heinrich, S, Edeling, F, Liebner, C, Hieronymus, H, Lange, T, Klemm, E, Catalyst as ignition source of an explosion inside a microreactor, Chem Eng Sci, 2012, 84(2012), 540–543.
[94] Chen, Y, Hone, CA, Gutmann, B, Kappe, CO, Continuous flow synthesis of carbonylated heterocycles via Pd-catalyzed oxidative carbonylation using CO and O2 at elevated temperatures and pressures, Org Process Res Dev, 2017, 21(7), 1080–1087.
[95] Wu, X-F, Neumann, H, Beller, M, Palladium-catalyzed oxidative carbonylation reactions, ChemSusChem, 2013, 6(2), 229–241.
[96] Yang, H, Martin, B, Schenkel, B, On-demand generation and consumption of diazomethane in multistep continuous flow systems, Org Process Res Dev, 2018, 22(4), 446–456.
[97] Lehmann, H, A scalable and safe continuous flow procedure for in-line generation of diazomethane and its precursor MNU, Green Chem, 2017, 19(6), 1449–1453.
[98] Dallinger, D, Kappe, CO, Lab-scale production of anhydrous diazomethane using membrane separation technology, Nat Protoc, 2017, 12(10), 2138–2147.
[99] Roydhouse, MD, Ghaini, A, Constantinou, A, Cantu-Perez, A, Motherwell, WB, Gavriilidis, A, Ozonolysis in flow using capillary reactors, Org Process Res Dev, 2011, 15(5), 989–996.
[100] Vasudevan, N, Sharma, MK, Reddy, DS, Kulkarni, AA, A multi-step continuous flow synthesis of the cystic fibrosis medicine ivacaftor, React Chem Eng, 2018, 3(4), 520–526.
[101] Lee, K, Lin, H, Jensen, KF, Ozonolysis of quinoline and quinoline derivatives in a Corning low flow reactor, React Chem Eng, 2017, 2(5), 696–702.
[102] Lau, S-H, Galván, A, Merchant, RR, Battilocchio, C, Souto, JA, Berry, MB, Ley, SV, Machines vs malaria: A flow-based preparation of the drug candidate OZ439, Org Lett, 2015, 17(13), 3218–3221.
[103] Inoue, T, Ohtaki, K, Adachi, J, Lu, M, Murakami, S, Direct synthesis of hydrogen peroxide using glass fabricated microreactor – Multichannel operation and catalyst comparison, Catal Today, 2015, 248, 169–176.
[104] Sun, B, Zhu, H, Liang, W, Zhang, X, Feng, J, Xu, W, A safe and clean way to produce H2O2 from H2 and O2 within the explosion limit range, Int J Hydrogen Energy, 2019, 44(36), 19547–19554.

[105] Mata, A, Cantillo, D, Kappe, CO, An integrated continuous-flow synthesis of a key oxazolidine intermediate to noroxymorphone from naturally occurring opioids, European J Org Chem, 2017, 2017(44), 6505–6510.
[106] Cantillo, D, Kappe, CO, Halogenation of organic compounds using continuous flow and microreactor technology, React Chem Eng, 2017, 2(1), 7–19.
[107] Keplinger, ML, Suissa, LW, Toxicity of Fluorine Short-Term Inhalation, Am Ind Hyg Assoc J, 1968, 29(1), 10–18.
[108] Harsanyi, A, Conte, A, Pichon, L, Rabion, A, Grenier, S, Sandford, G, One-step continuous flow synthesis of antifungal WHO essential medicine flucytosine using fluorine, Org Process Res Dev, 2017, 21(2), 273–276.
[109] Deng, Q, Shen, R, Ding, R, Zhang, L, Bromination of aromatic compounds using bromine in a microreactor, Chem Eng Technol, 2016, 39(8), 1445–1450.
[110] Steiner, A, Roth, PMC, Strauss, FJ, Gauron, G, Tekautz, G, Winter, M, Williams, JD, Kappe, CO, Multikilogram per Hour Continuous Photochemical Benzylic Brominations Applying a Smart Dimensioning Scale-up Strategy, Org Process Res Dev, 2020, 24(10), 2208–2216.
[111] Strauss, FJ, Cantillo, D, Guerra, J, Kappe, CO, A laboratory-scale continuous flow chlorine generator for organic synthesis, React Chem Eng, 2016, 1(5), 472–476.
[112] Chung Leung, GY, Ramalingam, B, Loh, G, Chen, A, Efficient and practical synthesis of sulfonamides utilizing SO 2 gas generated on demand, Org Process Res Dev, 2020, 24(4), 546–554.
[113] Chen, Y, Hone, CA, Gutmann, B, Kappe, CO, Continuous flow synthesis of carbonylated heterocycles via Pd-catalyzed oxidative carbonylation using CO and O 2 at elevated temperatures and pressures, Org Process Res Dev, 2017, 21(7), 1080–1087.
[114] Ramezani, M, Kashfipour, MA, Abolhasani, M, Minireview: Flow chemistry studies of high-pressure gas-liquid reactions with carbon monoxide and hydrogen, J Flow Chem, 2020, 10(1), 93–101.
[115] Hansen, SVF, Wilson, ZE, Ulven, T, Ley, SV, Controlled generation and use of CO in flow, React Chem Eng, 2016, 1(3), 280–287.
[116] Köckinger, M, Hone, CA, Kappe, CO, HCN on tap: On-demand continuous production of anhydrous HCN for organic synthesis, Org Lett, 2019, 21(13), 5326–5330.
[117] Vieira, T, Stevens, AC, Chtchemelinine, A, Gao, D, Badalov, P, Heumann, L, Development of a large-scale cyanation process using continuous flow chemistry en route to the synthesis of remdesivir, Org Process Res Dev, 2020, 24(10), 2113–2121.
[118] Picard, B, Gouilleux, B, Lebleu, T, Maddaluno, J, Chataigner, I, Penhoat, M, Felpin, F-X, Giraudeau, P, Legros, J, Oxidative neutralization of mustard-gas simulants in an on-board flow device with in-line NMR monitoring, Angew Chemie Int Ed, 2017, 56(26), 7568–7572.
[119] Emmanuel, N, Bianchi, P, Legros, J, Monbaliu, J-CM, A safe and compact flow platform for the neutralization of a mustard gas simulant with air and light, Green Chem, 2020, 22(13), 4105–4115.

Luuk T.C.G. Van Summeren, Floris P.J.T. Rutjes, Daniel Blanco-Ania and Tom G. Bloemberg

9 Diazo(diphenyl)methane synthesis in continuous flow: an experiment for the undergraduate teaching laboratory

9.1 Introduction

Although continuous-flow processes are a mainstay of industrial bulk chemistry, the advent of microreactors has constituted a paradigm change in laboratory-scale synthetic chemistry. Flow chemistry in microreactors has intrinsic advantages for many chemical reactions, because it is safe, reaction conditions can be precisely controlled – leading to a high reproducibility – and heat and mass transfer are very efficient. For photochemistry and photoredox chemistry, the advantages are even more pronounced: the same large surface-to-volume ratio of the reaction mixture in a flow cell that is the reason for the efficient heat transfer, also ensures that all reactants can be evenly illuminated. As a result of all these advantages, microreactor flow chemistry is of relevance for the fine-chemical and pharmaceutical industry. Therefore, it is important that this is also reflected in chemical education programs [1]. Although flow chemistry shares principles with industrial-scale continuous production and process chemistry, which are traditionally part of the chemical curriculum in technical universities, flow chemistry itself is still not widely taught. Students that have experience with flow chemistry may have increased chances of future employment.

At the same time, by its nature, flow chemistry also offers opportunities for integrating parts of chemical education in lab courses that are otherwise hard to incorporate. Flow chemistry facilitates higher level or meta-experimentation within the limited time frame of a (synthetic) laboratory period. Instead of just performing a single batch synthesis and determining the yield, students can perform multiple flow syntheses, for instance to:

- investigate the reproducibility of their results,
- learn from mistakes and improve on imperfect attempts,
- investigate the kinetics or thermodynamics of a reaction,
- optimize a reaction's yield,
- test hypotheses.

In other words, flow chemistry facilitates *doing actual chemical research* (or emulating that) within a limited amount of time. Such meta-experimentation compels students to combine, integrate and apply their theoretical knowledge from different

https://doi.org/10.1515/9783110693676-009

molecular subdisciplines and perceive chemistry as the broad and integrated science that it is.

This chapter presents the Swern-like oxidation of hydrazones to the corresponding diazo compounds (Fig. 9.1) to be performed as a model reaction in chemistry laboratories at universities across the world. An alternative version of this text has been published in the *Journal of Flow Chemistry* [2]. This reaction follows the same steps as the Swern oxidation [1]: activation of DMSO by reaction with an electrophile (TFAA) [2], reaction of the nucleophile to be oxidized (benzophenone hydrazone) with the activated DMSO intermediate, and [3] deprotonation of the resulting intermediate by a base (DIPEA) with subsequent elimination of dimethyl sulfide. Besides learning how to conduct a flow experiment, students will become acquainted with design of experiments (DoE) and reaction optimization, the organic chemistry of diazo compounds and the possible use of the flow products in subsequent reactions, oxidation chemistry, and quantitative VIS spectroscopy, used to monitor the reaction.

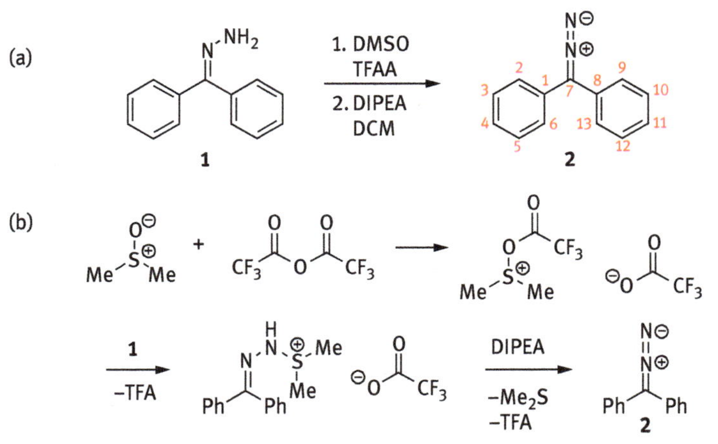

Fig. 9.1: The synthesis of diazo(diphenyl)methane (2) via the Omura–Sharma–Swern oxidation of benzophenone hydrazone (1). (a) The overall reaction and (b) the subsequent steps in the reaction. DMSO, dimethyl sulfoxide; TFAA, trifluoroacetic anhydride; DIPEA, *N,N*-diisopropylethylamine; DCM, dichloromethane. The numbers in the structure of **2** in the top right are used for referencing NMR signals in Section 9.7.2.

Flow chemistry has been part of our laboratory courses since 2011, when two *b330 Flow Start* education units from FutureChemistry, a university spin-off company specializing in flow chemistry, were acquired. From the start, the Omura–Sharma–Swern (OSS) oxidation has been the reaction that was used to acquaint students with flow chemistry. The first 2 years, students were provided with an application note provided by FutureChemistry, based on [3] and describing the straightforward synthesis of cinnamyl aldehyde by oxidation of cinnamyl alcohol.

In 2013, we realized that this same reaction could also be used to convert benzophenone hydrazone (**1**, Fig. 9.1) to diazo(diphenyl)methane (**2**) [4, 5]. This reaction is more appealing from a didactic viewpoint, for several reasons:

- the product is a relatively stable diazo compound: a chemically interesting type of compound [6, 7], attractive to students, but not frequently encountered in undergraduate education;
- the mechanism of the formation of the product (Fig. 9.2b) differs sufficiently from that of the commonly encountered aldehyde or ketone analogues (Fig. 9.2a) to be challenging for students to derive;
- last but not least, the product has an intense purple-red color, permitting observation of its formation by eye as well as allowing quantification by visible spectroscopy, for which equipment is readily available in most undergraduate laboratories.

To our delight, the synthesis of **2** could indeed also be performed in continuous flow at room temperature, even with relatively simple flow chemistry setups. A 2-day student project was formulated, in which students not only synthesize **2**, but actually perform a limited optimization of two reaction parameters by means of a formal *experimental design*, integrating the theory they have previously obtained about organic chemistry, analytical chemistry, DoE, statistics, and programming with the new concept of continuous-flow processes.

Since then, the synthesis of **2** in flow has been one of the two recurring instances of flow chemistry in our undergraduate synthetic lab courses; the other being the routine synthesis of ethyl diazoacetate in flow [8] on an as-needed basis for use in the synthesis of a bicyclononyl derivative [9].

In this chapter, the optimization of the synthesis of **2** both from both an experimental and a chemical education viewpoint will be discussed. A basic student manual for the experiment is provided.

9.2 Moffatt oxidations

Following the introduction of “activated DMSO” oxidations at the end of the 1950s, the direct oxidations of primary and secondary alcohols to the corresponding aldehydes and ketones by Moffatt-style DMSO-based activators at the beginning of the 1960s have become a standard tool in synthetic laboratories [10–15]. The general procedure involves activating DMSO by reaction with an “activator,” an electrophile that is attacked by a lone pair on the oxygen atom of DMSO. Two of the activators leading to the most reactive activated DMSO species are oxalyl chloride and trifluoroacetic anhydride, both introduced by Swern and coworkers [16–20].

There can be a cruel sense of humor in scientific progress: after Albright et al. in the 1960s had dismissed the use of trifluoroacetic anhydride as an activator [21, 22],

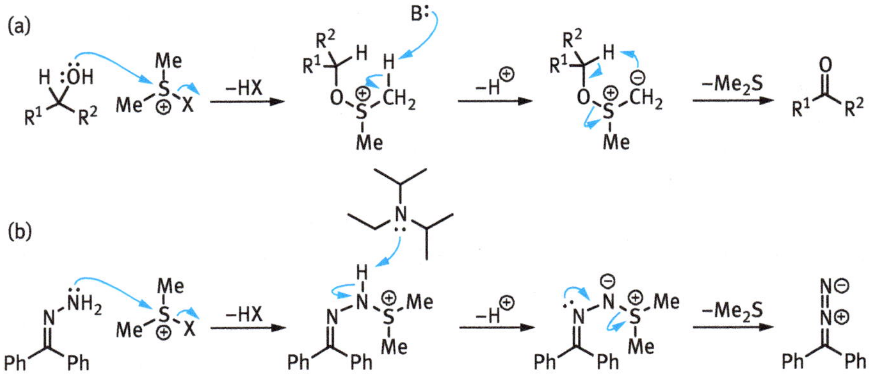

Fig. 9.2: (a) The general mechanism for the Moffatt oxidation of an alcohol by activated DMSO. (b) Proposed mechanism for the formation of **2** from **1** by activated DMSO. As opposed to the classical Moffatt mechanism under (a), in this case presumably the more acidic amine proton is directly abstracted by the base, giving rise to the corresponding ylide, which then collapses to the diazo functional group with concomitant loss of dimethyl sulfide.

because it was useless at room temperature, Swern and coworkers in the 1970s proved that it actually is a very good activator at low temperatures [18, 19, 23]. And with the advent of flow chemistry, the use of trifluoroacetic anhydride as an activator for DMSO at room temperature has been proven possible after all [3, 24, 25].

Contrary to their use in the oxidization of alcohols, Moffatt oxidations are far less widely used for the analogous conversion of amines into the corresponding imines [26–29], probably not because of inherent issues, but because the formation of an imine from an amine is simply less frequently required than the opposite pathway, as in reductive amination. Whereas the mechanism of imine formation starting from a secondary amine is probably analogous to that of carbonyl formation starting from an alcohol [30] (Fig. 9.2a), it is not immediately obvious that it is also possible to convert a hydrazone into a diazo group, since in that case, two hydrogen atoms need to be abstracted from the same nitrogen (Fig. 9.2b). Not surprisingly therefore, the discovery of this possibility has been described as serendipitous and provides an extension of the set of available ways to produce diazo compounds.

9.3 Diazo compounds

9.3.1 Synthesis

The experiment described in this chapter represents a further extension to the ways available for producing diazo compounds in a continuous-flow setup [8, 31–33].

There is a plethora of procedures to synthesize diazo compounds. Figure 9.3 depicts the five most common procedures:
Procedure **a**: This procedure is called *diazo transfer* and is used for molecules with two adjacent electron-withdrawing (EWG) groups. These compounds are treated with a base (e.g., potassium carbonate) and a diazo-transfer reagent with the structure RSO_2N_3, of which imidazole-1-sulfonyl azide salts are the safest to use [34, 35].

Procedure **b**: Compounds with α-amino EWG transform their amino group into a diazo group by diazotization using the standard conditions ($NaNO_2$, H^+) [36].

Procedure **c**: Hydrazones can be oxidized to diazo compounds with a variety of reagents, such as $Pb(OAc)_4$ [37] or with the procedure presented in this chapter [5]. This procedure is usually performed to synthesize diazo hydrocarbons.

Procedure **d**: *N*-Sulfonyl hydrazones eliminate sulfinic acids in the presence of bases (e.g., NaOEt) to form diazo compounds [38].

Procedure **e**: The most general procedure to synthesize simple diazo compounds [e.g., diazomethane, diazoethane, and diazo(phenyl)methane] is the base-promoted decomposition of *N*-nitroso ureas and *N*-nitroso sulfonamides [39, 40].

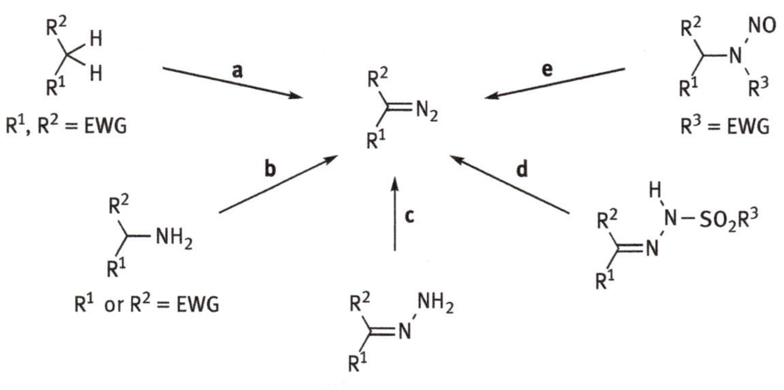

Fig. 9.3: Common procedures to synthesize diazo compounds.

9.3.2 Reactions

Thanks to their peculiar structure, diazo compounds can undergo numerous transformations. It should be noted that most of the reactions of diazo compounds entail the loss of dinitrogen because of the great stability of the nitrogen–nitrogen triple bond. The loss of dinitrogen from diazo compounds takes place with the aid of heat, UV light, or metal catalysis with the formation of the corresponding carbenes or carbenoids. These carbenes or carbenoids react further forming different products depending on their structure and the reaction conditions used. Figure 9.4 summarizes the most relevant reactions of diazo compounds.

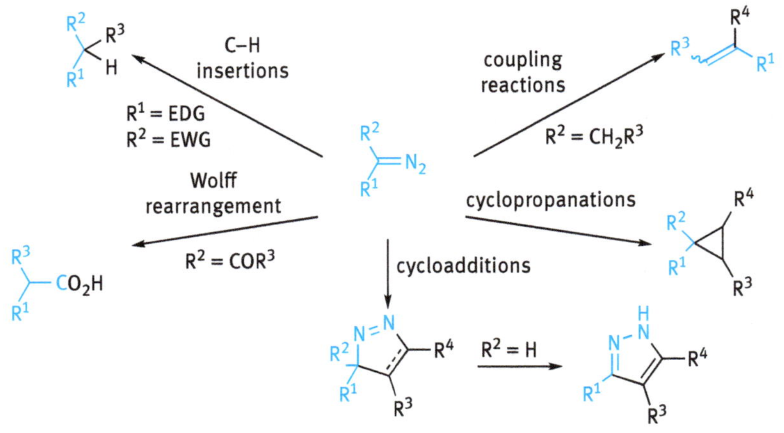

Fig. 9.4: Common reactions of diazo compounds.

Cycloadditions: Diazo compounds perform 1,3-dipolar cycloaddition reactions with alkenes and alkynes to form dihydropyrazoles, 3*H*-pyrazoles, or 1*H*-pyrazoles depending on the dipolarophile used (alkene or alkyne) and the substitution of the diazo compound [41, 42].

Wolff rearrangement: α-diazo ketones (readily available) lose dinitrogen promoted by heat, light, or metal catalysis (e.g., Ag_2O) with a subsequent rearrangement to form ketenes. These ketenes can then react with water, alcohols, or amines to form carboxylic acids (shown in Fig. 9.4), esters, and amides, respectively [43].

C–H insertions: Intermolecular metal-catalyzed (e.g., by rhodium complexes) C–H insertions of diazo compounds are predictable and regioselective when the diazo compound bears an EWG and electron-donating group at the same carbon atom [44]. Other insertion reactions are also possible, like O–H and N–H insertions.

Coupling reactions: Diazo compounds were added to the list of coupling partners of metal-catalyzed cross-coupling reactions at the beginning of this millennium. These reactions involve metal–carbene intermediates [45].

Cyclopropanations: Diazo compounds react with alkenes to form cyclopropanes. This reaction gives higher yields when it is catalyzed by transition metal complexes (e.g., $Rh_2(OAc)_4$) than when the reaction is promoted by heat or UV light [46].

9.4 The experiment – student protocol

In this experiment, starting from benzophenone hydrazone [1], you will synthesize diazo(diphenyl)methane [2], using the OSS oxidation and flow chemistry. You will optimize the conditions of the experiment by means of a formal experimental design [47, 48]. As a result, you typically have to carry out the synthesis seven or eight times in total; each individual synthesis will take about 15 min. You will analyze the product mixture of each separate synthesis with a UV–vis spectrometer, using a microcuvette.

9.5 Motivation

Why would you perform the OSS oxidation of benzophenone hydrazone in flow? The traditional oxidation described above [5] is carried out in a classical batch experiment in a flask at −78 °C. Such low temperatures, however, render a reaction almost by definition unsuitable for large-scale application (e.g., at 100–1,000 L scale) since the cooling capacity (and hence energy) that would be required would be enormous. Moreover, for an exothermic reaction like this one, it would be even more difficult to control the temperature and keep it at the required low value. Owing to the possibility to go to very low reaction times (i.e., residence times) in a flow experiment, providing more control over the reaction conditions, combined with the excellent heat transfer, it is often possible to run low-temperature batch reactions, at higher temperatures in a flow device. This is also seen in the current example, where the −78 °C batch experiment is replaced by a selective flow process that runs at room temperature. Also, when safety is an issue, like in this case with the formation of a diazo derivative, it may be advantageous to consider choosing for a flow process, which is inherently safe due to the small reactor volume. And you could even think of additional benefits such as better scalability and more reproducibility as a result of the well-controlled conditions. On the other hand, there might also be arguments against using a flow process, in particular when solids are formed in the process, or when you have to go to higher dilutions, or when the flow process becomes too expensive due to the use of dedicated (not multipurpose) equipment. A suitable way to make a proper comparison of a batch versus a flow experiment for a given process would be to set up a quantitative model in which all parameters costs (e.g., of the reactor, space), energy consumption, environmental impact, safety risks, chemicals, solvents, waste, and so on are taken into account. Such a model is called a life cycle assessment [49, 50] and is an increasingly common approach to make a judicious choice between one process and the other.

9.5.1 Outline

You will need to make three solutions in dichloromethane (DCM): Solution A contains *both* **1** and DMSO, solution B contains trifluoroacetic acid (TFAA), and solution Q (Q for quench) contains *N,N*-diisopropylethylamine (DIPEA) or triethylamine (TEA), both sterically hindered nitrogen bases.

During the actual experimentation, you are going to vary the flow rates of the three solutions in order to optimize the conversion of **1** into **2** as a function of two factors: (1) the "residence" time between mixing the TFAA and DMSO/**1** solutions and adding in the DIPEA, and (2) the ratio B/A between TFAA and **1** (the DMSO is left out of consideration). Note that this ratio is not simply the ratio of the concentrations of the two solutions, but the ratio of the "molar rates" at which the reactants are pumped through the microreactor. For an example, see the calculations below.

Another factor that could be varied easily is the temperature *T*, but it is clear from earlier experiments that this is more or less inversely related to the reaction time (higher *T* means shorter *t*, lower *T* means longer *t*), so in this experiment *T* is kept constant at 10 °C.

9.5.2 Experimental design

For this experiment, it is required to be familiar with *Experimental Design* or *Design of Experiments* (DoE) to some extent [47, 48]. These are methodologies to optimize a criterion as a function of a set of factors in a minimum number of experiments. Part of your preparation for the experiment is setting up an experimental design for the two factors. The basic option is to choose a *full factorial design* of two factors at two levels. In order to be able to estimate the significance of any observed effect, you will have to perform at least some combinations of factor settings in duplicate.

First, set up an experimental design in concept. For a full factorial design, this can be as simple as making a plot such as Fig. 9.5, in which you explicate the values you want to use. An example would be to place t_{rx} (s) at the *x*-axis and B/A at the *y*-axis.

The standard conditions (= standard factor settings) for this experiment are $t_{\text{rx}} = 4$ s, B/A = 2 and Q/B = 2. The ratio Q/B will remain constant throughout the experiment. The center point (the standard factor settings) would then be at the point (t_{rx}, B/A) = (4, 2). Choose settings for the other points that are not too close – since that will mean that effects will probably be too small to observe – nor too far away.

Next, set up a spreadsheet that calculates the flow rates that are needed to achieve the factor settings you chose, based on the following parameters:
- the reactor volume (1 µL in this case),
- the molarities of the reactant solutions,
- the factor settings themselves.

As an example, to calculate the flow rates (ϕ) for the center point, start by having your spreadsheet calculate the total flow rate in µL min^{-1} of solutions A and B combined, from the reactor volume and the required reaction time:

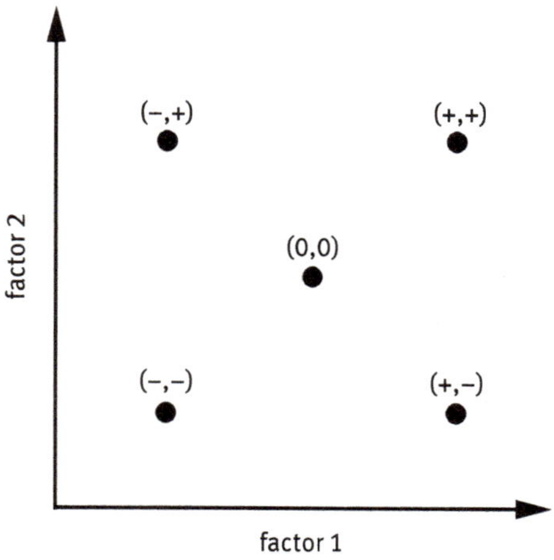

Fig. 9.5: Schematic depiction of a full factorial design of two factors at two levels: a high (+) and a low (–) level. Each point corresponds to an experiment with the given factor settings (temperature, pressure, pH, concentration of a reactant, etc.) in which the response (the quantity of interest, such as yield, ee, purity, etc.) is determined. The center point is not formally part of the full factorial design, but is often the starting point. It also allows for curvature detection of the response surface, which is vital if it is expected to have a maximum that is not an edge extreme.

$$\phi_{\text{tot}} = \phi_{\text{A}} + \phi_{\text{B}} = \frac{1\ \mu\text{L}}{(4/60)\ \text{min}} = 15\,\mu\text{L}\,\text{min}^{-1} \tag{9.1}$$

Furthermore, the ratio between the "molar rates" of A and B is

$$\frac{\phi_{\text{B}} \cdot [\text{B}]}{\phi_{\text{A}} \cdot [\text{A}]} = 2 \Rightarrow \phi_{\text{B}} \cdot [\text{B}] = 2 \cdot \phi_{\text{A}} \cdot [\text{A}] \tag{9.2}$$

From equations (9.1) and (9.2), you can proceed to calculate the flow rate of solution A by combining them:

$$\phi_{\text{A}} + \phi_{\text{B}} = \phi_{\text{A}} + \frac{2 \cdot \phi_{\text{A}} \cdot [\text{A}]}{[\text{B}]} = \phi_{\text{A}} \cdot \left(1 + \frac{2 \cdot [\text{A}]}{[\text{B}]}\right) = 15\ \mu\text{L}\,\text{min}^{-1} \tag{9.3}$$

$$\Rightarrow \phi_{\text{A}} = \frac{15}{\left(1 + \frac{2 \cdot [\text{A}]}{[\text{B}]}\right)} = \frac{15}{\left(1 + \frac{2 \cdot 0.2\text{M}}{0.4\text{M}}\right)} = \frac{15}{2} = 7.5\ \mu\text{L}\,\text{min}^{-1} \tag{9.4}$$

Now, it should be easy to have your spreadsheet calculate the flow rates of solutions B (7.5 µL min^{-1} as well) and Q (4.18 µL min^{-1}).

Using the spreadsheet, also calculate the flow rates for your other factor settings. Make sure to include an appropriate number of replicate measurements and to randomize the order of your syntheses.

9.5.3 Safety

The entire experiment should be conducted in a fume hood. All chemicals should be treated with normal precautions: look up their particular dangers in the appropriate Material Safety Data Sheet (MSDS). Be aware that DMSO increases uptake of other chemicals through the skin. Consider wearing appropriate gloves when filling, attaching, or detaching syringes. To take up liquids in the syringes used for dispensing the liquid in the experiment, consider using blunt needles. The produced diazo (diphenyl)methane is unstable and, like all diazo compounds, toxic and potentially explosive in pure form. Make sure to quench it before disposal.

9.5.4 Introductory experiment

You will have to prepare the following solutions:

A. About 20 mL of 0.2 M benzophenone hydrazone and 0.6 M DMSO (together!) in DCM
B. About 20 mL of 0.4 M trifluoroacetic anhydride (TFAA) in DCM
Q. About 10 mL of 1.435 M DIPEA or TEA in DCM

Carefully (in a fume hood!) add some droplets of the TFAA solution to a few milliliters of the mixture of **1** and DMSO, and observe. The mixture should turn red [51] for a brief moment; after this, the color disappears again. This is an interesting experimental observation: apparently, the intermediate that is formed *before* adding base (DIPEA) already has the typical color of the final product!

One of the objectives of the continuous-flow experiment is to find the settings that ensure the base is added *exactly* at the "temporal sweet spot": the moment the maximum amount of intermediate is present and can be converted into product.

9.5.5 Preparing the setup

Fig. 9.6 displays the schematic setup of the microreactor system. Ask a teacher or TA to help you setting up the real system.

Three pieces of tubing with "luer locks" for attaching syringes should be attached to the inlets for solutions A, B, and Q. A fourth piece of tubing should be attached to the reactor outlet. The other side of the tube can be put through the lid of a standard 10 mL vial in which (the lid, not the vial) a hole has been punched. This vial is for waste collection.

Fill two 5 mL syringes with solutions A and B, attach them to the tubing, and place them in the syringe pumps. Fill a 1 mL syringe with solution Q, attach it to the

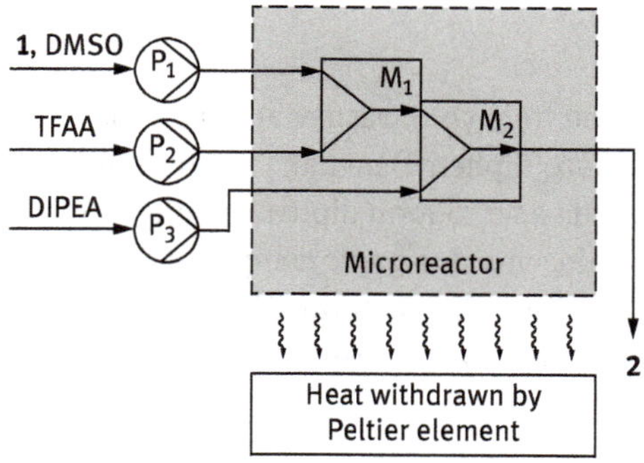

Fig. 9.6: Microreactor setup: the benzophenone hydrazone (**1**) & DMSO (A) and TFAA (B) solutions are combined and mixed in mixing channel M1 which has an internal volume of 1 μL. At M2 the DIPEA (Q) solution is added, after which the reaction mixture containing the product diazo(diphenyl) methane (**2**) leaves the microreactor.

correct tubing, and place it in the third syringe pump. Set the pumps to the correct syringe volumes. Set the temperature controller to 10 °C.

9.5.6 Carrying out the experiments

To begin with, the tubing must be filled with the solutions, so set the flow rates of all three pumps at a relatively high value (25 μL min^{-1} for instance). Start the pumps and wait for effluent to start dripping from the outlet tube. Then, set the required flow rates and start the pumps again. Wait 5 or 10 min until the effluent becomes red. Then, transfer the lid with the outlet tubing to a second vial in which 1 mL of DCM was pipetted. This is required to dilute the effluent to keep its absorbance at λ_{max} below 1. Make sure to collect effluent over several minutes (typically between 5 and 10 min should be sufficient) to "smooth out" any temporal variations in product formation.

Make sure to write down the total run time for the synthesis: if you want to know what the optimal settings are, you need to be able to compare the measured absorbance to the amount of benzophenone hydrazone that passed through the microreactor!

When you have collected enough effluent, place the lid with the end of the outlet tube on the waste vial once more, change the flow rates of the setup according to your experimental design (the order of which has preferably been randomized) and allow some time for equilibration of the system and removal of the effluent that is still in the outlet channel and corresponds to the previous factor settings. Meanwhile, transfer the collected effluent of the last synthesis run to a microcuvette and measure the absorption spectrum to determine λ_{max}, the wavelength at which absorbance is maximal. Also note down the absorbance of the solution at that wavelength. Since diazo(diphenyl) methane is an unstable compound that decomposes over time, it is important to measure the result of each combination of factors immediately after finishing the synthesis.

Wait long enough for the system to equilibrate and then start collecting effluent in another vial with a known amount of DCM in it again. Keep track of the time. When you have collected enough effluent, measure the absorbance of the diluted effluent.

Proceed to perform all other syntheses in the same manner.

9.5.7 Waste disposal and cleaning

Combine the effluents from the different runs and from your waste vial. Add some droplets of dilute acetic acid until the red color of diazo(diphenyl)methane has disappeared; diazo(diphenyl)methane will react with the acetic acid to form diphenylmethyl acetate. Dispose of the mixture in a waste canister for halogen-rich organic compounds.

After you finish your syntheses, clean the setup by each time first flushing the syringes and then the three different inlet channels with, consecutively:

- DCM
- acetone
- water
- 1 M NaOH
- water
- 1 M HCl
- water
- acetone
- air

The DCM washings should be disposed in a waste canister for halogenated organic compounds; the combined acetone washing should be disposed in a waste canister for halogen-poor solvents. The water, 1 M HCl and 1 M NaOH washings can be combined and flushed down the drain with some water.

9.5.8 Data analysis

After the experiment, analyze your data using either DoE effect estimation methods or multiple linear regression [47] in a program such as Matlab, R, or Microsoft Excel or by using a programming language such as Python. Write a short report about your findings, making sure to consider the following points:

1. Can you use the measured absorbances directly as input for your analysis? If not, what do you need to correct for? And how?
2. Calculate a pooled standard deviation for the replicate measurements.
3. Take into account the factors themselves, interactions between factors and quadratic terms.
4. Estimate which effects – if any – are significant.
5. Is there indication for curvature?
6. Can you estimate an optimum combination of factor settings and/or a direction for further inquiry? What would you advise as the next experiment or experiments to perform?

9.5.9 Extensions/variations

Instead of using the commercially available benzophenone hydrazone, you could synthesize it yourself first or even synthesize an alternative hydrazone [4, 5, 37, 52]. Acetophenone hydrazone or benzaldehyde hydrazone are obvious options. Take care with hydrazones that would lead to diazo compounds that are not resonance stabilized, for they may be very unstable and dangerous as a result!

You could also try to use your diazo compound in a subsequent reaction. The conversion into an ester by reaction with a carboxylic acid was already mentioned; similarly, an ether can be produced by reaction of **2** with an alcohol. But diazo chemistry is quite versatile; references [41–46] are good starting points to look for interesting reactions to consider.

An important drawback of the setup used in this experiment is the fact that there are only three pumps, but four reactants. This is why **1** and DMSO are dissolved together in a single solution. As a result, the ratio between TFAA and DMSO is always varied at the same time, and in the same manner, as the ratio of TFAA and **1**, so any effects due to the ratio TFAA/**1** are *confounded with* (cannot be distinguished from) effects due to the TFAA/DMSO ratio. If you have an extra pump available, as well as a T-piece or Y-piece, you can use the setup in Fig. 9.7a and separate solutions of **1** and DMSO. At the very least, this enables you to get rid of the confounding, but it also opens the possibility to consider more advanced experimental designs (or other optimization strategies), in which the flow rates of the solutions of **1** and DMSO are varied separately.

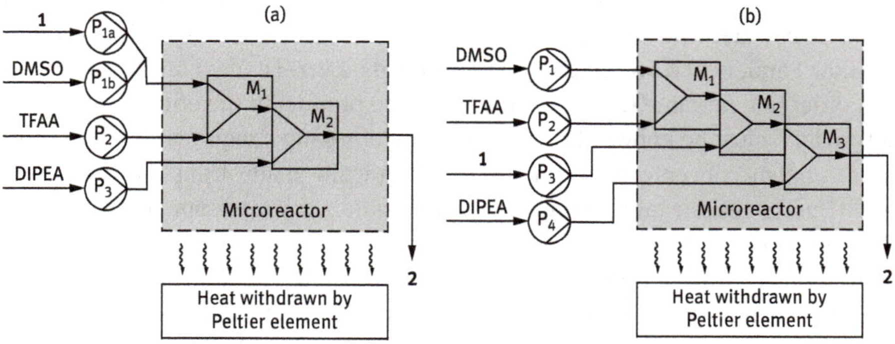

Fig. 9.7: Alternative microreactor setups. (a) With separate solutions and pumps for DMSO and **1**, removing the confounding and potentially allowing for more detailed optimization. (b) With separate solutions and pumps for all reactants as well and additionally combining the reactants in a different order. Because DMSO and TFAA react, proper mixing is required, whereas this is not strictly needed for DMSO and 1 in alternative setup (a).

Another option would be to try the setup from reference [24] (see Fig. 9.7b) in which the substrate is only added *after* the activated DMSO is produced. You can investigate the influence this has on the yield of the reaction: does it produce better results or doesn't it?

9.6 Results and discussion

Depending on the parameter settings chosen for the experimental design, the student experiment results in a limited number of measurements that indicate the direction of the optimal parameter settings and an estimate of the model describing the (relative) yield/conversion of the reaction as a function of the parameters. To provide a benchmark, a full characterization of the response surface of the relative conversion of **1** into **2** as a function of the two factors "residence time" and "TFAA/**1** ratio" (confounded with TFAA/DMSO ratio) was carried out. The results of this benchmark experiment can be compared with the student results.

The "visible" absorption spectrum of the product mixture shows a maximum around $\lambda_{max} = 521$ nm. Absorbance measurements to determine the relative conversion of **1** into **2** are carried out at that wavelength. In order to obtain good results, effluent was collected over some time to allow temporal variations in the conversion to average out. For each experiment, the effluent was collected in a glass vial over a precalculated time interval that was chosen such that the total amount of **1** that passed through the system was always equal. Afterward, either 0.5 or 1.0 mL DCM was added. In this way, the volume of the colored liquid became large enough to fill a semi-micro cuvette, while at the same time, its absorbance became sufficiently low to stay below 1 and, hence, was in the regime where the Beer–Lambert law is valid.

In order to determine the concentration of the product in the effluent, its molar absorptivity ε must be known. Unfortunately, determining ε requires standard solutions of known concentration and these are typically made from the pure compound. Due to the unstable nature of **2**, purification and assessment of purity is difficult (see below). Therefore, ε is as yet unknown and the absorbance can only be used to indicate concentrations *relative* to each other. Apart from that, the absorbance alone is not a good measure of the conversion, however, since – to a certain extent – increasing the amount of **1** by increasing flow rate 1 (Fig. 9.6) or decreasing the total volume of liquid by decreasing flows 2 and/or 3 may also result in higher concentrations of product in the effluent, even when *relatively* less **1** is converted into product. As an example, when the absorbance would increase 1.5-fold when flow rate 1 is doubled, the *relative* conversion of **1** into **2** is actually decreased to $2/1.5 = 0.75$ that of the previous setting.

Therefore, the measured absorbances were normalized by dividing them by *the theoretical concentration of* ***2*** *in case of total conversion* (or, the theoretical concentration of **1** in case of no conversion at all, which would amount to the same). The remaining differences in the corrected absorbance are either due to an actual effect (i.e., an actual change in the conversion) or to factors beyond control (i.e., experimental error).

Figure 9.8 shows two depictions of the response surface corresponding to the relative conversion of **1** into **2**. Some replicate measurements produced values that were suspected to be outliers. To reduce their influence in subsequent data fitting, the median values of the corrected absorbances were calculated and used as input for the model fit. It was also realized that the [TFAA]/[**1**] axis would be asymmetrical: starting from an equimolar situation (i.e., [TFAA]/[**1**] = 1), and *increasing* [TFAA] by a factor of 2 leads to [TFAA]/[**1**] = 2, whereas *decreasing* it with a factor of 2 leads to [TFAA]/[**1**] = 0.5. To correct for this, the natural logarithm of [TFAA]/[**1**] was used, so as to render the axis symmetrical around 0. This resulted in better

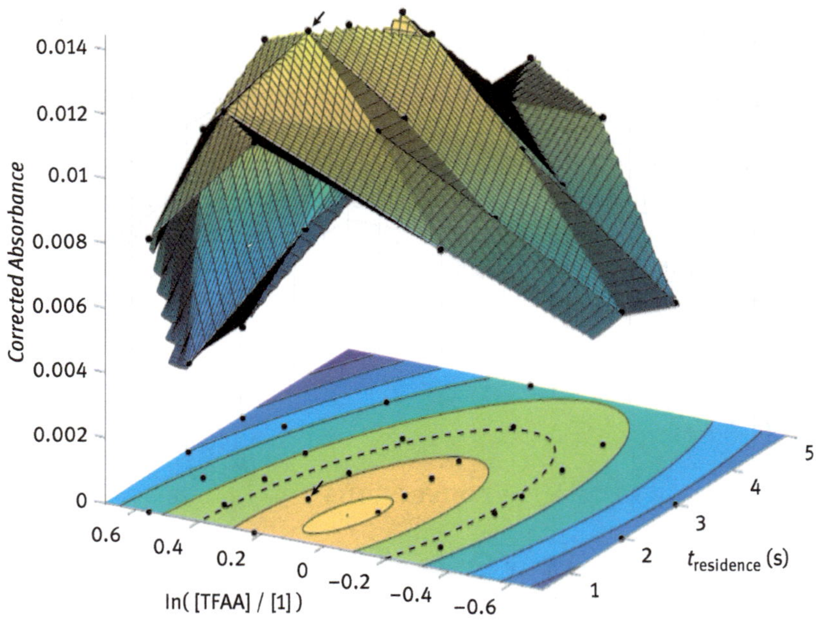

Fig. 9.8: Two depictions of the response surface of the corrected absorbance as a measure for relative conversion of **1** into **2**. The 3D surface simply "connects the dots", which are the median values of replicate measurements of the corrected absorbance for different combinations of $t_{residence}$ and ln([TFAA]/[**1**]). The contour lines at the bottom show the fit of the quadratic model to the data. The distance between the contour lines corresponds to steps of 0.02 in the *z* direction, but the innermost contour line is added for extra detail at a distance of 0.006 from the previous one. The dashed contour line corresponds to a distance of 1 pooled standard deviation – calculated from all replicate measurements – below the maximum. The arrows indicate the point that yielded optimal experimental results.

fits (lower RMS) than for the asymmetrical situation. The data were fit to the model $z = a + bx + cy + dxy + ex^2 + fy^2$, in which z is the corrected absorbance, x is the reaction time $t_{\text{residence}}$ (s) and y is $\ln([\text{TFAA}]/[\mathbf{1}])$ The resulting model was: $z = 0.012 + 7.115 \cdot 10^{-4} \cdot x + 0.004 \cdot y - 0.001 \cdot xy - 2.376 \cdot 10^{-4} \cdot x^2 - 0.015 \cdot y^2$. Although the fit of the quadratic model to the data is far from perfect, the addition of higher order terms resulted in overfitting, whereas the exchange of the quadratic terms for higher order terms resulted in fits that were actually worse, although the location of the optimum hardly changed. This observation conforms to the general one made in [47] that in experimental optimization, models do not normally benefit from incorporating terms beyond quadratic ones. For the given concentrations and the set-up used, at a temperature of 10 °C, the model predicts the optimal combination of factors to be around $t_{\text{residence}} = 1.75$ s and $\ln([\text{TFAA}]/[\mathbf{1}]) = 0.065$. Experimentally, optimal results were obtained at $t_{\text{residence}} = 1.5$ s and $\ln([\text{TFAA}]/[\mathbf{1}]) = 0.2231$.

9.6.1 Purification

Although assessing the yield and purity of diazo compounds is notoriously difficult [5, 53], an attempt was made to purify an amount of continuous-flow product for further characterization by NMR and IR spectroscopy. The small amount of product was obtained as a crystalline solid, and the NMR spectra did not show any starting material or major impurities. The NMR and IR spectra corresponded to literature [5, 54]. We conclude that the purified product should in principle be clean enough for further use in synthetic applications.

9.7 Experimental

9.7.1 Determination of relative conversions

Microreactor setup: All flow chemistry experiments were conducted using a FutureChemistry b330 FlowStart education setup, consisting of three syringe pumps, a temperature controller and a microreactor holder equipped with a Peltier element. The microreactor was a FutureChemistry "Short Quench" microreactor consisting of borosilicate glass, with an internal volume (measured between the point where solutions 1 and 2 are mixed and the point where solution 3 is added) of 1 µL. The syringes used for dispensing the reactant solutions were two gas-tight 5 mL borosilicate glass syringes and one gas-tight 1 mL borosilicate glass syringe (Integrated Lab Solutions GmbH). Connections between syringes and the microreactor consisted of FEP tubing, 1/16″ O.D., 0.50 mm I.D. (IDEX art. Nr. 1548 L).

All experiments were conducted with solutions in DCM (Fisher Scientific, purity ≥99%) made from the chemicals as obtained from the manufacturers without further purification or drying: TFAA (Alfa Aesar, ≥99%): 0.4 M; dimethyl sulfoxide (VWR Chemicals, technical grade): 0.6 M, together with benzophenone hydrazone (Acros Organics, purity ≥98%): 0.2 M; DIPEA (Fisher Scientific, purity ≥98%): 1.435 M. The solution of dimethyl sulfoxide and benzophenone hydrazone was freshly prepared on a daily basis, since after a day of standing it became slightly opaque and a water-soluble precipitate formed at the bottom of the flask.

UV analysis: All quantitative Vis spectroscopic measurements for determination of relative conversions were performed off-line on a Shimadzu UV-1800 UV–visible spectrophotometer using a semi-micro quartz cuvette with a pathlength of 1 cm.

9.7.2 Purification trial

For the "purification trial" of **2**, fresh bottles of the same chemicals were used, with the exception of: dimethyl sulfoxide (Acros Organics, purity ≥99.7%, "Extra Dry") and benzophenone hydrazone, which was recrystallized from a minimum volume of hot ethanol and dried in vacuo at 35 °C.

To the collected product solution in DCM (approximately 20 mL) an equal volume of pentane was added. The resulting organic phase was subsequently washed with saturated sodium bicarbonate solution (1 × 20 mL), water (3 × 20 mL), and brine (2 × 20 mL), and dried over sodium sulfate. The solvent was removed in vacuo at room temperature. The residual dark purple-red oil was dissolved in a minimal amount of pentane and filtered over basic alumina (approx. 3 cm). The solids were rinsed with pentane until the filtrate was colorless. The filtrate was evaporated in vacuo at room temperature yielding the product as a dark purple-red crystalline solid.

All NMR spectra were collected on a Bruker 300 MHz Avance III HD spectrometer equipped with a BBO probe at 298 K with a ^{1}H frequency of 300.13 MHz and a ^{13}C frequency of 75.48 MHz. Tetramethylsilane ($\delta = 0.00$ ppm) or the residual protons of the deuterated solvent were used as internal references. ^{1}H–^{13}C HSQC spectra were acquired using a 2,870.6 Hz spectral width in F2 and 12,500 Hz spectral width in F1 using 512 × 256 points and processed to 512 × 512 points, 4 scans per increment, relaxation delay of 1.5 s and 1-bond $J_{CH} = 145$ Hz. ^{1}H–^{13}C HMBC spectra were acquired using a 2,873.6 Hz spectral width in F2 and 16,778.5 Hz spectral width in F1 using 512 × 128 points and processed to 512 × 512 points, 8 scans per increment, a relaxation delay of 1.5 s and long-range $J_{CH} = 8$ Hz. Numbers of protons and carbons correspond to the numbers in Fig. 9.1.

^{1}H NMR (CD_2Cl_2, 300 MHz): δ_H 7.45–7.37 (4H, m, H3, H5, H10, and H12), 7.35–7.29 (4H, m, H2, H6, H9, and H13), 7.24–7.17 (2H, m, H4, and H11). ^{13}C{^{1}H} NMR (CD_2Cl_2, 75 MHz): δ_C 130.0 (C1 and C8), 129.5 (C3, C5, C10, and C12), 126.0 (C4 and C11), 125.5 (C2,

C6, C9, and C13), 62.7 (C7). FT-IR (diamond ATR): 3,057*w* (br, aromatic C–H), 3,030*w* (br, aromatic C–H), 2,032*s* (diazomethane N–N stretch), 1,592*m* (aromatic C–C), 1,492*m* (C–C), 1,262*w* (C–N), 1,075*w*, 1,030*w*, 747*s*, 688*s*, 649*m*, 472*m*.

9.7.3 Safety

All experiments should be conducted in a fume hood. All chemicals should be treated with normal precaution: we teach our students to look up the safety information for all chemicals and products in the appropriate MSDS themselves. They also have to come up with ways to quench and dispose any resulting reaction mixtures, products, solvents, and so on. This information is discussed with a teacher or TA prior to performing the experiment.

For this experiment in particular, note that DMSO increases uptake of other chemicals through the skin. The produced diazo(diphenyl)methane is unstable and, like all diazo compounds, toxic and potentially explosive in purified form; complete purification is not recommended for undergraduate students. The produced diazo compound can be quenched using small amounts of dilute acetic acid or another dilute acid.

Consider having students wear appropriate gloves when filling, attaching, or detaching syringes. To take up liquids in the syringes used for dispensing the liquid in the experiment, we used blunt needles (Sterican MIX 1.20 × 40 mm, B. Braun Melsungen AG).

References

[1] Blanco-Ania, D, Rutjes, FPJT, Continuous-flow chemistry in chemical education, J Flow Chem, 2017, 7(3–4), 157–158.

[2] Van Summeren, LTCG, Gerretzen, J, Rutjes, FPJT, Bloemberg, TG, Optimization of continuous-flow diphenyldiazomethane synthesis: an integrated undergraduate chemistry experiment, J Flow Chem, 2020, In Press.

[3] Nieuwland, PJ, Koch, K, Van Harskamp, N, Wehrens, R, Van Hest, JCM, Rutjes, FPJT, Flash chemistry extensively optimized: high-temperature Swern-Moffatt oxidation in an automated microreactor platform, Chem-Asian J, 2010, 5(4), 799–805.

[4] Javed, MI, Brewer, M, Diazo preparation via dehydrogenation of hydrazones with "activated" DMSO, Org Lett, 2007, 9(9), 1789–1792.

[5] Javed, MI, Brewer, M, Diphenyldiazomethane, Org Synth, 2008, 85, 189–195.

[6] Maas, G, New syntheses of diazo compounds, Angew Chem Int Edit, 2009, 48(44), 8186–8195.

[7] Hetterscheid, DGH, Hendriksen, C, Dzik, WI, Smits, JMM, Van Eck, ERH, Rowan, AE, et al., Rhodium-mediated stereoselective polymerization of "carbenes", J Am Chem Soc, 2006, 128(30), 9746–9752.

[8] Delville, MME, Van Hest, JCM, Rutjes, FPJT, Ethyl diazoacetate synthesis in flow, Beilstein J Org Chem, 2013, 9, 1813–1818.

[9] Dommerholt, J, Schmidt, S, Temming, R, Hendriks, LJA, Rutjes, FPJT, Van Hest, JCM, et al., Readily accessible bicyclononynes for bioorthogonal labeling and three-dimensional imaging of living cells, Angew Chem Int Ed, 2010, 49(49), 9422–9425.

[10] Kornblum, N, Jones, WJ, Anderson, GJ, A new and selective method of oxidation – the conversion of alkyl halides and alkyl tosylates to aldehydes, J Am Chem Soc, 1959, 81(15), 4113–4114.

[11] Kornblum, N, Powers, JW, Anderson, GJ, Jones, WJ, Larson, HO, Levand, O, et al., A new and selective method of oxidation, J Am Chem Soc, 1957, 79(24), 6562.

[12] Pfitzner, KE, Moffatt, JG, A new and selective oxidation of alcohols, J Am Chem Soc, 1963, 85(19), 3027–3028.

[13] Pfitzner, KE, Moffatt, JG, Sulfoxide-carbodiimide reactions. II. Scope of the oxidation reaction, J Am Chem Soc, 1965, 87(24), 5670–5678.

[14] Pfitzner, KE, Moffatt, JG, Sulfoxide-carbodiimide reactions. I. A facile oxidation of alcohols, J Am Chem Soc, 1965, 87(24), 5661–5670.

[15] Fenselau, AH, Moffatt, JG, Sulfoxide-carbodiimide reactions. III. Mechanism of the oxidation reaction, J Am Chem Soc, 1966, 88(8), 1762–1765.

[16] Tojo, J, Fernández, M, Tojo J, editor. Oxidation of Alcohols to Aldehydes and Ketones – A Guide to Current Common Practice, New York, Springer Science+Business Media, Inc, 2006.

[17] Mancuso, AJ, Huang, SL, Swern, D, Oxidation of long-chain and related alcohols to carbonyls by dimethyl-sulfoxide activated by oxalyl chloride, J Org Chem, 1978, 43(12), 2480–2482.

[18] Huang, SL, Omura, K, Swern, D, Oxidation of sterically hindered alcohols to carbonyls with dimethyl sulfoxide-trifluoroacetic anhydride, J Org Chem, 1976, 41(20), 3329–3331.

[19] Omura, K, Sharma, AK, Swern, D, Dimethyl sulfoxide-trifluoroacetic anhydride – new reagent for oxidation of alcohols to carbonyls, J Org Chem, 1976, 41(6), 957–962.

[20] Omura, K, Swern, D, Oxidation of alcohols by activated dimethyl-sulfoxide – preparative steric and mechanistic study, Tetrahedron, 1978, 34(11), 1651–1660.

[21] Albright, JD, Goldman, L, Dimethyl sulfoxide-acid anhydride mixtures. New reagents for oxidation of alcohols, J Am Chem Soc, 1965, 87(18), 4214–4216.

[22] Albright, JD, Goldman, L, Dimethyl sulfoxide-acid anhydride mixtures for oxidation of alcohols, J Am Chem Soc, 1967, 89(10), 2416–2423.

[23] Huang, SL, Omura, K, Swern, D, Further studies on oxidation of alcohols to carbonyl-compounds by dimethyl sulfoxide-trifluoroacetic anhydride, Synthesis-Stuttgart, 1978, 1978(4), 297–299.

[24] Kawaguchi, T, Miyata, H, Ataka, K, Mae, K, Yoshida, J, Room-temperature swern oxidations by using a microscale flow system, Angew Chem Int Edit, 2005, 44(16), 2413–2416.

[25] Van Der Linden, JJM, Hilberink, PW, Kronenburg, CMP, Kemperman, GJ, Investigation of the moffatt-swern oxidation in a continuous flow microreactor system, Org Process Res Dev, 2008, 12(5), 911–920.

[26] Keirs, D, Overton, K, Conversion of amines into imines by swern oxidation, J Chem Soc Chem Comm, 1987, 1987(21), 1660–1661.

[27] Jinbo, Y, Kondo, H, Taguchi, M, Sakamoto, F, Tsukamoto, G, Synthesis of new DNA gyrase inhibitors – application of the DMSO oxidation to the conversion of the amine into the imine, J Org Chem, 1994, 59(20), 6057–6062.

[28] Gentilucci, L, Grijzen, Y, Thijs, L, Zwanenburg, B, Convenient synthesis of optically-Active 2h-Azirine-2-carboxylic esters by swern oxidation of aziridine-2-carboxylic esters, Tetrahedron Lett, 1995, 36(26), 4665–4668.

[29] Kamal, A, Rao, NV, A New Route for the Synthesis of Pyrrolo[2,1-c][1,4]benzodiazepine Antibiotics via Oxidation of Cyclic Secondary Amine, Chem Commun, 1996, 1996(3), 385–386.

[30] Torssell, K, Mechanisms of dimethylsulfoxide oxidations, Tetrahedron Lett, 1966, 7(37), 4445–4451.

[31] Tran, DN, Battilocchio, C, Lou, SB, Hawkins, JM, Ley, SV, Flow chemistry as a discovery tool to access sp(2)-sp(3) cross-coupling reactions via diazo compounds, Chem Sci, 2015, 6(2), 1120–1125.

[32] Movsisyan, M, Delbeke, EIP, Berton, JKET, Battilocchio, C, Ley, SV, Stevens, CV, Taming hazardous chemistry by continuous flow technology, Chem Soc Rev, 2016, 45(18), 4892–4928.

[33] Hock, KJ, Koenigs, RM, The generation of diazo compounds in continuous-flow, Chem-Eur J, 2018, 24(42), 10571–10583.

[34] Goddard-Borger, ED, Stick, RV, An efficient, inexpensive, and shelf-stable diazotransfer reagent: Imidazole-1-sulfonyl azide hydrochloride, Org Lett, 2007, 9(19), 3797–3800.

[35] Fischer, N, Goddard-Borger, ED, Greiner, R, Klapötke, TM, Skelton, BW, Stierstorfer, J, Sensitivities of some imidazole-1-sulfonyl azide salts, J Org Chem, 2012, 77, 1760–1764.

[36] Holzwarth, MS, Alt, I, Plietker, B, Catalytic activation of diazo compounds using electron-rich, defined iron complexes for carbene-transfer reactions, Angew Chem Int Ed, 2012, 51, 5351–5354.

[37] Holton, T, Shechter, H, Advantageous synthesis of diazo compounds by oxidation of hydrazones with lead tetraacetate in basic environments, J Org Chem, 1995, 60, 4725–4729.

[38] Wu, -L-L, Ge, Y-C, He, T, Zhang, L, Fu, X-L, Fu, H-Y, Chen, H, Li, R-X, An efficient one-pot synthesis of 3,5-disubstituted 1H-Pyrazoles, Synthesis, 2012, 44(10), 1577–1583.

[39] Tuktarov, AR, Korolev, VV, Khalilov, LM, Ibragimov, AG, Dzhemilev, YM, Catalytic cyclopropanation of fullerene[60] with diazomethane, Russ J Org Chem, 2009, 45(11), 1594–1597.

[40] De Boer, TJ, Backer, HJ, Diazomethane, Org Synth, 1956, 36, 16.

[41] García-Ruano, JL, Alonso-de Diego, SA, Blanco-Ania, D, Martín-Castro, AM, Martín, MR, Rodríguez-Ramos, JH, (*Z*)-3-p-tolylsulfinylacrylonitrtiles as Chiral Dipolarophiles: Reactions with Diazoalkanes, Org Lett, 2001, 3(20), 3173–3176.

[42] He, S, Chen, L, Niu, Y-N, Wu, L-Y, Liang, Y-M, 1,3-dipolar cycloaddition of diazoacetate compounds to terminal alkynes promoted by Zn(OTf)2: An efficient way to the preparation of pyrazoles, Tetrahedron Lett, 2009, 50(20), 2443–2445.

[43] Kirmse, W, 100 years of the wolff rearrangement, Eur J Org Chem, 2002, 2193–2256.

[44] Davies, HML, Manning, JR, Catalytic CH Functionalization by Metal Carbenoid and Nitrenoid Insertion, Nature, 2008, 451, 417–424.

[45] Xiao, Q, Zhang, Y, Wang, J, Diazo compounds and N-tosylhydrazones: Novel cross-coupling partners in transition-metal-catalyzed reactions, Acc Chem Res, 2013, 46(2), 236–247.

[46] Blanco-Ania, D, Maartense, L, Rutjes, FPJT, Rapid production of *trans*-cyclooctenes in continuous flow, ChemPhotoChem, 2018, 2, 898–905.

[47] Massart, DL, Vandeginste, BGM, Buydens, LMC, De Jong, S, Smeyers-Verbeke, J (1997) Handbook of Chemometrics and Qualimetrics Part A. Massart DL, editor.

[48] Leardi, R, Experimental design in chemistry: A tutorial, Anal Chim Acta, 2009, 652(1–2), 161–172.

[49] Santos, A, Barbosa-Póvoa, A, Carvalho, A, Life cycle assessment in chemical industry – a review, Curr Opin Chem Eng, 2019, 26, 139–147.

[50] Kleinekorte, J, Fleitmann, J, Bachmann, M, Kätelhön, A, Barbosa-Póvoa, A, Von Der Assen, A, Bardow, A, Life cycle assessment for the design of chemical processes, products, and supply chains, Ann Rev Chem Biomol Eng, 2020, 11, 203–233.

[51] Of course, if you made a mistake in preparing your solutions, no red color will be visible. We have noticed however, that even when taking great care in preparing the solutions, in ~1 out

of 10 cases the reaction produces no product at all; preparing the **1**/DMSO solution anew in – as far as we are aware – exactly the same manner and restarting the reaction then results in product formation again. In our own experience with this particular experiment, a 10% failure (a 90% success rate, and with a simple remedy for the remaining 10%: just make one or two stock solutions again or use someone else's) is not prohibitive at all, but sometimes just delays the commencement of the experiment proper, by ~15 min.).

[52] Patrick, TB, Flory, PA, Direct Fluorination of aryl ketone hydrazones, J Fluorine Chem, 1984, 25(2), 157–164.

[53] Andrews, SD, Day, AC, Raymond, P, Whiting, MC, 2-Diazopropane, Org Synth, 1970, 50, 27.

[54] Yates, P, Shapiro, BL, Aliphatic Diazo Compounds. 3. Infrared Spectra, J Am Chem Soc, 1957, 79(21), 5756–5760.

Haruro Ishitani and Shu Kobayashi

10 Continuous flow catalysis

10.1 Introduction

Continuous-flow synthesis has attracted considerable attention in both academic research and industry because it is recognized as a key technology for the next-generation chemical production [1]. In comparison to batch systems, flow systems offer great advantages to the chemical industry in terms of sustainability, efficiency, and safety. Most research in synthetic organic chemistry has long been conducted using batch methods. The batch approach is currently, by far, the most commonly used synthetic method in almost all laboratories in academia and industry. For example, the production of fine chemicals, such as active pharmaceutical ingredients (APIs), agrochemicals, electronic chemicals, and fragrances, are mostly carried out by repeating-batch methods. On the other hand, for several other industrial products, continuous manufacturing is an important technique; it is used as a framework for the production of automobiles, electronic devices, steel, and food. For establishing continuous manufacturing in the chemical industry, continuous-flow systems operated at steady state are vital to achieve on-demand production of chemicals. In general, however, synthesis by flow methods seems to be more difficult than batch methods. It is considered that synthesis by flow methods may be applicable to chemical products through simple processes, as often seen in the petroleum industry, and that they are difficult to apply to the synthesis of more complicated molecules such as APIs. Nonetheless, in chemical industries, the number of continuous-flow methods in use has gradually begun to increase, even in fine chemicals industries, including the pharmaceutical industry. At the same time, investigations into organic reactions, including catalytic reactions using flow methods at the laboratory level, have become increasingly popular over the past decades [2].

10.2 Overview of catalytic processes in flow organic synthesis

Continuous-flow systems in synthetic organic chemistry can be classified into four different categories based on the styles of reactions carried out under flow conditions (Fig. 10.1) [3].

The most straightforward case is when all reagents are continuously fed into tubular reactors (consisting of a hollow tube or pipe through which reactants flow) and the product is continuously collected (Type I). Most homogeneous liquid-phase reactions will be appropriate for this process without the requirement of using special

https://doi.org/10.1515/9783110693676-010

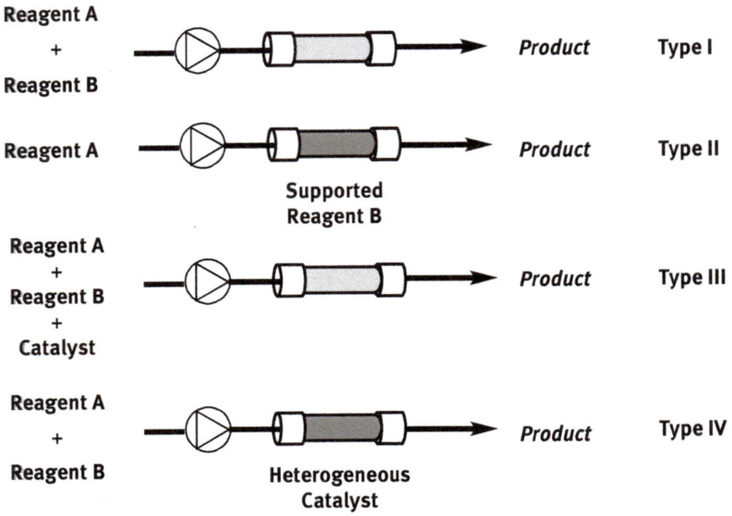

Fig. 10.1: Classification of flow reactions in fine organic synthesis.

techniques, whilst using rapid reactions enables one to draw maximum advantage of this process. This approach has, therefore, been developed mainly as microflow chemistry. Various remarkable milestones have been achieved to date. Several difficulties normally associated with the corresponding batch processes have been overcome, especially pertaining to the use of highly reactive and short-lived reagents or intermediates, such as aryl metal reagents and benzynes. Safer experimental systems are now available when one needs to handle hazardous, toxic, explosive, or flammable chemicals. However, unreacted starting materials and by-products are along with the product/s, and quenching and work-up processes are then required.

Type I flow reaction requires that all reagents, including catalysts, must dissolve in a reaction medium as a matter of course; however, this sometimes narrows the window of applicable reactions, especially reactions that use inorganic salt reagents. This type of insoluble reagents can be used directly in column reactors, although their immobilization and the use of the immobilized reagents with column flow reactors can reduce the coproduction of by-products (Type II flow reaction). Several solid support reagents are used in this Type II flow reaction. This is often evident in the telescoped multistep synthesis of target compounds. The drawback of this approach is that overreactions may occur and when the supported reagent is consumed, the reagent-packed reactor must be changed.

The other two reaction types are catalytic flow reactions. Because modern organic chemistry is heavily reliant upon catalysis to improve the efficiency and selectivity of reactions, flow reactions can be conducted by the feeding of homogeneous catalysts with substrates (Type III). The use of homogeneous catalysts for the desired reactions is a straightforward approach to achieve catalytic processes under flow. The usability of homogeneous catalysts for this process and the sustainability of product quality during the entire process are great advantages; however, quenching of catalytic reactions and processes to remove the catalyst residues is unavoidable, and this interrupts the multistep flow reactions from proceeding in a seamless fashion. Moreover, the advantages offered by the use of homogeneous catalytic reactions in batch systems may be limited in flow systems because the reaction rates of catalytic reactions are usually slow.

On the other hand, the use of heterogeneous catalysts with column-type fixed-bed reactors (catalyst-packed reactors) can be categorized as Type IV flow reactions. Compared with the similar catalytic flow Type III reaction, Type IV offers obvious advantages. One advantage is the elimination of any catalyst invalidation processes, and contamination of the reactions' output by catalyst residues can be avoided. From the viewpoint of the construction of a multistep sequential continuous-flow production of high-value-added compounds, Type IV is ideal. Another important aspect here is the enhancement of the catalytic efficiency in Type IV catalytic flow reactions. The [Substrate, S]/[Catalyst, C] ratio in a batch system changes with the progress of the reaction. High [S]/[C] in the initial stage is disadvantageous due to the undesirable interaction between the excess substrate and the catalyst, and low [S]/[C] in the later stage causes a decrement of the frequency of contact between the two. Hence, most catalytic reactions in batch require long reaction times to run to completion. In the case of Type IV catalytic flow reactions, reactants dissolved in the mobile phase are supplied to a catalyst-packed reactor, where a large amount of catalytic species exist. This is in clear contrast to the Type III flow with homogeneous catalysts as well as batch homogeneous/heterogeneous catalytic reactions, and clear evidence of reducing the reaction time to reaction completion within the residence time (Fig. 10.2).

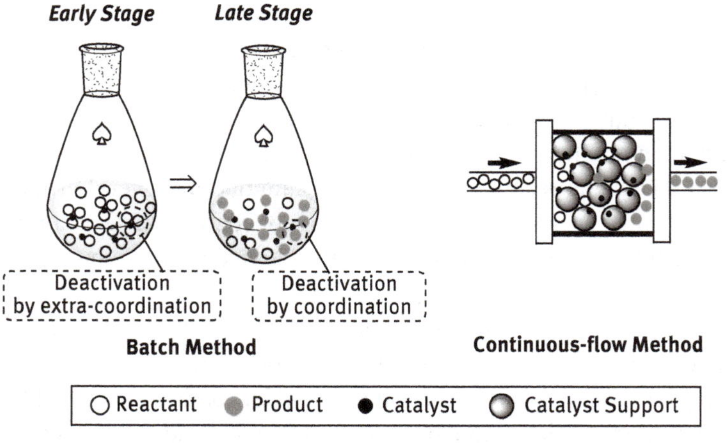

Fig. 10.2: Batch method and continuous-flow method; from the viewpoint of catalyst.

To date, a lot of effort has been spent and remarkable advances are seen in the field of heterogeneous catalysis, including fundamental research and industrial processes. With regard to the use of catalysis for fine organic synthesis into heterogeneous reactions, there still exists a great hurdle in investigation to find more efficient, stable, and reliable way, especially in continuous-flow systems. Highly active and stable heterogeneous catalysts allow making the flow system with high productivity and long life-time processes.

Progress in the Type I flow reaction is closely related to chemical engineering. Typical process parameters, such as temperature and pressure, affect the output of a reaction greatly; thus, differences in optimal reaction conditions between batch processes and flow processes can often be seen in literatures. The same is also true in catalytic flow reactions; however, the relatively slow reaction rate in such catalytic reactions motivates ones to start their investigation by just following the same reaction as in the batch system.

In the following sections, examples of catalytic flow processes, including recent achievements, will be described for each type of reaction (Types III and IV). Reactions taken up in the following part are all fundamental and general organic reactions, frequently seen in the research field of synthetic organic chemistry.

i *(For further details on this issue please see Volume 1, Chapter 1, Title: Fundamentals of flow chemistry)*

10.3 Acid/base-catalyzed bond-forming reactions

10.3.1 1,2- and 1,4-addition reactions

In the 1990s, results of early investigations into the catalytic flow carbon–carbon reactions were reported. Polymer-supported reagents/ligands were developed. A polystyrene-based chiral amino alcohol 1, used as a Lewis base catalyst for the addition of dialkylzinc to an aldehyde, was developed and used as a catalyst-packed reactor (Fig. 10.3) [4]. The same reaction was further examined by several groups using soluble/insoluble polymer-based amino alcohols under continuous-flow conditions. In 1999, a successful example was created; it demonstrated a continuous 275 h operation and satisfactory enantioselectivity [5].

As is evident in the early stage examples, catalysts generated by organic molecules are good motifs for demonstrating Type IV flow reactions [6]. One of the most fundamental and simple cases is the solid base-catalyzed aldol condensation of aldehydes and ketones, and Knoevenagel condensation of aldehydes and 1,3-dicarbonyl compounds. Solid base-catalyzed condensation reactions (e.g., with basic metal oxides, KF/alumina, and hydrotalcites) have been extensively investigated; however,

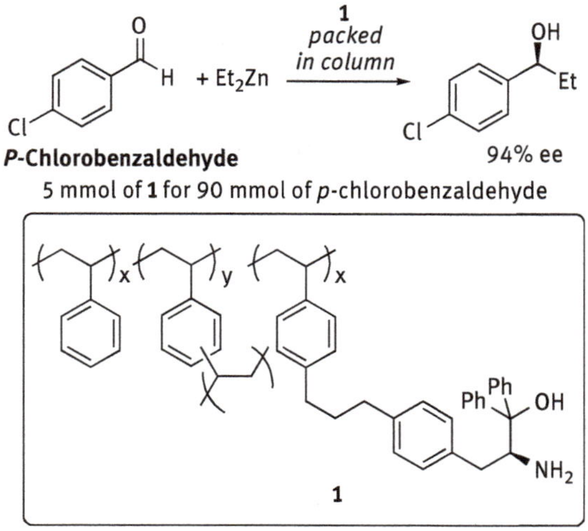

Fig. 10.3: Continuous-flow 1,2-addition using diethylzinc.

examples under continuous-flow mode are rather limited. In 2016, aldol condensation in the Type IV flow system was reported, where commercially available quaternary ammonium hydroxide resins (**2**) were used (Fig. 10.4) [7]. The target, φ-ionone, was obtained in good to high yield under the conditions of weight hourly space velocity (WHSV) = 7.90 $g_{feed}\ h^{-1}\ g_{catalyst}{}^{-1}$, using a nine-fold quantity of pronucleophile and acetone. Here, WHSV, one of the indexes of conditions of a flow reaction, means the weight ratio of the feed to catalyst per unit period. In this study, a 0.2 M 2-propanol solution of citral containing 9 equiv. acetone was fed at 1 mL min^{-1} flow rate into the catalyst-packed reactor containing 6.0 g of resin catalyst (Fig. 10.4). Later, the same type of aldol condensation of α-tetralone with a small excess of benzaldehyde was investigated (Fig. 10.5) [8]. In this study, typical solid/heterogeneous basic catalysts were first screened in batch mode, and it was determined that the quaternary ammonium hydroxide resin (**2**) exhibited the highest activity for the reaction. Here, a 0.1/0.105 M solution of the substrates was fed at 0.15 mL min^{-1} flow rate into a 10 mm × 100 mm stainless reactor, packed with 7.45 meq resin catalyst. The substrate/OH^- molar ratio per hour, calculated to be 0.12 h^{-1}, was rather low, but a 80–90% yield of the desired product was maintained over 100 h.

Knoevenagel condensation reactions, in which, tertiary amine-bound silica is used as a catalyst have also been reported. A 0.13 M toluene solution of the substrate was fed at 0.1 mL min^{-1} flow rate into a 2.4 mm × 6.6 mm reactor, packed with 1.364 g (1.5 mmol amine) silica catalyst. The overall yield, after 18.5 h, was 90% [9]. The substrate/amine molar ratio per hour was calculated to be 0.52 h^{-1}, meaning that the turnover number (TON) of the catalyst reached 10 after the 18.5 h operation. In a batch system, the use of 10 mol% catalyst for the same reaction required 20 h to reach 80% conversion, indicating a higher efficiency of the catalyst in the flow

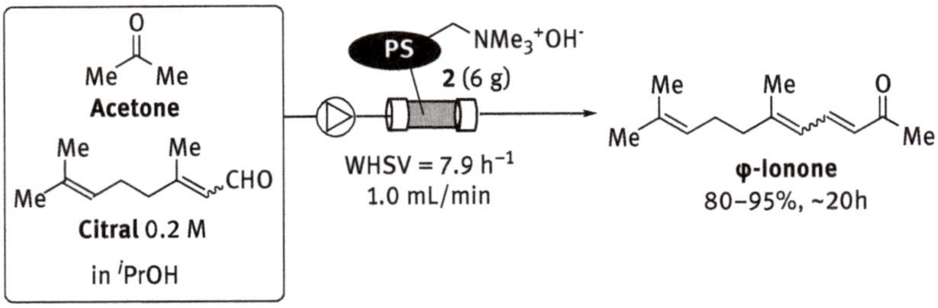

Fig. 10.4: Continuous-flow aldol condensation of citral and acetone.

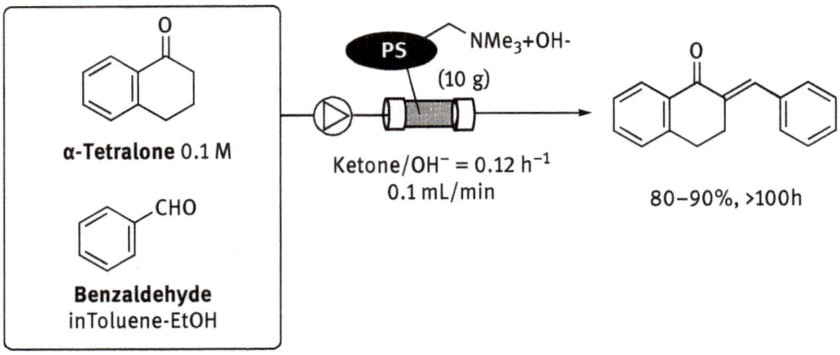

Fig. 10.5: Continuous-flow aldol condensation of α-tetralone and benzaldehyde.

mode. The applicability of solid/heterogeneous basic materials in the Knoevenagel condensation was examined by Ishitani et al. [10]. High activity of weakly basic primary amine-functionalized silica gel for the condensation between aldehyde and malonate was determined. Flow reaction between a variety of aldehydes and dimethyl malonate with an ~ 0.46 h^{-1} substrate/-NH_2 molar ratio proceeded well, to afford the desired vinylidene malonate in good to high yield, over >24 h (Fig. 10.6).

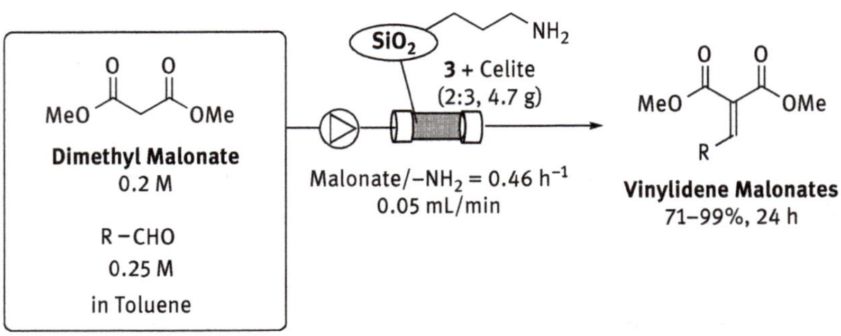

Fig. 10.6: Continuous-flow Knoevenagel condensation using amine-functionalized silica gel.

Supported chiral organocatalysts are useful tools for performing enantioselective flow reactions [6]. Examples of three-component enantioselective Mannich-type reactions under continuous-flow conditions, in which, immobilized natural primary amino acid catalysts (**4**) are used have been reported [11, 12]. A 0.5 M solution of the substrate was fed into the reactor with 300 mg of the supported catalyst (corresponding to 0.31 mmol of the amino acid). The reaction's progress was monitored at the exit of the reactor using in-line IR to identify the optimal flow rate. The best conversion was achieved at a flow rate of 0.03 mL min^{-1}, which corresponds to 3.0 h^{-1} substrate/amino acid molar ratio and 17 min residence time. After a 6 h operation, 3.31 mmol of the desired amino ketone derivative was obtained with satisfactory stereoselectivity (*anti*/*syn* = 88/12, 89% ee) [13]. Here, it was highlighted that the TON had doubled than under batch conditions (Fig. 10.7).

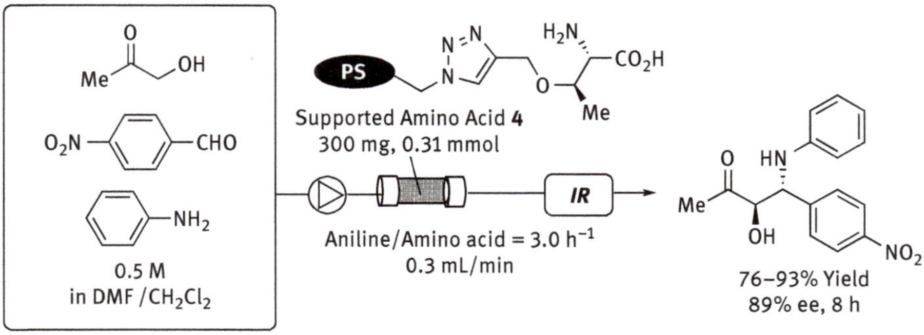

Fig. 10.7: Supported amino acid-catalyzed continuous-flow Mannich-type reaction.

BINOL-derived chiral phosphoric acid catalysts have also been used in organic acid-catalyzed continuous-flow reactions. Enantioselective 1,2-addition reactions of indoles to imines were examined; 6 h continuous-flow reaction with approximately 21 h^{-1} substrate/phosphoric acid molar ratio afforded the corresponding product in 3.6 g with 94% ee [14].

Immobilized metal catalysts are another category of materials to access Type IV acid-catalyzed flow reactions. Several heterogeneous systems, including chiral catalyst systems, have been developed to date [15]. Cyanation reactions of an aldehyde or imine also offer an important C–C bond-formation methodology, affording a secondary alcohol or amine. Ti(salen) immobilized on polystyrene and self-supported chiral Ti-alkoxide catalysts have been developed, and both have been used in the asymmetric 1,2-addition of cyanides [16, 17].

The Henry reaction is a reliable C–C bond-forming reaction that can be used to obtain 1,2-nitro alcohols, which are precursors of 1,2-amino alcohols [18]. The use of Nd/Na immobilized on multiwalled carbon nanotubes (MWNTs), as heterogeneous catalyst 5 for the Henry reaction under continuous-flow conditions, has been reported

[19]. The catalyst was prepared by mixing a chiral-amide ligand, $NdO_{1/5}(O^iPr)_{4/5}$, sodium hexamethyldisilazide (NaHMDS), and MWNTs in THF. The obtained solid was packed into a column reactor and the Henry reaction between the aldehyde and the excess nitroethane was carried out under continuous-flow conditions of an aldehyde/bimetallic catalyst molar ratio of 7.7 h^{-1}. The desired Henry adduct was obtained at an excellent yield and with excellent stereoselectivity during the 30 h operation without any loss of either reactivity or selectivity (Fig. 10.8).

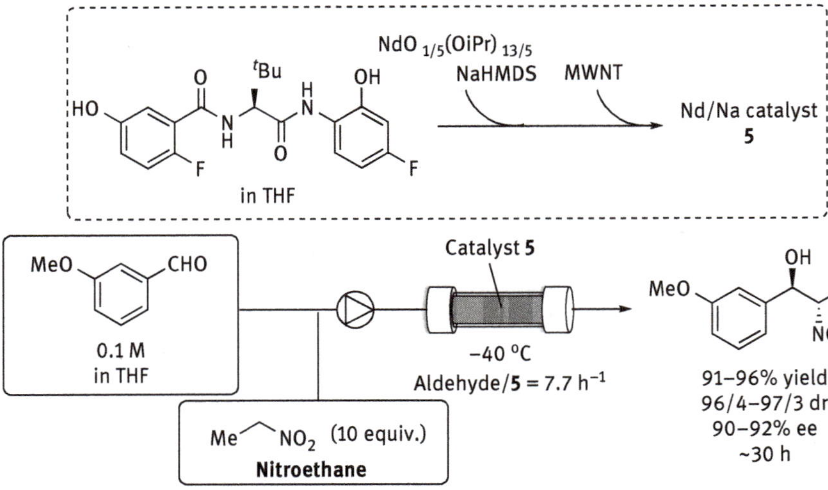

Fig. 10.8: Enantioselective Henry reaction under continuous-flow conditions.

Activated olefin moieties in enones, vinylidene malonates, nitroolefins, and so on are good acceptors of nucleophiles, affording 1,4-addition products. Because of their accessibility (as is evident in the example of aldol condensations and Knoevenagel condensation), preparation of these classes of electrophilic reactants and their use in 1,4-addition can be connected in one-flow. For example, a two-step sequential flow reaction comprising nitroolefin synthesis and 1,4-addition reactions with several catalyst-packed reactors has been reported [20]. Weakly basic primary amine-functionalized silica gel (**3**) was used for the first reactor and the resulting nitroolefin stream was directly fed to the second reactor. Various solid/immobilized bases, including MgO, CaO, and secondary amine-functionalized mesoporous silica, were used in the 1,4-addition of amine, 1,3-dicarbonyl, and thiolate, respectively (Fig. 10.9).

In 2012, a continuous-flow asymmetric 1,4-addition reaction with a homogeneous chiral catalyst (Type III) was reported [21]. In this reaction, an aryl bromide was employed as a starting material and directly converted to the corresponding aryl boronate in flow, before feeding into the final reactor containing Rh-QuinoxP (**6**). As a result, the desired 1,4-adduct (**7**) was obtained at a good to excellent yield and with excellent enantioselectivity (Fig. 10.10).

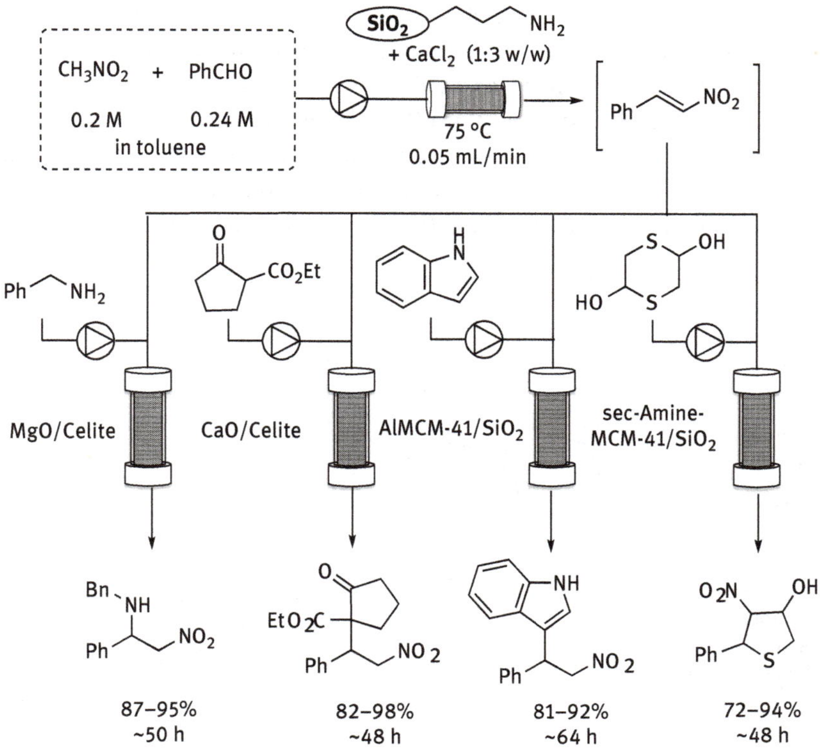

Fig. 10.9: Continuous-flow 1,4-addition reactions using heterogeneous catalysts.

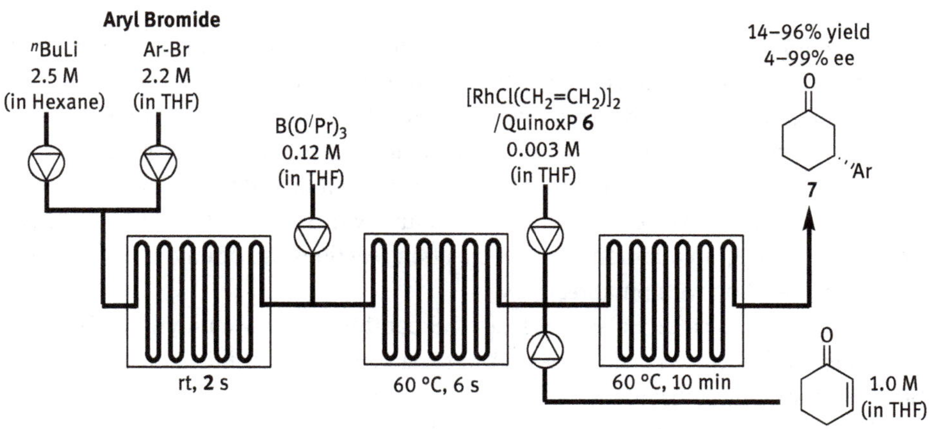

Fig. 10.10: Type III flow asymmetric 1,4-addition sequence.

In the same year, yet another successful example of an enantioselective 1,4-addition reaction was reported [22]. Here, the Type IV strategy was used. Polystyrene-supported chiral Ca-PyBOX (**8**) catalyst was used as a chiral catalyst and a substrate/PyBOX molar ratio of 1.2 h^{-1} was used. With this system, the desired 1,4-adduct of dimethyl malonate and 2-nitrostyrene was obtained at an excellent yield and with excellent enantioselectivity. Notably, the catalyst remained active for >200 h without loss of either activity or selectivity, and the TON reached was 228 (Fig. 10.11).

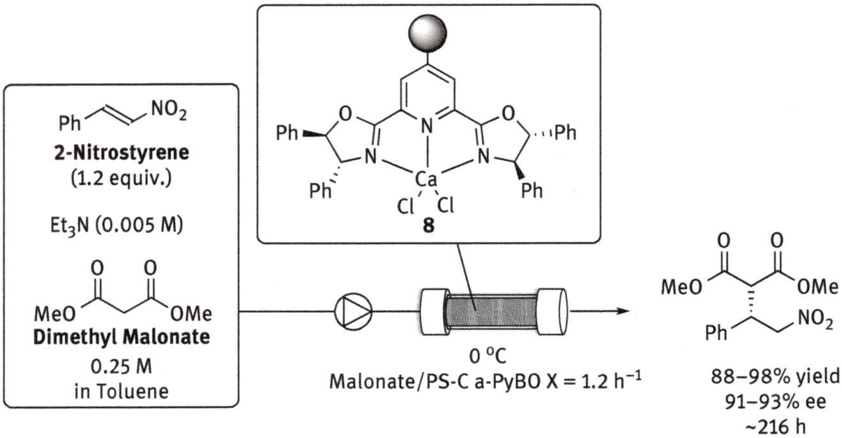

Fig. 10.11: Enantioselective 1,4-addition reaction catalyzed by supported chiral calcium catalyst.

10.3.2 Cycloaddition reactions

Cycloaddition reactions are one of the most useful tools for constructing cyclic structures in organic synthesis. However, besides the Diels–Alder cycloaddition and the related 1,3-dipolar cycloaddition strategies, they often require harsh or extreme reaction conditions to generate reactive intermediates such as carbenes, nitrenes, benzynes, ketenes, and singlet oxygen. Flow chemistry can provide acceptable solutions to manage these troublesome synthetic techniques with Type I or Type II noncatalytic flow reactions [23]. In addition, photocyclization reactions are often employed in the Type I flow reaction with photoreactors.

With regard to catalytic systems, several continuous-flow catalytic Diels–Alder reactions are known. The first Diels–Alder reaction under continuous-flow conditions with an immobilized catalyst was reported in 1996 [24]; here, a chiral borane amide immobilized on polystyrene was employed. The reactant solution containing cyclopentadiene with methacrolein was provided via a dropping funnel. The reaction proceeded well to afford a cycloadduct at an excellent yield and with good stereoselectivity. Achiral versions of this reaction have also been reported [25]. According to the literature, there are numerous candidate solid-acid catalysts that are useful for the reaction under flow conditions: aluminosilicate catalysts, such as aluminum-grafted silica monoliths, USY zeolite, and Beta zeolite. In the case of the H-Beta zeolite-catalyzed continuous-flow Diels–Alder reaction between methyl acrylate and cyclopentadiene, the reaction with a substrate/zeolite ratio of 17.6 mmol h^{-1} $g_{cat}{}^{-1}$ proceeded to afford the desired product at a high yield during an operation of at least 7 h (Fig. 10.12). In 2019, a new type of Lewis acidic metal–organic framework (MOF) was introduced: trifluoromethanesulfonated Zr-benzene tricarboxylate MOF, Zr-OTf-BTC. It was employed in the continuous-flow Diels–Alder reaction of benzoquinone and cyclohexadiene at a substrate/catalyst molar ratio of 100 h^{-1}. Although the stopped-flow mode was used to test the activity, an almost quantitative yield of the cycloadduct was attained after 17 sequences, and the turnover frequency (TOF) and TON were calculated to be 100 h^{-1} and 1,700, respectively [26].

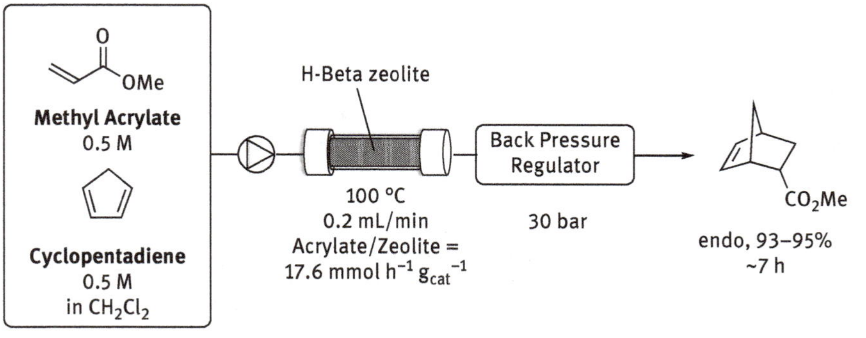

Fig. 10.12: Zeolite-catalyzed continuous-flow Diels–Alder reaction.

Examples of enantioselective organocatalytic Diels–Alder reactions in Type IV flow have been reported. In 2014, a triazole-anchored polystyrene-supported chiral imidazolidinone (**9**) was examined [27a]. An enantioselective Diels–Alder reaction of cinnamaldehyde and cyclopentadiene with a 0.43 h^{-1} cinnamaldehyde/catalyst molar ratio proceeded to afford the desired product at 68–73% yield with 90% ee during 26 h (Fig. 10.13a). Later, another type of chiral imidazolidinone organocatalyst (**10**), was employed for the reaction of the same catalyst. Although the reaction conditions were relatively mild, with a cinnamaldehyde/catalyst molar ratio of 0.1 h^{-1}, a decrease in enantioselectivities during a 81 h operation was observed from 90% to 62% ee (Fig. 10.13b) [27b]. Cycloaddition reactions are well known to have negative activation volume and pressure effects on reactivity, and regioselectivity in azide–alkyne cycloaddition reaction was seen [28]. In this paper, the performance of two reactors, a specialized autoclave batch reactor for high-pressure operation up to 1,800 bar and a capillary flow reactor (up to 400 bar) were compared.

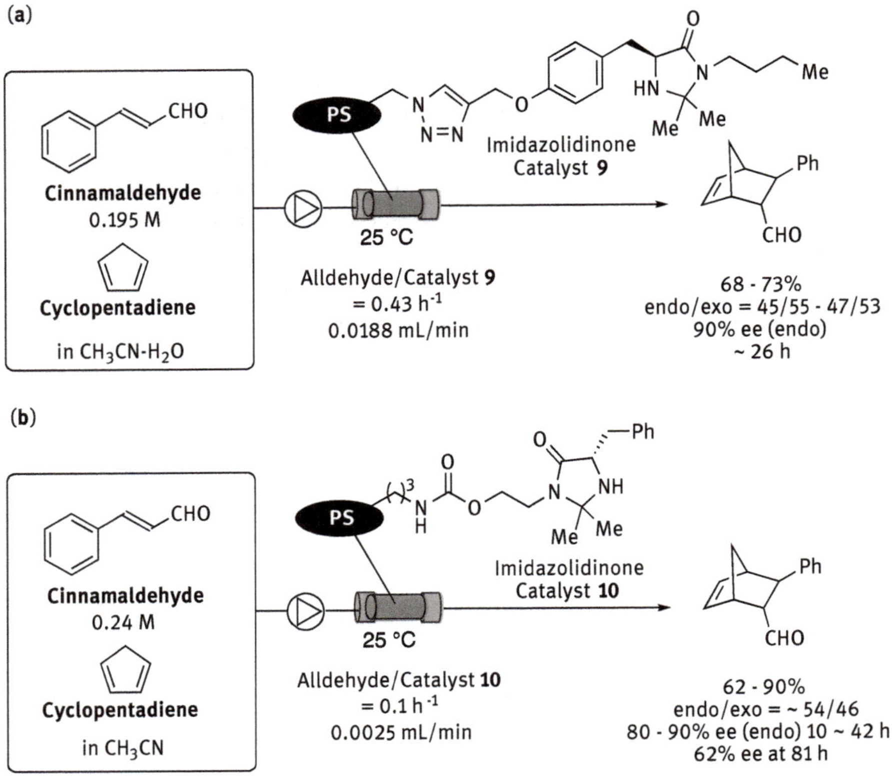

Fig. 10.13: Organocatalytic continuous-flow enantioselective Diels–Alder reactions.

10.3.3 Friedel–crafts-type reactions

A Friedel–Crafts reaction is a straightforward method for the synthesis of alkylarenes and aromatic ketones, which are important classes of compounds that can function as precursors in the manufacture of fine chemicals. Industrial processes that employ zeolites have been developed for the acylation of aromatic moieties using acetic anhydride as an acylating reagent in a recycle fixed-bed reactor [29]. However, deactivation of these heterogeneous catalysts is a serious problem that limits the application of the acylation reaction under continuous flow. Alkylation of benzene with benzyl alcohol was examined in a flow manner using sulfonic acid resins [30]. A sulfonic acid resin, the polymer backbone of which consists of fluorinated porous organic polymer, exhibited efficient activity in the reaction with a 2.8 h^{-1} benzyl alcohol/-SO_3H molar ratio and afforded the product diphenylmethane at a yield >90% over 5 h. Friedel–Crafts acylation with acid anhydride was carried out with Type IV continuous-flow systems using zirconium-exchanged Beta zeolite as a catalyst [31]. The acylation of anisole with propionic anhydride, with 0.9 $mol_{sub}\ h^{-1}\ g_{cat}{}^{-1}$, which corresponds to 1.8 h^{-1} anisole/total acid amount molar ratio proceeded to afford the desired ketone at a high yield over 120 h (Fig. 10.14). Zr-OTf-BTC was also used in the reaction between acetic anhydride and 2-methoxynaphthalene in a stopped-flow mode [26]. Continuous-flow Friedel–Crafts acylation reactions are reported for the telescoped synthesis of ibuprofen, a nonsteroidal anti-inflammatory drug. Type III homogeneous flow reactions with stoichiometric amounts of acids were used [32].

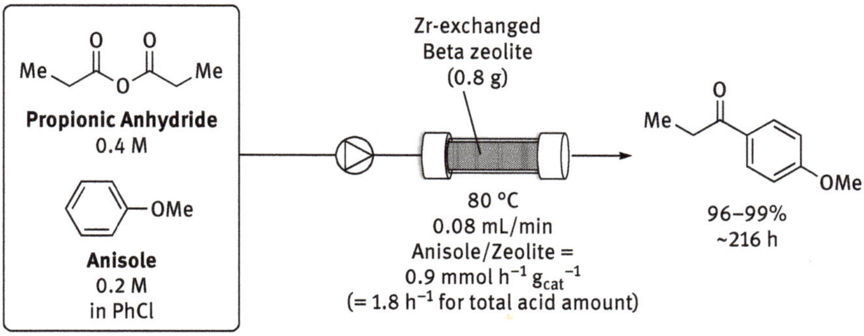

Fig. 10.14: Zr-exchanged Beta zeolite-catalyzed continuous-flow Friedel–Crafts acylation.

10.4 Hydrogenation/reduction

10.4.1 Metal-catalyzed nonchiral hydrogenation and reduction

Catalytic hydrogenation reactions of unsaturated C–C or carbon–heteroatom bonds are very important in chemistry. These reactions are widely used, not only on a laboratory scale but also at an industrial scale, when gaseous hydrogen is used as a reductant under catalytic conditions. Recent demands in this area are for safer, cheaper, and smaller systems – thus, there are high expectations to use continuous-flow Type IV hydrogenation. Hydrogenation reactions using catalyst-packed reactors require liquid–gas–solid catalyst systems, the so-called trickle-bed systems/reactors. Here, the liquid-phase substrate and hydrogen gas are well mixed when they pass through a catalyst in the static phase, to achieve a high diffusion efficiency.

Numerous excellent review articles summarize the many contributions made over the past decade. A useful summary from one such review [33] is now presented. See Tab. 10.1. Here, not only are precious metals such as Pd, Pt, Ru, and Rh included, but base metals such as Ni and Co are also often considered, in combination with a variety of solid supports.

An example of the continuous-flow hydrogenation of organic molecules such as organic nitro compounds and nitrile compounds using a designed ternary Pd

Tab. 10.1: Summary of recent examples of continuous-flow hydrogenation with catalysts, taken from Ref. 2k.

Catalyst	Modification/support	Substrates
Commercial catalysts		
Pd/C		Azide [34], imine [35], nitroarene [36], pyridine [37]
Pt/C	Sulfided [38], V-doped [39], Fe-doped [40]	Alkene [41], nitroarene [38–40]
Ru/C		Arene [42]
Rh/C		Alkene [43], arene [42]
Raney Ni		Alkene [44, 45], nitroarene [46]
Raney Co		Nitroarene [39, 47]
PtO_2		Nitroarene [48]
Designed catalysts		
Pd	TiO_2 [49, 50], Bi-TiO_2 [51], Cation-exchange resin [52], Al_2O_3 [53, 54], ZrO_2-PVA [55], TiS [56], polysilane [57–60], borate-PS [61], ZnO [62], maghemite [63], mesocellular SiO_2 [64], sulfonated SiO_2, [65, 66], monolithic SiO_2 [67]	Alkene [49, 52, 55, 58, 63–65, 67], Alkyne [50, 51, 55, 56, 58, 61, 62, 66, 67], ketone [52, 67], pyridine [49], nitroarene [55, 57, 63], nitroalkene [59], nitrile [60], azide [63], debenzylation [67]
Pt	TiO_2 [68], SiO_2 [69], Al_2O_3 [70], PS-PEG [71], ZnO [72]	Nitroarene [68, 71], aldehyde [69], alkene [70–72]
Au	TiO_2 [73, 74], Al_2O_3 [75–77]	Nitroarene [73, 75–77], alkyne [74]
Ag	SiO_2 [74]	Alkyne [74]
Ru	Hyperbranched PS [78]	Glucose [78]
Rh		Alkene [79]
Cu_2O	PEG-NH_2 [80]	Nitrophenol [80]
Ni	PEG-NH_2 [81]	Alkene [81]
CeO_2	TiO_2 [82]	Alkyne [82]
Co_3O_4	Nanographene-CNT [83]	Nitroarene [83]

heterogeneous catalyst under ambient H_2 pressure is shown in Fig. 10.15 [57, 60]. In the nitrile hydrogenation study, using PDMSi-Pd/SiO_2 (PDMSi: polydimethylsilane), a benzonitrile/Pd molar ratio of 33 h^{-1} was used and quantitative yields were maintained over 300 h without leaching of metal species. In the case of nitropropane hydrogenation with PDMSi-Pd/bone charcoal, a nitropropane/Pd molar ratio of 32 h^{-1} was used and quantitative yields with no leaching were maintained over 120 h. As is common in heterogeneous catalysis, the choice of solid support material plays a crucial role in controlling the reactivity of the catalyst for the reduction of various types of functional groups. Although the actual role of the PDMSi modifier is unclear here, highly efficient and highly robust catalysis could be achieved. Dual-pore-type monolithic silica was used as a Pd support and employed in the continuous-flow hydrogenation of C–C multiple bonds [67]. Under the conditions of 12.8 h^{-1} substrate/Pd molar ratio, hydrogenation of C–C double and triple bonds to single bonds, and azide to primary amine proceeded to afford the corresponding product at a high yield (Fig. 10.16). The lifetime for the hydrogenation of cinnamyl alcohol reported in the literature is 72 h. A continuous-flow hydrogenation is also a desirable technique in the highly oxidized biomass conversion to bio-oils. Glucose hydrogenation to sorbitol was examined using a Ru nanoparticle (NP) catalyst on hyperbranched polystyrene. Under continuous-flow conditions, with a highly concentrated glucose solution, the designed catalyst exhibited competitive results with an industrial Raney Ni catalyst [78].

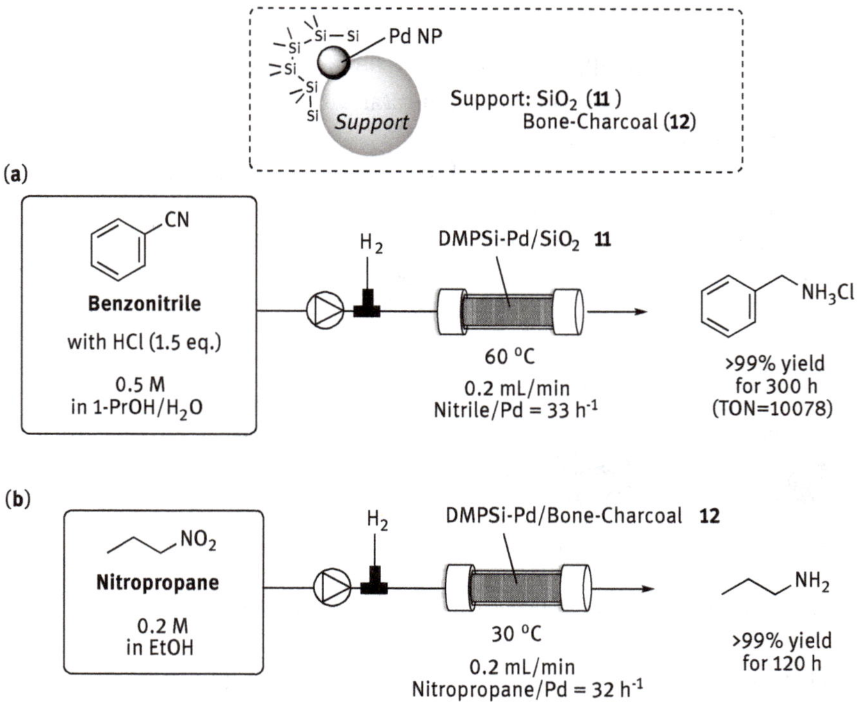

Fig. 10.15: PDMSi-Pd catalysts for continuous-flow hydrogenation.

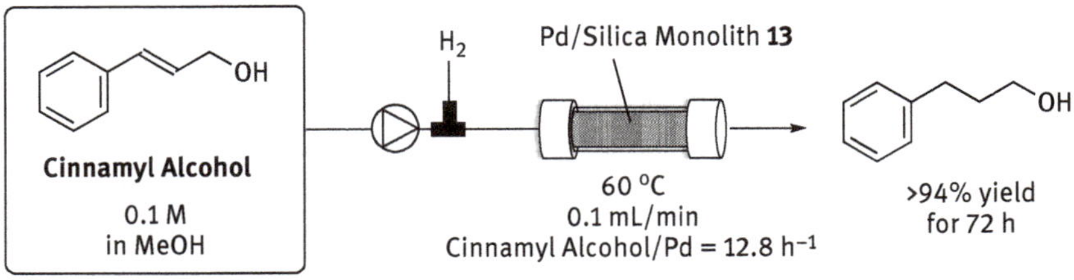

Fig. 10.16: Continuous-flow hydrogenation using Pd/silica monolith.

10.4.2 Enantioselective hydrogenation

Asymmetric enantioselective hydrogenation reactions of prochiral C–C or carbon–heteroatom double bonds are among the most important transformations for obtaining chiral organic compounds, and numerous enantioselective hydrogenations under Type III and Type IV continuous-flow conditions have been reported. This is not surprising because chiral-modified Pt-heterogeneous catalysts for enantioselective hydrogenations are among the most classical catalysts in the history of asymmetric catalysis. In the field of chiral heterogeneous catalysts, this classical approach, involving surface catalytic sites and modifiers, has certainly dominated over alternatives (Fig. 10.17, Type A). In efforts to enable the fine-tuning of asymmetric environments around the metal center, several chiral-catalyst-anchored inorganic materials have also been developed (Type B).

One of the impressive characteristics of this avenue of investigation is the use of supercritical fluids as the mobile phase. This medium has a high affinity for molecular hydrogen and opens new possibilities for the management of molecular catalysts, which has indeed attracted attention (Type C).

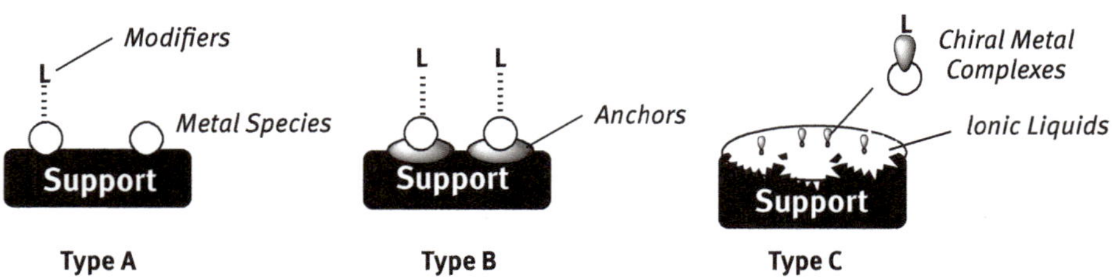

Fig. 10.17: Type of immobilized catalysts for enantioselective continuous-flow hydrogenation.

Enantioselective hydrogenation of activated ketones such as pyruvate esters over cinchona alkaloid-modified Pt on solid supports has been widely studied, ever since the pioneering findings of Orito's group several decades ago (Type A in Fig. 10.17) [84]. A preliminary investigation of the reaction with a flow reactor was first reported in 1991 [85]. The hydrogenation of methyl pyruvate using Pt/SiO_2 as a catalyst was investigated using cinchonidine as a chiral modifier. Although the conversion of pyruvate was < 10%, the product was obtained with 80% ee under flow conditions (Fig. 10.18). Later, some improvement in flow-asymmetric hydrogenation of keto-pantolactone Pt/Al_2O_3 was achieved and the corresponding alcohol was obtained with 83.4% ee, with 94 mmol/g_{cat}^{-1} h^{-1} [86, 87].

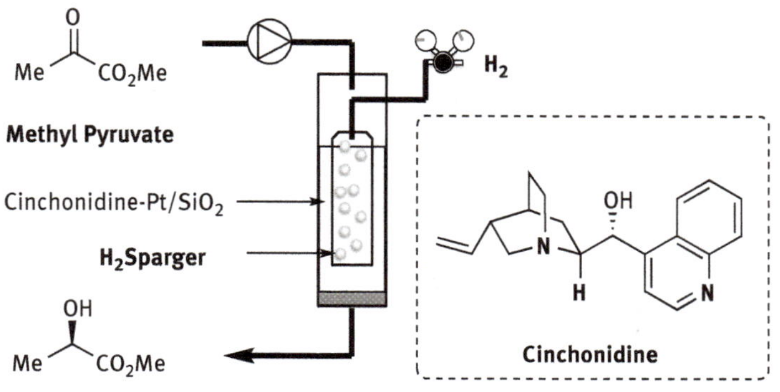

Fig. 10.18: Flow enantioselective hydrogenation using cinchonidine-Pt/SiO_2.

The use of cationic Rh complexes immobilized on solid supports was investigated with the Type B method, shown in Fig. 10.17, which consisted of phosphotungstic acid and alumina as a base support [88]. Continuous-flow asymmetric hydrogenation of itaconates or enamides under continuous-flow conditions was investigated by various research groups, and a high level of enantioselectivity as well as chemical yield were finally attained [89]. According to the literature, a TON of 2638, with 99% conversion and 98% ee, was attained during a 23 h operation, yielding 68 g of enantiomerically pure product (Fig. 10.19). Quite recently, this approach was used in continuous-flow enantioselective hydrogenation using a Rh-QuinoxP*/silicotungstic acid/base-functionalized mesoporous silica KIT-6 system. Enantioselective hydrogenation of enamide to the corresponding amide by this system with an enamide/Rh molar ratio of 120 h^{-1}, proceeded to afford the corresponding amide quantitatively, with 99% ee during a 90 h operation [90].

Another approach that takes advantage of both homogeneous and heterogeneous catalysis was recently examined, based on the concept of metal catalysis in supported ionic-liquid phases (SILPs; Type C in Fig. 10.17). The high affinity of a chiral Rh-naphthyl-QUINAPHOS complex with an ionic liquid or a supported ionic liquid phase on silica enables maintaining the homogeneous catalyst inside a tubular reactor. Furthermore, an important factor pertaining to immobilization is the use of supercritical CO_2 ($scCO_2$) as the mobile phase (Fig. 10.20). In a typical hydrogenation experiment using dimethyl itaconate, high ee values (>95% ee) are evident while the TON reached 70,000 [91, 92].

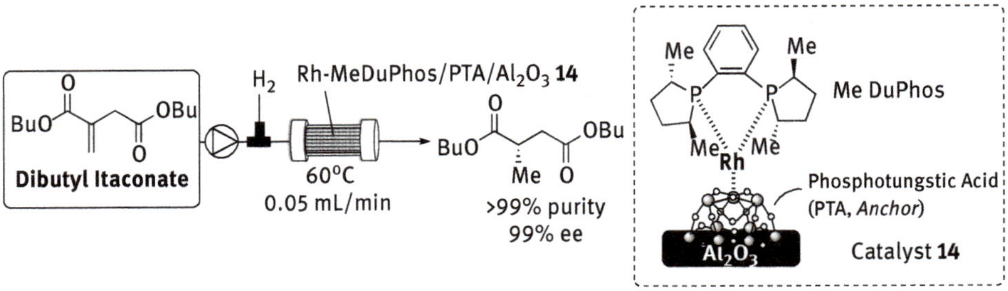

Fig. 10.19: Enantioselective continuous-flow hydrogenation over Rh-MeDuPhos/PTA/Al_2O_3.

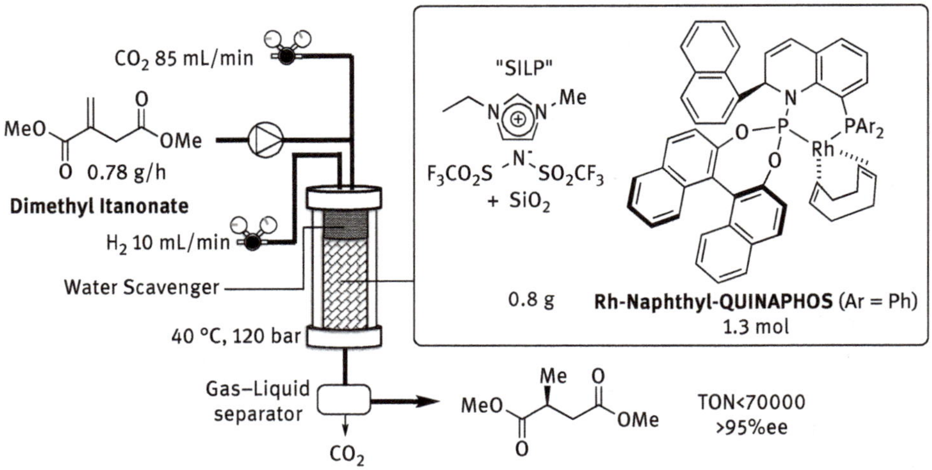

Fig. 10.20: Enantioselective continuous-flow hydrogenation using $scCO_2$.

(For related issues please see Volume 1, Chapter 11, Title: Gaseous reagents in flow chemistry).

10.5 Oxidation

10.5.1 Aerobic oxidation of alcohols

Several oxidation processes under continuous-flow conditions are reported in the literature; however, progress in the controlled oxidation of organic compounds, especially by catalytic processes with wide substrate scope lags somewhat behind other flow processes. This is due to lack of suitable catalyst/oxidant systems for enabling suitable transformations in flow systems. Traditionally, many stoichiometric oxidants, such as heavy metals, peroxo compounds, and hypervalent halogens, are commonly used in synthetic organic chemistry laboratories, but none of them are suitable for continuous-flow synthesis, especially telescoped multistep synthesis, due to unavoidable by-products derived from the oxidants. A process that includes the use of more suitable oxidants, such as molecular oxygen, air, and hydrogen peroxide, is desirable. Some flow-oxidation reactions of primary and secondary alcohols to form the corresponding carbonyl compounds with a catalyst-packed reactor have been examined using immobilized precious metal catalysts. In an early example of 2005, a polymer-incarcerated Ru catalyst/*N*-methyl morpholine *N*-oxide (NMO) system was examined for the oxidation of benzyl alcohol to benzaldehyde with an alcohol/Ru molar ratio of 7.0 h^{-1} [93]. The reaction proceeded to afford the aldehyde at a high yield in 8.4 h (Fig. 10.21). Later, the use of Ru/alumina and $Ru(OH)_X$/alumina were reported [94, 95]. Both catalytic systems afforded the corresponding oxidation product at a high yield when using molecular oxygen as an oxidant. However, in the former case, a recirculation system was required to obtain a high yield, while in the latter case, a rather low WHSV of 0.008 h^{-1}, which corresponds to a 0.3 h^{-1} 2-hydroxymethyl thiophene/Ru molar ratio, was required to achieve a satisfactory yield in a long-time operation (Fig. 10.22).

Supported chiral salen complexes have also been employed in kinetic resolution and asymmetric epoxidation in flow [96, 97].

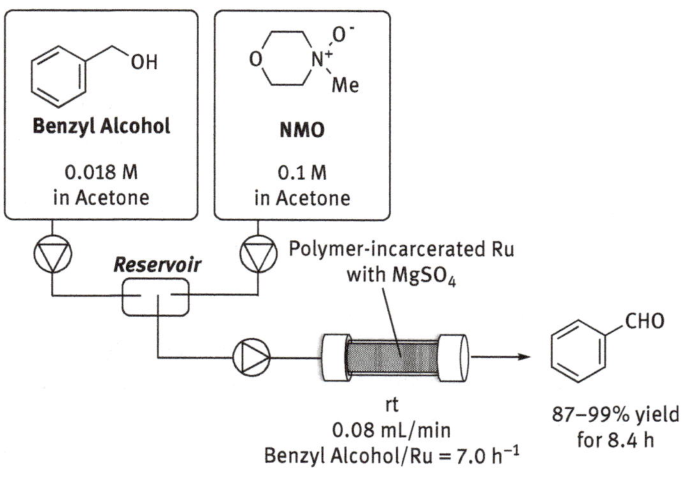

Fig. 10.21: Polymer-incarcerated Ru-catalyzed continuous-flow oxidation of benzyl alcohol.

10.5.2 Oxidative transformation

Other types of oxidation of organic compounds are also reported. In this context, photoredox catalytic processes through single electron transfer, promoted by light sources, have recently attracted great attention among the global chemistry community

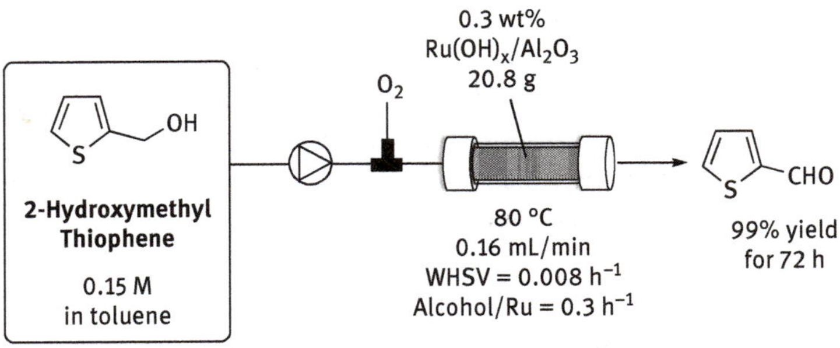

Fig. 10.22: Alumina-supported Ru-catalyzed continuous-flow oxidation with molecular oxygen.

[98]. In addition, due to the availability of an infinite energy supply, the application of continuous-flow processes has revealed tremendous advantages due to the efficacy of light penetration, irradiation control, and heat control [99]. Several transition-metal complexes and organic dyes have been used as photocatalysts in Type III and Type IV manners and demonstrated great utility in continuous-flow systems.

C–H bond functionalization is one of the essential oxidative catalytic processes in modern organic chemistry. Among other approaches, C–H activation by photocatalysts is possible through hydrogen-atom transfer (HAT) mechanisms. Several methods have been developed in flow using polyoxometalates as HAT catalysts [100]. For example, a tetrabutylammonium decatungstate (TBADT) (**15**), a typical polyoxometalate, has proven to be an efficient hydrogen abstraction catalyst under near-UV irradiation and its use in continuous Type III Csp^3–H oxidation reactions are reported (Fig. 10.23) [101]. The versatility of this method is highlighted by a large substrate scope (30 substrates), including activated and inactivated aliphatic C–H bonds, as well as natural product scaffolds. Moreover, the flow system was designed to make this transformation safe and readily scalable, and it exhibited reaction accelerations comparable to the batch method (45 min instead of 4 h). In addition, another C–H activation reaction, late-stage C–H fluorination, was also investigated [102].

Natural organic dyes (NODs) have also been widely used in continuous flow as they are inexpensive, nontoxic, and readily available photocatalysts. Over the past five years, in particular, eosin Y has been used as an efficient catalytic photosensitizer [103]. Common methylene blue and rose Bengal have also been used in isolated cases [104]. Although NODs are "green" and are readily available, their limited photophysical properties produces compatibility issues with certain transformations, especially with supposedly inert reactants. One of the most versatile types of synthetic photosensitizers is metal-free porphyrinoid photocatalysts, such as tetra-phenylporphyrins (TPPs) [105]. The reduction potential of TPPs is generally higher than NODs, which makes them very attractive for photoredox transformations. A simple protocol for the photooxygenation of activated naphthols to naphthoquinones in continuous flow was explored by screening several porphyrinoid photocatalysts [106].

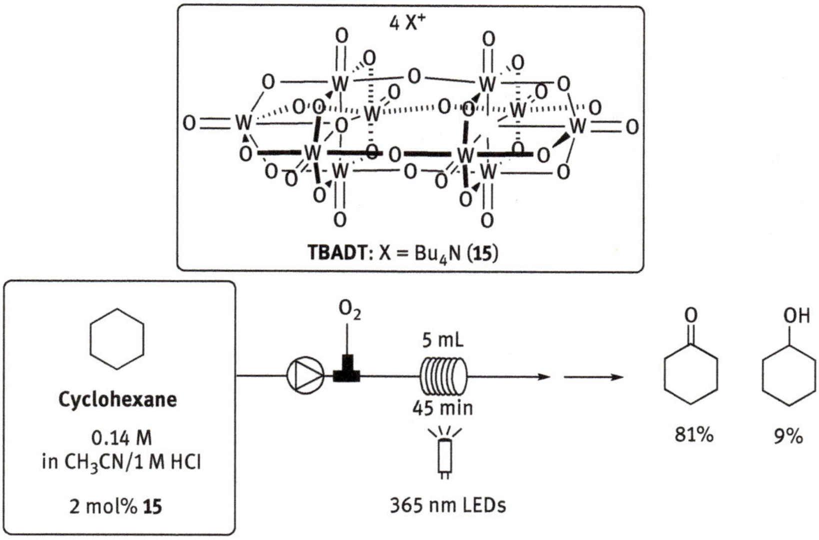

Fig. 10.23: Continuous-flow photooxidation with TBADT as a homogeneous catalyst.

One of the most attractive reagents to emerge over the past few years is the abundant gas CO_2. In recent research, it was determined that by generating the single electron reduction of CO_2, through a combination of *p*-terphenyl (**16**) as compatible photosensitizer ($E^0 = -2.63$ V vs SCE in acetonitrile) and $KOCOCF_3$ as base (Fig. 10.24a), the α-carboxylation of amines, such as 1-benzyl piperidine with CO_2, was achieved at a high yield [107]. This CO_2-mediated photocatalytic combination was further used by the same group for the continuous β-selective hydrocarboxylation of styrene (Fig. 10.24b) [108].

By analogy, with homogeneous photocatalysis, photosensitizer immobilization is the most intuitive strategy to promote heterogeneous photocatalysis in the Type IV manner. The first immobilized photosensitizer employed in continuous flow was used for the production of peroxides via $^1O_2$ generation in $scCO_2$, using immobilized TPP sensitizers, through covalent bonding on PVC beads [109]. More recently, the same authors employed a similar strategy for the synthesis of the antimalarial drug artemisinin (Fig. 10.25) [110]. In this work, the advantages of flow chemistry were directly related to green chemistry principles, in addition to serving the pharmaceutical industry. One of the keys to the success relied on the use of $scCO_2$, whose main property is to solubilize gases such as O_2 fully. $scCO_2$ is also abundant, an inert solvent, and is considered nontoxic, making it ideal for continuous processes. A second important point was the immobilization of *meso*-TPP and *meso*-TPFPP (tetrakis (pentafluorophenyl)porphyrin) on Amberlyst-15 (sulfonic acid-functionalized polystyrene, Amb-15) through ionic anchoring, such as catalyst (**17**). This resulted in bifunctional heterogeneous photocatalysts, bearing both acidic and photosensitizer functionalities on the same support material. This catalyst was packed into a fixed-bed

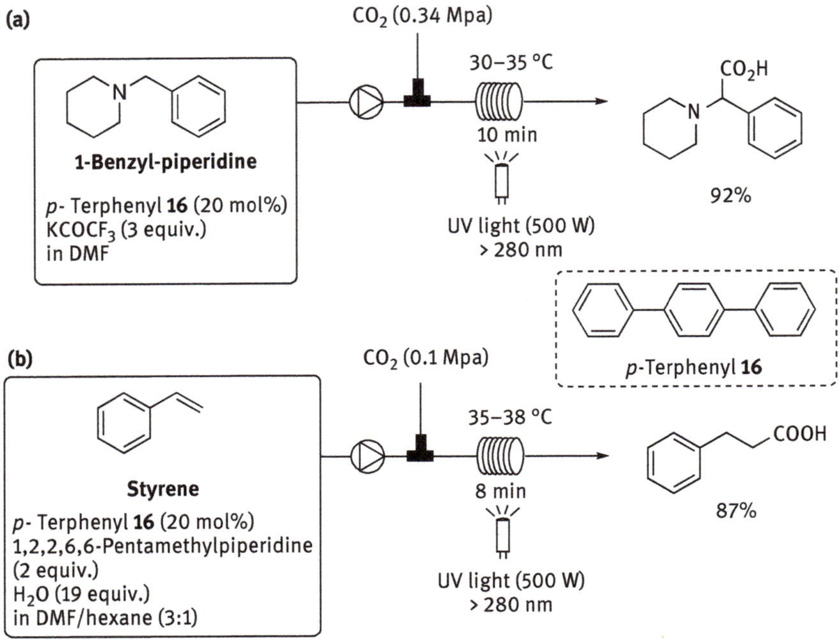

Fig. 10.24: Photoactivation of CO_2 in flow and use in oxidative transformations.

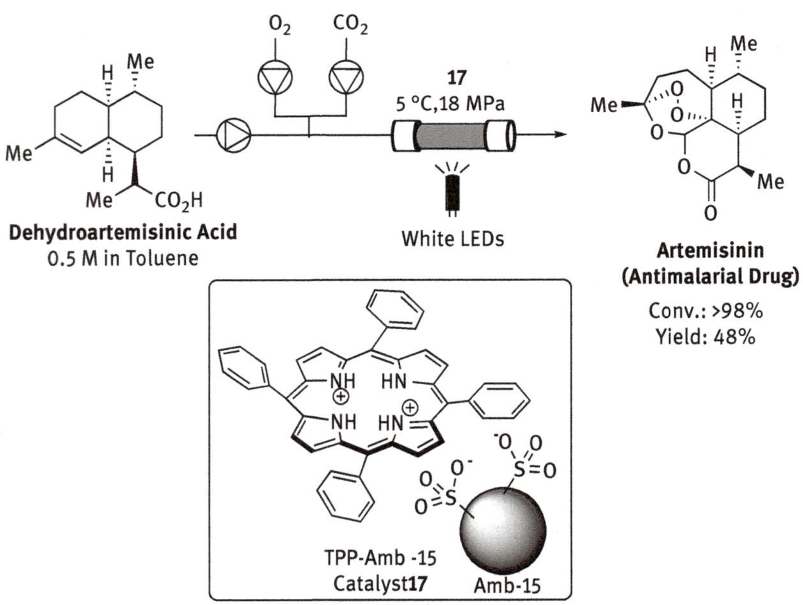

Fig. 10.25: Singlet oxygen manipulation in flow for the synthesis of Artemisinin.

photoreactor and provided with ~50% of pure artemisinin (within a single column), directly from dehydroartemisinic acid. Furthermore, immobilized 2,1,3-benzothiadiazole was used as a photoactive cross-linking monomer in visible light to generate singlet oxygen and the subsequent conversion of α-terpinene to ascaridole [111].

10.6 Coupling reactions

Organometallic reagents, which can be generated by transmetalation from nucleophilic reagents and a transition-metal precatalyst, are widely used in cross-coupling reactions. Several name reactions are widely known, based on the class of organometallic reagents, such as organoboron (Suzuki–Miyaura), vinyl metal (Mizoroki–Heck), organozinc (Negishi), and acetylide (Sonogashira) [112]. One of the most important reactions is the Pd-catalyzed cross-coupling reaction, such as the Suzuki–Miyaura coupling reaction. The use of homogeneous Pd catalysis, derived from precatalyst (**18**), in a Type III continuous-flow reaction is shown in Fig. 10.26. In this process, lithiation of arylhalide and borylation of the resulting aryllithium reagents were combined before the coupling reaction [113]. Later, the use of heterogeneous Pd catalyst, monolithic Pd catalyst, was used in a similar system [114].

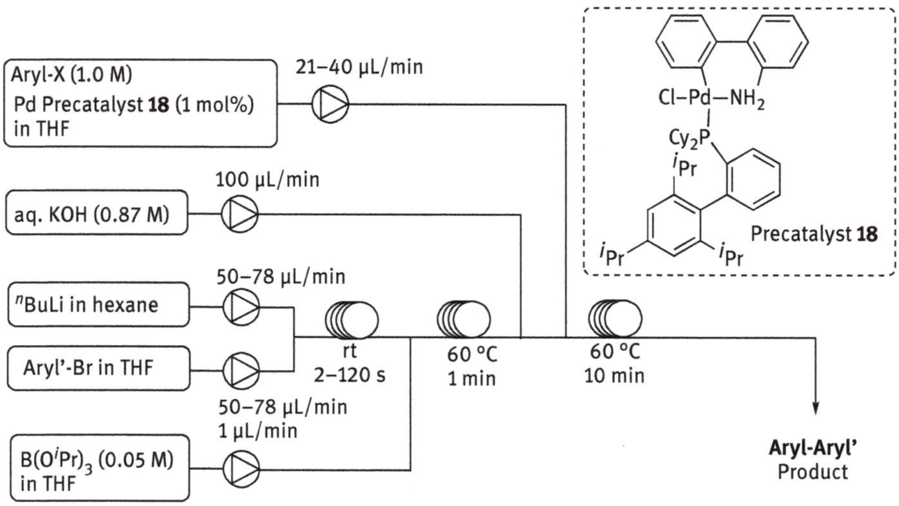

Fig. 10.26: Lithiation/borylation/cross-coupling sequential flow reaction.

As described in the previous sections, the use of heterogeneous catalysts for these types of reactions (Type IV flow) is attracting much attention – nowadays, there are numerous contributions related to heterogeneous catalysis in this class of reactions. Several immobilized Pd catalysts, including commercially available and laboratory-prepared ones, have been employed in continuous-flow Suzuki–Miyaura cross-coupling reactions as one of the benchmark investigations of continuous-flow cross-coupling reaction. A typical schematic image of this reaction and select results are illustrated in Figs. 10.27 and 10.28.

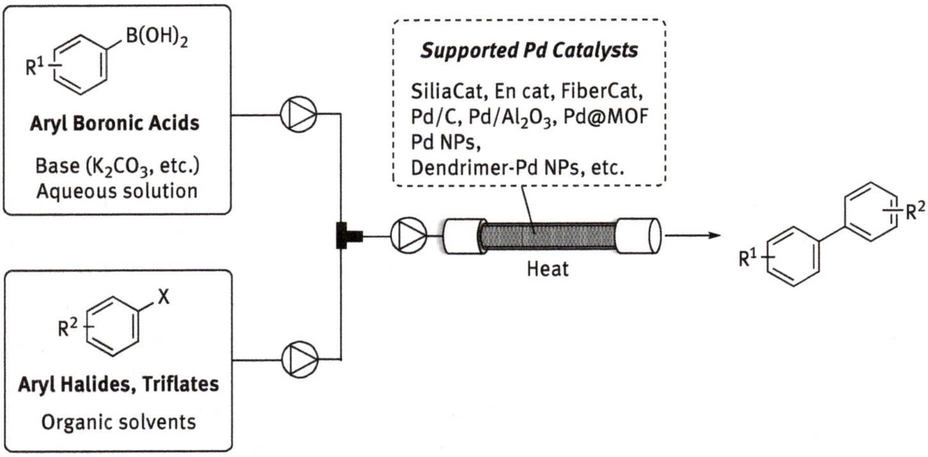

Fig. 10.27: Continuous-flow Suzuki–Miyaura cross coupling with supported Pd catalysts.

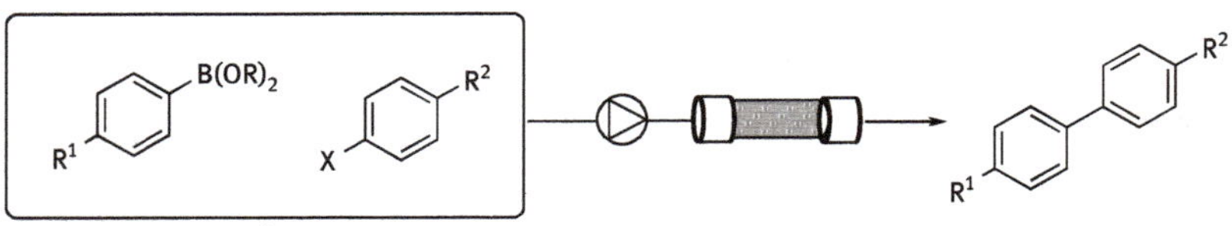

Catalyst:	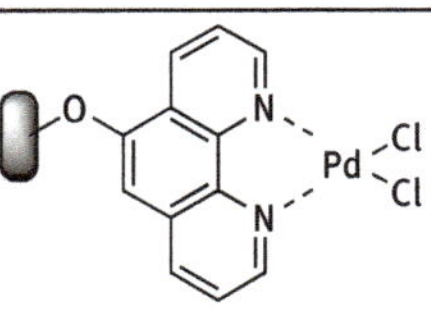	R^1 = Me,R = H R^2 = COMe, X = Br	**Substrate/Pd = 3.1 h^{-1}** 80 °C	**59% yield** **TOF = 1.8 h^{-1}**	ref. 115
Catalyst:	SiliaC at DPP-Pd	R^1 = OMe, R = H R^2 = 2-Me, X = Br	**Substrate/Pd = 7.5 h^{-1}** 60 °C	**99% yield** **for 8 h**	ref. 116
Catalyst:	Pd@MIL-101(Cr)-NH_2	R^1 =H, R = Pin R^2 = CHO, X = Br	**Substrate/Pd = 0.6 h^{-1}** 20 °C	**95% yield**	ref. 117
Catalyst:	Pd/C	R^1 =H, R = H R^2 = COMe, X = I	**Substrate/Pd = 60 h^{-1}** 25 °C	**>99% yield**	ref. 118

Fig. 10.28: Selected examples of continuous-flow Suzuki–Miyaura cross coupling.

Apart from the problem of metal leaching, an issue to be addressed is the unavoidable by-products that accompany the desired coupling products, such as residues of metal species and residues of halides. In the telescoped flow synthesis of complex organic molecules through multistep transformation, carryovers from upstream should be minimized to avoid undesirable influences downstream. Thus, a combination of a flow reaction system with an in-line purification system, including in-line extraction of inorganic materials, will become an important task toward the successful use of this chemistry in continuous manufacturing. Another way in which low-waste cross coupling could be addressed would be the use of redox-mediated C–H activation methodology, including a cross-dehydrative coupling reaction. Photoredox catalytic coupling reactions should also be useful for this purpose. Figure 10.29 is an interesting example of the photocatalytic cross coupling of a heteroarene with an arene diazonium salt. A semiconductor, TiO_2, was used as a heterogeneous photocatalyst, under irradiation with blue LED light [119].

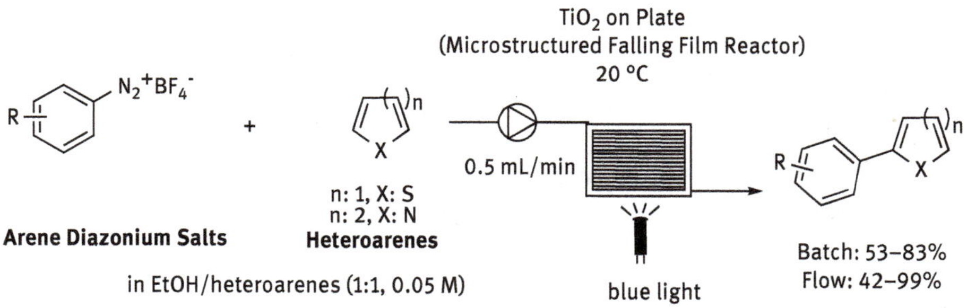

Fig. 10.29: Photocatalytic cross-coupling reaction under continuous-flow conditions.

10.7 Carbene transfer reactions

Carbene transfer reactions, such as cyclopropanation, are important processes in synthetic organic chemistry. This is because the specific structures are often observed in biologically active molecules. However, cyclopropanation reactions require a specified methodology to overcome the high barrier associated with forming the strained ring. Among the several types of transformations developed to date, carbene transfer reactions, including enantioselective ones, catalyzed by transition metals such as Rh and Cu, are used in catalytic continuous-flow reactions. The first cyclopropanation reaction catalyzed by an immobilized metal complex under continuous-flow conditions was reported in 2007 [120]. The immobilized Cu-BOX complex (**19**), using a polystyrene backbone, was employed for the reaction. Selected examples of enantioselective cyclopropanation of styrene with ethyl diazoacetate, using supported copper catalysts, are shown in Fig. 10.30. Later, an example of cyclopropanation, using Ru-PyBOX immobilized on polystyrene, was reported by the same group in the same year [121]. Although the use of a high substrate/Ru molar ratio per hour decreased both yield and selectivity, the use of $scCO_2$ as solvent dramatically improved both attributes. They then also reported the use of the Cu-AzaBOX complex, immobilized on siliceous mesocellular foam as a modified chiral copper catalyst, for the enantioselective cyclopropanation reaction [122].

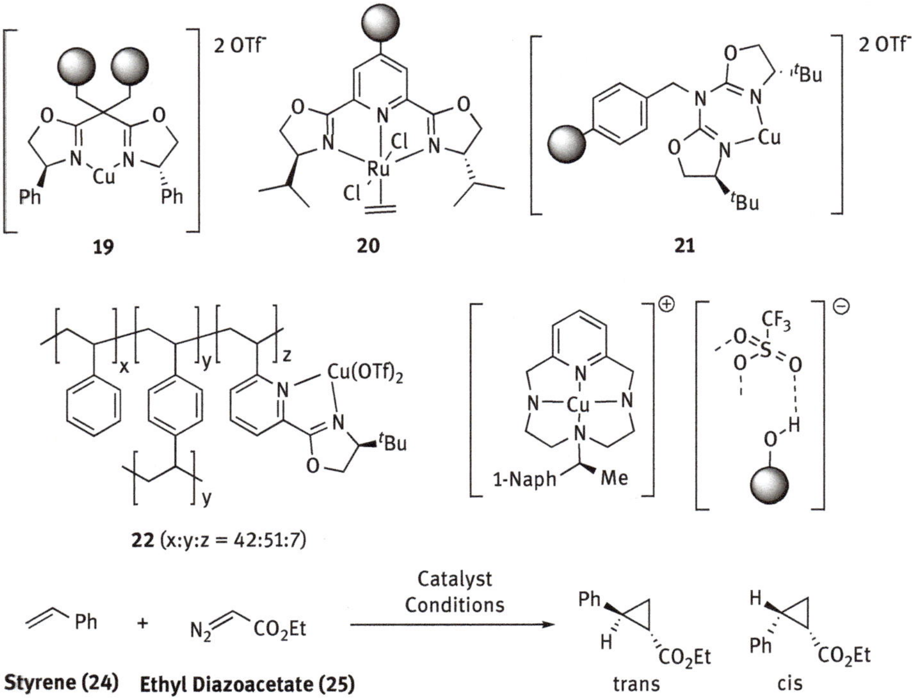

Catalyst	Conditions	Results	refs
19	**24**: 2 M, **25**: 0.5 M (in CH_2Cl_2), 0.002 mL/min	61% yield 71% ee (trans)	120
20	**24**: 1.73 M 0.05 mL/min $scCO_2$8 Mpa: 0.55 mL/min, 40 °C	22% yield 89% ee (trans)	121
21	circulation (in CH_2Cl_2) 5 mL/min	80% yield 93% ee (trans)	122
22	**24**: 1.5 M, **25**: 0.5 M (in CH_2Cl_2), 0.023 mL/min	24% yield 60% ee (cis)	123
23	**24**: 87.8 mmol, **25**: 17.5 mmol, 0.02 mL/min $scCO_2$ 130 bar: 0.5 mL/min, 40 °C	65.5% yield 62% ee (cis)	124

Fig. 10.30: Supported chiral catalysts for continuous-flow cyclopropanation.

Another interesting catalysis system that involves a Rh dimer complex to achieve diazo compound activation has been used to perform carbonyl ylide cycloaddition reactions. In 2011, the cycloaddition of styrene with carbonyl ylide formed in situ from diazo ester under continuous-flow conditions was reported with immobilized Rh dimer catalyst (**26**) [125]. The corresponding product was obtained in high yield and with excellent enantioselectivity for as long as 60 h, and the level of Rh complex leaching was 0.013% (Fig. 10.31).

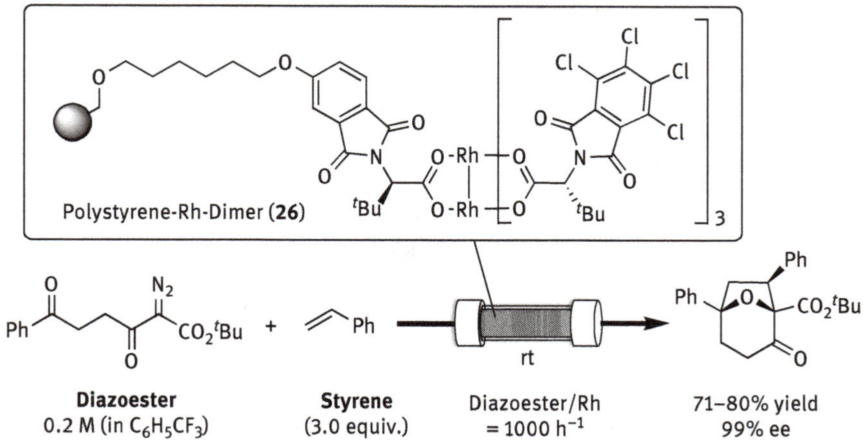

Fig. 10.31: Carbene-mediated 1,3-dipolar cycloaddition reaction in flow.

10.8 Olefin metathesis

Olefin metathesis is one of the most important C–C bond-formation reactions. It is widely used not only for the preparation of small organic molecules but also for large-membered ring systems. It is also used in polymer chemistry, in ring-opening metathesis polymerization (ROMP). The reactions are catalyzed by group VI metal compounds (organometallic complexes), such as Grubbs catalysts and Hoveyda–Grubbs catalysts, which have been well-developed as homogeneous catalysts and group VI metal oxides. Because of their potent activity in this class of reactions, there have been good developments in continuous-flow olefin metathesis, accompanied by progress in reaction engineering.

Homogeneous Type III and heterogeneous Type IV catalytic continuous-flow reactions of the ring-closing metathesis reaction have been developed. In an example reported in 2010, a continuously stirred tank reactor with homogeneous catalysts was employed [126]. It has also been demonstrated that the pervaporation approach with a membrane reactor can be used to overcome mass transport limitation [127]. The use of homogeneous Ru catalyst **27** with a membrane sheet-in-frame ring-closing metathesis module yielded a TON > 7,500 for the reaction of diethyl diallylmalonate (Fig. 10.32). In the heterogeneous Type IV catalytic continuous-flow reaction, immobilized Ru-based catalysts, through electrostatic interaction between ligand and support, were used in ring-closing metathesis reactions [128]. In early investigations, the TON was low. This was caused by the instability of the immobilized catalytic species. Although the activity was unsatisfactory here, the low stability was later improved by using a mesoporous silica support (Fig. 10.33) [129]. A TON of 35,500 was achieved in the ring-closing metathesis of *N,N*-diallyl-*p*-tosylamide when using the Ru catalyst 28/SBA-15 composite.

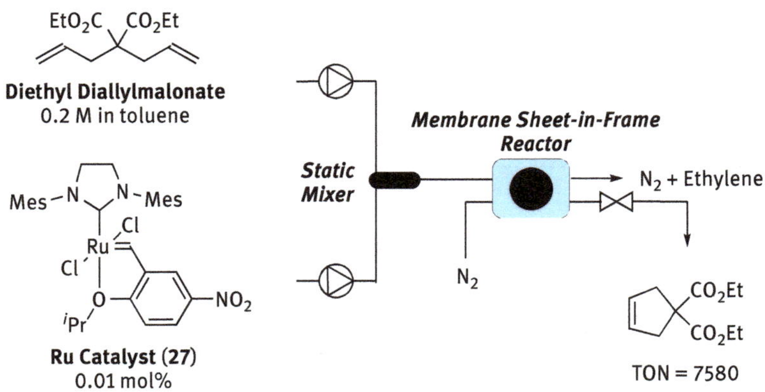

Fig. 10.32: Continuous-flow ring-closing metathesis using homogeneous Ru catalyst.

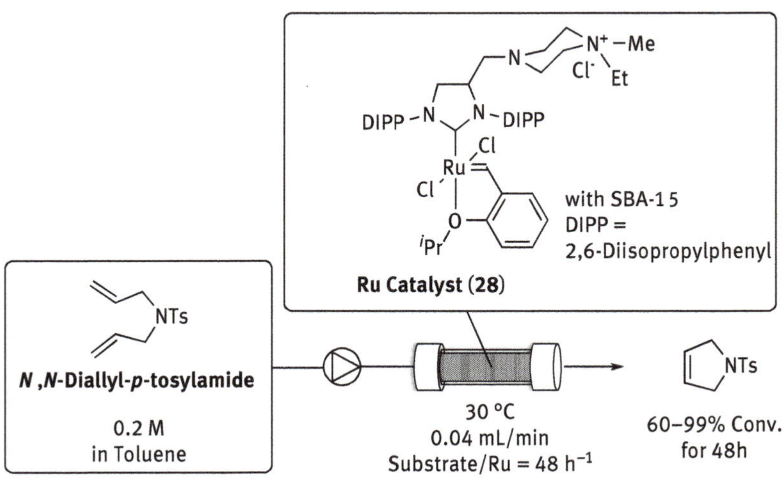

Fig. 10.33: Continuous-flow ring-closing metathesis using immobilized Ru catalyst.

The immobilized methyl oxorhenium approach was also investigated. Chlorinated alumina was employed as a solid support for methyltrioxorhenium and used in propene metathesis to 2-butene; a TON of 100,000 was achieved [130].

10.9 Catalytic multistep reactions

As described in the earlier sections, attention to continuous-flow chemistry has increased rapidly, particularly over the past decade, specifically in the manufacture of fine chemicals. To achieve multistep continuous-flow systems, catalyst invalidation processes and contamination of the output of reactions by catalyst residues should be avoided. Type IV flow reactions are ideal for this purpose – several telescoped syntheses of complex molecules involving catalytic Type IV flow reactions have been developed. A benchmark work is the continuous-flow synthesis of rolipram, an important γ-aminobutyric acid (GABA) derivative, from commercially available aldehyde (**29**), using only a Type IV flow reaction (Fig. 10.34) [131]. The enantioselective sequential flow synthesis of rolipram was achieved by telescoping an immobilized amine-catalyzed nitro-aldol condensation with a chiral calcium-catalyzed asymmetric 1,4-addition reaction, a Pd-catalyzed nitro reduction step, and a solid acid-catalyzed hydrolysis/decarboxylation/lactonization sequence. Of great importance is that both enantiomers of rolipram could be obtained exclusively by changing the column-bearing PyBOX-calcium catalyst (**8**).

Other GABA derivatives from commercially available chemicals, prepared via a three-step continuous-flow process involving only Type IV flow reactions, have also been reported: for example, (±)-pregabalin intermediate (Fig. 10.35) and baclofen (Fig. 10.36) [10, 132].

A continuous-flow processing platform permits the integration of microanalytics into the reaction process, thus providing vital information about the reaction's progress in real time, and ultimately reducing the time required for reaction screening. In 2014, a machine-assisted multistep flow process for the preparation of pyrazine-2-carboxamide and its reduced derivative piperazine-2-carboxamide was reported, where an open-source software package and a computer-controlled multistep continuous-flow chemistry devices were used [133]. Flow IR was used in the first step in the process as an analytic tool and NMR was used in the second step (Fig. 10.37). Notably, the use of microanalytic tools with automated continuous-flow systems simplifies the reaction processes. This semicontinuous protocol is useful for reaction steps that are not easy to telescope, require different flow rates, and/or include an easy separation or purification step such as solvent extraction.

Many engineering problems are yet to be addressed, such as long lead time, productivity, and costs for development, and this new technique must open up a new era of sustainable chemical industry for fine and specialty chemicals.

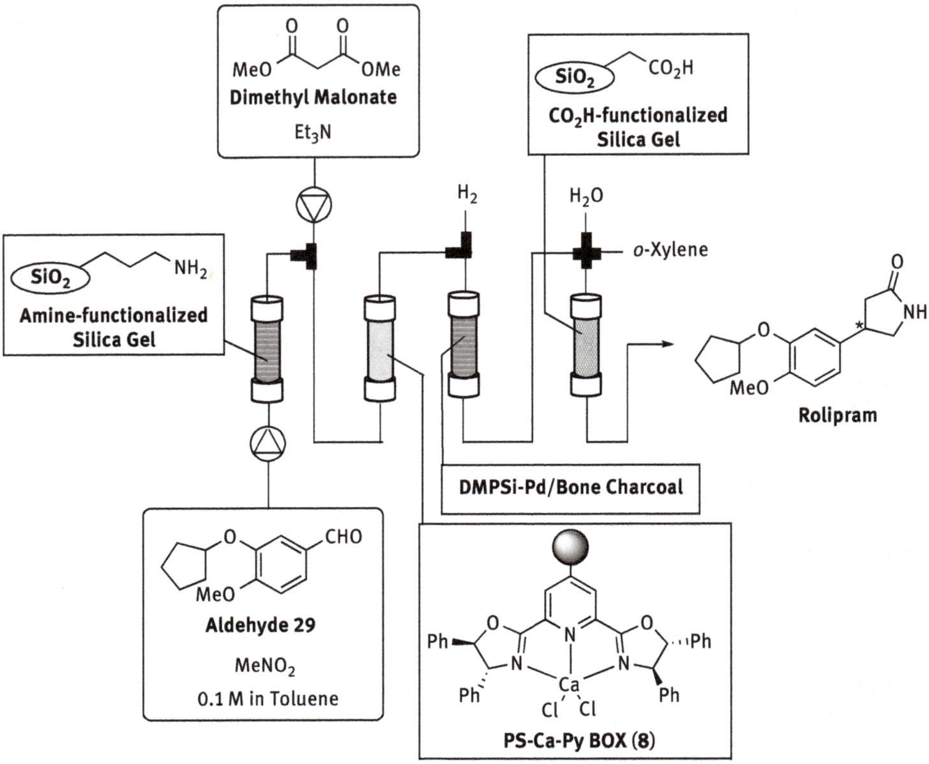

Fig. 10.34: Enantioselective sequential flow synthesis of rolipram.

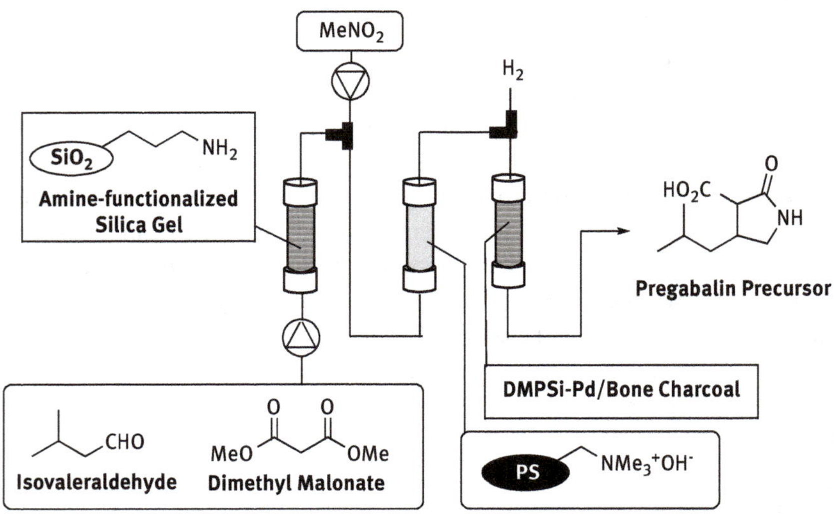

Fig. 10.35: Sequential flow synthesis of pregabalin precursor.

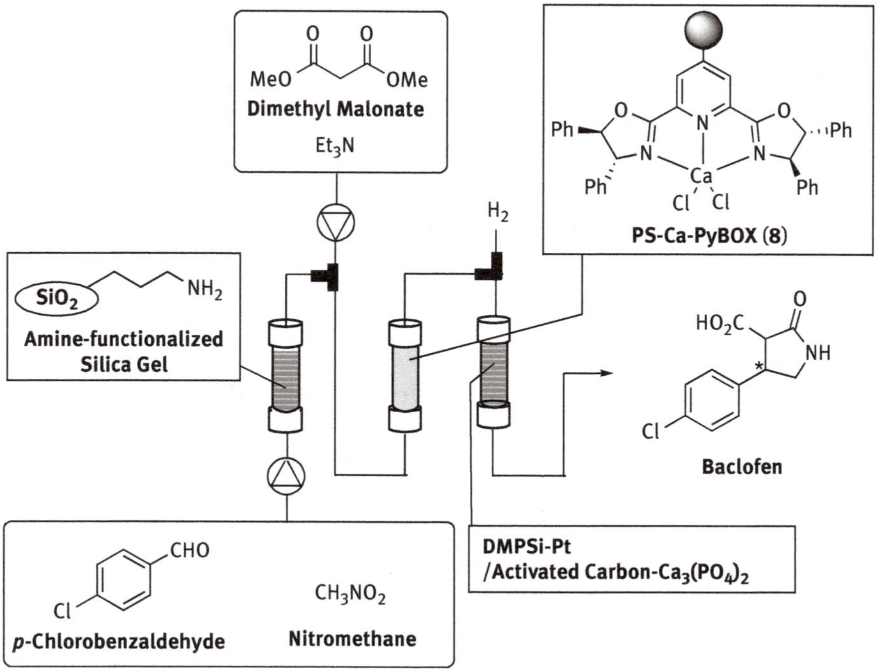

Fig. 10.36: Enantioselective sequential flow synthesis of baclofen.

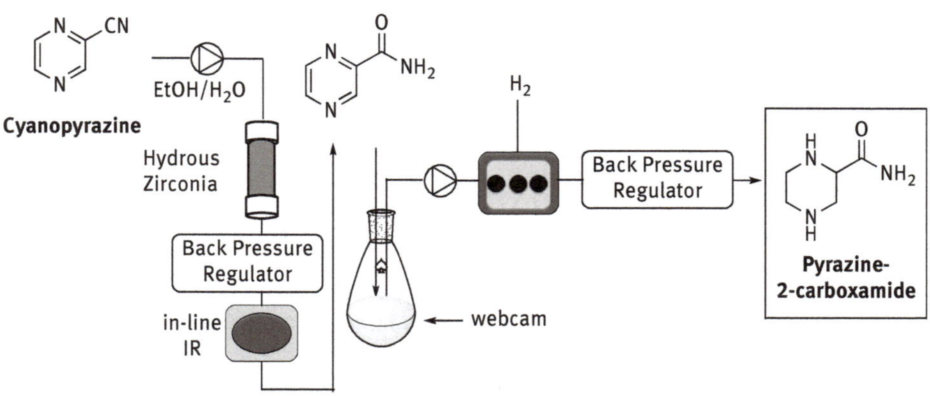

Fig. 10.37: Reaction monitoring and sequential flow synthesis.

10.10 Conclusion and outlook

Continuous-flow chemistry is currently a key technology in the manufacture of not only bulk or common chemicals but also fine and specialty chemicals – it offers more convenient on-demand and on-site production. In particular, the application of catalytic reactions with heterogeneous catalysts in continuous flow is most promising because the catalyst and the product are physically separated and waste by-products are minimized. However, compared with the noncatalytic Type I or homogeneous catalytic Type III flow reactions, the use of truly efficient heterogeneous catalysts in flow has only recently commenced. New synthetic methodologies, based on addition and condensation reactions with heterogeneous catalysis in place of substitution reactions, will become increasingly important, particularly in efforts to avoid the carryover of undesirable by-products in multistep flow systems. To attain this goal, more efficient heterogeneous catalysts for various atom-economical organic transformations under continuous-flow conditions must be developed and, furthermore, a fair comparison and evaluation of catalysis is necessary.

In various reactions described in the above sections, some of the results cited from the original literature were reevaluated using a parameter of the substrate/catalyst molar ratio per unit period (h). This is because the yield in a flow reaction is considered a controllable outcome by changing the substrate/catalyst molar ratio. In addition to the fair evaluation of catalytic activity, the productivity per time, productivity per catalyst, space-time yield, and TON may also be important outcomes.

i

Further readings

A selection of reviews is provided in the following references.

- Mak, X. Y., Laurio, P., Seeberger, P. H. *Beilstein J. Org. Chem.*, 2009, 5, 19.
- Webb, D., Jamison, T. F. *Chem. Sci.*, 2010, 1, 675.
- Wenger, J., Ceylan, S., Kirschning, A. *Chem. Commun.*, 2011, 47, 4583.
- Hessel, V., Cortese, B., de Croon, M. H. J. M. *Chem. Eng. Sci.*, 2011, 66, 1426.
- Wiles, C., Watts, P. *Green Chem.*, 2012, 14, 38.
- Wenger, J., Ceylan, S., Kirschning, A. *Adv. Synth. Catal.*, 2012, 354, 17.
- Anderson, N. G. *Org. Process. Res. Dev.*, 2012, 16, 852.
- Baxendale, I. R. *J. Chem. Technol. Biotechnol.*, 2013, 88, 519.
- Bieringer, T., Buchholz, S., Kockmann, N *Chem. Eng. Technol.*, 2013, 36, 900.
- Pastre, J. C., Browne, D. L., Ley, S. V. *Chem. Soc. Rev.*, 2013, 42, 8849.
- Masuda, K., Ichitsuka, T., Koumura, N., Sato, K., Kobayashi, S. *Tetrahedron*, 2018, 74, 1705.

?

Study questions

10.1 Divide catalytic flow reactions into two major categories.

10.2 Provide an advantage and a disadvantage of the two major categories of catalytic flow reactions.

10.3 Answer the following questions with respect to the image of a flow reaction with a catalyst-packed reactor that is provided.

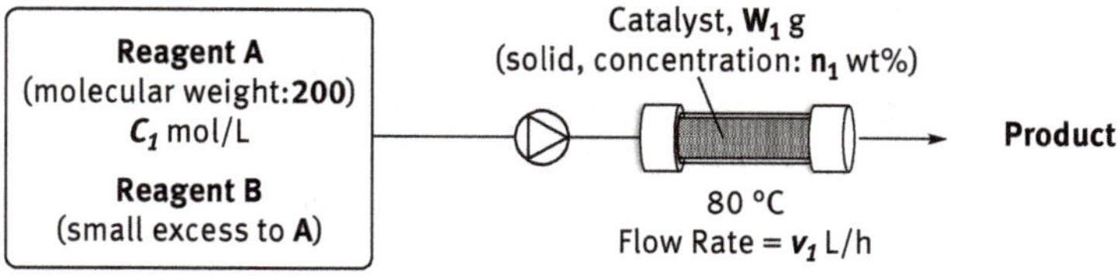

(a) Provide a general formula for the weight hourly space velocity (WHSV).
(b) Describe methods to reduce the above WHSV to half of that value.
(c) What is the effect of reducing the WHSV on the reaction outcome, that is, the yield of the product?
(d) When C_1, W_1, n_1, and v_1 are 0.5, 1.0, 5, and 0.5, respectively, calculate the value of WHSV [h^{-1}].
(e) Let us assume that above system (4) affords 8 g of product within 1 h. Maintaining the conversion rate from the reagents to the product, now describe at least one set of conditions that will afford 80 g of the product within 1 h.

10.4 From the following given types of reactions (a–f), state the advantageous reactions for conducting a multistep "one-flow" reaction:
(a) S_N1 substitution reaction
(b) pericyclic reaction
(c) addition reaction
(d) Wittig olefination
(e) Jones oxidation
(f) hydrogenation reaction

References

[1] Kirschning, A, Guest Editor, Chemistry in flow system, Beilstein J Org Chem, 2009, 5, 15.

[2] (a) Mak, XY, Laurio, P, Seeberger, PH, Asymmetric reactions in continuous flow, Beilstein J Org Chem, 2009, 5, 19. (b) Webb, D, Jamison, TF, Continuous Flow Multi-step Organic Synthesis Chem Sci, 2010, 1, 675–680; (c) Wenger, J, Ceylan, S, Kirschning, A, Ten Key Issues in Modern Flow Chemistry, Chem Commun, 2011, 47, 4583–4592; (d) Hessel, V, Hessel, Cortese, B, de Croon, M H J. M, Novel Process Windows Concept, Proposition and Evaluation Methodology, and Intensified Superheated Processing. Chem Eng Sci, 2011, 66, 1426–1448; (e) Wiles, C., Watts, P. Continuous Flow Reactors: a Perspective. Green Chem, 2012, 14, 38–54; (f) Wenger, J, Ceylan, S, Kirschning, A. Flow Chemistry – A Key Enabling Technology for Multistep Organic Synthesis. Adv Synth Catal, 2012, 354, 17–57; (g) Anderson, NG, Using Continuous Processes to Increase Production, Org. Process. Res. Dev., 2012, 16, 852–869; (h) Baxendale, I. R. The Integration of Flow Reactors into Synthetic Organic Chemistry, J Chem Technol Biotechnol, 2013, 88, 519–552; (i) Bieringer, T., Buchholz, S., Kockmann, N. Future Production Concepts in the Chemical Industry: Modular – Small-Scale – Continuous, Chem Eng Technol, 2013, 36, 900–910; (j) Pastre, JC, Browne, DL, Ley, SV, Flow Chemistry Syntheses of Natural Products. Chem Soc Rev, 2013, 42, 8849–8869; (k) Masuda, K, Ichitsuka, T., Koumura, N, Sato, K, Kobayashi, S, Flow Fine Synthesis with Heterogeneous Catalysts, Tetrahedron, 2018, 74, 1705–1730.

[3] Kobayashi, S, Flow "fine" synthesis: high yielding and selective organic synthesis by flow methods, Chem Asian J, 2016, 11, 425–436.

[4] Itsuno, S, Sakurai, Y, Ito, K, Maruyama, T, Nakahama, S, Frechet, JMJ, New solid-phase catalysts for asymmetric synthesis: cross-linked polymers containing a chiral schiff base-zinc complex, J Org Chem, 1990, 55, 304–310.

[5] Hodge, P, Sung, DWL, Stratford, PW, Asymmetric synthesis of 1-phenylpropanol using polymer-supported chiral catalysts in simple bench-top flow systems, J Chem Soc, Perkin Trans, 1, 1999, 2335–2342.

[6] (a) Finelli, FG, Miranda, LSM, De Souza, ROMA, Expanding the toolbox of asymmetric organocatalysis by continuous-flow process, Chem Commun, 2015, 51, 3708–3722. (b) Rodriguez-Escrich, C., Pericas, M. A. Organocatalysis on Tap: Enantioselective Continuous Flow Processes Mediated by Solid-Supported Chiral Organocatalysts. Eur. J. Org. Chem., 2015, 1173–1188; (c) Puglisi, A., Benaglia, M., Porta, R., Coccia, F. Organocatalysis Chemistry in Flow. Curr. Organocat., 2015, 2, 79–101; (d) Guillena, G., Alonso, D., Baeza, A., Chinchilla, R., Flores-Ferrandiz, J., Gomez-Martinez, M., Trillo, P. Pursuing Chemical Efficiency by Using Supported Organocatalysts for Asymmetric Reactions under Aqueous Conditions. Curr. Organocat., 2015, 2, 102–123; (e) Atodiresei, I., Vila, C., Rueping, M. Asymmetric organocatalysis in continuous flow: opportunities for impacting industrial catalysis. ACS Catal., 2015, 5, 1972–1985.

[7] Bonrath, W, Pressel, Y, Schutz, J, Ferfecki, J, Topp, K-D, Aldol condensations catalyzed by basic anion-exchange resins, ChemCatChem, 2016, 8, 3584–3591.

[8] Laroche, B, Saito, Y, Ishitani, H, Kobayashi, S, Basic anion-exchange resin-catalyzed aldol condensation of aromatic ketones with aldehydes in continuous flow, Org Process Res Dev, 2019, 23, 961–967.

[9] Pandarus, V, Gingras, G, Beland, F, Ciriminna, R, Pagliaro, M, Enhanced catalysis under flow conditions using siliabond functionalized silica gels, Catal Sci Technol, 2011, 1, 1600–1604.

[10] Ishitani, H, Kanai, K, Saito, Y, Tsubogo, T, Kobayashi, S, Synthesis of (±)-pregabalin by utilizing a three-step sequential-flow system with heterogeneous catalysts, Eur J Org Chem, 2017, 6491–6494.

[11] (a) Alza, E, Rodiriguez-Escrich, C, Sayalero, S, Bastero, A, Pericas, MA, A solid-supported organocatalyst for highly stereoselective, batch, and continuous-flow mannich reactions, Chem Eur J, 2009, 15, 10167–10172. (b) Martin-Rapun, R., Satyalero, S., Pericas, M. A. Asymmetric Anti-Mannich Reactions in Continuous Flow. Green Chem., 2013, 15, 3295–3301; (c) Sagamanova, I., Rodiriguez-Escrich, C., Molnar, I. G., Sayalero, S., Gilmore, R., Pericas, M. A. Translating the enantioselective michael reaction to a continuous flow paradigm with an immobilized, fluorinated organocatalyst. ACS Catal., 2015, 5, 6241–6248.

[12] Nakashima, E, Yamamoto, H, Process catalyst mass efficiency by using proline tetrazole column-flow system, Chem Eur J, 2018, 24, 1076–1079.

[13] Ayats, C, Henseler, AH, Dibello, E, Pericas, MA, Continuous flow enantioselective three-component anti-mannich reactions catalyzed by a polymer-supported threonine derivative, ACS Catal, 2014, 4, 3027–3033.

[14] Osorio-Planes, L, Rodiriguez-Escrich, C, Pericas, MA, Enantioselective continuous-flow production of 3-indolylmethanamines mediated by an immobilized phosphoric acid catalyst, Chem Eur J, 2014, 20, 2367–2372.

[15] Ishitani, H, Saito, Y, Kobayashi, S, Enantioselective organometallic catalysis in flow, Top Organomet Chem, 2016, 57, 213–248.

[16] Lundgren, S, Ihre, H, Moberg, C, Enantioselective cyanation of benzaldehyde. An asymmetric polymeric catalyst in a microreactor, ARKIVOC, 2008, 6, 73–80.

[17] Sayad, AM, Ramalingam, B, Chai, CLL, Li, C, Garland, MV, Yoshinaga, K, Self-Supported Chiral Titanium Cluster (SCTC) as a robust catalyst for the asymmetric cyanation of imines under batch and continuous flow at room temperature, Chem Eur J, 2012, 18, 5693–5700.

[18] (a) He, Y, Jawad, A, Li, X, Atanga, M, Rezaeim, F, Rownaghi, AA, Direct aldol and nitroaldol condensation in an aminosilane-grafted Si/Zr/Ti composite hollow fiber as a heterogeneous catalyst and continuous-flow reactor, J Catal, 2016, 341, 149–159. (b) Tsukanov, S. V., Johnson, M. D., May, S. A., Rosemeyer, M., Watkins, M. A., Kolis, S. P., Yates, M. H., Johnston, J. N. Development of an intermittent-flow enantioselective aza-Henry reaction using an arylnitromethane and homogeneous brønsted acid–base catalyst with recycle. Org. Process Res. Dev., 2016, 20, 215–226. (c) Yamada, T., Kuwata, M., Takakura, R., Monguchi, Y., Sajiki, H., Sawama, Y. Organocatalytic nitroaldol reaction associated with deuterium-Labeling. Adv. Synth. Catal., 2018, 360, 637–641.

[19] Hashimoto, K, Kumagai, N, Shibasaki, M, Self-assembled asymmetric catalyst engaged in a continuous-flow platform: An anti-selective catalytic asymmetric nitroaldol reaction, Org Lett, 2014, 16, 3496–3499.

[20] Ishitani, H, Saito, Y, Tsubogo, T, Kobayashi, S, Synthesis of nitro-containing compounds through multistep continuous flow with heterogeneous catalysts, Org Lett, 2016, 18, 1346–1349.

[21] Shu, W, Bachwald, SL, Enantioselective β-arylation of ketones enabled by lithiation/borylation/1,4-addition sequence under flow conditions, Angew Chem Int Ed, 2012, 51, 5355–5358.

[22] Tsubogo, T, Yamashita, Y, Kobayashi, S, Toward efficient asymmetric carbon-carbon bond formation: continuous flow with chiral heterogeneous catalysts, Chem Eur J, 2012, 18, 13624–13628.

[23] Garcia-Lacuna, J, Dominguez, G, Perez-Castells, J, Flow chemistry for cycloaddition reactions, ChemSusChem, 2020, 13, 5138–5163.

[24] Kamahori, K, Ito, K, Itsuno, S, Asymmetric diels–alder reaction of methacrolein with cyclopentadiene using polymer-supported catalysts: Design of highly enantioselective polymeric catalysts, J Org Chem, 1996, 61, 8321–8324.

[25] (a) Sachse, A, Hulea, V, Finiels, A, Coq, B, Fajula, F, Galarneau, A, Alumina-grafted macro-/mesoporous silica monoliths as continuous flow microreactors for the Diels–Alder reaction, J Catal, 2012, 287, 62–67. (b) Seghers, S., Protasova, L., Mullens, S., Thybaut, J. W., Stevens, C. Improving the efficiency of the Diels–Alder process by using flow chemistry and zeolite catalysis. Green Chem., 2017, 19, 237–248.

[26] Ji, P, Feng, X, Oliveres, P, Li, Z, Murakami, A, Wang, C, Lin, W, Strongly Lewis Acidic metal–organic frameworks for continuous flow catalysis, J Am Chem Soc, 2019, 141, 14878–14888.

[27] (a) Chiroli, V, Benaglia, M, Puglisi, A, Porta, R, Jumde, R. P., Mandoli, A, A chiral organocatalytic polymer-based monolithic reactor, Green Chem, 2014, 16, 2798–2806. (b) de Oliveira, P. H. R., Santos, B. M. daS., Leão, R. A. C., Miranda, L. S. M., San Gil, R. A. S., de Sauza, R. O. M. A., Finelli, F. G. From immobilization to catalyst use: a complete continuous-flow approach towards the use of immobilized organocatalysts. ChemCatChem, 2019, 11, 5553–5561.

[28] Borukhova, S, Seeger, AD, Noel, T, Wang, Q, Busch, M, Hessel, V, Pressure-accelerated azide–alkyne cycloaddition: Micro capillary versus autoclave reactor performance, ChemSusChem, 2015, 8, 504–512.

[29] (a) Beers, AEW, Nijhuis, TA, Kapteijn, F, Moulijn, JA, Zeolite coated structures for the acylation of aromatics, Microporous Mesoporous Mater, 2001, 48, 279–284. (b) Guidotti, M., Canaff, M., Coustard, J.-M., Magnoux, P., Guisnet, M. Acetylation of aromatic compounds over H-BEA zeolite: the influence of the substituents on the reactivity and on the catalyst stability. J. Catal., 2005, 230, 375–383; (c) Dejaegere, E. A., Thybaut, J. W., Marin, G. B., Baron, G. V., Denayer, J. F. M. Modeling of toluene acetylation with acetic anhydride on H-USY zeolite. Ind. Eng. Chem. Res., 2011, 50, 11822–11832; (d) Chen, Z., Chen, W., Tong, T., Zeng, A. Continuous

liquid phase acylation of toluene over hbea zeolite: solvent effects and origin of the deactivation. J. Mol. Catal. A, 2015, 396, 231–238.

[30] Sun, Q, Hu, K, Leng, K, Yi, X, Agulia, B, Sun, Y, Zheng, A, Meng, X, Ma, S, Xiao, F-S, A porous Brønsted superacid as an efficient and durable solid catalyst, J Mater Chem A, 2018, 6, 18712–18719.

[31] Rao, X, Ishitani, H, Yoo, W-J, Kobayashi, S, Zirconium-β zeolite-catalyzed continuous-flow friedel-crafts acylation reaction, Asian J Org Chem, 2019, 8, 316–319.

[32] (a) Snead, DR, Jamison, TF, A three-minute synthesis and purification of ibuprofen: Pushing the limits of continuous-flow processing, Angew Chem Int Ed, 2015, 54, 983–987. (b) Ghislieri, K., Gilmore, D., Seeberger, P. H. Chemical assembly systems: layered control for divergent, continuous, multistep syntheses of active pharmaceutical ingredients. Angew. Chem. Int. Ed., 2015, 54, 678–682.

[33] Yu, T, Jiao, J, Song, P, Nie, W, Yi, C, Zhang, Q, Li, P, Recent progress in continuous-flow hydrogenation, ChemSusChem, 2020, 13, 2876–2893.

[34] Gutmann, B, Roduit, J-P, Roberge, D, Kappe, CO, A two-step continuous-flow synthesis of N-(2-Aminoethyl)acylamides through ring-opening/hydrogenation of oxazolines, Chem Eur J, 2011, 17, 13146–13150.

[35] DallaVechia, L, Reichart, B, Glasnov, T, Miranda, LSM, Kappe, CO, De Souza, ROMA, A three step continuous flow synthesis of the biaryl unit of the HIV protease inhibitor atazanavir, Org Biomol Chem, 2013, 11, 6806–6813.

[36] Chen, M, Buchwald, SL, Continuous-flow synthesis of 1-substituted benzotriazoles from chloronitrobenzenes and amines in a C–N bond formation/hydrogenation/diazotization/cyclization sequence, Angew Chem Int Ed, 2013, 52, 4247–4250.

[37] Cantillo, D, Wolf, B, Goetz, R, Kappe, CO, Continuous flow synthesis of a Key 1,4-benzoxazinone intermediate via a nitration/hydrogenation/cyclization sequence, Org Process Res Dev, 2017, 21, 125–132.

[38] Barwinski, B, Migowski, P, Gallou, F, Franciò, G, Leitner, W, Continuous-flow hydrogenation of 4-phenylpyridine to 4-phenylpiperidine with integrated product isolation Using a CO2 switchable system, J Flow Chem, 2017, 7, 41–45.

[39] Hizartzidis, L, Cossar, PJ, Robertson, MJ, Simone, M, Young, KA, McCluskey, A, Gordon, CP, Expanding the utility of flow hydrogenation – a robust protocol restricting hydrodehalogenation, RSC Adv, 2014, 4, 56743–56748.

[40] Loos, P, Alex, H, Hassfeld, J, Lovis, K, Platzek, J, Steinfeldt, N, Hubner, S, Selective hydrogenation of halogenated nitroaromatics to haloanilines in batch and floworg, Process Res Dev, 2016, 20, 452–464.

[41] Baramov, T, Hassfeld, J, Gottfried, M, Bauer, H, Herrmann, T, Ben Said, M, Roggan, S, Continuous platinum-mediated hydrogenation of refametinib iodo-nitroaniline key intermediate DIM-NA: The combined challenges of selectivity and catalyst deactivation, Eur J Org Chem, 2017, 3921–3928.

[42] Rubulotta, G, Luska, KL, Urbina-Blanco, CA, Eifert, T, Palkovits, R, Quadrelli, EA, Theuleux, C, Leitner, W, Highly selective hydrogenation of R-(+)-limonene to (+)-p-1-menthene in batch and continuous flow reactors, ACS Sustain Chem Eng, 2017, 5, 3762–3767.

[43] Hattori, T, Ida, T, Tsubone, A, Sawama, Y, Monguchi, Y, Sajiki, H, Facile arene hydrogenation under flow conditions catalyzed by rhodium or ruthenium on carbon, Eur J Org Chem, 2015, 2492–2497.

[44] Fink, MJ, Schön, M, Rudroff, F, Schnürch, M, Mihovilovic, MD, Single operation stereoselective synthesis of aerangis lactones: Combining continuous flow hydrogenation and biocatalysts in a chemoenzymatic sequence, ChemCatChem, 2013, 5, 724–727.

[45] Viviano, M, Glasnov, TM, Reichart, B, Tekautz, G, Kappe, CO, A scalable two-step continuous flow synthesis of nabumetone and related 4-Aryl-2-butanones, Org Process Res Dev, 2011, 15, 858–870.
[46] Martinez, ST, Pinto, AC, Glasnov, T, Kappe, CO, Chemistry of pyrrolizidine alkaloids revisited – semi-synthetic microwave and continuous-flow approaches toward crotalaria-alkaloids, Tetrahedron Lett, 2014, 55, 4181–4184.
[47] Chen, J, Przyuski, K, Roemmele, R, Bakale, RP, Improved continuous flow processing: Benzimidazole ring formation via catalytic hydrogenation of an aromatic nitro compound, Org Process Res Dev, 2014, 18, 1427–1433.
[48] Said, MB, Baramov, T, Herrmann, T, Gottfried, M, Hassfeld, J, Roggan, S, Continuous selective hydrogenation of refametinib iodo-nitroaniline key intermediate DIM-NA over raney cobalt catalyst at kg/day scale with online UV–visible conversion control, Org Process Res Dev, 2017, 21, 705–714.
[49] Cantillo, D, Damm, M, Dallinger, D, Bauser, M, Berger, M, Kappe, CO, Sequential nitration/hydrogenation protocol for the synthesis of triaminophloroglucinol: Safe Generation and use of an explosive intermediate under continuous-flow conditions, Org Process Res Dev, 2014, 18, 1360–1366.
[50] Linares, N, Hartmann, S, Galarneau, A, Barbaro, P, Continuous partial hydrogenation reactions by Pd@unconventional bimodal porous titania monolith catalysts, ACS Catal, 2012, 2, 2194–2198.
[51] Linares, N, Moreno-Marrodan, C, Barbaro, P, PdNP@Titanate nanotubes as effective catalyst for continuous-flow partial hydrogenation reactions, ChemCatChem, 2016, 8, 1001–1011.
[52] Cherkasov, N, Ibhadon, AO, Rebrov, EV, Novel synthesis of thick wall coatings of titania supported Bi poisoned pd catalysts and application in selective hydrogenation of acetylene alcohols in capillary microreactors, Lab Chip, 2015, 15, 1952–1960.
[53] Moreno-Marrodan, C, Barbaro, P, Catalano, M, Taurino, A, Green production of polymer-supported pdnps: application to the environmentally benign catalyzed synthesis of Cis-3-Hexen-1-ol under flow conditions, Dalton Trans, 2012, 41, 12666–12669.
[54] Ouchi, T, Battilocchio, C, Hawkins, JM, Ley, SV, Process intensification for the continuous flow hydrogenation of ethyl nicotinate, Org Process Res Dev, 2014, 18, 1560–1566.
[55] Lévêque, J, Philippe, R, Zanota, ML, Meille, V, Sarrazin, F, Baussaron, L, De Bellefon, C, Hydrodynamics and mass transfer in a tubular reactor containing foam packings for intensification of G-L-S Catalytic reactions in co-current up-flow configuration, Chem Eng Res Des, 2016, 109, 686–697.
[56] Liguori, F, Barbaro, P, Sawa, H, Continuous flow hydrogenation reactions by Pd catalysts onto Hybrid ZrO2/PVA materials, Appl Catal A: Gen, 2014, 488, 58–65.
[57] Vilé, G, Almora-Barrios, N, Mitchell, S, López, N, Pérez-Ramírez, J, From the lindlar catalyst to supported ligand-modified palladium nanoparticles: Selectivity patterns and accessibility constraints in the continuous-flow three-phase hydrogenation of acetylenic compounds, Chem Eur J, 2014, 20, 5926–5937.
[58] Ueno, M, Morii, Y, Uramoto, K, Oyamada, H, Mori, Y, Kobayashi, S, Catalytic flow hydrogenation of aromatic nitro compounds using polysilane-supported palladium, J Flow Chem, 2014, 4, 160–163.
[59] Kobayashi, S, Okumura, M, Akatsuka, Y, Miyamura, H, Ueno, M, Oyamada, H, Powerful continuous-flow hydrogenation by using Poly(dimethyl)silane-Supported palladium catalysts, ChemCatChem, 2015, 7, 4025–4029.
[60] Saito, Y, Ishitani, H, Kobayashi, S, Catalytic hydrogenation of aliphatic nitro compounds with polysilane/bone charcoal-supported palladium catalysts under continuous-flow conditions, Asian J Org Chem, 2016, 5, 1124–1127.

[61] Saito, Y, Ishitani, H, Ueno, M, Kobayashi, S, Selective hydrogenation of nitriles to primary amines catalyzed by a Polysilane/SiO2-supported palladium catalyst under continuous-flow conditions, ChemistryOpen, 2017, 6, 211–215.
[62] Liguori, F, Barbaro, P, Green semi-hydrogenation of alkynes by Pd@Borate monolith catalysts under continuous flow, J Catal, 2014, 311, 212–220.
[63] Cherkasov, N, Al-Rawashdeh, M, Ibhadon, AO, Rebrov, EV, Scale up study of capillary microreactors in solvent-free semihydrogenation of 2-Methyl-3-butyn-2-ol, Catal Today, 2016, 273, 205–212.
[64] Rathi, AK, Gawande, MB, Ranc, V, Pechousek, J, Petr, M, Cepe, K, Varma, RS, Zboril, R, Continuous flow hydrogenation of nitroarenes, azides and alkenes using maghemite–Pd nanocomposites, Catal Sci Technol, 2016, 6, 152–160.
[65] Nagendiran, A, Sörensen, H, Johansson, MJ, Tai, C-W, Bäckvall, J-E, Nanopalladium-catalyzed conjugate reduction of michael acceptors – application in flow, Green Chem, 2016, 18, 2632–2637.
[66] Furuta, A, Hirobe, Y, Fukuyama, Y, Ryu, I, Manabe, Y, Fukase, K, Flow dehydration and hydrogenation of allylic alcohols: Application to the waste-free synthesis of pristane, Eur J Org Chem, 2017, 1365–1368.
[67] Liguori, F, Barbaro, P, Said, B, Galarneau, A, Santo, VD, Passaglia, E, Feis, A, Unconventional Pd@Sulfonated Silica Monoliths catalysts for selective partial hydrogenation reactions under continuous flow, ChemCatChem, 2017, 9, 3245–3258.
[68] Kataoka, S, Takeuchi, Y, Harada, A, Takagi, T, Takenaka, T, Fukaya, N, Yasuda, H, Ohmori, T, Endo, A, Microreactor Containing platinum nanoparticles for nitrobenzene hydrogenation, Appl Catal A: Gen, 2012, 427–428, 119–124.
[69] Durndell, LJ, Wilson, K, Lee, AF, Platinum-catalysed Cinnamaldehyde Hydrogenation in Continuous flow, RSC Adv, 2015, 5, 80022–80026.
[70] Hsieh, C-T, Ötvös, SB, Wu, Y-C, Mándity, IM, Chang, F-R, Fülöp, F, Highly selective continuous-flow synthesis of potentially bioactive deuterated chalcone derivatives, ChemPlusChem, 2015, 80, 859–864.
[71] Osako, T, Torii, K, Tazawa, T, Uozumi, Y, Continuous-flow hydrogenation of olefins and nitrobenzenes catalyzed by platinum nanoparticles dispersed in an amphiphilic polymer, RSC Adv, 2015, 5, 45760–45766.
[72] Rehm, TH, Reinhard, D, Kost, H-J, Hofmann, C, Zapf, R, Löb, P, Laribi, Y, Perrichon, P, Berguerand, C, Kiwi-Minsker, L, Sulman, E, Szirbik, G, Richert, H, Lang, J, Gottfried, M, Roggan, S, Chemoselective three-phase hydrogenation of an ombrabulin nitro-stilbene intermediate in a continuous-flow mobile platform, Chem Eng J, 2017, 316, 1069–1077.
[73] Wang, X, Cárdenas-Lizana, F, Keane, MA, Toward sustainable chemoselective nitroarene hydrogenation using supported gold as catalyst, ACS Sustain Chem Eng, 2014, 2, 2781–2789.
[74] Vilé, G, Pérez-Ramírez, MJ, Beyond the Use of modifiers in selective alkyne hydrogenation: Silver and gold nanocatalysts in flow mode for sustainable alkene production, Nanoscale, 2014, 6, 13476–13482.
[75] Artiukha, EA, Nuzhdin, AL, Bukhtiyarova, GA, Zaytzev, SY, Plyusnin, PE, Shubin, YV, Bukhtiyarov, VI, One-pot reductive amination of aldehydes with nitroarenes over an Au/Al2O3 Catalyst in a continuous flow reactor, Catal Sci Technol, 2015, 5, 4741–4745.
[76] Nuzhdin, AL, Moroz, BL, Bukhtiyarova, GA, Reshetnikov, SI, Pyrjaev, PA, Aleksandrov, PV, Bukhtiyarov, VI, Selective liquid-phase hydrogenation of a nitro group in substituted nitrobenzenes over Au/Al2O3 Catalyst in a packed-bed flow reactor, ChemPlusChem, 2015, 80, 1741–1749.
[77] Nuzhdin, AL, Artiukha, EA, Bukhtiyarova, GA, Zaytzev, SY, Plyusnin, PE, Shubin, YV, Bukhtiyarov, VI, Synthesis of unsaturated secondary amines by direct reductive amination of

aliphatic aldehydes with nitroarenes over Au/Al2O3 catalyst in continuous flow mode, RSC Adv, 2016, 6, 88366–88372.

[78] Gericke, D, Ott, D, Matveeva, VG, Sulman, E, Aho, A, Murzin, DY, Roggan, S, Danilova, L, Hessel, V, Loeb, P, Kralisch, D, Green catalysis by nanoparticulate catalysts developed for flow processing? Case study of glucose hydrogenation, RSC Adv, 2015, 5, 15898–15908.

[79] Haywood, T, Miller, PW, Microfluidic hydrogenation reactions by using a channel-supported rhodium catalyst, ChemCatChem, 2014, 6, 1199–1203.

[80] Paun, C, Giziński, D, Zienkiewicz-Machnik, M, Banaś, D, Kubala-Kukuś, A, Sá, J, p-Nitrophenol flow hydrogenation with Nano-Cu2O grafted on polymeric resin, Catal Commun, 2017, 92, 61–64.

[81] Giziński, D, Goszewska, I, Zieliński, M, Lisovytskiy, D, Nikiforov, K, Masternak, J, Zienkiewicz-Machnik, M, Srębowata, A, Sá, J, Chemoselective flow hydrogenation of α,β – unsaturated aldehyde with nano-nickel, Catal Commun, 2017, 98, 17–21.

[82] Vilé, G, Wrabetz, S, Floryan, L, Schuster, ME, Girgsdies, F, Teschner, D, Perez-Ramirez, J, Stereo- and chemoselective character of supported CeO2 catalysts for continuous-flow three-phase alkyne hydrogenation, ChemCatChem, 2014, 6, 1928–1934.

[83] Baramov, T, Loos, P, Hassfeld, J, Alex, H, Beller, M, Stemmler, T, Meier, G, Gottfried, M, Roggan, S, Encapsulated cobalt oxide on carbon nanotube support as catalyst for selective continuous hydrogenation of the showcase substrate 1-Iodo-4-nitrobenzene, Adv Synth Catal, 2016, 358, 2903–2911.

[84] Orito, Y, Imai, S, Niwa, S, Nguyen, GH, Asymmetric hydrogenation of methyl benzoylformate using platinum-carbon catalysts modified with cinchonidine, J Synth Org Chem Jpn, 1979, 37, 173–174.

[85] Meheux, PA, Ibbotson, A, Wells, PB, Enantioselective Hydrogenation: II. variation of activity and optical yield with experimental variables in methyl pyruvate hydrogenation catalyzed by cinchona-modified platinum/silica (EUROPT-1), J Catal, 1991, 128, 387–396.

[86] (a) Künzle, N, Hess, R, Mallat, T, Baiker, A, Continuous enantioselective hydrogenation of activated ketones, J Catal, 1999, 186, 239–241. (b) Meier, D. M., Ferri, D., Mallat, T., Baiker, A., Molecular Insight into the Dynamics of Chiral Modification of Pt/alumina. J. Catal., 2007, 248, 68–76.

[87] (a) Künzle, N, Mallat, T, Baiker, A, Enantioselective Hydrogenation of isopropyl-4,4,4-trifluoroacetoacetate in a continuous flow reactor, Appl Catal A: Gen, 2003, 238, 251–257. (b) Wandeler, R., Künzle, N., Schneider, M. S., Mallat, T., Baiker, A. Continuous Platinum-catalyzed Enantioselective Hydrogenation in 'Supercritical' Solvents. Chem. Commun., 2001, 673–674; (c) Szöllösi, G., Cserényi, S., Balázski, K., Fülöp, F., Bartók, M. New Data in the Enantioselective Hydrogenation of Ethyl Pyruvate on Pt-cinchona Chiral Catalyst Using Continuous-flow Fixed-bed Reactor System: The Origin of Rate Enhancement. J. Mol. Catal. A: Chem., 2009, 305, 155–160; (d) Szöllösi, G., Cserényi, S., Bucsi, I., Bartók, T., Fülöp, F., Bartók, M. Origin of the Rate Enhancement and Enantiodifferentiation in the Heterogeneous Enantioselective Hydrogenation of 2,2,2-trifluoroacetophenone over Pt/alumina Studied in Continuous-flow Fixed-bed Reactor System. Appl. Catal. A: Gen., 2010, 382, 263–271; (e) Szöllösi, G., Makra, Z., Fekete, M., Fülöp, F., Bartók, M. Heterogeneous Enantioselective Hydrogenation in a Continuous-flow Fixed-bed Reactor System: Hydrogenation of Activated Ketones and Their Binary Mixtures on Pt–Alumina–Cinchona Alkaloid Catalysts. Catal. Lett., 2012, 142, 889–894; (f) Hermán, B., Szöllösi, G., Fülöp, F., Bartók, M. Enantioselective Hydrogenation of α,β-Unsaturated Carboxylic Acids in Fixed-bed Reactor. Appl. Catal. A: Gen., 2007, 331, 39–43; (g) Szöllösi, G., Cserényi, S., Bartók, M. Novel Evidence on the Role of the Nucleophilic Intermediate Complex in the Orito-Reaction: Unexpected Inversion

in the Enantioselective Hydrogenation of 2,2,2-Trifluoroacetophenone on Pt-Cinchona Chiral Catalyst Using Continuous-Flow Fixed-Bed Reactor. Catal. Lett., 2010, 134, 264–269.
[88] (a) Augustine, RL, Tanielyan, SK, Mahata, N, Gao, Y, Zsigmond, A, Yang, H, Anchored homogeneous catalysts: the role of the heteropoly acid anchoring agent, Appl Catal A: Gen, 2003, 256, 69–76. (b) Stephenson, P., Licence, P., Ross, S. K., Poliakoff, M. Continuous Catalytic Asymmetric Hydrogenation in Supercritical CO_2. Green Chem. 2004, 6, 521–523; (c) Al Herz, M. A., Tsoligkas, A. N., Simmons, M. J. H., Wood, J. Enantioselective Hydrogenation of Dimethyl Itaconate with Immobilised Rhodium-duphos Complex in a Recirculating Fixed-bed Reactor. Appl. Catal. A: Gen., 2011, 396, 148–158; (d) Madarász, J., Farkas, G., Balogh, S., Szöllösy, A., Kovács, J., Darvas, F., Urge, L., Bakos, J. A Continuous-Flow System for Asymmetric Hydrogenation Using Supported Chiral Catalysts. J. Flow Chem., 2011, 2, 62–67.
[89] Duque, R, Pogorzelec, PJ, Cole-Hamilton, DJ, A single enantiomer (99 %) directly from continuous-flow asymmetric hydrogenation, Angew Chem Int Ed, 2013, 52, 9805–9807.
[90] Saito, Y, Kobayashi, S, Development of Robust heterogeneous chiral rhodium catalysts utilizing acid–base and electrostatic interactions for efficient continuous-flow asymmetric hydrogenations, J Am Chem Soc, 2020, 142, 16546–16551.
[91] Hintermair, U, Francio, G, Leitner, W, A fully integrated continuous-flow system for asymmetric catalysis: Enantioselective hydrogenation with supported ionic liquid phase catalysts using supercritical CO2 as the Mobile Phase, Chem Eur J, 2013, 19, 4538–4547.
[92] (a) Theuerkauf, J, Franciò, G, Leitner, W, Continuous-flow asymmetric hydrogenation of the β-Keto ester methyl propionylacetate in ionic liquid–supercritical carbon dioxide biphasic systems, Adv Synth Catal, 2013, 355, 209–219. (b) Hintermair, U., Höfener, T., Pullmann, T., Franciò, G., Leitner, W. Continuous Enantioselective Hydrogenation with a Molecular Catalyst in Supported Ionic Liquid Phase under Supercritical CO2 Flow. ChemCatChem, 2010, 2, 150–154.
[93] Kobayashi, S, Miyamura, H, Akiyama, R, Ishida, T, Highly active, immobilized ruthenium catalysts for oxidation of alcohols to aldehydes and ketones. Preparation and use in both batch and flow systems, J Am Chem Soc, 2005, 127, 9251–9254.
[94] Zotova, N, Hellgardt, K, Kelsall, GH, Jessiman, AS, Hii, KK, Catalysis in flow: the practical and selective aerobic oxidation of alcohols to aldehydes and ketones, Green Chem, 2010, 12, 2157–2163.
[95] Mannel, DS, Stahl, SS, Root, TW, Continuous flow aerobic alcohol oxidation reactions using a heterogeneous Ru(OH)x/Al2O3 Catalyst, Org Process Res Dev, 2014, 18, 1503–1508.
[96] Liu, H, Feng, J, Zhang, J, Miller, PW, Chen, L, Su, C-Y, A catalytic chiral gel microfluidic reactor assembled via dynamic covalent chemistry, Chem Sci, 2015, 6, 2292–2296.
[97] (a) Beigi, M, Haag, R, Liese, A, Continuous application of polyglycerol-supported salen in a membrane reactor: Asymmetric epoxidation of 6-Cyano-2,2-dimethylchromene, Adv Synth Catal, 2008, 350, 919–925. (b) Cho, S.-H., Walther, N. D., Nguyen, S. T., Hupp, J. T. Anodic aluminium oxide catalytic membranes for asymmetric epoxidation. Chem. Commun., 2005, 5331–5333; (c) Sandel, S., Weber, S. K., Trapp, O. Oxidations with Bonded Salen-catalysts in Microcapillaries. Chem. Eng. Sci., 2012, 83, 171–179.
[98] (a) Romero, NA, Nicewicz, DA, Organic photoredox catalysis, Chem Rev, 2016, 116, 10075–10166. (b) Shaw, M. H., Twilton, J., MacMillan, D. W. C. Photoredox Catalysis in Organic Chemistry. J. Org. Chem., 2016, 81, 6898–6926; (c) Prier, C. K., Rankic, A., MacMillan, D. W. C. Visible Light Photoredox Catalysis with Transition Metal Complexes: Applications in Organic Synthesis. Chem. Rev., 2013, 113, 5322–5363; (d) Narayanam, J. M. R., Stephenson, C. R. J. Visible Light Photoredox Catalysis: Applications in Organic Synthesis. Chem. Soc. Rev., 2011, 40, 102–113.

[99] (a) Morse, PD, Beingessner, RL, Jamison, TF, Enhanced Reaction Efficiency in Continuous Flow, Isr J Chem, 2017, 57, 218–227. (b) Gilmore, K., Seeberger, P. H. Continuous flow photochemistry. Chem. Record, 2014, 14, 410–418; (c) Schuster, E. M., Wipf, P. Photochemical Flow Reactions. Isr. J. Chem., 2014, 54, 361–370; (d) Garlets, P. J., Nguyen, D., Stephenson, C. R. J. The Development of Visible-Light Photoredox Catalysis in Flow. Isr. J. Chem., 2014, 54, 351–360; (e) Su, Y., Straathof, N. J. W., Hessel, V., Noël, T. Photochemical Transformations Accelerated in Continuous-Flow Reactors: Basic Concepts and Applications. Chem. Eur. J., 2014, 20, 10562–10589.

[100] Bonassi, F, Ravelli, D, Protti, S, Fagnoni, M, Decatungstate photocatalyzed acylations and alkylations in flow via hydrogen atom transfer, Adv Synth Catal, 2015, 357, 3687–3695.

[101] Laudadio, G, Govaerts, S, Wang, Y, Ravelli, D, Koolman, HF, Fagnoni, M, Djuric, SW, Noël, T, Selective C(sp3)–H aerobic oxidation enabled by decatungstate photocatalysis in flow, Angew Chem Int Ed, 2018, 57, 4078–4082.

[102] Halperin, SD, Kwon, D, Holmes, M, Regalado, EL, Campeau, L-C, DiRocco, DA, Britton, R, Development of a direct photocatalytic C–H fluorination for the preparative synthesis of odanacatib, Org Lett, 2015, 17, 5200–5203.

[103] (a) Straathof, N, Osch, D, Schouten, A, Wang, X, Schouten, J, Hessel, V, Noël, T, Visible light photocatalytic metal-free perfluoroalkylation of heteroarenes in continuous flow, J Flow Chem, 2014, 4, 12–17. (b) Talla, A., Driessen, B., Straathof, N. J. W., Milroy, L.-G., Brunsveld, L., Hessel, V., Noël, T. Metal-Free Photocatalytic Aerobic Oxidation of Thiols to Disulfides in Batch and Continuous-Flow. Adv. Synth. Catal., 2015, 357, 2180–2186; (c) Bottecchia, C., Rubens, M., Gunnoo, S. B., Hessel, V., Madder, A., Noël, T. Visible-Light-Mediated Selective Arylation of Cysteine in Batch and Flow. Angew. Chem. Int. Ed., 2017, 56, 12702–12707.

[104] (a) Wu, G, Lv, T, Mo, W, Yang, X, Gao, Y, Chen, H, One-pot Synthesis of Tricyclo-1,4-Benzoxazines via Visible-light Photoredox Catalysis in Continuous Flow, Tetrahedron Lett, 2017, 58, 1395–1398. (b) Penders, I. G. T. M., Amara, Z., Horvath, R., Rossen, K., Poliakoff, M., George, M. W. Photocatalytic Hydroxylation of Arylboronic Acids Using Continuous Flow Reactors. RSC Adv., 2015, 5, 6501–6504.

[105] Barona-Castrano, JC, Carmona-Vargas, CC, Brocksom, TJ, De Oliveira, KT, Porphyrins as catalysts in scalable organic reactions, Molecules, 2016, 21, 310.

[106] De Oliveira, KT, Miller, LZ, McQuade, DT, Exploiting photooxygenations mediated by porphyrinoid photocatalysts under continuous flow conditions, RSC Adv, 2016, 6, 12717–12725.

[107] Seo, H, Katcher, MH, Jamison, TF, Photoredox activation of carbon dioxide for amino acid synthesis in continuous flow, Nat Chem, 2017, 9, 453–456.

[108] Seo, H, Liu, A, Jamison, TF, Direct β-selective hydrocarboxylation of styrenes with CO2 enabled by continuous flow photoredox catalysis, J Am Chem Soc, 2017, 139, 13969–13972.

[109] Han, X, Bourne, RA, Poliakoff, M, George, MW, Immobilised photosensitisers for continuous flow reactions of singlet oxygen in supercritical carbon dioxide, Chem Sci, 2011, 2, 1059–1067.

[110] Amara, Z, Bellamy, JF, Horvath, R, Miller, SJ, Beeby, A, Burgard, A, Rossen, K, Poliakoff, M, George, MW, Applying green chemistry to the photochemical route to artemisinin, Nat Chem, 2015, 7, 489–495.

[111] Tobin, JM, McCabe, TJD, Prentice, AW, Holzer, S, Lloyd, GO, Paterson, MJ, Arrighi, V, Cormack, PAG, Vilela, F, Polymer-supported photosensitizers for oxidative organic transformations in flow and under visible light irradiation, ACS Catal, 2017, 7, 4602–4612.

[112] (a) Cantillo, D, Kappe, CO, Immobilized transition metals as catalysts for cross-couplings in continuous flow – A Critical assessment of the reaction mechanism and metal leaching, ChemCatChem, 2014, 6, 3286–3305. (b) Munirathinam, R., Huskens, J., Verboom, W.

Supported Catalysis in Continuous-Flow Microreactors. Adv. Synth. Catal., 2015, 357, 1093–1123; (d) Len, C., Bruniaux, S., Delbecq, F., Parmar, V. Palladium-Catalyzed Suzuki–Miyaura Cross-Coupling in Continuous Flow. Catalysts, 2017, 7, 146; (e) Plutschack, M. B., Pieber, B., Gilmore, K., Seeberger, P. H. The Hitchhiker's Guide to Flow Chemistry, Chem Rev, 2017, 117, 11796–11893; (f) Frost, C. G., Mutton, L. Heterogeneous Catalytic Synthesis Using Microreactor Technology, Green Chem, 2010, 12, 1687–1703.

[113] Shu, W, Pellegatti, L, Oberli, MA, Buckwald, SL, Continuous-flow synthesis of biaryls enabled by multistep solid-handling in a lithiation/borylation/suzuki–miyaura cross-coupling sequence, Angew Chem Int Ed, 2011, 50, 10665–10669.

[114] Nagaki, A, Hirose, K, Moriwaki, Y, Mitamura, K, Matsukawa, K, Ishizuka, N, Yoshida, J, Integration of borylation of aryllithiums and suzuki–miyaura coupling using monolithic pd catalyst , Catal Sci Technol, 2016, 6, 4690–4694.

[115] Bolton, KF, Canty, AJ, Deverell, JA, Guijt, RM, Hilder, EF, Rodemann, T, Smith, JA, Macroporous monolith supports for continuous flow capillary microreactors, Tetrahedron Lett, 2006, 47, 9321–9324.

[116] Munoz, JM, Alcazar, J, De La Hoz, A, Diaz-Ortiz, A, Cross-coupling in flow using supported catalysts: Mild, clean, efficient and sustainable Suzuki–Miyaura coupling in a single pass, Adv Synth Catal, 2012, 354, 3456–3460.

[117] Pascanu, V, Hansen, PR, Bermejo Gomez, A, Ayats, C, Platero-Prats, AE, Johansson, MJ, Pericas, MA, Martin-Mature, B. Highly Functionalized Biaryls via Suzuki–Miyaura Cross-Coupling Catalyzed by Pd@MOF under Batch and Continuous Flow Regimes, ChemSusChem, 2015, 8, 123–130.

[118] Hattori, T, Tsubone, A, Sawama, Y, Monguchi, Y, Sajiki, H, Palladium on carbon-catalyzed suzuki-miyaura coupling reaction using an efficient and continuous flow system, Catalysts, 2015, 5, 18.

[119] Fabry, DC, Ho, YA, Zapf, R, Tremel, W, Panthofer, M, Rueping, M, Rehm, TH, Blue light mediated C–H Arylation of heteroarenes Using TiO2 as an immobilized photocatalyst in a continuous-flow microreactor, Green Chem, 2017, 19, 1911–1918.

[120] Burguete, MI, Cornejo, A, García-Verdugo, E, García, J, Gil, MJ, Luis, SV, Martínez-Merino, V, Mayoral, JA, Sokolova, M, Bisoxazoline-functionalised Enantioselective Monolithic Mini-flow-reactors: Development of efficient processes from batch to flow conditions, Green Chem, 2007, 9, 1091–1096.

[121] Burguete, MI, Cornejo, A, García-Verdugo, E, Gil, MJ, Luis, SV, Mayoral, JA, Martínez-Merino, V, Sokolova, M, Pybox monolithic miniflow reactors for continuous asymmetric cyclopropanation reaction under conventional and supercritical conditions, J Org Chem, 2007, 72, 4344–4350.

[122] Lim, J, Riduan, SN, Lee, SS, Ying, JY, Siliceous mesocellular foam-supported Aza (bisoxazoline)-copper catalysts, Adv Synth Catal, 2008, 350, 1295–1308.

[123] Aranda, C, Cornejo, A, Fraile, JM, García-Verdugo, E, Gil, MJ, Luis, SV, Mayoral, JA, Martínez-Merino, V, Ochoa, Z, Efficient enhancement of copper-pyridineoxazoline catalysts through immobilization and process design, Green Chem, 2011, 13, 983–990.

[124] Castano, B, Gallo, E, Cole-Hamilton, DJ, Dal Santo, V, Psaro, R, Caselli, A, Continuous flow asymmetric cyclopropanation reactions using Cu(I) Complexes of Pc-L* Ligands supported on silica as catalysts with carbon dioxide as a carrier, Green Chem, 2014, 16, 3202–3209.

[125] Takeda, K, Oohara, T, Shimada, N, Nambu, H, Hashimoto, S, Continuous flow system with a polymer-supported dirhodium(II) Catalyst: Application to enantioselective carbonyl ylide cycloaddition reactions, Chem Eur J, 2011, 17, 13992–13998.

[126] Monfette, S, Eyholzer, M, Roberge, DM, Fogg, DE, Getting Ring-closing metathesis off the bench: reaction-reactor matching Transforms metathesis efficiency in the assembly of large rings, Chem Eur J, 2010, 16, 11720–11725.
[127] Breen, CP, Parrish, C, Shangguan, N, Majumdar, S, Murnen, H, Jamison, TF, Bio, MM, A scalable membrane pervaporation approach for continuous flow olefin metathesis, Org Process Res Dev, 2020, 24, 2298–2303.
[128] Michrowska, A, Mennecke, K, Kunz, U, Kirschning, A, Grela, K, A new concept for the noncovalent binding of a ruthenium-based olefin metathesis catalyst to polymeric phases: Preparation of a catalyst on raschig rings, J Am Chem Soc, 2006, 128, 13261–13267.
[129] Skowerski, K, Pastva, J, Czarnocki, S, Janoscova, J, Exceptionally stable and efficient solid supported hoveyda-type catalyst, Org Process Res Dev, 2015, 19, 872–877.
[130] Gallo, A, Fong, A, Szeto, KC, Reib, J, Delevoye, L, Gauvin, RM, Taoufik, M, Peters, B, Scott, SL, Ligand exchange-mediated activation and stabilization of a re-based olefin metathesis catalyst by chlorinated alumina, J Am Chem Soc, 2016, 138, 12935–12947.
[131] Tsubogo, T, Oyamada, H, Kobayashi, S, Multistep continuous-flow synthesis of (R)- and (S)-rolipram using heterogeneous catalysts, Nature, 2015, 520, 329–332.
[132] Ishitani, H, Furiya, Y, Kobayashi, S, Enantioselective sequential-flow synthesis of baclofen precursor via asymmetric 1,4-Addition and chemoselective hydrogenation on platinum/carbon/calcium phosphate composites, Chem Asian J, 2020, 15, 1688–1691.
[133] Ingham, R, Battilocchio, C, Hawkins, JM, Ley, SV, Integration of enabling methods for the automated flow preparation of piperazine-2-carboxamide, *Beilstein J Org Chem*, 2014, 10, 641–652.

Klára Lövei and Gellért Sipos

11 Gaseous reagents in flow chemistry

11.1 Introduction

11.1.1 Introduction

Small gaseous reagents are essential building blocks in chemical industries and are utilized on a daily basis in research laboratories on mmol scale and in multiton production facilities as well. The "Nobel winning" Haber–Bosch process and the fact that approximately 25% of the marketed drugs require at least one hydrogenation step during their synthesis elegantly demonstrate the importance of hydrogenations [1]. The multistep semisynthetic synthesis of artemisinin, a drug used against malaria, commercialized by Sanofi, incorporated both an asymmetric hydrogenation and a photochemical oxidation step [2]. Carbon monoxide, carbon dioxide, and diazomethane are important C1 synthons. Chlorinated compounds, frequently applied in cross-coupling reactions, are potentially accessed through reactions of their respective precursors with anhydrous HCl or chlorine gas.

The use of gaseous reagents in synthetic chemistry laboratories is usually not preferred or is avoided. This is most likely due to regulatory and technical challenges when it comes to gases. Quite a few gaseous reagents (e.g., CO) are toxic and pose explosion or fire risk (e.g., H_2, acetylene). Even if none of these issues are present, running reactions under pressure is not always among the capabilities of research laboratories.

The objectives of this chapter are (a) to show the potential advantages of utilizing gaseous reagents in organic synthesis under continuous flow conditions; (b) to provide an overview on the current status of the field; and (c) to encourage synthetic organic chemists to make gaseous continuous flow chemistry part of their daily laboratory routine.

In this chapter, we focus on practical applications of gas–liquid and gas–liquid–solid reactions for the synthesis of small organic molecules. In Sections 11.1.2–11.1.6, we briefly overview some important technological aspects and highlight the key advantages of handling gases under flow conditions. Specific aspects are covered in more detail in relevant chapters of this book as well as in the previous edition and in excellent reviews [3, 4]. In Sections 11.2–11.12, the applications of individual gases are summarized. Engineering aspects, modelling, gas phase (gas-solid phase) reactions, and industrial processes are out of scope.

https://doi.org/10.1515/9783110693676-011

11.1.2 Reactor technology and flow regimes

Any flow reactor can be utilized for gas–liquid reactions. Therefore, when one decides to set up a laboratory for gas–liquid reactions or starts developing methodology for a certain gas–liquid reaction, it might be cumbersome to make a decision on the reactor technology to be used without deep knowledge of the process. The most straightforward and probably the most budget-friendly setup for gas–liquid reactions consists of a piston (e.g., an HPLC pump) or syringe pump, a mass flow controller (MFC), a T- or Y-mixer, a coil reactor, and a backpressure regulator (BPR) (Fig. 11.27). Apart from the MFC, all of these components are essential parts of a "flow chemistry starter pack" and can be also found in every laboratory that is equipped with an HPLC instrument.

When gas–liquid reactions are implemented in small inner diameter tubing (0.25–0.75 mm), Taylor flow becomes dominant; however, in case of a large difference between the gas and liquid flow rates, annular flow regime might be observed. If the gas is highly soluble in the reaction media or can be liquefied under the operating pressure of the system, homogeneous reaction conditions might occur. Bubble or bubbly flow conditions can be obtained in reactors with complex mixing geometry (e.g., the Corning glass plates). Immediate consumption of the diffused gas is often observed in tube-in-tube reactors equipped with Teflon AF-2400 membrane. Falling film microreactors have been operated both in co-current and counter-current modes. Finally, gas–liquid–solid reactions, mostly catalytic hydrogenations, are typically performed on fixed-bed catalysts.

i *For further details on this issue, please see Volume 1, Chapter 3, Title: Technology overview/overview of the devices.*

11.1.3 Source of the gas – cylinders and on-demand production

The conventional approach is to source the gas from a cylinder. Infusion of the reagent can be performed using a gas-tight syringe or an MFC. The latter provides more accurate dosing, and continuous processing is only limited by the capacity of the cylinder. Gases with low vapor pressure might be condensed and used in the liquid phase or in solution (e.g., NH_3, $CClF_2H$). On-demand generation of various gases (by electrolysis or using chemical generators) has been described [5]. In this case, the gaseous reagent is generated from stable or benign precursors and utilized shortly after production. The disadvantage of this approach is the reduced atom economy. In exchange, stable but hazardous (e.g., H_2 can be generated by electrolysis of water) or otherwise inaccessible unstable reagents (e.g., CH_2N_2 generated chemically from *N*-methyl-nitroso amine precursors) can be utilized in a well-controlled and relatively safe manner.

11.1.4 Residence time considerations

For homogeneous reactions in solution and for liquid-liquid biphasic reactions, quite often, residence time calculations can be performed with good accuracy by dividing the volume of the reactor by the sum of the flow rates of the individual streams. However, residence time calculations in systems employing gases are more difficult. For example, the residence time might be affected by the solubility of the gas in the liquid phase, which depends on the pressure and temperature applied. Furthermore, in an ideal situation, the gas can be applied in stoichiometric ratio and will get consumed during the reaction. Not only the consumption of the gas but also its in situ formation might alter the residence time. Overall, it is advisable to experimentally measure the residence time under steady-state conditions [6].

11.1.5 Units and notations

Most often, authors report pressure values in bar or psi (1 bar = 14.504 psi), and sometimes in Pa. The notation barg or bar(g) represents gauge pressure, which means, the pressure above ambient or atmospheric pressure. The "g" notation is oftentimes omitted; therefore "0 bar" displayed on an instrument or written in article will actually refer to "0 barg".

Liquid flow rates are typically given in $\mu Lmin^{-1}$ or $mLmin^{-1}$ units. Gas flow rates are typically given for normal or standard conditions (such as standard cubic centimeters per minute [sccm] or normal milliliters per minute [$NmLmin^{-1}$]).

11.1.6 Advantages of flow chemistry in the utilization of gaseous reagents

Increased safety is one of the key reasons for applying gaseous reagents under continuous flow conditions. Chapter 8 focuses on risk mitigation; therefore, we will mention only two safety aspects in this section: (a) the chemical inventory of a flow reactor is smaller than that of a batch reactor with the same productivity; (b) generally, the materials used for flow chemistry (even polymer tubes) have high-pressure resistance. Operating in high-pressure resistance equipment not only improves safety but can also lead to enhanced reaction rates. Eliminating the headspace results in larger interfacial area, which can lead to higher mass transfer rates. Also, the precise control of stoichiometry often allows running reactions with stoichiometric amounts of the gaseous reagent.

11.2 Hydrogen (H_2)

11.2.1 Introduction

Catalytic hydrogenations represent a versatile synthetic tool in organic chemistry. A plethora of valuable compounds might be accessed through the reduction of various functional groups (e.g., NO_2, C=O, C=C), de-aromatization of aromatic rings, by the removal of protecting groups (e.g., Bn, Cbz), or via hydrodehalogenation and deuteration reactions. Within the flow chemistry literature, a large body of the disclosed work focuses on hydrogenations. To provide a comprehensive overview of flow hydrogenations is out of the scope of this chapter. In this section, we will go over the evolution of flow hydrogenations and then provide a few selected examples for the transformations performed in this area. The field has been reviewed intensively, and the reader is directed to these reports for more information [3, 4, 7, 8].

11.2.2 Evolution of continuous flow hydrogenations

One of the major challenges of multiphase reactions is to achieve sufficient mass transfer rates between the different phases. This is especially true for gas–liquid–solid reactions, as gases tend to have low solubility in solvents. In their influential contribution, Kobayashi et al. reasoned that efficient multiphase mixing might be achieved using microreactor technology in which the solid catalyst would be immobilized on the reactor walls [9]. A glass "microchannel" reactor was put in use, and the channels were coated with a Pd catalyst. The reactor had a very high interfacial area per unit of volume compared to conventional reactors used in chemical processes. The efficiency of this technology was demonstrated on the hydrogenation of alkenes and alkynes, as well as in deprotection reactions (Fig. 11.1). Nearly quantitative yields were achieved under annular flow conditions. Although the throughput of the system was rather low (0.1 M, 0.1 mLh^{-1}), no Pd leaching was observed, and the reactor could be reused several times without the loss of activity.

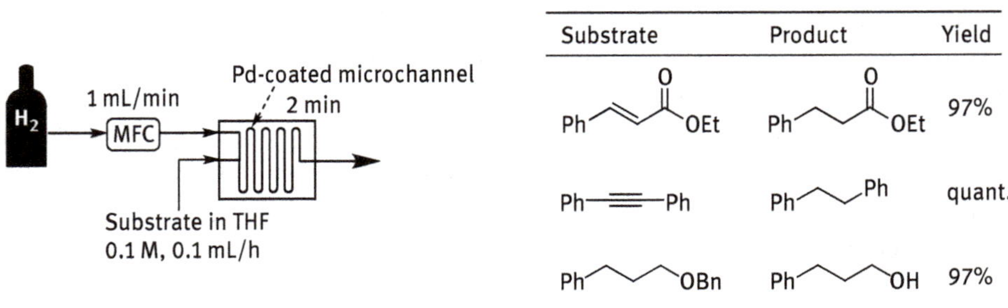

Fig. 11.1: Continuous flow hydrogenation in a Pd-coated microreactor [9].

Hazards associated with hydrogenations are mainly fires and explosions due to the pyrophoric nature of the catalyst, the formation of unstable intermediates, and the use of flammable solvents in combination with hydrogen gas and pressure [10]. To address these concerns, commercially available benchtop flow reactors have been developed. The first of such instruments, ThalesNano's H-Cube® system, specifically designed for hydrogenations, eliminates the need for gas cylinders, since it features a high-pressure electrocatalytic cell in which the H_2 gas is produced via the electrolysis of water. The total reactor volume is usually limited to a few milliliters, therefore, only a small amount of H_2 inventory is present at a certain time. Furthermore, the (pyrophoric) catalyst is contained in a cartridge, minimizing human exposure and the potential for fire during the reaction and workup [1, 11].

Schematic representation of the H-Cube® system is shown on Fig. 11.2. as part of an autonomous process. The H_2 gas, generated by the electrolytic cell, is introduced into the reaction line through a porous titanium frit mixer to ensure gas–liquid mixing by dispersing the gas bubbles. The substrate is delivered by an HPLC pump. The gas–liquid mixture is then passed through a bubble detector, which determines whether there is any hydrogen in the line, and then through the catalyst cartridge (CatCart®), which is located in a heating block. A pressure sensor and a BPR allow the user to perform reactions at elevated pressures up to 100 bar. The system is operated through a touch screen.

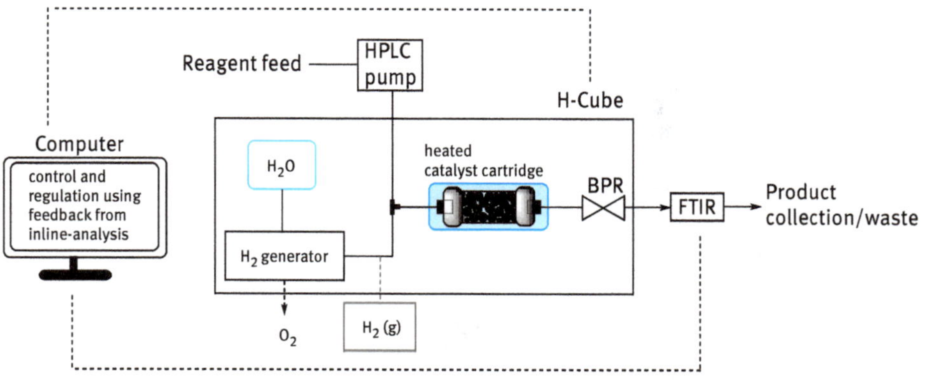

Fig. 11.2: Schematics of instrumentation for autonomous hydrogenation [16].

Shortly after its introduction in 2004, the technology has generated significant interest in the pharmaceutical industry. Academic and industrial research groups developed systems for automated synthesis using liquid handlers connected to the hydrogenation reactor [11–13]. Researchers at Amgen carried out investigations to identify the key parameters affecting reproducibility and performance during routine work [14, 15]. A fully automated heterogeneous flow-hydrogenation system was described in 2019 (Fig. 11.2) [16]. The computer-controlled system utilized the simplex algorithm for

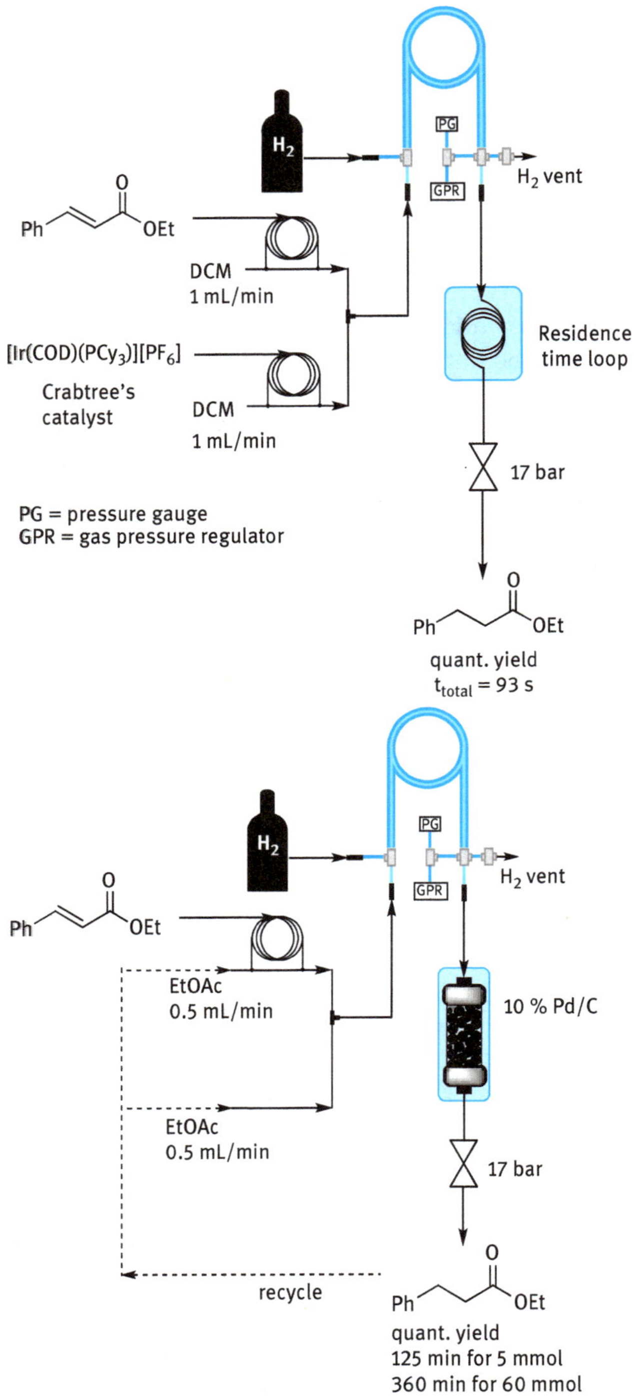

Fig. 11.3: Tube-in-tube reactors for homogeneous and heterogeneous hydrogenations [19].

autonomous optimization. The main hardware components of the platform were an autosampler, an H-Cube®, and an FTIR instrument for in-line analysis.

Tube-in-tube reactors featuring the Teflon AF-2400 gas permeable membrane were developed by the Ley group [17–19], and have become commercially available as well (Gas Addition Module from Uniqsis and Gas–Liquid reactor from Vapourtec). In Ley's prototype, the semi-permeable Teflon AF-2400 inner tube was placed inside a larger diameter PTFE tube. The gas was introduced into the outer tube, while the liquid was fed into the AF-2400 tube. Diffusion of the reactive gas into the substrate stream allowed the chemical reaction to take place. The tube-in-tube reactor proved to be suitable for processes with a broad range of gases, including homogeneous and heterogeneous hydrogenations (Fig. 11.3) [19]. Homogenous hydrogenation of alkenes with Crabtree's catalyst afforded quantitative yields in around 93 s. When the tube-in-tube reactor was adapted for heterogeneous hydrogenation, low conversions were obtained in single pass runs. To get high yield, the authors created a closed-loop system in which the substrate was circulated in the reactor, while hydrogen was continuously fed and consumed until full conversion was reached. It was observed that outgassing of hydrogen was lost 3 min after commencing the reaction and returned when all ethyl cinnamate was consumed. In the same study, measurement of hydrogen dissolution via degassing was carried out either using a simple gas burette or by a bubble counting technique, based on computer-assisted image processing.

The combination of supported ionic liquid phase (SILP) with supercritical fluids represents an attractive catalyst immobilization strategy for flow chemistry. SILP is a noncovalent immobilization strategy mostly used for homogeneous cationic transition metal catalysis. In the stationary phase, the catalyst is bound to the ionic liquid (IL) solvent film on the surface of a support material. The mobile phase contains the substrate dissolved in supercritical CO_2 ($scCO_2$). SILP technology proved to be an excellent fit for continuous flow asymmetric hydrogenations of olefins [20].

For more information on SILP, please see Volume 1, Chapter 10, Title: Continuous flow catalysis.

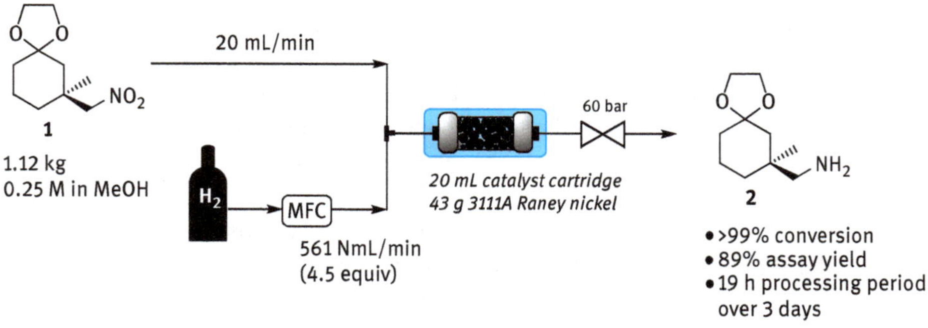

Fig. 11.4: Kilo-scale reduction of intermediate **1**. Simplified reactor setup is shown [22].

Kilo-scale fixed-bed hydrogenations have received increasing interest in peer-reviewed literature in the last few years. During their explorations, Ley et al. and Williams and coworkers followed similar strategies [21, 22]. Gram scale studies were carried out in the H-Cube® before moving to the HEL FlowCat reactor. Benchtop instruments, with surprisingly small reactor volumes can afford $kgday^{-1}$ productivities; for example, 1.12 kg of **1** was reduced in a 20 mL reactor over a period of 19 h (Fig. 11.4). Importance of the physical properties of the catalyst (e.g., mechanical strength and particle size distribution) becomes more pronounced at a higher scale [22]. Custom-built systems were also utilized for kg-scale aromatic nitro reductions. Notably, iodo-nitroaniline, a key intermediate of refametinib, was reduced to the corresponding aniline with up to 144 gh^{-1} productivity. The isolated material contained only 0.09 wt% desiodo side-products [23]. The above examples sourced hydrogen from high pressure tanks. Recently, a larger scale standalone high pressure hydrogen generator suitable for batch and flow chemistry was also developed [24].

11.2.3 Functional group transformations

Catalytic hydrogenations in microreactors are routinely performed on several different functional groups. A majority of these reactions can be carried out with Pd/C or Raney Ni catalysts. Carbonyl reductions and hydrogenation of aromatic rings usually require Pt, Ru, or Rh catalysts. Nonetheless, the Holy Grail of catalysts does not exist, since every process has different requirements. For example, economic considerations are profoundly different for laboratory-scale synthesis than for industrial production. Furthermore, the presence of certain functional groups, for example a reduction-sensitive aryl-iodide, might necessitate the careful tuning of reaction conditions, including the choice of the catalyst.

For the synthesis of pharmaceutically relevant compounds, reductions have been successfully incorporated into sequential cascade reactions [25], as well as into processes where the reduction of multiple functional groups takes place at the same time [26]. In this vein, the work of McCluskey et al. beautifully demonstrates the versatility of flow hydrogenations [27, 28]. α,β-Unsaturated nitriles, some of which also featured a furan moiety, were reduced in chemoselective fashion (Fig. 11.5). Mild reaction conditions in combination with 10% Pd/C resulted in alkene hydrogenation only. Simultaneous reduction of the alkene and nitrile groups was achieved with Raney Ni catalyst under more forcing conditions (50 °C, 10 bar). It was also possible to reduce the furan together with, or independently from, the nitrile.

It should be noted that for small scale applications the hydrogen flow rate (or any gas, for that matter) is not always reported or, except when an MFC is used, not controlled precisely. The reactions shown on Fig. 11.5 were carried out in a commercial reactor that produces ca. 60 NmL H_2 per a minute. "H_2 (10%)," in this case, refers to an approximately 10% hydrogen production rate (Fig. 11.5 top right). Furthermore, when planning a fixed-bed catalytic reaction, one should take into account the amount of catalyst to be used, especially on larger scales, where reaching high turnover numbers is crucial. In small scale applications, it is common to select a cartridge that will accommodate enough catalyst to reach practically useful production rates, by controlling the liquid flow rate (and, by that, the residence time). Therefore, except for cases where catalyst performance (e.g., high turnover numbers are required) is vital, relatively little attention is given to the substrate/catalyst ratio.

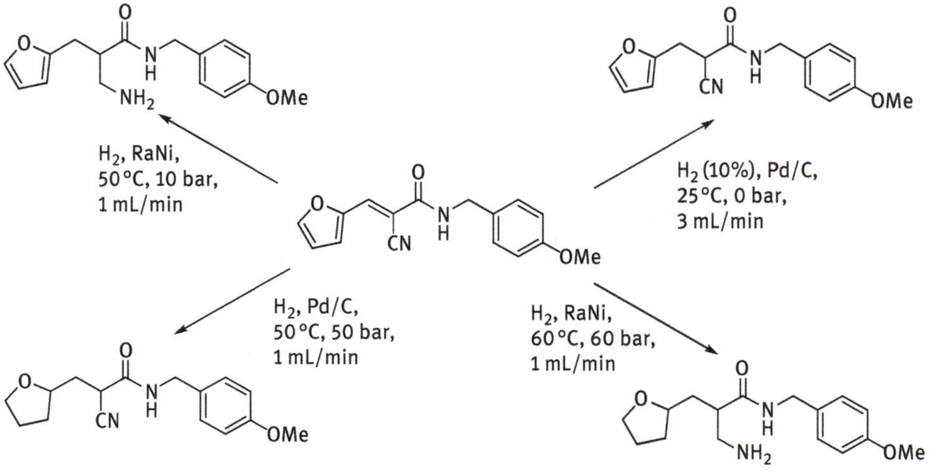

Fig. 11.5: Selective hydrogenations of alkene, nitrile, and furan moieties [27].

Hydrogenation of halogenated nitroarenes can benefit from flow conditions. Loos and coworkers studied the selective hydrogenation of halogenated nitroaromatics in batch and flow [29]. After a broad screening of commercially available catalysts, Pt-V/C and Raney Co were selected for further studies. Although good results were obtained with Pt-V/C in batch, moving to flow led to a fall in catalyst performance. On the other hand, the Raney Co catalyst delivered halogenated anilines in good yields (78–98%), with less than 2% dehalogenation taking place (Fig. 11.6).

Deuterated compounds are of significant interest for physical organic chemistry (e.g., kinetic isotope effect), pharmaceutical (e.g., drug metabolism studies), and analytical applications (e.g., internal standards in mass spectrometry), as well. Instead of bypassing the hydrodehalogenation reaction, researchers at the University of Szeged, exploited this reaction for the deuterium labelling of benzene derivatives (Fig. 11.7) [30]. Deuterium gas was generated from heavy water, and the Pd-catalyzed process was performed under harsh conditions (100 bar, 100 °C).

Protecting group manipulations are done daily in medicinal chemistry laboratories. Removal of *N*- and *O*-benzyl (Bn) and carbobenzyloxy (Cbz) groups can be performed with relative ease, even with the possibility of chemoselective deprotections.

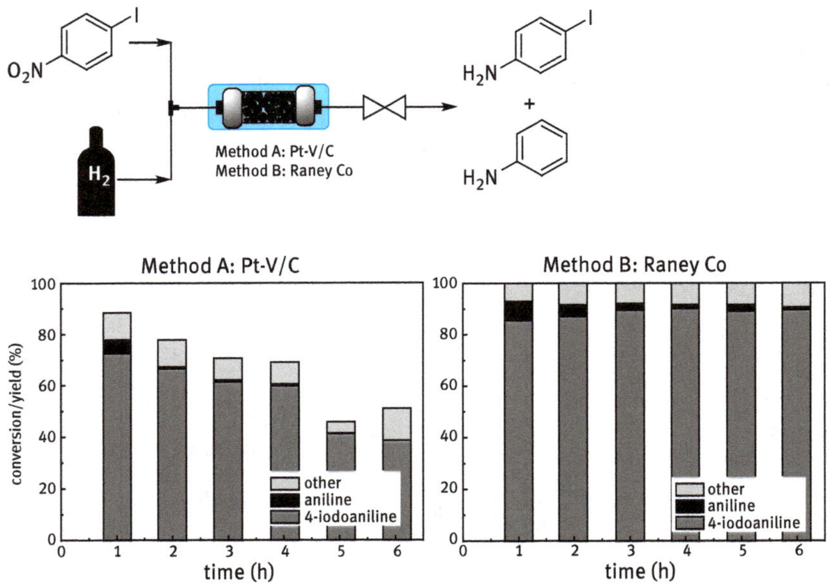

Fig. 11.6: Avoiding hydrodehalogenation during nitro reductions. Charts are adopted from Ref. [29].

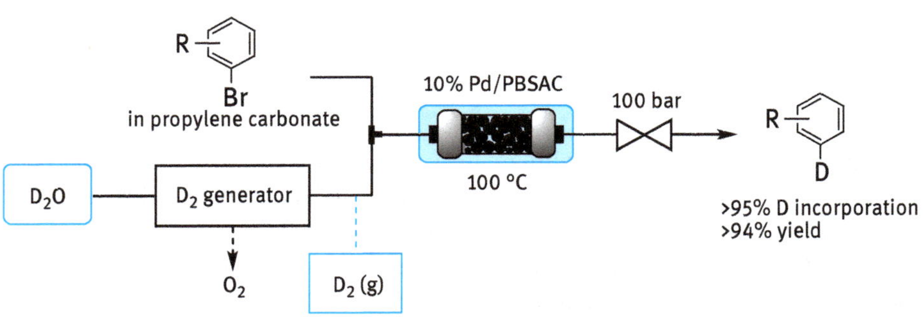

Fig. 11.7: Exploiting the potential of hydrodehalogenation for deuterium labeling [30].

For example, as shown in Fig. 11.8, *O*-debenzylation was affected at 40 bar and 45 °C, while an increase in pressure to 50 bar resulted in *N*-debenzylation, albeit with low yield [31].

Finally, although not discussed here, direct reductive aminations and imine hydrogenations, selective or complete alkyne hydrogenations, amide, azide and azo reductions, and the incorporation of hydrogenations in multistep flow syntheses have been described [7, 8, 32].

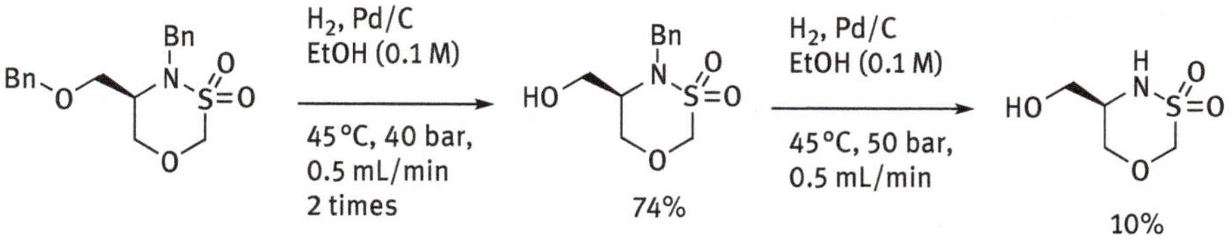

Fig. 11.8: Sequential *O*- and *N*-debenzylation [31].

11.3 Carbon monoxide (CO)

Carbon monoxide is regularly used in organic chemistry due to its versatility as a C_1 building block. However, various risks are associated with the traditional batch processes, since carbon monoxide is a colorless, odorless, toxic, and highly flammable gas. Since its solubility is low in organic solvents, elevated temperature and high pressure are often required for carbonylation reactions.

The Ley group developed effective methods for the palladium-catalyzed carbonylation of aryl iodides and bromides, and among others, for the synthesis of aminocarbonyls, using carbon monoxide and gaseous dimethyl amine simultaneously with the combination of two tube-in-tube reactors. The challenges posed by the effect of different solvents, pressure control, as well as the removal of the catalyst, and the excess carbon monoxide were studied thoroughly [33].

Continuous flow systems provide a safe way to extreme process conditions, such as excessively high temperatures. The palladium-catalyzed reductive carbonylation of nitrobenzene to phenyl isocyanate was carried out in segmented flow, at 220 °C and 10 bar, with significantly higher conversion and cleaner product than in the batch process [34].

In a similar vein, intensified conditions were applied for the palladium-catalyzed methoxycarbonylation of heteroaryl chlorides by the Kappe group. Carbon monoxide and methanol were dosed into the flow reactor at high temperature and pressure in a safe and accurately controlled manner. The methyl ester products were formed significantly faster in flow than under batch conditions [35]. An oxidative carbonylation was performed by the same group to synthesize carbonylated heterocycles [36]. Gaseous oxygen and carbon monoxide were utilized simultaneously in a custom flow chemical setup, which ensured the safe handling of the potentially explosive gas mixture (Fig. 11.9).

On-demand generation of carbon monoxide offers an elegant alternative to the use of pressurized cylinders. Dehydration of formic acid in the presence of sulfuric acid is an effective method for CO generation. The Ryu group has reported the Koch-Haaf reaction of adamantols [37] and palladium-catalyzed Heck carbonylations [38], based on this strategy. The Heck reactions were implemented in a tube-in-tube reactor, proving that ex situ CO generation can be safely performed with this technique (Fig. 11.10).

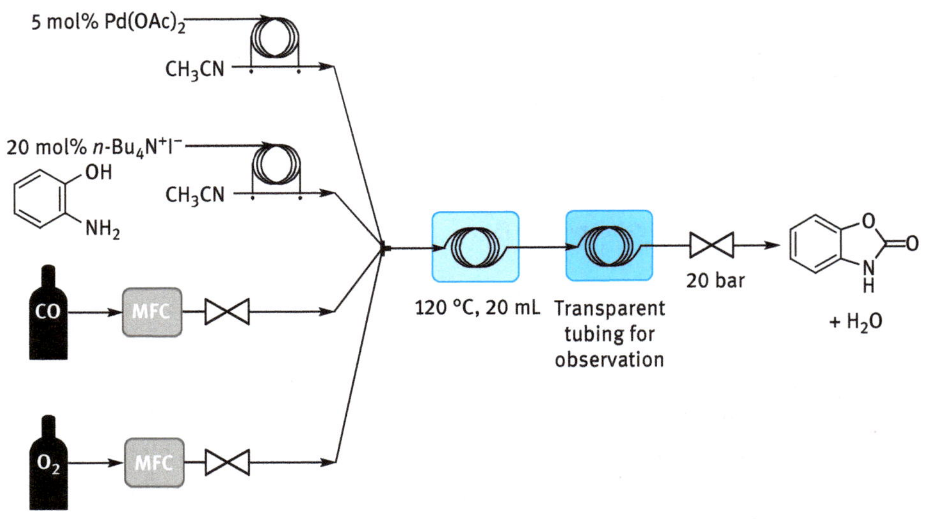

Fig. 11.9: Oxidative carbonylation protocol for the synthesis of carbonylated heterocycles [36].

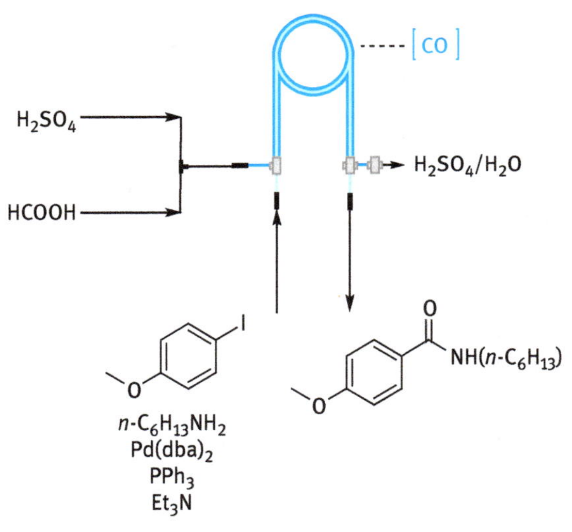

Fig. 11.10: Heck reaction with ex situ CO generation [38].

11.4 Synthesis gas (CO/H_2)

The application of synthesis gas (syngas), a mixture of carbon monoxide and hydrogen, is an effective and economic way to perform formylation reactions. The Kappe group has reported a safe, sustainable, and scalable continuous flow process to apply syngas in the palladium-catalyzed formylation of aryl bromides for the synthesis of several key APIs [39]. Carbon monoxide and hydrogen were accurately dosed with MFCs into the liquid stream via a four-way mixer, resulting in segmented flow. The group has found that the formation of undesirable palladium carbonyl clusters could be avoided by applying CO and H_2 in 1:3 stoichiometric ratio. A similar setup was used for the reductive carbonylation of aryl fluorosulfonates (Fig. 11.11) [40].

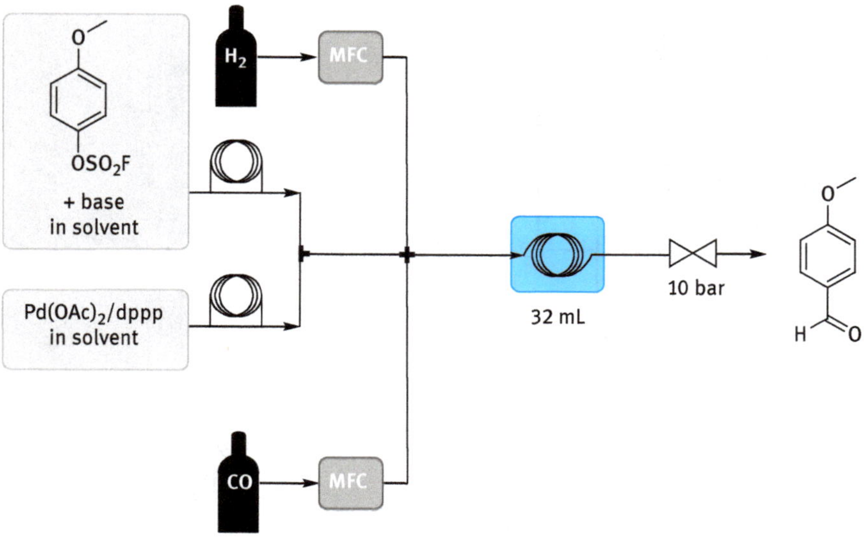

Fig. 11.11: Reductive carbonylation of aryl fluorosulfonates [40].

Tube-in-tube reactors offer a convenient alternative for the application of syngas in continuous flow. Branched aryl aldehydes were synthesized in the rhodium-catalyzed hydroformylation of styrenes [41], while linear aldehydes were prepared from 1-octene with low-pressure syngas and a suitable Rh-catalyst in a cost-effective, scalable process [42].

11.5 Carbon dioxide (CO_2)

Since carbon dioxide is a non-flammable, inexpensive and easily accessible gas, which can be used as a valuable C_1 synthon, several research groups have investigated its application both in batch and continuous flow chemistry. Due to its thermodynamic stability, high pressure and elevated temperature are often required for reactions with carbon dioxide.

Traditionally, CO_2 is mainly used as a weak electrophile, reacting with strong nucleophiles, like organomagnesium or organolithium compounds. Carboxylation reactions with Grignard substrates were implemented by the Ley group in a tube-in-tube system to synthesize various carboxylic acids [17]. The flow process included the application of carbon dioxide at a relatively low pressure and the in-line purification of the carboxylic acid through a catch-and-release protocol (Fig. 11.12).

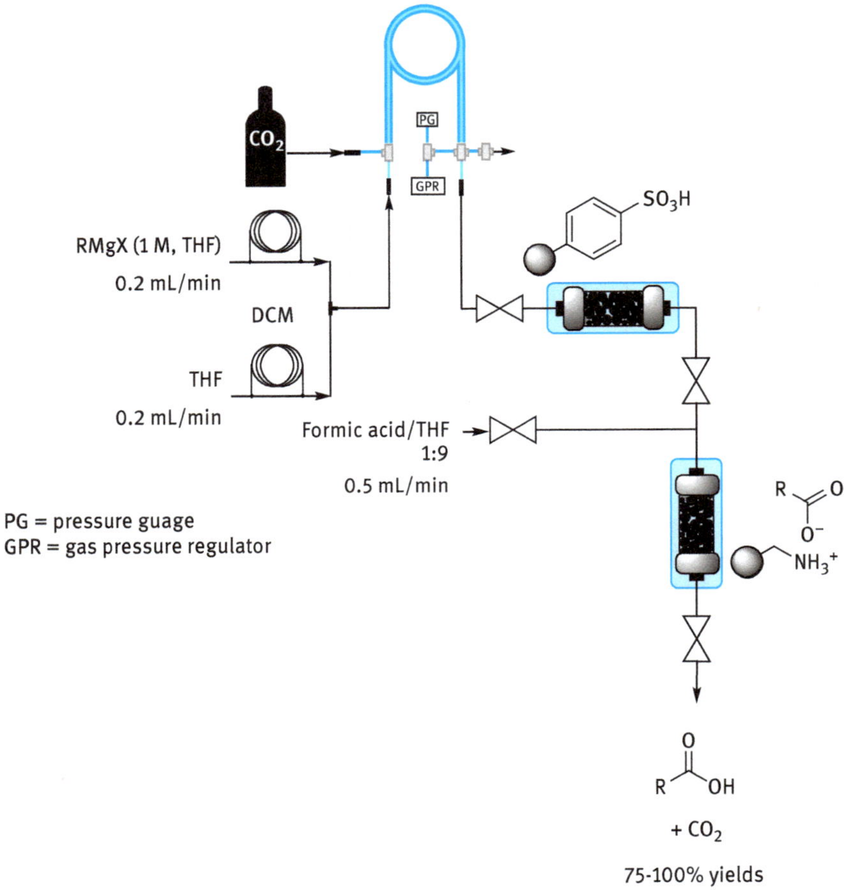

Fig. 11.12: Synthesis of carboxylic acids in a tube-in-tube reactor with in-line purification [17].

The Kirschning group developed an elegant flow strategy for the preparation of dibenzosuberone, a key intermediate in the synthesis of amitriptyline [43]. The reaction sequence involves multiple lithiation steps: a Wurtz-type dimerization is followed by carboxylation with CO_2, before the Parham cyclisation affords dibenzosuberone. A tube-in-tube reactor was used for the introduction of carbon dioxide into the reaction stream. Subsequently, CO_2 was removed by degassing through a Teflon® AF2400 tube to avoid side reactions with *n*BuLi during the Parham cyclization step.

A simple T-mixer was applied by the Kappe group for mixing carbon dioxide into the liquid stream in their lithiation/carboxylation protocol to synthesize carboxylic acids [44]. The authors reported that to stabilize the carbon dioxide gas flow, a preheater before the MFC needed to be applied.

The formation of C–O bonds in the synthesis of cyclic carbonates from epoxide precursors has been reported by several research groups. Earlier studies have evaluated the use of *sc*CO_2, acting both as the solvent and the carbon source of the reaction [45]. The Jamison group generated bromine from catalytic amounts of *N*-bromosuccinimide (NBS) and benzoyl peroxide (BPO) to activate the epoxide, which reacted with the CO_2 to form the carbonate (Fig. 11.13) [46].

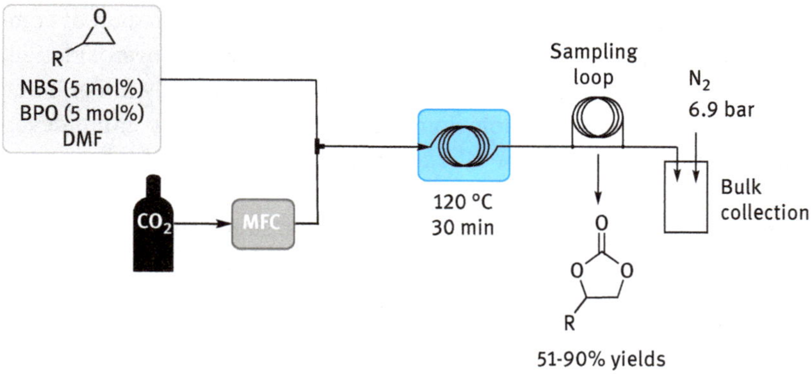

Fig. 11.13: Synthesis of cyclic carbonates in the presence of NBS and BPO [46].

Photoredox catalysis and carboxylation were effectively combined by the Jamison group. α-Amino acids were synthesized in the reaction of carbon dioxide and various amines. The reported process involved the in-line generation of carbon dioxide radical anion in a photochemical continuous flow setup. Carbon dioxide was mixed into the liquid stream with a T-mixer, and the segmented flow of the gas–liquid mixture was pumped directly into the photochemical reactor [47].

11.6 Oxygen (O_2)

Oxygen is a colorless, tasteless, odorless gas which is inexpensive, readily available, and harmless to the environment as well as to humans. Molecular oxygen is widely used in commodity chemical manufacturing, which is in sharp contrast to the fine chemical industry, where aerobic oxidations are less developed. This difference likely originates from the particular approaches followed by the respective industries, since commodity chemicals are typically manufactured in purpose-built plants, while the fine chemical sector traditionally employs multipurpose batch reactors. Nevertheless, the field of aerobic oxidations has grown exponentially since the early 1990s, including a plethora of methodologies for continuous flow oxidations [48, 49]. The key challenges of aerobic oxidations lie in the low solubility of oxygen in organic solvents and in oxygen's capability to form combustible mixtures with organic vapors. Flow chemistry can address both of these issues by eliminating the headspace, by pressurization, and by providing increased surface area between the gas and liquid phases.

Aerobic oxidations might be divided into five major categories: (1) transition metal catalysis (including oxidative coupling reactions); (2) photocatalysis; (3) biocatalysis; (4) organocatalysis; and (5) uncatalyzed reactions. Selected examples from categories 1, 2, and 5 are discussed below.

The Tsuji–Wacker oxidation is the palladium-catalyzed transformation of complex olefins to aldehydes or ketones. When there is a chance for overoxidation, using oxygen gas in this reaction can be rather challenging, especially under batch conditions where the control of gas stoichiometry is difficult. The Ley group developed a Pd-catalyzed anti-Markovnikov Wacker oxidation in a tube-in-tube reactor (Fig. 11.14 left) [50]. The process proceeded with stoichiometric amount of oxygen, and a back-pressure of 25 bar was necessary to maintain homogenous conditions (i.e., avoid out-gassing of O_2) in the reaction stream (note that the O_2 gas pressure was only 8 bar!). In a similar vein, the Kappe group developed a complementary approach for the Pd-catalyzed oxidative cleavage of olefins (Fig. 11.14 right) [51]. Terminal alkenes afforded aldehydes, while the reaction with gem-disubstituted alkenes led to ketones. The reactor operated in a segmented flow regime.

Copper-catalyzed aerobic oxidations were thoroughly studied by Stahl and coworkers [52]. A homogenous (bpy)Cu/TEMPO catalyst showed excellent performance in

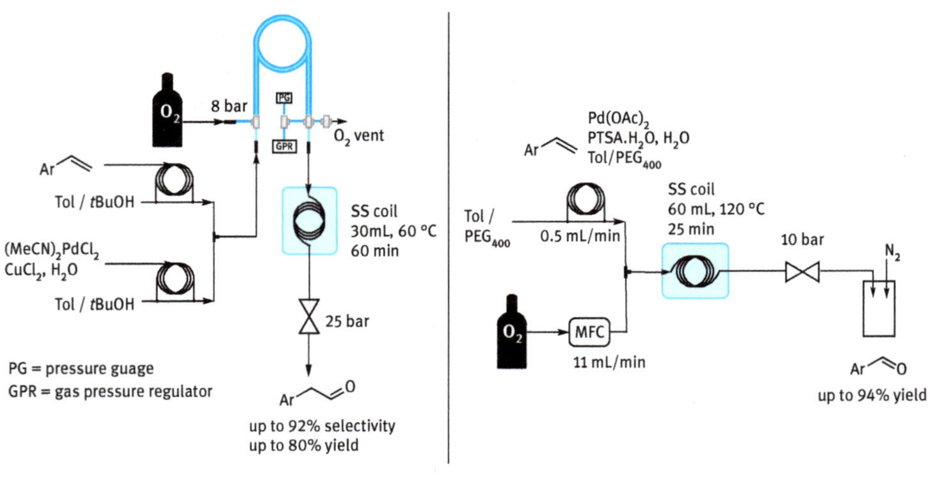

Fig. 11.14: Pd-catalyzed anti-Markovnikov Wacker oxidation (left) and oxidative cleavage (right) of styrenes [50, 51].

the oxidation of alcohols (*see case study 4.6 in Volume 1, Chapter 4, Title: Practical aspects of performing continuous flow chemistry*) [52]. Later on, the same group developed a tube-in-shell (tube-in-flask) reactor for the safe introduction of oxygen into the liquid phase [89]. The authors found that PTFE tubing is permeable for oxygen; therefore, it was adopted for this application, instead of the commonly used, but, fragile and expensive Teflon AF-2400 tube. The aerobic oxidation of alcohols using heterogeneous transition metal catalysts has received significant attention as well. For example, Stahl, Root, and coworkers described the oxidation of alcohols with $Ru(OH)_x/Al_2O_3$ in a packed-bed reactor as well as in a tube-in-shell reactor [54, 55].

Triplet oxygen ($^3O_2$) is converted to singlet oxygen ($^1O_2$) upon light irradiation in the presence of a suitable photosensitizer. The Seeberger group developed flow methods employing singlet oxygen, a highly reactive species, for the synthesis of artemisinin (Fig. 11.15) [56, 57]. The key step of this reaction sequence is the singlet oxygen-induced ene reaction to form **3**, subsequent acid catalyzed Hock cleavage, reaction with triplet oxygen and, finally a series of condensation steps generate the ring system of artemisinin.

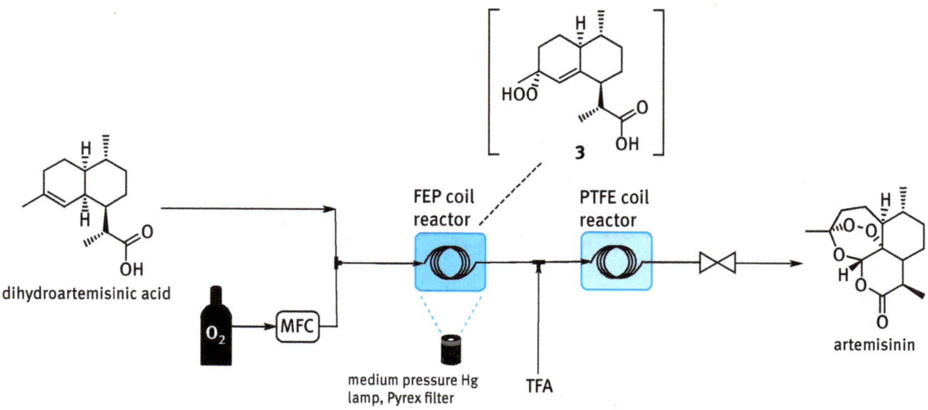

Fig. 11.15: The first flow photochemical oxidation approach for the synthesis of artemisinin [56].

Selective photochemical oxidation of C(sp^3)-H bonds was developed by the Noel group [58]. The method employs oxygen as an oxidant and decatungstate photocatalyst, which performs hydrogen abstraction on the C(sp^3)-H fragment. A simple PFA tube reactor was applied, and the reaction operated in a Taylor flow regime (Fig. 11.16). *ortho*-Functionalized phenols were prepared through a three-step integrated flow approach. Grignard reagents were generated via a benzyne intermediate in the first two capillary reactors and then directly oxidized to phenols with compressed air in the third unit [59].

Oxidation of hydrazine monohydrate ($N_2H_4{\cdot}H_2O$) affords diimide (N_2H_2), a highly reactive transfer hydrogenation agent. To speed up the otherwise relatively slow

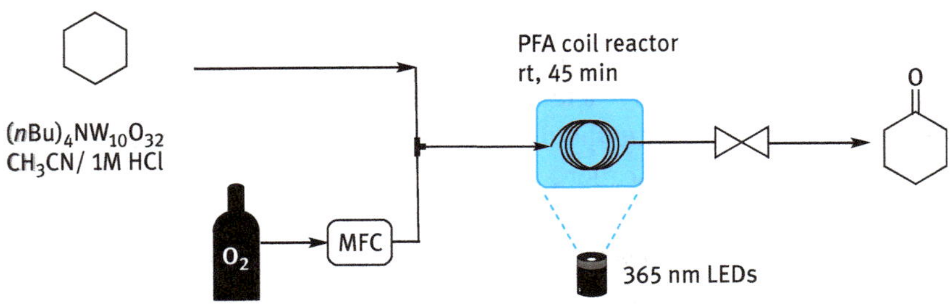

Fig. 11.16: Photocatalytic oxidation of C(sp^3)-H bonds [58].

oxidation, the Kappe group employed intensified process conditions in a microfluidic reactor [60]. The reduction of simple olefins afforded good to excellent yields; however, transfer hydrogenation of less reactive olefins provided moderate yields. The insufficient reaction rates were due to overoxidation and disproportionation of the diimide. To ensure that sufficient amount of the diimide reagent is available, a multi-injection protocol was developed (Fig. 11.17) [61].

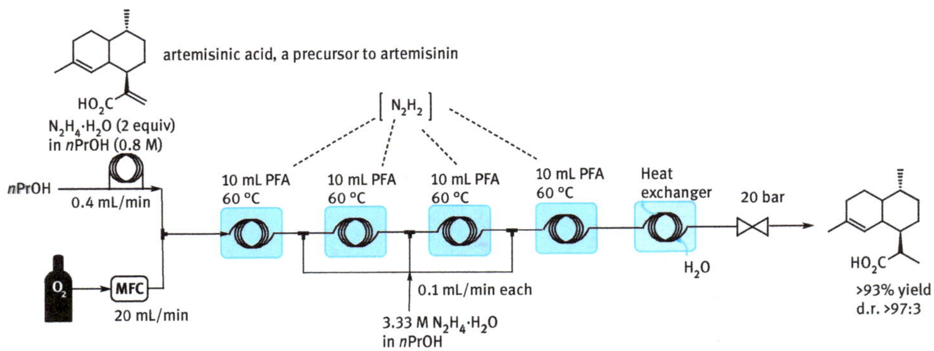

Fig. 11.17: Multi-injection protocol for utilizing diimide reagent N_2H_2 [61].

11.7 Ozone (O_3)

Ozonolysis is an excellent approach to prepare alcohols, aldehydes, ketones, or acids from the corresponding alkenes. On one hand, ozone is an economical reagent that leaves behind no waste (generated from and decomposes to oxygen). On the other hand, along with the ozonides generated during ozonolysis, ozone itself is toxic and explosive. Microreactors for ozonolysis were first used by Jensen and coworkers [62]. A multichannel reactor was fabricated from silicon and Pyrex wafers. The study focused on engineering aspects; nevertheless the oxidation of a phosphite, an amine and an olefin were demonstrated. Since Jensen's report, continuous flow ozonolysis has been demonstrated in a variety of reactors (e.g., falling film and cyclone mixer) [3, 48].

Independently from each other, the groups of Gavriilidis [63] and Kappe [64] performed ozonolysis in similar experimental setups, consisting of an ozone generator and a tube reactor unit (Fig. 11.18). The reactor unit was built-up from three subunits: (1) reagent pre-cooling; (2) cooled reaction loop; (3) and a cooled quench loop. Quenching of the residual ozone gas might be performed in a cartridge filled with manganese dioxide/copper oxide [63]. In a second contribution, Gavriilidis et al. performed ozonolysis on a broad range of alkenes and furans to prepare aldehydes and carboxylic acids, respectively [65].

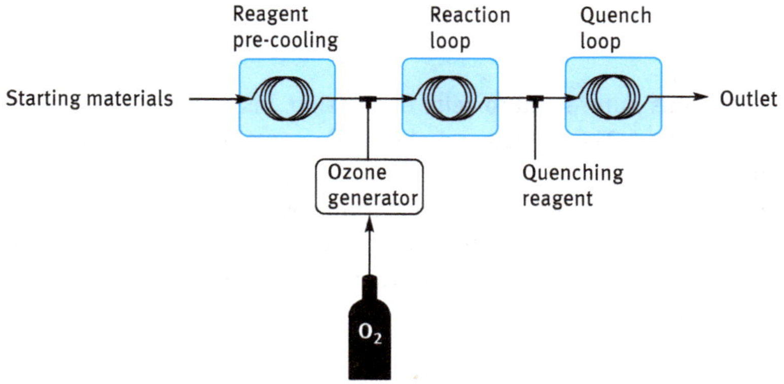

Fig. 11.18: Schematic representation of an ozonolysis setup [63, 64].

The flow ozonolysis step during the synthesis of 2-aminoadamantane-2-carboxylic acid was carried out with a droplet spray method (Fig. 11.19) [66]. The substrate feed was introduced into the stream of fast moving O_3/O_2 gas. The droplet spray at the reactor outlet was collected into a purge vessel flushed with argon or nitrogen. A secondary pump was then used to take the collected reaction mixture for further processing.

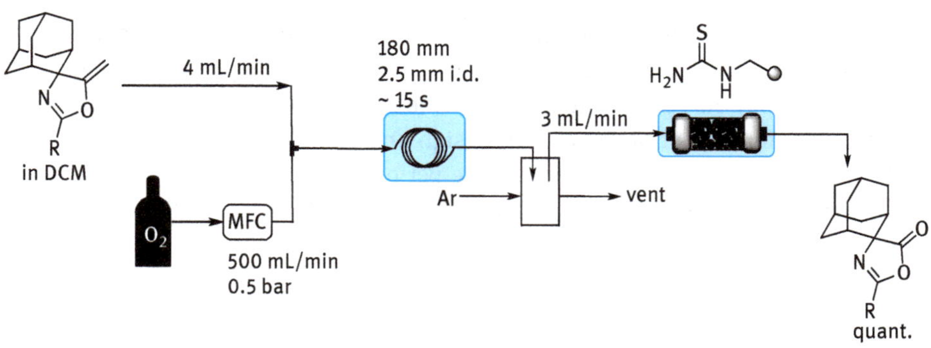

Fig. 11.19: Droplet spray ozonolysis [66].

11.8 Chlorine (Cl_2) and hydrogen chloride (HCl)

With the aid of continuous flow chemistry, the strong oxidizing power of the highly toxic and corrosive chlorine gas can be exploited safely. The application of gaseous Cl_2 enables the atom economic synthesis of various chlorinated compounds. One of the first examples was performed by the Jähnisch group. A falling film microreactor was applied for the photochlorination of toluene-2,3-diisocyanate under UV irradiation. Higher selectivity and production rate were achieved compared to batch reactions [67]. Similarly, the Ryu group reported the photo-initiated radical chlorination of various cycloalkanes with chlorine gas, yielding the monochlorinated products with high selectivity [68].

Due to the hazardous nature and toxicity of chlorine, handling and storing of the gas is strictly regulated. The on-demand chemical generation of Cl_2 is a viable alternative to the implementation of chlorination reactions (Fig. 11.20). The Kappe group developed an effective method in which chlorine was generated in situ from the reaction of NaOCl and HCl. Cl_2 was then extracted with a suitable organic solvent, while the aqueous waste was eliminated from the reaction mixture by an in-line membrane separator [69].

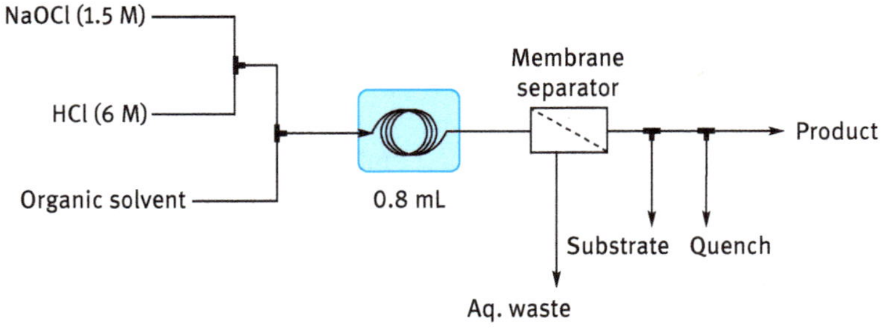

Fig. 11.20: In situ generation of chlorine [69].

Gaseous HCl can be efficiently utilized for the synthesis of chlorinated compounds; however, hydrogen chloride is highly corrosive in the presence of moisture. The Hessel group has overcome the difficulties associated with the application of this gas in continuous flow by careful design and complete purging of the system with nitrogen. Notably, the reactions were performed at elevated temperature and high pressure, yielding the corresponding chlorides in 10–15 min [70].

11.9 Gaseous reagents for the introduction of fluorine or fluorocarbon motifs

While most of the gaseous reagents can be handled using equipment available in almost all flow chemistry laboratories, working with fluorine requires specialized equipment and training. It is probably not surprising, that only a handful of groups study direct fluorinations utilizing 10% F_2/N_2 mixtures. Most notably, the Durham Fluorine group published numerous reports on selective direct fluorinations [71]. Fluorination of 1,3-dicarbonyls and subsequent cyclization with hydrazine were carried out in a continuous gas/liquid–liquid/liquid flow process [72]. Flucytosine was prepared via direct fluorination in a silicon carbide reactor (Fig. 11.21) [73]. Hypofluorous acid acetonitrile complex (HOF·MeCN), a reagent used for epoxidation of alkenes and oxidation of amines, was generated from fluorine, water, and acetonitrile in flow (Fig. 11.22) [74, 75].

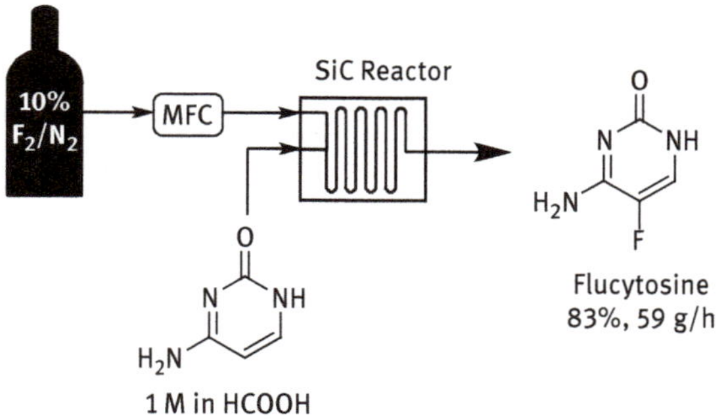

Fig. 11.21: Continuous synthesis of flucytosine [73].

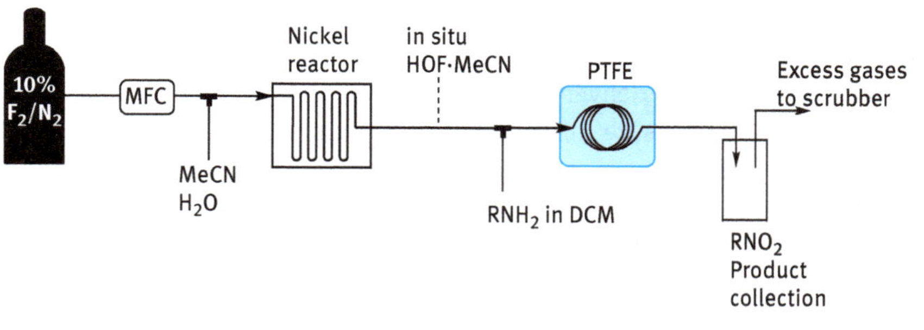

Fig. 11.22: Oxidation of amines by in situ formed HOF·MeCN complex [75].

Many of the small fluoroalkanes, halofluoroalkenes, and perfluoroalkanes (C1–C3) are gases at room temperature; nevertheless, these gaseous reagents often have good solubility in organic solvents and can be handled in solution or liquefied before introduction into the reactor [76, 77].

The Noel group applied CF_3I as gaseous CF_3 source in a range of photochemical trifluoromethylations of thiols [78], heterocycles [79], and styrenes [80]. Notably, the authors were able to drive the reaction between styrenes and CF_3I toward the trifluoromethylation or hydrotrifluoromethylation by careful tuning of the reaction conditions. Trifluoromethylation was achieved in the presence of CsOAc base in DMF/MeOH, while hydrotrifluoromethylation occurred with 4-hydroxythiophenol as an H-atom donor in DCE/EtOH. This methodology required a simple flow setup. The liquid feed was delivered by a syringe pump, and the flowrate of the CF_3I gas was controlled by an MFC. The gas and liquid feeds were mixed in a simple T-mixer, before passing through the irradiated PFA reactor and BPR before product collection. The trifluoromethylation reaction was performed within 90 min, and high *E*/*Z* ratios were obtained (Fig. 11.23).

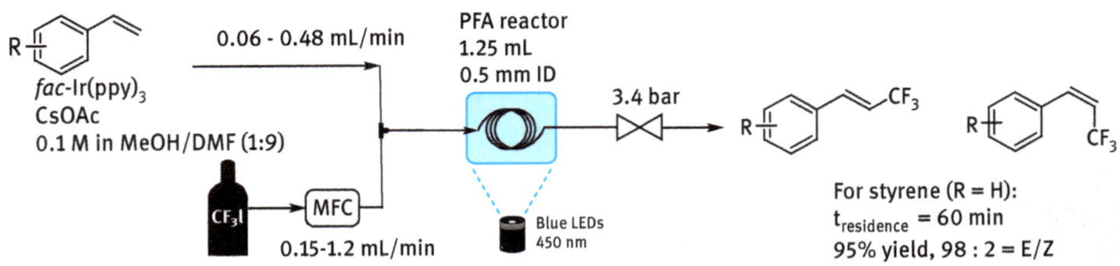

Fig. 11.23: Continuous flow photocatalytic trifluoromethylation of styrenes [80].

Trifluromethylation of ketones and aldehydes by fluoroform (CF_3H) was investigated by Ley and coworkers [81]. Fluoroform is a potent greenhouse gas; therefore its release into the atmosphere should be avoided. By integrating in-line analytics (IR, NMR), the authors were able to ensure full consumption of fluoroform under optimized conditions. Shibata et al. disclosed a microflow system for the trifluoromethylation of aldehydes, ketones, chalcones, and *N*-sulfinylimines [82]. All reactions could be carried out in the same flow system consisting of syringe pumps for liquid, an MFC for fluoroform delivery, a mixer with two or three inlets, and a stainless steel reactor tubing to ensure sufficient residence time (Fig. 11.24). Even though large excess of CF_3H (11 equiv., 25 mL min^{-1} gas flow rate vs. 1.43 mL min^{-1} liquid flow rate) was used, the reaction proceeded in a slug flow regime.

Urakawa, Grushin et al. developed a continuous flow cupration method to transform CF_3H into $CuCF_3$, a potent trifluoromethylating reagent [83]. Kappe and coworkers used fluoroform and chlorodifluoromethane ($ClCF_2H$) for the C^{α}-difluoromethylation of malonates, esters, and protected α-amino acids [85, 86]. These investigations focused on the scalable synthesis of eflornithine [86]. The authors devised a telescoped flow procedure utilizing fluoroform, which afforded eflornithine in 86% isolated yield over two steps, with a throughput of 24 $mmolh^{-1}$. The reaction sequence starts with the deprotonation of the protected amino acid with LiHMDS and is followed by the reaction of the activated amino acid with fluoroform. Finally, deprotection with HCl at high

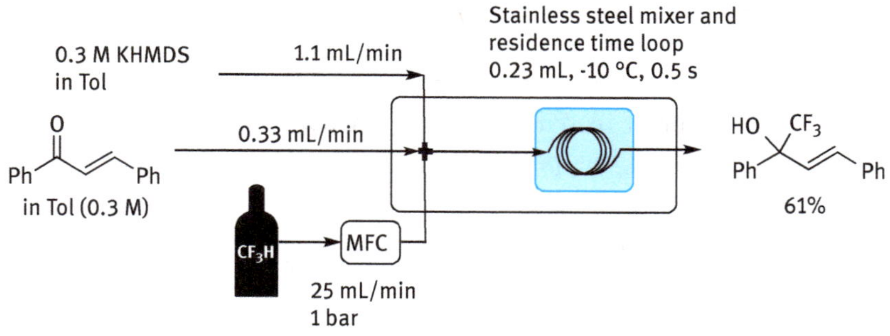

Fig. 11.24: Trifluoromethylation of chalcones with fluoroform [82].

temperature affords eflornithine hydrochloride monohydrate. The process shown in Fig. 11.25 was significantly improved compared to previous approaches, where pressure fluctuation and clogging were identified as key challenges. Selection of an appropriate BPR and the use of stoichiometric amount of CF_3H proved to be crucial to develop a scalable process.

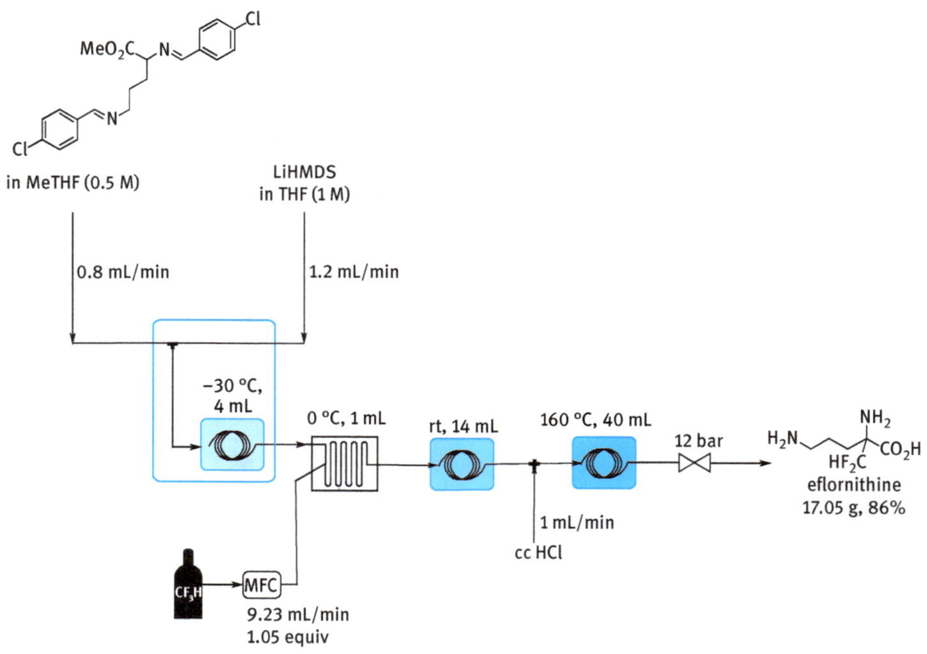

Fig. 11.25: Synthesis of eflornithine: difluoromethylation with fluoroform [86].

The use of chlorodifluoromethane for deuteriodifluoromethylation and gem-difluoroalkenylation of aldehydes was reported by Fu and Jamison [87]. Fluorous analogues of diazomethane, namely, trifluoromethyl diazomethane (CF_3CHN_2)

and difluoromethyl diazomethane (CF_2HCHN_2) were prepared in situ and engaged in [3 + 2] cycloadditions to afford pyrazoles or pyrazolines in a telescoped coil reactor [88]. Methyl(trifluoromethyl)dioxirane [$(CF_3)(CH_3)CO_2$] gas was also generated in situ and used in the direct oxidation of C(sp^3)-H bonds [89]. Photochemical fluorination of allylic alcohols was performed with the inexpensive and inert SF_6 gas as fluorinating agent (Tab. 11.1) [90].

11.10 Light hydrocarbons (alkanes, alkenes, and alkynes)

The selective functionalization of gaseous alkanes through the direct activation of intrinsically inert C(sp^3)-H bonds represents a formidable challenge in chemistry. Zuo et al. applied cerium photocatalysis for the C–H amination, alkylation, and arylation of light alkanes [91]. Although reaction optimization was performed in batch pressure vessels, later on, flow experiments were carried out using a commercially available Corning photoreactor to investigate the scale-up potential. Flow reactions

Tab. 11.1: Gaseous organofluorine compounds utilized in flow chemistry.

Reagent gas	Product/reaction	Reference
CF_3I	RCF_3	[76–80]
CF_3H	RCF_2H; RCF_3	[81–84,86]
CF_2ClH	RCF_2H	[85]
CF_3CHN_2	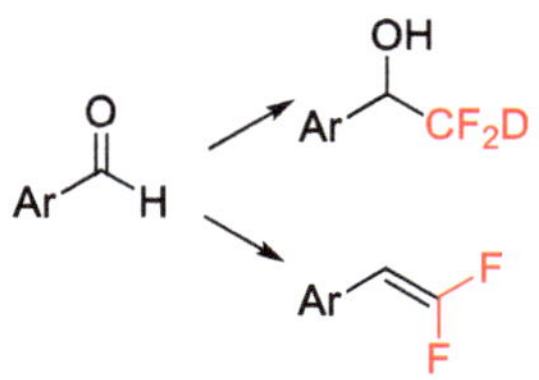	[87]
CF_2HCHN_2	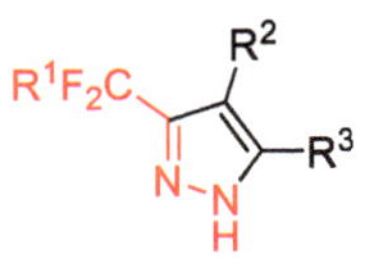 R^1 = H, F	[88]
SF_6	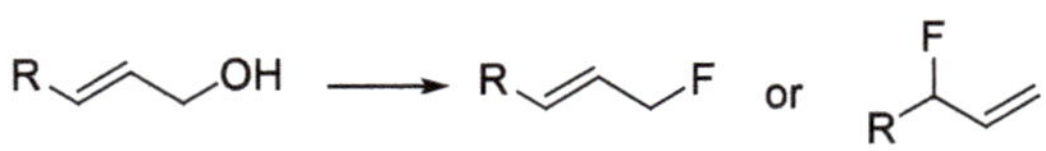	[90]
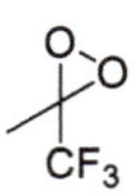	C-H oxidation C(sp^3)-H ⟶ R_3C-OH or R_2C=O	[89]

proceeded with good efficiency; in most cases, 1–2 $mmolh^{-1}$ productivity was achieved. Noel and coworkers expanded the scope of light alkane functionalization. In their report, decatungstate photocatalysis was employed to activate $C(sp^3)$–H bonds in simple alkanes (Fig. 11.26) [92]. Using olefins as radical traps, functionalization of methane, ethane, propane, and isobutane was carried out. Not surprisingly, C–H activation of methane (since its bond dissociation energy is higher than other alkanes) proved to be the most difficult. The reactions were carried out under continuous flow conditions, using several slightly different reactor configurations (homemade, Vapourtec, and HANU photoreactors were applied). To avoid gas-to-liquid mass transfer limitations, relatively high pressures were applied, ensuring liquefaction of the gaseous alkane.

The continuous flow application of gaseous alkenes and alkynes were successfully demonstrated in Pd-catalyzed coupling reactions. Noel and coworkers developed homogeneous and gas–liquid Catellani-type reactions in continuous flow [93]. Sterically hindered ortho-disubstituted styrene building blocks were prepared using gaseous ethylene, propylene, and 3,3,3-trifluoropropene as coupling partners (Fig. 11.27). The reactions were carried out in a stable Taylor flow regime. The importance of precise parameter control was highlighted through a batch control experiment, in which the yield of the desired product was smaller (66% vs 12% for the substrate shown) and formation of several by-products was observed.

Wu et al. developed a so-called stop-flow microtubing (SFMT) reactor to accelerate the discovery process of gas–liquid reactions [94]. The SFMT reactor consists of a gas feed controlled by an MFC, two pumps for the delivery of solutions, mixers, a microtubing reactor, a pressure gauge, a BPR, and a number of shut-off valves. For condition screening, the microtubing reactor is filled with the reaction mixture (typically a gas–liquid slug) and then sealed using the shut-off valves. The sealed reactor tube can then be removed from the system and placed in an oil bath. The same process might be repeated with another reactor unit under different experimental conditions (e.g., at higher pressure) (Fig. 11.28). This stop-flow approach saves time when screening of multiple parameters is required. It should be noted, that mixing efficiency is expected to be smaller than in batch reactors or under Taylor flow

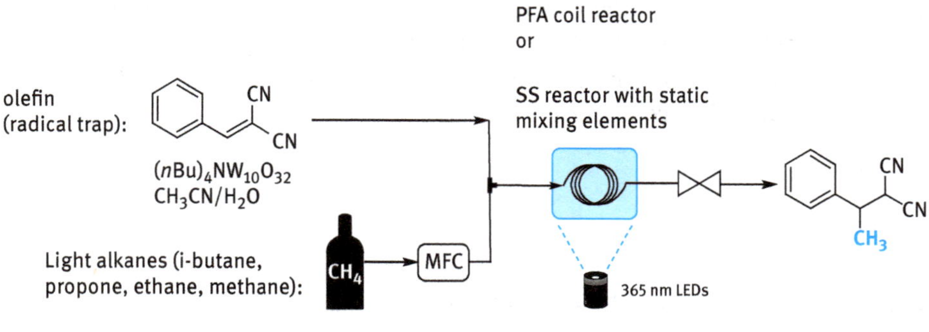

Fig. 11.26: Continuous flow $C(sp^3)$–H bond functionalization with light alkanes [92].

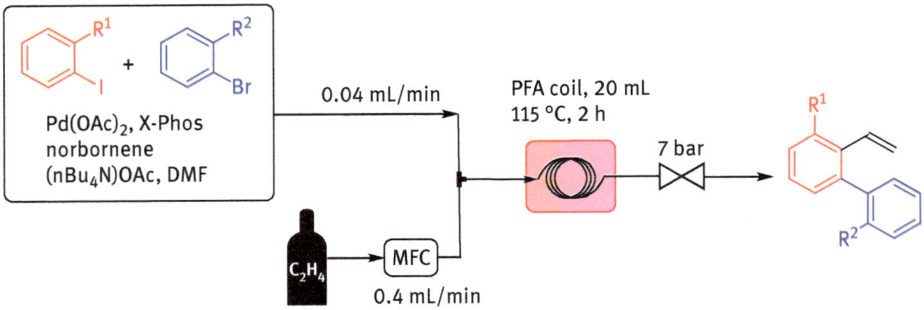

Fig. 11.27: Continuous flow Catellani-type reaction with ethylene gas [93].

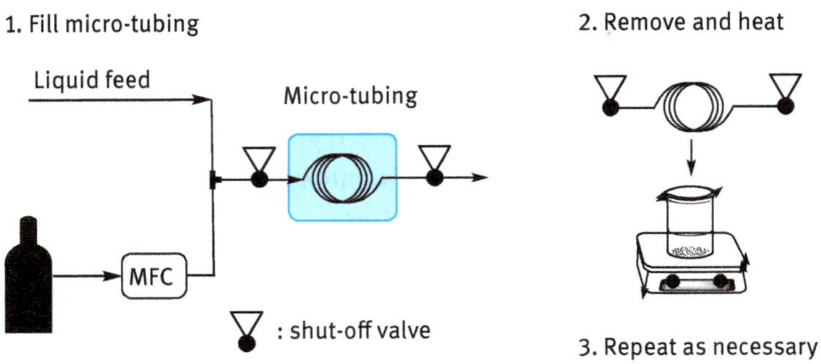

Fig. 11.28: "Stop-flow" microtubing (SFMT) reactor [94].

conditions. Nevertheless, the authors found that the results obtained with the SFMT reactor were comparable with those of a continuous flow system. In the proof-of-concept study, acetylene gas was used in Pd-catalyzed reactions for the synthesis of aryl fulvenes and aryl alkynes; furthermore, fluorinated styrenes were obtained under photoredox conditions.

The reaction of acetylene and EtMgBr afforded the ethynyl-Grignard reagent in a falling film microreactor. This reaction step was integrated with the addition of the Grignard reagent to carbonyl compounds, followed by in-line quenching to afford propargylic alcohols [95]. Finally, a number of groups, mainly with industrial interest, explored the reactivity of ethylene in photochemical [2 + 2] cycloadditions under continuous flow conditions [96–98].

11.11 Diazomethane (CH_2N_2)

Diazomethane is a versatile C1 building block for the introduction of methyl or methylene groups. The widespread use of diazomethane is hampered by its explosive and highly toxic nature. Diazomethane is mostly used in ethereal solution (typically in diethyl ether) and is usually prepared as needed, in the laboratory, through the hydrolysis of *N*-methyl-nitroso amine precursors (e.g., Diazald®, *N*-methyl-*N*-nitrosourea) under basic conditions. Traditional batch procedures, especially when the reagent is needed in anhydrous form, include a distillation step. Note, however, that diazomethane has been reported to explode upon contact with sharp edges (ground glass joints, starches). To lower the risk of explosion, custom kits for diazomethane distillation are available. Nevertheless, diazomethane has been prepared commonly on industrial scale [99].

In an early attempt, the Stark group demonstrated that microreactor technology provides a safe method for the utilization of diazomethane on laboratory scale [100]. Diazomethane was generated from Diazald® and KOH in residence time unit 1 (RTU1) and, quenched with benzoic acid in RTU2 (Fig. 11.29). The esterification of benzoic acid with diazomethane is quantitative and almost instantaneous; therefore, this reaction is frequently used to determine the efficiency of diazomethane generators. Even though good yields (up to 75%) were obtained, the system suffered from a major drawback: diazomethane was not separated from the aqueous base. Not only diazomethane decomposition can occur, but also reactions featuring water sensitive intermediates might suffer from undesired formation of hydrolysed products (e. g., the Arndt-Eistert reaction and the Wolf-rearrangement). To solve this issue, Kim et al. designed a dual-channel diazomethane generator featuring a PDMS membrane (Fig. 11.30) [101]. While diazomethane can diffuse through the membrane, KOH, water, and potassium *p*-toluenesulfonate are retained in the bottom channel.

The use of membrane technology for diazomethane generation and separation has become common. Although *N*-methyl-*N*-nitrosourea (NMNU) is instable and carcinogenic, Lehman at Novartis chose it as a precursor for diazomethane generation, because it provided cleaner and faster reaction compared to Diazald [102].

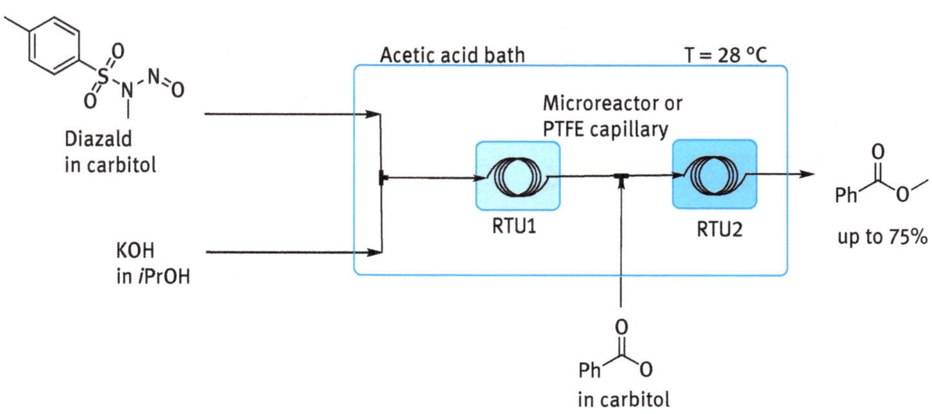

Fig. 11.29: Diazomethane generation without separation of the aqueous and organic phases [100].

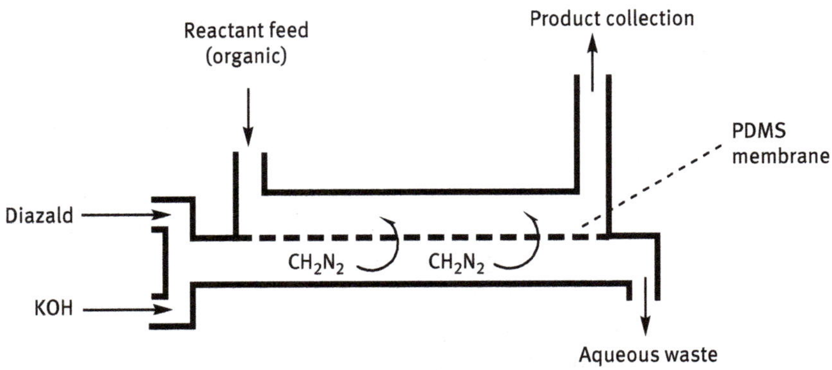

Fig. 11.30: Diazomethane generation with membrane separation [101].

For further details on this issue, please see Volume 1, Chapter 8, Title: Mitigation of chemical hazards under continuous flow conditions.

The Kappe group carried out detailed studies using the Teflon AF-2400 membrane. Tube-in-tube [103], tube-in-flask [104], and, more recently, CSTR cascade [105] configurations were tested as well.

A three-step continuous flow synthesis of α-chloro ketones– important intermediates of antiretroviral agents – was developed using a tube-in-tube diazomethane generator (Fig. 11.31). The synthesis starts with the formation of mixed anhydride in RTU1. Next, the anhydride is combined with diazomethane in the outer tube of the tube-in-tube reactor, and a second reactor unit ensures that diazo ketone formation can go to completion. Finally, the diazo ketone is quenched with HCl to afford the α-chloro ketone in 87% yield, with a 1.25 $mmolh^{-1}$ throughput [103]. Following a similar approach, researchers at Abbvie created an automated system in which a library of cyclopropyl boronic esters were prepared [106].

For further details on this issue, please see Volume 2, Chapter 7, Title: Continuous Gas–Liquid and Liquid–Liquid Separation.

11.12 Other gases (NH_3, HCN, H_2S, CH_2O, $COCl_2$)

Ammonia is a colorless gas with a distinctive pungent odor, used mainly in aqueous solution in traditional chemistry. Since its volatility limits its use in batch, the application of gaseous ammonia in continuous flow can be an effective alternative resulting in higher productivity. As an excellent example, Paal–Knorr pyrrole synthesis was performed by the Ley group in continuous flow. Gaseous ammonia was

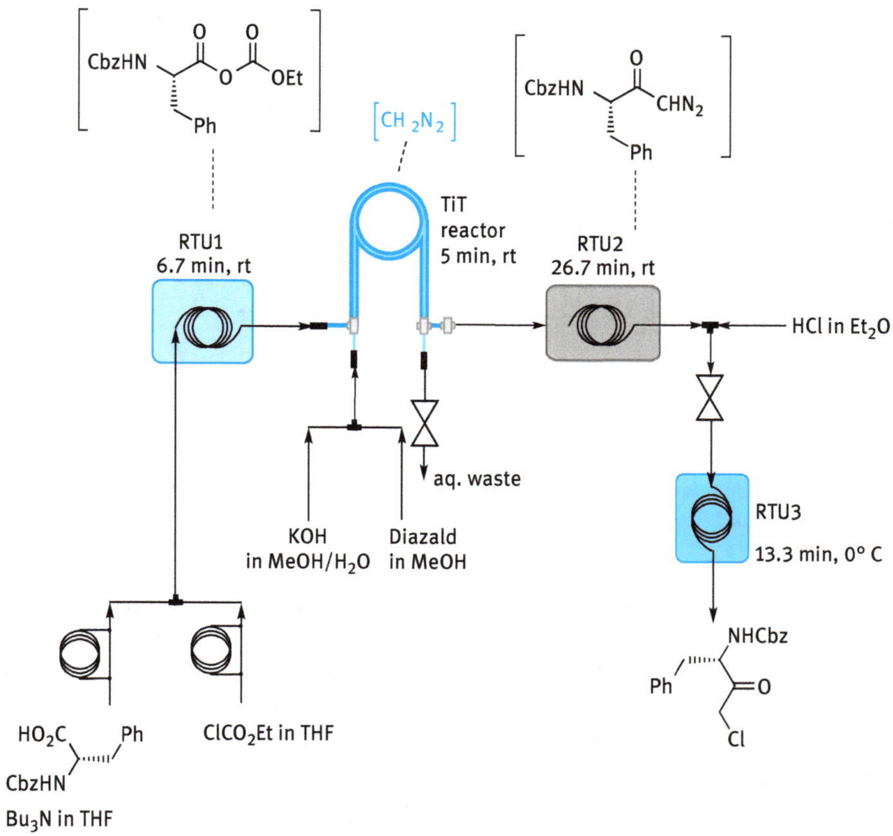

Fig. 11.31: Continuous flow synthesis of α-halo ketones [103].

mixed at low temperature with 1,4-diketones in the tube-in-tube reactor, and the reaction mixture was heated to 110 °C at 20 bar to achieve complete conversion. The authors reported that lower mixing temperature, longer residence time, and the application of methanol as solvent increased the concentration of ammonia in the liquid stream [107]. Further optimization of the protocol was later reported by the same group, enabling the mesoscale synthesis of fanetizole [108].

Aqueous ammonia was used as an ammonia source in a tube-in-tube reactor by Xue et al. for the continuous flow amination of aromatic and heteroaromatic halides, in a safe, simple, and scalable process (Fig. 11.32) [109].

Since the traditional batch application of **hydrogen cyanide** is considered extremely dangerous, safer alternatives need to be implemented, like the chemical generator concept, which was applied successfully by the Kappe group. Hydrogen cyanide was generated on-demand from aqueous NaCN and H_2SO_4 inside a non-permeable tube, and then transferred to a tube-in-tube reactor, where it was dissolved in an organic solvent, and used for hydrocyanation reactions [110].

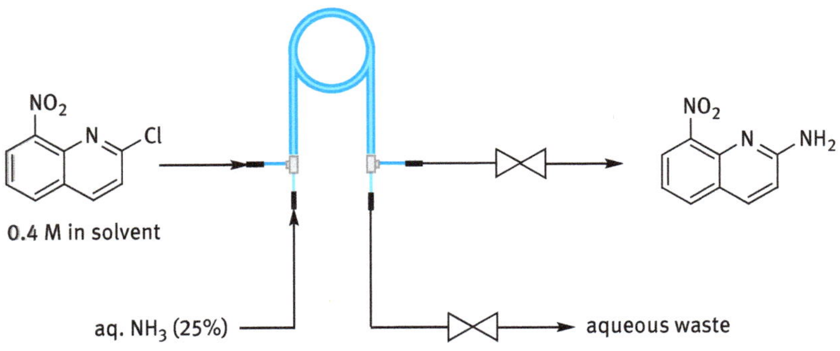

Fig. 11.32: Continuous flow reaction with aqueous ammonia [109].

(For further details on this issue, please see Volume 1, Chapter 8, Title: Mitigation of Chemical Hazards Under Continuous Flow Conditions)

The Kappe group has implemented the application of **hydrogen sulfide** in continuous flow for the synthesis of 2-oxopropanethioamide. Preliminary experiments showed that the presence of moisture in the MFC could cause serious corrosion in the gas line. Thus, a nitrogen purge line was installed, or alternatively a peristaltic pump was used for gas dosing instead of the MFC. Both solutions minimized the corrosion problem [111].

Gaseous **formaldehyde** can be generated in situ by heating paraformaldehyde in a pressure-resistant vessel and then used in a tube-in-tube reactor for the synthesis of N-Fmoc oxazolidinones, as Buba et al. have reported [112].

Due to its high toxicity and the hazards associated with its application, **phosgene** is usually generated and consumed in situ during chemical transformations. The Jensen group investigated the reaction of chlorine and carbon monoxide for phosgene generation [113], while Takahashi et al. applied triphosgene during the synthesis of amides [114].

Further readings

Reviews by Baxendale and Mallia [3] and by Gilmore, Seeberger et al. [4, 115] provide an excellent overview of the field. For a practical starting point, the reader is directed to articles in Nature Protocols, where detailed instructions on the assembly of gas–liquid photoreactors were published by the Noel group [116], while Kappe et al. described the step-by-step construction of tube-in-flask reactors for diazomethane generation [104].

Study questions

11.1 How many moles of $ArNO_2$ can be reduced in a 1 h process with a given hydrogen flow rate of 60 sccm?

11.2 Which factor will limit the throughput of an alkene hydrogenation process with the following parameters: 100 mL min^{-1} H_2, 0.5 M substrate, 0–10 mL min^{-1} HPLC pump flow rate, 0–150 °C temperature, and 1–100 pressure range?
11.3 How would you determine the residence time in a continuous flow reactor used for a gas–liquid–solid reaction?
11.4 Provide a description for a flow regime that might be observed during a gas–liquid reaction.
11.5 What is the chemical generator concept? Provide an example.
11.6 Design a tube-in-tube reactor in which the following transformation might be realized:

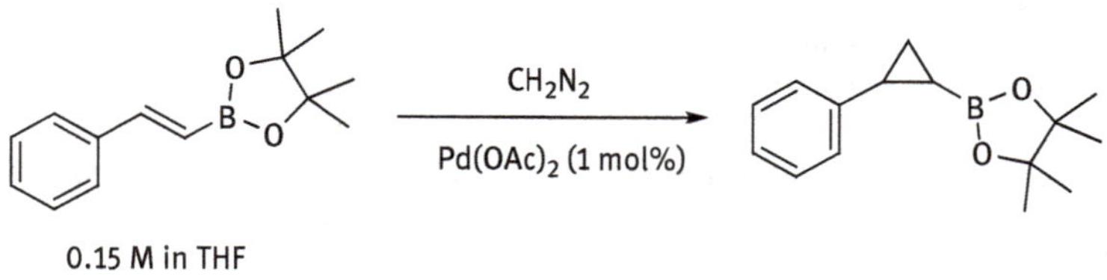

References

[1] Dormán, G, Kocsis, L, Jones, R, Darvas, F, A benchtop continuous flow reactor: A solution to the hazards posed by gas cylinder based hydrogenation, J Chem Health Safety, 2013, 20, 3–8.
[2] Turconi, J, Griolet, F, Guevel, R, et al., Semisynthetic artemisinin, the chemical path to industrial production, Org Process Res Dev, 2014, 18, 417–422.
[3] Mallia, CJ, Baxendale, IR, The use of gases in flow synthesis, Org Process Res Dev, 2016, 20, 327–360.
[4] Plutschack, MB, Pieber, B, Gilmore, K, Seeberger, PH, The Hitchhiker's guide to flow chemistry, Chem Rev, 2017, 117, 11796–11893.
[5] Dallinger, D, Gutmann, B, Kappe, CO, The concept of chemical generators: on-site on-demand production of hazardous reagents in continuous flow, Acc Chem Res, 2020, 53, 1330–1341.
[6] Hopley, A, Doyle, BJ, Roberge, DM, Macchi, A, Residence time distribution in coil and plate micro-reactors, Chem Eng Sci, 2019, 207, 181–193.
[7] Cossar, PJ, Hizartzidis, L, Simone, MI, McCluskey, A, Gordon, CP, The expanding utility of continuous flow hydrogenation, Org Biomol Chem, 2015, 13, 7119–7130.
[8] Yu, T, Jiao, J, Song, P, et al., Recent progress in continuous-flow hydrogenation, ChemSusChem, 2020, 13, 2876–2893.
[9] Kobayashi, J, Mori, Y, Okamoto, K, et al., A microfluidic device for conducting gas-liquid-solid hydrogenation reactions, Science, 2004, 304, 1305–1308.
[10] Chandra, T, Zebrowski, JP, Hazards associated with laboratory scale hydrogenations, J Chem Health Saf, 2016, 23, 16–25.
[11] Jones, RV, Godorhazy, L, Varga, N, Szalay, D, Urge, L, Darvas, F, Continuous-flow high pressure hydrogenation reactor for optimization and high-throughput synthesis, J Comb Chem, 2006, 8, 110–116.
[12] Clapham, B, Wilson, NS, Michmerhuizen, MJ, et al., Construction and validation of an automated flow hydrogenation instrument for application in high-throughput organic chemistry, J Comb Chem, 2008, 10, 88–93.

[13] Knudsen, KR, Holden, J, Ley, SV, Ladlow, M, Optimisation of conditions for O-benzyl and N-benzyloxycarbonyl protecting group removal using an automated flow hydrogenator, Adv Synth Catal, 2007, 349, 535–538.
[14] Bryan, MC, Wernick, D, Hein, CD, Petersen, JV, Eschelbach, JW, Doherty, EM, Evaluation of a commercial packed bed flow hydrogenator for reaction screening, optimization, and synthesis, Beilstein J Org Chem, 2011, 7, 1141–1149.
[15] Eschelbach, JW, Wernick, D, Bryan, MC, Doherty, EM, Characterization of dispersion effects on reaction optimization and scale-up for a packed bed flow hydrogenation reactor, Aust J Chem, 2013, 66, 165–171.
[16] Fabry, DC, Heddrich, S, Sugiono, E, Liauw, MA, Rueping, M, Adaptive and automated system-optimization for heterogeneous flow-hydrogenation reactions, React Chem Eng, 2019, 4, 1486–1491.
[17] Polyzos, A, O'Brien, M, Petersen, TP, Baxendale, IR, Ley, SV, The continuous-flow synthesis of carboxylic acids using CO_2 in a tube-in-tube gas permeable membrane reactor, Angew Chem Int Ed, 2011, 50, 1190–1193.
[18] Brzozowski, M, O'Brien, M, Ley, SV, Polyzos, A, Flow chemistry: intelligent processing of gas–liquid transformations using a tube-in-tube reactor, Acc Chem Res, 2015, 48, 349–362.
[19] O'Brien, M, Taylor, N, Polyzos, A, Baxendale, IR, Ley, SV, Hydrogenation in flow: Homogeneous and heterogeneous catalysis using Teflon AF-2400 to effect gas–liquid contact at elevated pressure, Chem Sci, 2011, 2, 1250–1257.
[20] Geier, D, Schmitz, P, Walkowiak, J, Leitner, W, Franció, G, Continuous flow asymmetric hydrogenation with supported ionic liquid phase catalysts using modified CO_2 as the mobile phase: From model substrate to an active pharmaceutical ingredient, ACS Catal, 2018, 8, 3297–3303.
[21] Ouchi, T, Battilocchio, C, Hawkins, JM, Ley, SV, Process intensification for the continuous flow hydrogenation of ethyl nicotinate, Org Process Res Dev, 2014, 18, 1560–1566.
[22] Carangio, A, Edwards, LJ, Fernandwez-Puertas, E, et al., Evaluation of sponge metal catalysts in a trickle bed reactor for the continuous hydrogenation of an aliphatic nitro intermediate, Org Process Res Dev, 2020, 24, 1909–1915.
[23] Said, MB, Baramov, T, Herrmann, T, Gottfried, M, Hassfield, J, Roggan, S, Continuous selective hydrogenation of Refametinib iodo-nitroaniline key intermediate DIM-NA over Raney cobalt catalyst at kg/day scale with online UV-VIS conversion control, Org Process Res Dev, 2017, 21, 705–714.
[24] Lebl, R, Zhu, Y, Ng, D, Hornung, CH, Cantillo, D, Kappe, CO, Scalable continuous flow hydrogenations using Pd/Al_2O_3-coated rectangular cross-section 3D-printed static mixers, Catal Today, 2020, in press, https://doi.org/10.1016/j.cattod.2020.07.046.
[25] Dimitrou, E, Jones, RH, Pritchard, RG, Miller, GJ, O'Brien, M, Gas-liquid flow hydrogenation of nitroarenes: Efficient access to a pharmaceutically relevant pyrrolobenzo[1,4]diazepine scaffold, Tetrahedron, 2018, 74, 6795–6803.
[26] Day, JP, et al., Elucidation of a structural basis for the inhibitor-driven, p62 (SQSTM1)-dependent intracellular redistribution of cAMP phosphodiesterase-4A4 (PDE4A4), J Med Chem, 2011, 54, 3331–3342.
[27] Tarleton, M, McCluskey, A, A flow chemistry route to 2-phenyl-3-(1H-pyrrol-2-yl)propan-1-amines, Tetrahedron Lett, 2011, 52, 1583–1586.
[28] Hizartzidis, L, Tarleton, M, Gordon, CP, McCluskey, A, Chemoselective flow hydrogenation approaches to isoindole-7-carboxylic acids and 7-oxa-bicyclio[2.2.1]heptanes, RSC Adv, 2014, 4, 9709–9722.
[29] Loos, P, Alex, H, Hassfeld, J, et al., Selective hydrogenation of halogenated nitroaromatics to haloanilines in batch and flow, Org Process Res Dev, 2016, 20, 452–464.

[30] Orsy, G, Fülöp, F, Mándity, IM, Continuous-flow catalytic deuterodehalogenation carried out in propylene carbonate, Green Chem, 2019, 21, 956–961.
[31] Borcard, F, Baud, M, Bello, C, et al., Synthesis of new oxathiazinane dioxides and their in vitro cancer cell growth inhibitory activity, Bioorg Med Chem Lett, 2010, 20, 5353–5356.
[32] Fülöp, Z, Szemesi, P, Bana, P, Éles, J, Greiner, I, Evolution of flow-oriented design strategies in the continuous preparation of pharmaceuticals, React Chem Eng, 2020, 5, 1527–1555.
[33] Gross, U, Koos, P, O'Brien, M, Polyzos, A, Ley, SV, A general continuous flow method for palladium catalysed carbonylation reactions using single and multiple tube-in-tube gas-liquid microreactors, Eur J Org Chem, 2014, 6418–6430.
[34] Takebayashi, Y, Sue, K, Yoda, S, Furuya, T, Mae, K, Direct carbonylation of nitrobenzene to phenylisocyanate using gas–liquid slug flow in microchannel, Chem Eng J, 2012, 180, 250–254.
[35] Mata, A, Hone, CA, Gutmann, B, Moens, L, Kappe, CO, Continuous-flow Pd-catalyzed carbonylation of aryl chlorides with carbon monoxide at elevated temperature and pressure, ChemCatChem, 2019, 11, 997–1001.
[36] Chen, Y, Hone, CA, Gutmann, B, Kappe, CO, Continuous flow synthesis of carbonylated heterocycles via Pd-catalyzed oxidative carbonylation using CO and O_2 at elevated temperatures and pressures, Org Process Res Dev, 2017, 21, 1080–1087.
[37] Fukuyama, T, Mukai, Y, Ryu, I, Koch–Haaf reaction of adamantanols in an acid-tolerant Hastelloy-made microreactor, Beilstein J Org Chem, 2011, 7, 1288–1293.
[38] Brancour, C, Fukuyama, T, Mukai, Y, Skrydstrup, T, Ryu, I, Modernized low pressure carbonylation methods in batch and flow employing common acids as a CO source, Org Lett, 2013, 15, 2794–2797.
[39] Hone, CA, Lopatka, P, Munday, R, O'Kearney-McMullan, A, Kappe, CO, Continuous-flow synthesis of aryl aldehydes by Pd-catalyzed formylation of aryl bromides using carbon monoxide and hydrogen, ChemSusChem, 2019, 12, 326–337.
[40] Köckinger, M, Hanselmann, P, Hu, G, Hone, CA, Kappe, CO, Continuous flow synthesis of aryl aldehydes by Pd-catalyzed formylation of phenol-derived aryl fluorosulfonates using syngas, RSC Adv, 2020, 10, 22449–22453.
[41] Kasinathan, S, Bourne, SL, Tolstoy, P, et al., Syngas-mediated C-C bond formation in flow: Selective rhodium-catalysed hydroformylation of styrenes, Synlett, 2011, 18, 2648–2651.
[42] Raghuvanshi, K, Zhu, C, Ramezani, M, et al., Highly efficient 1-octene hydroformylation at low syngas pressure: from single-droplet screening to continuous flow synthesis, ACS Catal, 2020, 10, 7535–7542.
[43] Kupracz, L, Kirschning, A, Multiple organolithium generation in the continuous flow synthesis of amitriptyline, Adv Synth Catal, 2013, 355, 3375–3380.
[44] Pieber, B, Glasnov, T, Kappe, CO, Flash carboxylation: fast lithiation–carboxylation sequence at room temperature in continuous flow, RSC Adv, 2014, 4, 13430–13433.
[45] Lu, XB, Xiu, JH, He, R, Jin, K, Luo, LM, Feng, XJ, Chemical fixation of CO_2 to ethylene carbonate under supercritical conditions: continuous and selective, Appl Catal Gen, 2004, 275, 73–78.
[46] Kozak, JA, Wu, J, Su, X, Simeon, F, Hatton, TA, Jamison, TF, Bromine-catalyzed conversion of CO_2 and epoxides to cyclic carbonates under continuous flow conditions, J Am Chem Soc, 2013, 135, 18497–18501.
[47] Seo, H, Katcher, MH, Jamison, TF, Photoredox activation of carbon dioxide for amino acid synthesis in continuous flow, Nat Chem, 2017, 9, 453–456.
[48] Gemoets, HPL, Su, Y, Shang, M, Hessel, V, Luque, R, Noel, T, Liquid phase oxidation chemistry in continuous-flow microreactors, Chem Soc Rev, 2016, 45, 83–117.
[49] Hone, CA, Kappe, CO, The use of molecular oxygen for liquid phase aerobic oxidations in continuous flow, Top Curr Chem, 2019, 377, 1–44. doi: https://doi.org/10.1007/s41061-018-0226-z.

[50] Bourne, SL, Ley, SV, A continuous flow solution to achieving efficient aerobic anti-markovnikov wacker oxidation, Adv Synth Catal, 2013, 355, 1905–1910.

[51] Hone, C, O'Kearney-McMullan, A, Munday, R, Kappe, CO, Development of a continuous-flow process for a Pd-catalyzed olefin cleavage using oxygen within the explosive regime, ChemCatChem, 2017, 9, 3298–3302.

[52] McCann, SD, Stahl, SS, Copper-catalyzed aerobic oxidations of organic molecules: Pathways for two-electron oxidation with a four-electron oxidant and a one-electron redox-active catalyst, Acc Chem Res, 2015, 48, 1756–1766.

[53] Greene, JF, Hoover, JM, Mannel, DS, Root, TW, Stahl, SS, Continuous-flow aerobic oxidation of primary alcohols with a copper(I)/TEMPO catalyst, Org Process Res Dev, 2013, 17, 1247–1251.

[54] Greene, JF, Preger, Y, Stahl, SS, Root, TW, PTFE-membrane flow reactor for aerobic oxidation reactions and its application to alcohol oxidation, Org Process Res Dev, 2015, 19, 858–864.

[55] Mannel, DS, Stahl, SS, Root, TW, Continuous flow aerobic alcohol oxidation reactions using a heterogeneous $Ru(OH)_x/Al_2O_3$ catalyst, Org Process Res Dev, 2014, 18, 1503–1508.

[56] Lévesque, F, Seeberger, PH, Continuous-flow synthesis of the anti-malaria drug artemisinin, Angew Chem Int Ed, 2012, 51, 1706–1709.

[57] Kopetzki, D, Lévesque, F, Seeberger, PH, A continuous-flow process for the synthesis of artemisinin, Chem Eur J, 2013, 19, 5450–5456.

[58] Laudadio, G, Govaerts, S, Wang, Y, et al., Selective $C(sp^3)$-H aerobic oxidation enabled by decatungstate photocatalysis in flow, Angew Chem Int Ed, 2018, 57, 4078–4082.

[59] He, Z, Jamison, TF, Continuous-flow synthesis of functionalized phenols by aerobic oxidation of Grignard reagents, Angew Chem Int Ed, 2014, 53, 3353–3357.

[60] Pieber, B, Martinez, ST, Cantillo, D, Kappe, CO, In situ generation of diimide from hydrazine and oxygen: continuous-flow transfer hydrogenation of olefins, Angew Chem Int Ed, 2013, 52, 10241–10244.

[61] Pieber, B, Glasnov, T, Kappe, CO, Continuous flow reduction of artemisinic acid utilizing multi-injection strategies – closing the gap towards a fully continuous synthesis of antimalarial drugs, Chem Eur J, 2015, 21, 4368–4376.

[62] Wada, Y, Schmidt, MA, Jensen, KF, Flow distribution and ozonolysis in gas-liquid multichannel microreactors, Ind Eng Chem Res, 2006, 45, 8036–8042.

[63] Roydhouse, MD, Ghaini, A, Constantinou, A, Cantu-Perez, A, Motherwell, WB, Gavriilidis, A, Ozonolysis in flow using capillary reactors, Org Process Res Dev, 2011, 15, 989–996.

[64] Irfan, M, Glasnov, TN, Kappe, CO, Continuous flow ozonolysis in a laboratory scale reactor, Org Lett, 2011, 13, 984–987.

[65] Roydhouse, MD, Motherwell, WB, Constantinou, A, et al., Ozonolysis of some complex organic substrates in flow, RSC Adv, 2013, 3, 5076–5082.

[66] Battilocchio, C, Baxendale, IR, Biava, M, Kitching, MO, Ley, SV, A flow-based synthesis of 2-aminoadamantane-2-carboxylic acid, Org Process Res Dev, 2012, 16, 798–810.

[67] Ehrich, H, Linke, D, Morgenschweis, K, Baerns, M, Jahnisch, K, Application of microstructured reactor technology for the photochemical chlorination of alkylaromatics, Chimia, 2002, 56, 647–653.

[68] Matsubara, H, Hino, Y, Tokizane, M, Ryu, I, Microflow photo-radical chlorination of cycloalkanes, Chem Eng J, 2011, 167, 567–571.

[69] Strauss, FJ, Cantillo, D, Guerra, J, Kappe, CO, A laboratory-scale continuous flow chlorine generator for organic synthesis, React Chem Eng, 2016, 1, 472–476.

[70] Borukhova, S, Noel, T, Hessel, V, Hydrogen chloride gas in solvent-free continuous conversion of alcohols to chlorides in microflow, Org Process Res Dev, 2016, 20(2), 568–573.

[71] Sandford, G, Continuous flow selective direct fluorination using fluorine gas, In: Groult, H, Leroux, FR, Tressaud, A, ed, Modern Synthesis Processes and Reactivity of Fluorinated Compounds, Elsevier, 2017, 339–348.
[72] Breen, JR, Sandford, G, Yufit, DS, Howard, JAK, Fray, J, Patel, B, Continuous gas/liquid–liquid/liquid flow synthesis of 4-fluoropyrazole derivatives by selective direct fluorination, Beilstein J Org Chem, 2011, 7, 1048–1054.
[73] Harsanyi, A, Conte, A, Pichon, L, Rabion, A, Grenier, S, Sandford, G, One-step continuous flow synthesis of antifungal WHO essential medicine flucytosine using fluorine, Org Process Res Dev, 2017, 21, 273–276.
[74] McPake, CB, Murray, CB, Sandford, G, Epoxidation of alkenes using HOF·MeCN by a continuous flow process, Tetrahedron Lett, 2009, 50, 1674–1676.
[75] McPake, CB, Murray, CB, Sandford, G, Sequential continuous flow processes for the oxidation of amines and azides by using HOF·MeCN, ChemSusChem, 2012, 5, 312–319.
[76] Rosso, C, Williams, JD, Filippini, G, Prato, M, Kappe, CO, Visible-light-mediated iodoperfluoroalkylation of alkenes in flow and its application to the synthesis of a key fulvestrant intermediate, Org Lett, 2019, 21, 5341–5345.
[77] Monteiro, JL, Carneiro, PF, Elsner, P, et al., Continuous flow homolytic aromatic substitution with electrophilic radicals – a fast and scalable protocol for trifluoromethylation, Chem Eur J, 2017, 23, 176–186.
[78] Staathof, NJW, Tegelbeckers, BJP, Hessel, V, Wang, X, Noel, T, A mild and fast photocatalytic trifluoromethylation of thiols in batch and continuous-flow, Chem Sci, 2014, 5, 4768–4773.
[79] Su, Y, Kuijpers, KPL, König, N, Shang, M, Hessel, V, Noel, T, A mechanistic investigation of the visible-light photocatalytic trifluoromethylation of heterocycles using CF_3I in flow, Chem Eur J, 2016, 22, 12295–12300.
[80] Staathof, NJW, Cramer, SE, Hessel, V, Noel, T, Practical photocatalytic trifluoromethylation and hydrotrifluoromethylationof styrenes in batch and flow, Angew Chem Int Ed, 2016, 55, 15549–15553.
[81] Musio, B, Gala, E, Ley, SV, Real-time spectroscopic analysis enabling quantitative and safe consumption of fluoroform during nucleophilic trifluoromethylation in flow, ACS Sustainable Chem Eng, 2018, 6, 1489–1495.
[82] Hirano, K, Gondo, S, Punna, N, Tokunaga, E, Shibata, N, Gas/liquid-phase micro-flow trifluoromethylation using fluoroform: Trifluoromethylation of aldehydes, ketones, chalcones, and N-sulfinylimines, ChemistryOpen, 2019, 8, 406–410.
[83] Mazloomi, Z, Bansode, A, Benavente, P, Lishchynskyi, A, Urakawa, A, Grushin, VA, Continuous process for production of $CuCF_3$ via direct cupration of fluoroform, Org Process Res Dev, 2014, 18, 1020–1026.
[84] Köckinger, M, Ciaglia, T, Bersier, M, Hanselmann, P, Gutmann, B, Kappe, CO, Utilization of fluoroform for difluoromethylation in continuous flow: A concise synthesis of a-difluoromethyl-amino acids, Green Chem, 2018, 20, 108–112.
[85] Gutmann, B, Hanselmann, P, Bersier, M, Roberge, D, Kappe, CO, Continuous-flow difluoromethylation with chlorodifluoromethane under biphasic conditions, J Flow Chem, 2017, 7, 46–51.
[86] Köckinger, M, Hone, CA, Gutmann, B, et al., Scalable continuous flow process for the synthesis of eflornithine using fluoroform as difluoromethyl source, Org Process Res Dev, 2018, 22, 1553–1563.
[87] Fu, WA, Jamison, TF, Deuteriodifluoromethylation and gem-difluoroalkenylation of aldehydes using $ClCF_2H$ in continuous flow, Angew Chem Int Ed, 2020, 59, 13885–13890.
[88] Britton, J, Jamison, TF, A unified continuous flow assembly-line synthesis of highly substituted pyrazoles and pyrazolines, Angew Chem Int Ed, 2017, 129, 8949–8953.

[89] Lesieur, M, Battilocchio, C, Labes, R, et al., Direct oxidation of Csp^3-H bonds using in situ generated trifluoromethylated dioxirane in flow, Chem Eur J, 2019, 25, 1203–1207.

[90] McTeague, TA, Jamison, TF, Photoredox activation of SF_6 for fluorination, Angew Chem Int Ed, 2016, 55, 15072–15075.

[91] Hu, A, Guo, JJ, Zuo, Z, Selective functionalization of methane, ethane, and higher alkanes by cerium photocatalysis, Science, 2018, 361, 668–672.

[92] Laudadio, G, Deng, Y, Wal, KVD, et al., C(sp^3)–H functionalizations of light hydrocarbons using decatungstate photocatalysis in flow, Science, 2020, 369, 92–96.

[93] Castani, A, Gemoets, HPL, Motti, E, Ca', ND, Noel, T, Homogeneous and gas–liquid Catellani-type reaction enabled by continuous-flow chemistry, Chem Eur J, 2018, 24, 14079–14083.

[94] Xue, F, Deng, H, Xue, C, Mohamed, DKB, Tang, KY, Wu, J, Reaction discovery using acetylene gas as the chemical feedstock accelerated by the "stop-flow" micro-tubing reactor system, Chem Sci, 2017, 8, 3623–3627.

[95] Deng, Q, Shen, R, Ding, R, Zhang, L, Generation of ethynyl-Grignard reagent in a falling film microreactor: An expeditious flow synthesis of propargylic alcohols and analogues, Adv Synth Catal, 2014, 356, 2931–2936.

[96] Corcoran, EB, Lévesque, F, McMullen, JP, Naber, JR, Studies toward the scaling of gas-liquid photocycloadditions, ChemPhotoChem, 2018, 2, 931–937.

[97] Williams, JD, Nakano, M, Gérardy, R, et al., Finding the perfect match: A combined computational and experimental study toward efficient and scalable photosensitized [2+2] cycloadditions in flow, Org Process Res Dev, 2019, 23, 78–87.

[98] Beaver, MG, Zhang, E, Liu, Z, et al., Development and execution of a production-scale continuous [2 + 2] photocycloaddition, Org Process Res Dev, 2020, 24, 2139–2146.

[99] Yang, H, Martin, B, Schenkel, B, On-demand generation and consumption of diazomethane in multistep continuous flow systems, Org Process Res Dev, 2018, 22(4), 446–456.

[100] Struempel, M, Ondruschka, B, Daute, R, Stark, A, Making diazomethane accessible for R&D and industry: Generation and direct conversion in a continuous micro-reactor set-up, Green Chem, 2008, 10, 41–43.

[101] Maurya, RA, Park, CP, Lee, JH, Kim, DP, Continuous in situ generation, separation, and reaction of diazomethane in a dual-channel microreactor, Angew Chem Int Ed, 2011, 50, 5952–5955.

[102] Lehmann, H, A scalable and safe continuous flow procedure for in-line generation of diazomethane and its precursor MNU, Green Chem, 2017, 19, 1449–1453.

[103] Pinho, VD, Gutmann, B, Miranda, LSM, De Souza, ROMA, Kappe, CO, Continuous Flow synthesis of α-halo ketones: Essential building blocks of antiretroviral agents, J Org Chem, 2014, 79, 1555–1562.

[104] Dallinger, D, Kappe, CO, Lab-scale production of anhydrous diazomethane using membrane separation technology, Nat Prot, 2017, 12, 2138–2147.

[105] Wernik, M, Poechlauer, P, Schmoelzer, C, Dallinger, D, Kappe, CO, Design and optimization of a continuous stirred tank reactor cascade for membrane-based diazomethane production: Synthesis of α-chloroketones, Org Process Res Dev, 2019, 23, 135–1368.

[106] Koolman, HF, Kantor, S, Bogdan, AR, Wang, Y, Pan, JY, Djuric, SW, Automated library synthesis of cyclopropyl boronic esters employing diazomethane in a tube-in-tube flow reactor, Org Biomol Chem, 2016, 14, 6591–6595.

[107] Cranwell, PB, O'Brien, M, Browne, DL, et al., Flow synthesis using gaseous ammonia in a Teflon AF-2400 tube-in-tube reactor: Paal–Knorr pyrrole formation and gas concentration measurement by inline flow titration, Org Biomol Chem, 2012, 10, 5774–5779.

[108] Pastre, JC, Browne, DL, O'Brien, M, Ley, SV, Scaling up of continuous flow processes with gases using a tube-in-tube reactor: Inline titrations and fanetizole synthesis with ammonia, Org Process Res Dev, 2013, 17, 1183–1191.

[109] Xue, C, Li, J, Lee, JP, Zhang, P, Wu, J, Continuous amination of aryl/heteroaryl halides using aqueous ammonia in a Teflon AF-2400 tube-in-tube micro-flow reactor, React Chem Eng, 2019, 4, 346–350.
[110] Köckinger, M, Hone, CA, Kappe, CO, HCN on tap: On-demand continuous production of anhydrous HCN for organic synthesis, Org Lett, 2019, 21, 5326–5330.
[111] Cantillo, D, Inglesby, PA, Boyd, A, Kappe, CO, Hydrogen sulfide chemistry in continuous flow: Efficient synthesis of 2-oxopropanethioamide, J Flow Chem, 2017, 7, 29–32.
[112] Buba, AE, Koch, S, Kunz, H, Löwe, H, Fluorenylmethoxycarbonyl-*N*-methylamino acids synthesized in a flow tube-in-tube reactor with a liquid–liquid semipermeable membrane, Eur J Org Chem, 2013, 4509–4513.
[113] Ajmera, SK, Losey, MW, Jensen, KF, Microfabricated packed-bed reactor for phosgene synthesis, AIChE J, 2001, 47, 1639–1647.
[114] Fuse, S, Tanabe, N, Takahashi, T, Continuous in situ generation and reaction of phosgene in a microflow system, Chem Commun, 2011, 47, 12661–12663.
[115] Guidi, M, Seeberger, PH, Gilmore, K, How to approach flow chemistry, Chem Soc Rev, 2020, 49, 8910–8932.
[116] Straathof, NJW, Su, Y, Hessel, V, Noel, T, Accelerated gas-liquid visible light photoredox catalysis with continuous-flow photochemical microreactors, Nat Prot, 2016, 11, 10–21.

Answers to the study questions

1.1 The typical arrangement of a continuous-flow apparatus contains reagent pumps, mixing units (T-mixer), reaction zone (typically tube/coil or packed-bed reactors that can be heated, cooled, or irradiated), quenching inlet, back-pressure regulator and product collection vessel.

1.2 The major advantageous features of the heterogeneous phase, packed-bed flow reactors are: Particle movement is mainly restricted; high interfacial area facilitates better mass transfer and improved heat distribution, and packed beds have improved lifetime due to decreased exposure to the environment; high reagent/catalyst excess could drive the reactions to completion; immobilized reagents are removed from reactions (placed in cartridges); catalyst poisoning and side reactions are minimized due to the immediate removal of the product from the reactor zone (no backmixing).

1.3 Diffusion is a passive transport due to a concentration gradient, while convection is a transport due to pressure gradient or in other words, the collective movement of molecules within fluids.

1.4 Micromixers can be characterized as passive (without external energy) or active (external energy added such as oscillation or vibration). The most widely used micromixers are passive, utilizing mixer geometry (increasing the interfacial area between fluidic layers) to generate effective contact between reagents.
For macromixing in large-scale flow equipments static mixers are used. Static mixers generate turbulence and intense radial mixing by applying tube inserts such as blades/helices or stretching of the fluid.

1.5 In a laminar flow regime radial mixing is strictly due to diffusion. Damkoehler number (Da) shows if a reaction is under diffusion or kinetic control in laminar flow. It expresses the reaction rate relative to the mass transport rate (diffusion). Chemical transformations are reaction rate limited when $Da < 1$. When $Da > 1$, experimentally measured reaction rates are controlled by diffusion or mass transport.

1.6 There are two dispersion effects due to axial and radial diffusion. Rapid convection (pressure-driven fluid delivery) increases the axial dispersion.

1.7 In batch processes, stoichiometry is defined by the concentration of chemical reagents and the ratio of their molar quantities. Stoichiometry in flow reactors is defined by the concentration of reagents and the ratio of their flow rate.

1.8 Reaction time in batch is the time spent under the defined condition, while in flow, reaction time equals the residence time spent in the reaction zone, depending on the flow rate and reaction volume.

1.9 Reaction progress can be characterized by the time spent in the flask, while in flow, it is characterized by the distance traveled in the channel.

https://doi.org/10.1515/9783110693676-012

1.10 Surface-to-volume ratio: area per unit volume of the reactor is a crucial factor for heat ad mass transfer. Generally, volume is equal to the length cubed, while surface area is equal to the length squared. When the length is shortened, the surface-to-volume ratio increases. Thus, microreactors have higher surface-to-volume ratio than macroreactors. If the size is reduced 100-fold, the surface-to-volume increases 100-fold.

1.11 Continuously flowing microreactors allow for rapid and homogeneous mixing because of their small dimensions (channel or capillary diameter is 0.05–0.5 cm) and laminar flow is predominant. Microreactors can achieve complete mixing in microseconds, whereas classical reactors mix on a timescale of seconds or longer. The fast mixing relies on the short diffusion path in microreactors. A molecule in the center of a typical microfluidic channel can reach the wall of that channel in a few seconds. The same molecule in the middle of a reaction flask (batch) would require hours to diffuse to the side wall (without mixing).

The time (t_d) needed for molecular diffusion is proportional to the square of the length of the diffusion path. A marked shortening of the diffusion path in flow (in a microreactor) results in a mixing speed that is unobtainable in batch.

1.12 Because microreactors have a greater surface area per unit volume than macroreactors, heat transfer occurs rapidly in a flow microreactor, enabling fast cooling/heating and hence, precise temperature control.

1.13 Hydrodynamic pumping exploits conventional or microscale pumps, notably syringe-type pumps, to deliver solutions around the channel network. The static nature of the fluid at the boundary produces a parabolic velocity profile within the channel.

1.14 The segmented flow system delivers pulses of reactants, segregated by an immiscible solvent, into the flow reactor. Each segment can consist of a different combination of reactants for different reactions to occur.

1.15 Reynolds number determines the flow regimes in the channels and is defined as the ratio of momentum (inertial) forces to viscous forces.

1.16 The flow is
truly laminar when Re < 100
transient when 100 < Re < 2,500
turbulent flow 2,500 < Re

1.17
- Lab scale: 1–10 mL min^{-1}
- Pilot plant scale: 50–150 mL min^{-1}
- Industrial scale: 200–600 mL min^{-1}

1.18 Continuous-flow (micro)reactors consist of a network of miniaturized channels with diameters typically in the range of 10–500 μm, and constructed of glass, quartz, silicon, polymers, or stainless steel.

1.19 Space–time yield (kg m^{-3} s^{-1}) represents the mass of a product *P* formed per volume of the reactor and time.

1.20 Flash chemistry is a field of chemical synthesis where extremely fast reactions involving short-lived highly reactive intermediates are conducted without deceleration in a highly controlled manner to produce desired compounds with high selectivity by virtue of high-resolution reaction control.

1.21 High-resolution reaction time control is simply adjusting the residence time to a particular step in a multistep reaction sequence. Thus, the isolation of the intermediate products could be resolved by the residence time (partial reaction time), which is not possible in batch reactors.

1.22 Novel process windows (NPW) operates at process conditions that are beyond the usual condition; that considerably speed up conversion rates while maintaining selectivity. This can be achieved by an increase in temperature, pressure, or concentration (solvent-free operation), by a simplification of process protocols or by function integration. Combined high-temperature and high-pressure flow regime, that is, >200 °C and >50 bar, is one of the main directions, with many applications focusing on the generation of high-temperature or supercritical water (scH_2O).

1.23 Process intensification (PI) can be defined as the ability to obtain equivalent or better results in terms of purity, selectivity, and yield of the desired product in a reduced period of time and, therefore, with an enhanced throughput, by increasing parameters such as temperature and pressure.

1.24 Flow reactors can be interconnected to form a multiple line of reactors in different configurations, including parallel and consecutive setups. Continuous-flow devices can be linked with in-line analytical tools such as UV, IR, HPLC, GLC, mass spectrometry, and/or NMR for real-time analysis and feedback, which allows for rapid automated optimization of the reaction parameters.

1.25 Production can only be initiated if the products are needed or the precursors are available. On-site–on-demand production of hazardous starting materials that are immediately used after generation saves transport, storage, and handling of these agents. On-demand production is also feasible at remote or multiple locations.

2.1

(a)

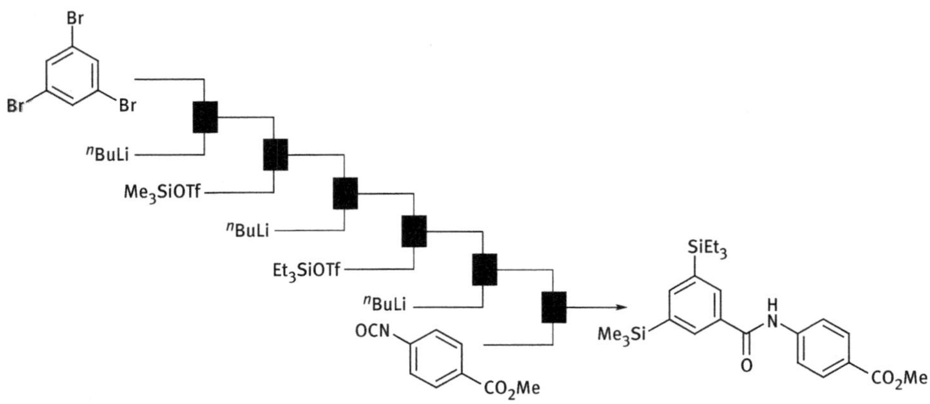

(b)

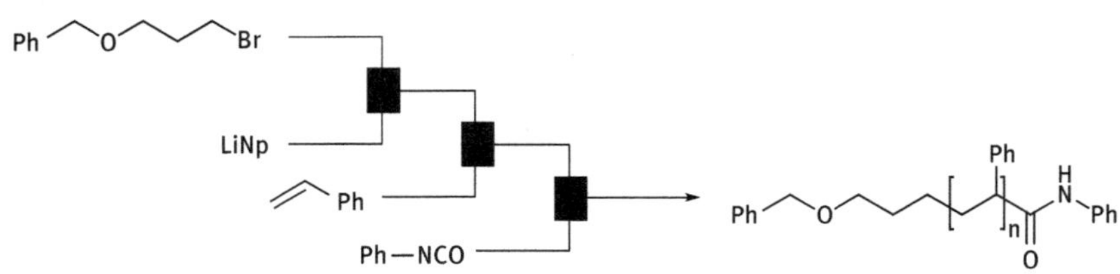

2.2

(a)

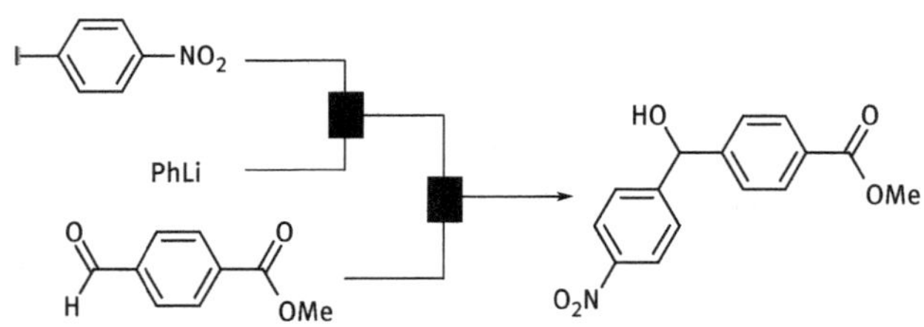

A 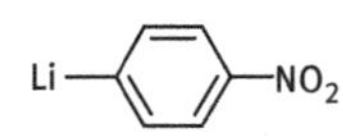

(b)

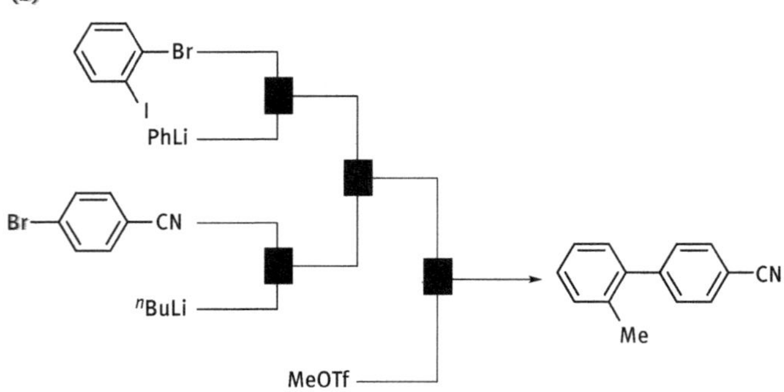

B 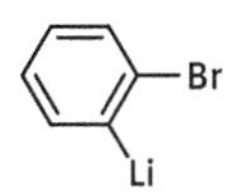

C 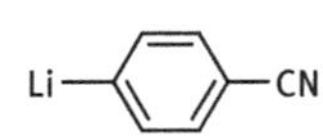

D 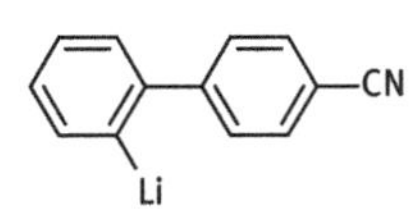

3.1 Pump, reactor, for example, coil or chip, heating/cooling, and backpressure regulator.

3.2 The reaction conditions: residence/reaction time, temperature regime, pressure regime, strong basic or acidic conditions, possible chemical interactions with the used reactor materials, solubility of all reagents, products and side/by-products, physical and health hazards of the reagents, products, and side/by-products.

3.3 a) Minutes
  b) 1:1.5 A:B
  c) 3.2 min
  d) You can prepare a more concentrated solution of B, so that the required equivalents of B can be delivered at a lower flow rate (e.g. 1 mL min of a 0.15 M solution of B). If this is not feasible (e.g. if compound B has limited solubility), you can increase the length of the reaction loop: a 10 mL loop would provide 4 min residence time.
  e) M of quenching reagent at a flow rate of 3 mL min

4.1 Introducing fluorine in organic molecules increases their stability and lipophilicity, which is relevant for drugs, where metabolic stability and membrane permeability are important properties.

4.2 Superheating of solvent (THF at 120 °C) is exploited to accelerate kinetics. Contained flow environment minimizes the impact of moisture on reagents and reactive carbene intermediate. The process is readily scaled to produce multigram quantities of product.

4.3 Avoiding high (partial) pressure oxygen is desirable to minimize the risk of solvent ignition. High oxygen pressure may give faster reactions, but a compromise between safety and efficiency is targeted in this application.

4.4 Further examples include fluorine, chlorine, hydrogen, carbon monoxide, and ammonia.

4.5 Apart from mechanical agitation, sonication can be exploited in flow mode.

4.6 Solids and slurries offer facile product isolation. Flow also allows for controlled and reproducible crystallization to give a specific crystal morphology, which is important in API synthesis.

4.7 High temperature batch setups need efficient insulation strategies, often requiring unattractive high-boiling solvents and monitoring (p, T) to guarantee safe operations.

4.8 In-line temperature and pressure control as well as automated shutdown processes are necessary in the interest of process safety.

4.9 Precooling loops are important to ascertain that reagents are mixed at an appropriately low temperature to minimize degradation of reactive species.

4.10 Metals are characterized by high heat capacity, and plate reactors can easily accommodate advanced static mixers to provide for better mixing.

4.11 Immobilization of enzymes allows for easy separation (post reaction), increased mass transfer, reduced product inhibition, and increased stability (mechanically and chemically). Immobilization can be time-consuming and may reduce the enzyme activity.

4.12 Contact between substrate and enzyme is increased, thus leading to faster reactions. The substrate experiences a local excess of enzyme if packed in a column.

4.13 In flow setups, a short and constant distance between lamp and tubing (or chip) is provided in combination with narrow tubing/channel diameters, accounting for uniformity.

4.14 If direct irradiation of substrates is not feasible or effective, a photosensitizer or photocatalyst can be employed to absorb and transfer the light energy to the substrate.

4.15 Control of interelectrode distance is challenging in batch, where electrolytes are often needed. Scalability of batch systems can be inefficient and may lead to safety concerns.

4.16 Electrons (and photons) are traceless reagents, obviating the need for removal as in analogous reagent-based scenarios. This increases the overall atom-economy.

4.17 Faster kinetics result in bimolecular reactions. Harmful solvents are avoided in the synthesis (not the purification), which also improves the PMI and e-factor of the process. In flow mode, this can lead to higher viscosity (pump issues) and heat dissipation may be affected.

4.18 Sufficient optimization of individual steps is necessary along with evaluation of whether spent reagents and impurities affect downstream reactions (need for in-line purification). Increasing the dilution will impact kinetics and can lead to inefficient late stages.

4.19

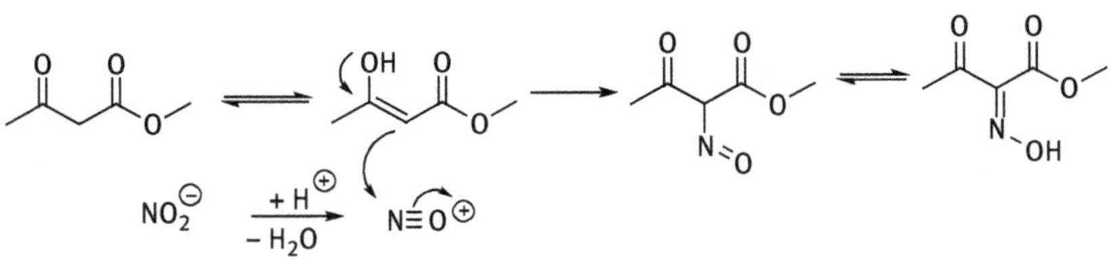

4.20 THF, chlorinated solvents, and toluene are classified as harmful due to their environmental impact. Ethyl acetate is more susceptible towards hydrolysis (NaOH is used).

4.21

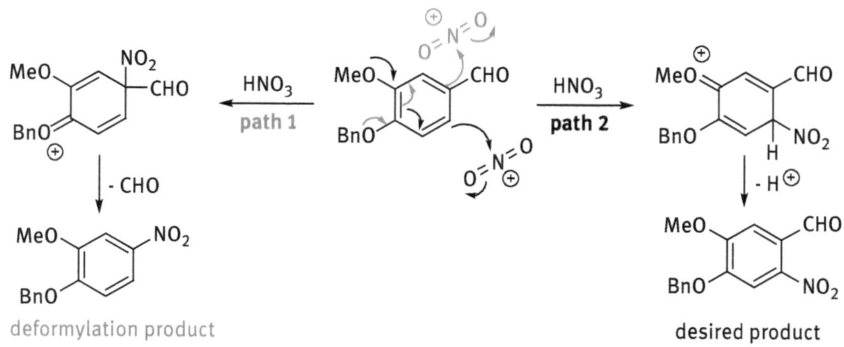

4.22 Fuming nitric acid is very corrosive and is a strong oxidant (e.g. toward stainless steel). Harmful nitrous oxides are liberated and there is a need for effective ventilation or scrubbing.

5.1 Total flow rate = 5 mL/40 min = 0.125 mL min
$X = 0.125/(1 + 1.5) \times 1 = 0.05$ mL min
$Z = 0.125/(1 + 1.5) \times 1.5 = 0.075$ mL min

5.2 Replace the reactor with a bigger one, for example, 10 mL so that X becomes 0.1 and Z 0.15 mL min

5.3 Total flow rate = 5 mL/15 min = 0.33 mL min
The solution is a system with two equations and two unknowns:
$X + Z = 0.133$ (addition of relative flow rates gives total flow rate) and $1.5 \times Z$: $2 \times X = 2{:}1$ (the molar concentration of each feed multiplied for its flow rate has to give the correct stoichiometry).
Solution: $X = 0.09$ and $Z = 0.24$.

5.4 2 mmol/ mL × 0.4 mL min × 10 min = 8 mmol

6.1 Configurable flow-paths on the rotor (slots) are symmetrically etched, similar to the rotor of Fig. 6.4. Configurable flow-paths of the rotor can also be etched unsymmetrically to maintain fluid communication to certain ports and disengage fluid communication to certain other ports. For example, the rotor of the valve in Fig. A1 is turned 20° counter-clockwise. The rotor positions before (left) and after (center) the turn are shown. In this configuration, ports 1 and 2 maintain fluid communication through a configurable flow-path that is slightly elongated, in comparison to the remaining two configurable flow-paths in both positions. The configurable flow-path that is in fluid communication with port 3 is disengaged from port 4, in the position at the center. In this position, fluid in the configurable flow-path connecting port 3 can be compressed or decompressed by a pump connected to port 3, without affecting fluids downstream of port 4. Once the pressure of the fluid in the flow-path connecting to port 3 is adjusted, the valve can be rotated by another 100°

counter-clockwise (right) to bring newly pressurized fluid in communication with the high pressure fluids at ports 1 and 2. This principle of pressure mitigation is used in modern PAT injections as well as in flow chemistry involving pressurized reactors.

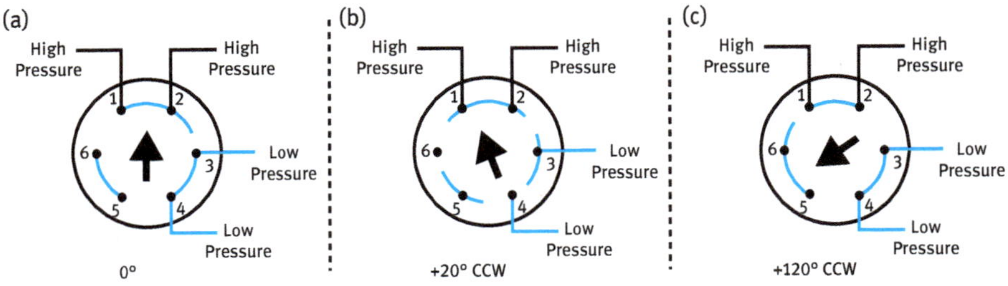

Fig. A.1: Extended rotor slot for pressure management in sampling valves.

6.2 First, one must identify the "load" configuration of the sampling valve, which is shown in Fig. 6.6 (left). In the "load" configuration, flowed sample travels via filter through five ports (1, 2, 5, 6, and 9) and two rotor slots in order to overfill the 1 μL sampling loop between ports 6 and 9. The net volume that the reactor sample needs to travel to overfill sampling loop L1 is [0.5 + 1 + 0.5 + 10 + 0.5 + 20 + 0.5 + 10 + 0.5 + 1 + 0.5 + 1 + 0.5] μL, which adds up to 46 μL. At a flow rate of 3 μL min, this would take 15.3 min. Assuming that the sampling valve is capable of switching to the "load" configuration immediately after a GC injection is made, one must wait for 15.3 min for a second sampling. At a frequency of 15.3 min per sampling, one can thus acquire 94 samples per day. If the sample does not contain particulates, the filtration step is redundant. In that case, a six-port valve equipped with a 1 μL sampling loop (similar to Fig. 6.6) will suffice for the sampling operation. In that case, the sample would travel through three ports and one rotor slot, which would equate to 3 μL. At 3 μL min flow rate, sample loading would take only 1 minute using a 6-port sampler. In that case, the GC analysis time would control the sampling frequency. Consequently, the team can acquire up to 411 data points per day.

6.3 **A.** PAT for the enolization step would ideally come with fast scan rate and a noninvasive detection method. An in-line IR PAT can be used for such measurements. If the diastereomeric distribution in the final product is considered, a CQA for the process in question, an in-line/on-line NMR PAT, or an at-line NMR PAT equipped with a solvent-switching mechanism may be used. In-line/on-line NMR PAT is best suited for the second step to monitor the conversion of the enol to the corresponding anion of the aldol product. The diastereomeric distribution between the syn-/anti-aldol products can be also investigated using this PAT. 19F NMR F-NMR can be used to readily monitor the conversion of the aldehyde into the anionic aldol product. The final product formed after protonation can be monitored by various in-line/on-line/at-line PATs, including LC-UV or LC-MS. One must exercise caution for possible elimination in the aldol product to the corresponding α,β-unsaturated ketone during analysis. Readers are encouraged to revisit this exercise after reading the recirculatory chromatographic analysis, which is presented later in this section.

The quality of a microlot can be validated by temporarily isolating microlots in one of two sample holding loops downstream, as shown in Fig. A2 [35]. The downstream eight-port valve isolates a segment of the product stream that is to be analyzed from an integrated PAT downstream, while allowing the reactor matter to flow through the second sample, holding the loop in a continuous manner.

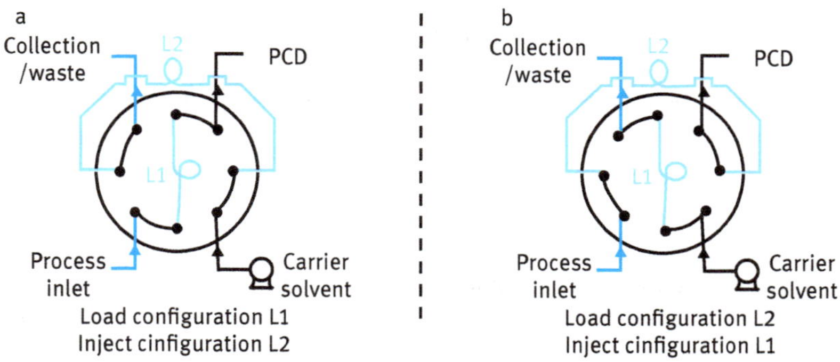

Fig. A2: Eight-port rotary valve with two holding loops (L1 and L2). (a) Load configuration for L1 and inject configuration for L2. (b) Inject configuration for L1 and load configuration for L2. PCD, pressure creating device.

6.4 **A.** Results from experiment II and I are the best and worst cases, respectively. Hence, the process condition (20 μL min flow rate at 5 °C) of experiment III would be reflected across a virtual line formed by connecting the process coordinates of the worst (50, 50) and the best (100, 100) experiments. This will set the flow rate at 5 μL min flow rate and the reaction temperature at 20 °C in the next experiment. **B.** If the yield from the fourth experiment is between 2% and 80%, the process conditions of the best and the worst outcome would remain unchanged. The Nelder–Mead strategy would require search propagation methods other than reflection.

6.5 The matrix multiplication rule shown below can be used to solve this exercise:

$$\begin{bmatrix} a & b \\ c & d \end{bmatrix} * \begin{bmatrix} e \\ f \end{bmatrix} = \begin{bmatrix} g \\ h \end{bmatrix} \tag{9}$$

$$g = a^*e + b^*f \tag{10}$$

$$h = c^*e + d^*f \tag{11}$$

Considering the second (x, y) pair, which is $\left(\begin{bmatrix} 6 \\ 10 \end{bmatrix}, \begin{bmatrix} 2 \\ 2 \end{bmatrix}\right)$, from the exercise, one can get the equation as follows:

$$\begin{bmatrix} z_1 & 2 \\ 2 & z_2 \end{bmatrix} * \begin{bmatrix} 6 \\ 10 \end{bmatrix} = \begin{bmatrix} 2 \\ 2 \end{bmatrix}$$

Solving the equation shown above, one can derive values of z_1 and z_2 from the following equations:

$$z_1{}^*6 + 2^*10 = 2 \text{ and } 2^*6 + z_2{}^*10 = 2$$

$$z_1 = -3 \text{ and } z_2 = -1$$

It is important to recognize that parameters z_1 and z_2 are not always experimentally measurable quantities. Matrix Z can be considered as the trained neural network in this example.

The same solution can be obtained using any of the (x, y) pairs from the exercise.

Using the trained network of matrix Z (shown below), one can construct a similar matrix multiplication operation for the concentrations of C and D, when concentrations of A and B are 1 and 1.8 M, respectively.

$$\begin{bmatrix} -3 & 2 \\ 2 & -1 \end{bmatrix} * \begin{bmatrix} 10 \\ 18 \end{bmatrix} = \begin{bmatrix} [C] \\ [D] \end{bmatrix}, \text{ where } Z = \begin{bmatrix} -3 & 2 \\ 2 & -1 \end{bmatrix}$$

$$[C] = -3^*10 + 2^*18 = 6$$

$$[D] = 2^*10 + (-1)^*18 = 2$$

Consequently, the machine can learn from any of the supplied (x, y) pair, each of which uses different concentrations of A and B and arrive at the identical value for z_1 and z_2 in matrix Z, irrespective of which (x, y) pair it chooses.

When concentrations of A and B are 1 M and 1.8 M respectively, the machine will predict concentrations of C and D will be 0.6 M and 0.2 M respectively in the current example.

7.1 Gas is used as a reagent and a carrier in many flow syntheses. It is put in excess; so it needs to be separated from the liquid stream after the flow synthesis. The example is hydrogenation, in which hydrogen gas is added. In some cases, the gas is used as a carrier or a diluent. As discussed in the chapter, the examples include the use of supercritical carbon dioxide for photo-oxidations.

7.2 There are many applications of the continuous liquid–liquid separation. It can be used to separate the biphasic liquid–liquid reactive stream, particularly those involving organic–aqueous reactions. Also, it can be used as a separator for the flow-based solvent extraction. In flow chemistry, many intermediates and products need to be purified via the solvent extraction. The continuous liquid–liquid separation allows this operation to be in flow and simply implemented in flow synthesis.

7.3 Sensors for gravity-based separators include a capacitance sensor and an impedance probe. A camera-based approach can also be used to detect the liquid–liquid interface.

7.4 The lower limit is $\Delta P_{per} = \frac{8\mu_{org} Q_{org}}{n\pi R_{mem}^4} L_{mem}$

The upper limit is $\Delta P_c = \frac{2\gamma}{R_{mem}} \cos\theta$

where y is the interfacial tension, R_{mem} is the pore size, and θ is the wetting angle, μ_{org} is the viscosity of the organic phase, Q_{org} is the flow rate of the organic (permeate) phase), and L_{mem} is the thickness of the membrane.

7.5 The downstream pressures can be fluctuating; so the backpressure regulators need to be adjusted real time. Otherwise, the separation will not be completed.

8.1 Global harmonized system (GHS).

8.2 Eliminate or reduce the hazard at source.

8.3 Transport intensification, robust structure, and small volume under continuous flow.

8.4 The intensification of chemical processes by extreme/harsh process conditions. High temperature, high pressure, high concentration, and conditions within the explosive regime (forbidden chemistry).

8.5 Thermal runaways can lead to accidents when heat generation is faster than heat removal; the high surface area of microreactors facilitates heat removal.

8.6 In-line generation of a reactive intermediate from less dangerous precursors; allows forbidden chemicals to be used as only small amounts are present at a given time.

8.7 HN_3: azide + acid; CH_2N_2: *N*-methyl-*N*-nitrosocompounds + base; ArLi: RLi + ArH/ArX.

8.8 Hydrogen, peroxides, halogens (fluorine, chlorine, bromine), carbon monoxide, sulfur dioxide, and hydrazine.

8.9 Use appropriate PPE and CPE (personal and collective protection equipment); quench dangerous intermediates in-line to avoid accumulation; monitor the reactor's operating parameters (pressure, temperature, flow rate) to ensure conditions are nominal and shutdown if necessary; develop clear SOPs (standard operation procedures).

8.10 Can result in reactor blockage and leaks of hazardous materials; use higher reaction temperature or use sonicator to avoid reactor precipitation.

8.11 Transport and storage of chemicals, and waste management.

10.1 Category 1. Reactions with homogeneous catalysts
Category 2. Reactions with heterogeneous catalysts

10.2 Category 1. Advantage: the possibility of using batch reaction conditions
Category 1. Disadvantage: requirement for a quenching step to terminate reactions
Category 2. Advantage: quenching not required, no contamination of catalyst residue to product solution.
Category 2. Disadvantage: requirement of immobilization, immobilization study, and technique

10.3 (1) WHSV $= 2 \times 10^4 C_1 v_1 / n_1 W_1$
(2) Reduce C_1 by half, reduce v_1 by half, or increase W_1 by doubling it
(3) Examples: Upon reducing WHSV, the number of contact opportunities between substrate and catalyst will increase, hence resulting in improvements in conversion and yield (however, there is the possibility of overreactions)
(4) WHSV $= 2 \times 10^4 C_1 v_1 / n_1 W_1 = 2 \times 10^4 \times 0.5 \times 0.5/(5 \times 1) = 1{,}000\ \mathrm{h}^{-1}$
(5) Increase C_1 to $10 \times C_1$ with $10 \times W_1$, etc.

10.4 Advantageous reactions are b, c, and f. Reactions a, d, and e are accompanied by the formation of by-products.

11.1 Approximately 36 mmol, assuming that nitro reduction is performed with 4.5 equiv. H_2.

11.2 The maximum throughput based on H_2 amount is 268 mmol h. The maximum throughput at 10 mL/min flow rate and 0.5 mmol substrate concentration is 300 mmol h. Temperature and pressure data are not relevant for answering the question (at least without the structure being specified). Therefore, the hydrogen flow rate will limit the throughput, assuming a 1:1 = H_2: alkene stoichiometric ratio.

11.3 It is advisable to measure the residence time with a tracer compound (e.g. some kind of a dye, or a compound which can be traced by UV spectrophotometry, and TLC).

11.4 For example, annular, Taylor flow. See Section 11.2.

11.5 On-demand generation of hazardous or unstable substance from benign precursors. Examples: diazomethane, chlorine, carbon monoxide, and HCN. Also, see Ref. [5].

11.6 See Fig. 11.3 in Ref. [106].

Index

https://doi.org/10.1515/9783110693676-013

Bei Fragen zur Produktsicherheit wenden Sie sich bitte an:
If you have any questions regarding product safety, please contact:

Walter de Gruyter
Genthiner Straße 13
10785 Berlin
productsafety@degruyterbrill.com